Growth Patterns in Physical Sciences and Biology

NATO ASI Series

Advanced Science Institutes Series

A series presenting the results of activities sponsored by the NATO Science Committee, which aims at the dissemination of advanced scientific and technological knowledge, with a view to strengthening links between scientific communities.

The series is published by an international board of publishers in conjunction with the NATO Scientific Affairs Division

A	**Life Sciences**	Plenum Publishing Corporation
B	**Physics**	New York and London
C	**Mathematical and Physical Sciences**	Kluwer Academic Publishers
D	**Behavioral and Social Sciences**	Dordrecht, Boston, and London
E	**Applied Sciences**	
F	**Computer and Systems Sciences**	Springer-Verlag
G	**Ecological Sciences**	Berlin, Heidelberg, New York, London,
H	**Cell Biology**	Paris, Tokyo, Hong Kong, and Barcelona
I	**Global Environmental Change**	

Recent Volumes in this Series

Series B: Physics

Growth Patterns in Physical Sciences and Biology

Edited by

Juan Manuel Garcia-Ruiz

CSIC–Universidad de Granada
Granada, Spain

Enrique Louis

Universidad de Alicante
Alicante, Spain

Paul Meakin

University of Oslo
Oslo, Norway

and

Leonard M. Sander

University of Michigan
Ann Arbor, Michigan

SPRINGER SCIENCE+BUSINESS MEDIA, LLC

Proceedings of a NATO Advanced Research Workshop on
Growth Patterns in Physical Sciences and Biology,
held October 7–11, 1991,
in Granada, Spain

NATO-PCO-DATA BASE

The electronic index to the NATO ASI Series provides full bibliographical references (with keywords and/or abstracts) to more than 30,000 contributions from international scientists published in all sections of the NATO ASI Series. Access to the NATO-PCO-DATA BASE is possible in two ways:

—via online FILE 128 (NATO-PCO-DATA BASE) hosted by ESRIN, Via Galileo Galilei, I-00044 Frascati, Italy

—via CD-ROM "NATO-PCO-DATA BASE" with user-friendly retrieval software in English, French, and German (©WTV GmbH and DATAWARE Technologies, Inc. 1989)

The CD-ROM can be ordered through any member of the Board of Publishers or through NATO-PCO, Overijse, Belgium.

Library of Congress Cataloging-in-Publication Data

Growth patterns in physical sciences and biology / edited by Juan
  Manuel García-Ruiz ... [et al.].
       p.   cm. -- (NATO ASI series. Series B, Physics ; vol. 304)
    "Published in cooperation with NATO Scientific Affairs Division."
    "Proceedings of a NATO Advanced Research Workshop on Growth
  Patterns in Physical Sciences and Biology, held October 7-11, 1991,
  in Granada, Spain"--T.p. verso.
    Includes bibliographical references and index.
    ISBN 978-1-4613-6235-7   ISBN 978-1-4615-2852-4 (eBook)
    DOI 10.1007/978-1-4615-2852-4
    1. Pattern perception--Congresses. 2. Pattern formation
  (Biology)--Congresses. I. García-Ruiz, Juan Manuel. II. North
  Atlantic Treaty Organization. Scientific Affairs Division.
  III. NATO Advanced Research Workshop on Growth Patterns in Physical
  Sciences and Biology (1991 : Granada, Spain) IV. Series: NATO ASI
  series. Series B, Physics ; v. 304.
  Q327.G76   1993
  500--dc20   93-23117

Additional material to this book can be downloaded from http://extra.springer.com.

ISBN 978-1-4613-6235-7

©1993 Springer Science+Business Media New York
Originally published by Plenum Press in 1993
Softcover reprint of the hardcover 1st edition 1993

PREFACE

During the past decade interest in the formation of complex disorderly patterns far from equilibrium has grown rapidly. This interest has been stimulated by the development of new approaches (based primarily on fractal geometry) to the quantitative description of complex structures, increased understanding of non-linear phenomena and the introduction of a variety of models (such as the diffusion-limited aggregation model) that provide paradigms for non-equilibrium growth phenomena. Advances in computer technology have played a crucial role in both the experimental and theoretical aspects of this enterprise. Substantial progress has been made towards the development of comprehensive understanding of non-equilibrium growth phenomena but most of our current understanding is based on simple computer models.

Pattern formation processes are important in almost all areas of science and technology, and, clearly, pattern growth pervades biology. Very often remarkably similar patterns are found in quite diverse systems. In some case (dielectric breakdown, electrodeposition, fluid-fluid displacement in porous media, dissolution patterns and random dendritic growth for example) the underlying causes of this similarity is quite well understood. In other cases (vascular trees, nerve cells and river networks for example) we do not yet know if a fundamental relationship exists between the mechanisms leading the formation of these structures.

This NATO Advanced Research Workshop was organized with the objective of bringing together physicists and biologists with a common interest in pattern growth and in applying new tools across the areas of their disciplines to explore the similarities and differences in their subjects. In general, the community of physicists is interested in the most simple pattern formation processes and focuses its attention on the similarities (universalities) that are associated with different processes. Biologists, on the other hand, must necessarily work with extremely complex systems and are primarily concerned with the development of a detailed (but generally qualitative) understanding of specific systems (organisms). Nevertheless there is a large "common ground" between these divergent approaches. For example, theoretical biologists have long studied simple growth models that are quite similar to those developed by physicists for quite different purposes. It was apparent to us that both communities should benefit substantially from an exchange of ideas and techniques.

The theory of pattern formation is still under development, and part of our task was to review the advances made. In some cases (such as the Eden model for the growth of cell colonies) the relationship between the growth algorithm and the pattern generated by the model is now quite will understood. In other cases (such as the diffusion-limited aggregation model) we are still quite far from developing a comprehensive analytical understanding.In this case a quite wide range of phenomena can nevertheless be understood in terms of this model. At present there is a growing acceptance of the idea that physical phenomena can be understood equally well in terms of simple algorithms or continuum equations.

Both the Eden model and diffusion-limited aggregation played a large role in many of the discussions in the workshop. Despite their simplicities both seem to seize some of the essential aspects of biological growth for some systems. However much remains to be learned about generalizations and more realistic (but more complicated) models.

The workshop was held in Granada Spain during the period 7-11 October, 1991. Despite the rather large fraction of physicists the main objectives of the workshop were achieved. In some cases, as the contributions to this volume attest, the confrontation between the point of view of physics and biology was very fruitful indeed. In other cases a large gaps remain, but a start was made.We think that substantial progress was made towards establishing a common language.The beautiful city of Granada provided a delightful environment for the workshop that was conducive to informal exchange of ideas.

The organizers would like to thank the NATO Division of Scientific Affairs for sponsorship. We had additional support from the Universidad de Granada, the Junta de Andalucia, the Comision Interministerial de Ciencia Tecnologia, the Ayuntamiento de Granada, and the Consejo Superior de Inverstigaciones Cientificas.

The Editors

CONTENTS

CELL COLONIES

SURFACE AND INTERFACES

CELLULAR PATTERNS

DYNAMICAL SYSTEMS

SELF REPLICATION

SELF ORGANIZATION

MEASUREMENT AND CHARACTERIZATION

FRACTAL GROWTH AND MORPHOLOGICAL CHANGE IN

BACTERIAL COLONY FORMATION

Mitsugu Matsushita, Masahiro Ohgiwari and Tohey Matsuyama[†]

Department of Physics, Chuo University
Kasuga, Bunkyo-ku, Tokyo 112, Japan
[†]Department of Bacteriology, Niigata University
School of Medicine, Asahimachidori, Niigata 951, Japan

INTRODUCTION

Pattern formation in biological systems is believed to be much more complicated than that in physical and/or chemical systems. Biological phenomena usually take place through complex intertwinement between inherently complex biological factors and environmental (physico-chemical) conditions. However, setting morphogenesis of individual organisms aside, pattern formation of the population of simple biological objects may sometimes be dominated by purely physical conditions. Let us here pay attention to the colony formation of bacteria which are one of the simplest biological objects.

Usual bacteria such as *Escherichia* (*E.*) *coli* are single cell organisms. Even very small number of bacteria (parent cells), once they are inoculated on the surface of approapriate medium such as solid agar which contains enough nutrient and incubated for a while, repeat the growth and cell division many times. Eventually the cell number of the progeny bacteria becomes huge, and usually swarm on the solid medium to form a visible colony. The colony differs in size, form and color according to bacterial species. It also changes its form with the variation of environmental conditions.[1] The investigation of bacterial colony growth seems to bring us quite a treasure house for the study of pattern formation. For instance, Matsuyama et al.[2] showed recently that bacteria called *Serratia marcescens* exerts fractal morphogenesis in the process of spreading growth on an agar surface.

Here we would like to discuss a part of this rich-in-variety bacterial colony formation from the viewpoint of physics of random pattern formation which has recently progressed considerably.[3-5]

EXPERIMENTAL PROCEDURES

In order to investigate the pattern change in bacterial colonies due to environmental variation, we varied only two parameters; concentrations of nutrient (peptone) C_n and agar C_a in a thin agar plate as the incubation medium. Other parameters such as temperature were kept in constant.

Throughout the present experiments we used bacterial species called *Bacillus* (*B.*) *subtilis*. Our strain OG-01 is the same as that used in the former experiments.[6-8] This strain is rod-shaped with flagella and motile in

Growth Patterns in Physical Sciences and Biology, Edited
by J. M. Garcia-Ruiz *et al.*, Plenum Press, New York, 1993

water by collectively rotating them. We also used immotile mutants with no flagella in order to study the effect of bacterial movement on the surface of solid agar medium. They were obtained from the wild type strain (OG-01) by nitrosoguanidine-mutagenesis. Otherwise the experimental procedures used here were exactly the same as those in our previous reports.[6-8]

DIFFUSION-LIMITED COLONY GROWTH

Bacterial colonies grow two-dimensionally on the agar surface because average pore size of the network of prepared agar gel is smaller than the size of bacteria used. Bacterial cells cannot move into the agar medium.

First the agar concentration C_a was fixed at some value in the range of 10-15 g/l. Agar plates obtained thus are rather hard, and bacterial cells cannot move around on the surface of such an agar plate. They only grow and perform cell division locally by feeding on nutrient peptone.

The effect of nutrient was first investigated. We observed colony patterns which had been incubated for three weeks after inoculation at various initial nutrient concentrations C_n in the range of 0-1 g/l. An increase of the nutrient concentration was found to enhance the colony growth. In particular, the growth does not take place without nutrient. This means that the local bacterial growth and cell division (local growth process) at the interface of colonies is governed primarily by the presence of nutrient.

Now we fixed the initial nutrient concentration at 1 g/l. Note that this concentration is much less than usual, e.g., 15 g/l. In Fig. 1(a) a typical example is shown of colony patterns incubated one month after inoculation at this nutrient concentration. It really looks like a two-dimensional DLA cluster, an example of which is also shown in Fig. 1(b), except that branches are thicker in the colony than in the DLA cluster. The DLA (diffusion-limited aggregation) model[4,5,9-11] was first proposed by Witten and Sander[9] as a prototype model for the diffusion-limited growth. Averaged over about 25 samples of colony patterns from the same strain, we obtained a

a   b

Fig. 1. (a) A typical example of DLA-like colony patterns of B. *subtilis* incubated at 35°C for a month after inoculation on the surface of agar plates with C_n=1 g/l and C_a=10 g/l. (b) An example of two-dimensional DLA clusters with the particle number N=10,000 obtained by computer simulations on a square lattice.

a   b

Fig. 2. The growth of a DLA-like colony pattern photographed (a) 12 and (b) 35 days after inoculation. C_n=1 g/l and C_a=10 g/l. Note that many inner branches, typical examples of which are pointed by triangles, are seen to stop growing afterwards.

fractal dimension of D=1.72±0.02. This is in good agreement with that of two-dimensional DLA patterns.[10] But this does not immediately mean that bacterial colonies are formed by the DLA mechanism. Additional evidence is clearly needed to claim that.

The next problem is whether these colony patterns are really growing through DLA processes or not. Some years ago, Meakin[12] demonstrated by computer simulations that when biological growth is governed by the DLA processes, the growing patterns show characteristic features such as screening, repulsion, and so on. Conversely, the existence of these behaviors enables one to confirm the DLA growth in biological systems. In order to examine these behaviors, we first tried to observe the existence of the screening effect during the colony growth. This effect is characteristic for the pattern formation in a Laplacian field. As clearly seen in Fig. 2, many inner branches were found to stop growing afterwards in spite of their open neighborhood. This evidences the existence of the screening effect of protruding main branches against inner ones in a colony.

We next observed the behavior of two neighboring colonies. Figure 3(a) shows the colony pattern which was incubated one month after having inoculat-

a   b

Fig. 3. (a) Two neighboring colonies inoculated simultaneously at two points and then incubated for a month. C_n=1 g/l and C_a=10 g/l. (b) Two-dimensional DLA clusters grown from two seed particles to whole number of particles N=10,000.

ed at two points simultaneously. As clearly seen in the figure, two neighboring colonies repel each other and never fuse together. This behavior is another feature of the pattern formation in a Laplacian field. To see this more convincingly, we show two-dimensional DLA clusters grown from two seeds in Fig. 3(b). Note the similarity in the repulsion behavior of bacterial colonies and DLA clusters.

Taking all these experimental results into account, it is now clear that the present bacterial colonies grow through DLA processes. Then, there still remains one more problem: What makes the Laplacian field, or what diffuses? Since we are treating biological organisms, there are two possibilities: (1) Nutrient diffuses in toward the colonies, and/or (2) some waste material discharged by bacteria themselves diffuses out from the colonies. Both cases are equally possible to produce the same DLA-like patterns, although the effect of nutrient concentration described before strongly suggests that the nutrient diffusion is more important. In order to determine which one is mainly taking place, we put the nutrient at a corner of a dish (elsewhere no nutrient initially), and inoculated the bacteria at the center. Remember that in the present case bacterial cells cannot move around on the surface of agar plates. Nevertheless, the colony showed a clear tendency to grow toward where the nutrient was initially placed.[7,8] This result inevitably leads us to take the first possibility. We can now conclude that the present bacterial colonies grow through DLA mechanism in the nutrient concentration field.

MORPHOLOGICAL CHANGE IN BACTERIAL COLONY FORMATION

Let us now turn our attention to the morphological change of bacterial colonies under the variation of environmental conditions. Colony patterns were found to change drastically when varying the concentrations of nutrient C_n and agar C_a. Figure 4 shows the phase diagram of various colony patterns observed in the present experiments. The abscissa is indicated by the in-

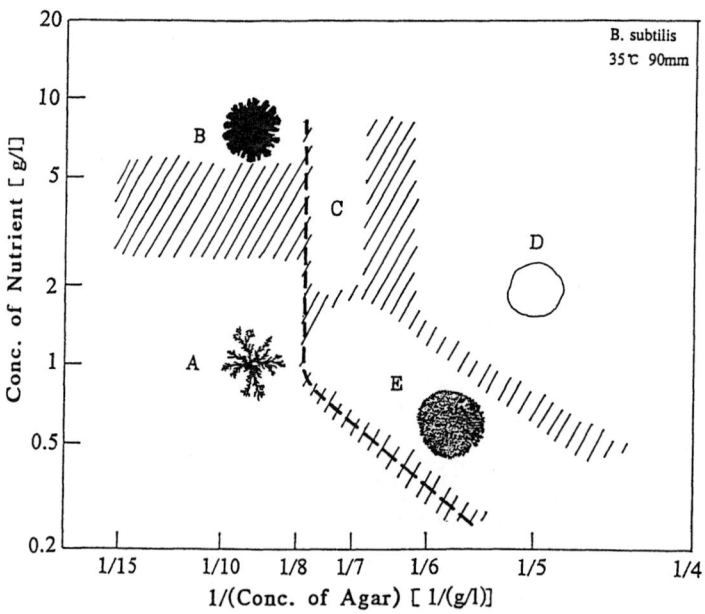

Fig. 4. Phase diagram of pattern change in colonies of *B. subtilis* (wild type). Thick broken line indicates the boundary of observing bacterial cell movement inside colonies.

Fig. 5. Typical Eden-like colony pattern of *B. subtilis* in the region B. C_n=4.5 g/l and C_a=9 g/l. The photo was taken 5 days after inoculation.

verse of agar concentration, which represents the softness of the agar medium. Note also that the ordinate is indicated by a logarithmic scale. Namely, agar plates become softer as we follow the diagram from left to right, while the nutrient richness increases from bottom to top. Describing more relevantly to pattern formation, the softer an agar medium becomes, the faster the nutrient diffuses. This is one way of reducing the nutrient diffusion length ℓ. Another way is to increase the nutrient concentration. Making the agar medium softer induces the active movement of individual bacterial cells as well, as shown in the following section. This turns out to enhance the growth rate of a colony very much.

Here we classify colony patterns into five types, each of which was observed in the region labeled as A-E in Fig. 4, respectively. However, absolute numerical values of the abscissa and ordinate should not be taken too seriously. Phase boundaries of colony patterns have broad characteristic, and also change from strain to strain.

At low C_n (poor nutrient medium) and high C_a (usual hard agar plates), i.e., in the region A in the diagram, we obtained DLA-like colony patterns, as described in detail in the previous section. Figures 1(a) and 2(b) are the typical DLA-like colony patterns.

Increasing C_n with C_a more or less fixed, i.e., moving from the region A to B, colony patterns showed gradual crossover from DLA-like to Eden-like (Fig. 5) at high C_n (rich nutrient medium). The branch thickness of the colony increased gradually as the nutrient concentration was increased. Eventually branches fused together to form a compact pattern, as seen in Fig. 5. However, outwardly growing surface looks still rough. Moreover, two colonies

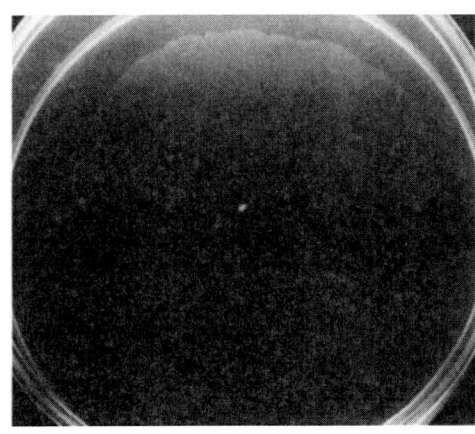

Fig. 6. Typical simply spreading colony pattern of *B. subtilis* in the region D. C_n=1 g/l and C_a=5 g/l. This photo was taken 16 hours after inoculation. Compare the elapsed time with that of DLA-like in Fig. 1(a).

5

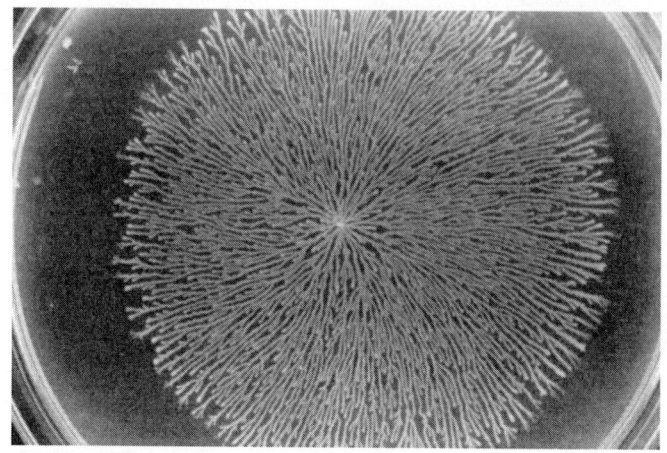

Fig. 7. Typical DBM-like colony pattern of *B. subtilis* in the region E. $C_n=1$ g/l and $C_a=7$ g/l. This photo was taken one day after inoculation.

inoculated simultaneously at two points grow and come close together in the region B, in contrast to the region A (see the former section). This means that the diffusion length ℓ of the nutrient peptone is very small. These properties are characteristic to the Eden growth.[13] However, fjord-like sharp cuts seen in Fig. 5 are not found in the usual Eden growth. They may be due to some biological effects such as the remnant of waste material discharged by bacteria.

In the wide region D with low C_a (soft agar medium) colonies spread confluently with smooth growing surface frontline and no branching at all, as shown in Fig. 6. They grew very fast, and looked too transparent to photograph clearly.

In the narrow region E between A and D where the nutrient is poor and the medium softness is intermediate, the colony morphology (Fig. 7) took patterns clearly reminiscent of the so-called dense-branching morphology (DBM).[14,15] Colonies in this region branch densely, but the advancing envelope (outline) looks characteristically smooth, compared with DLA-like colonies. In this sense the colony patterns seen in the region D look as if dense branches fused together to form homogeneously spreading patterns.

In the region C the morphology took something intermediate (Fig. 8). Colonies grew with tip-splitting and narrow branching just as DBM. But the

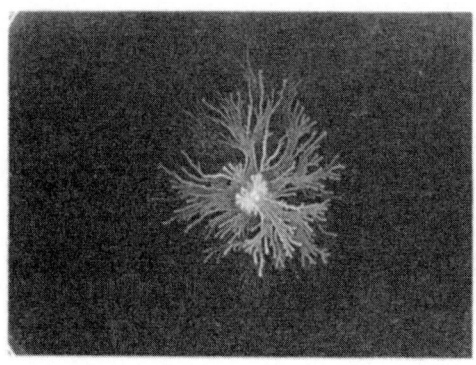

Fig. 8. Typical colony pattern of *B. subtilis* in the region C. $C_n=3$ g/l and $C_a=7$ g/l. This photo was taken one day after inoculation.

6

outline was rather rough as DLA- or Eden-like patterns. In this region colony patterns became denser as the nutrient concentration C_n was increased.

EFFECT OF BACTERIAL CELL MOVEMENT

We notice that the morphological change is conspicuous from the region A (DLA-like patterns) to E (DBM-like), compared with very gradual change from A to B (Eden-like). The growth rate in the region E is also much higher than A. It takes a colony about a month in A and a week in B to grow to the size of about 5 cm (half a diameter of a petri dish containing agar medium), while it takes only a day in C and E and even a shorter time in D. This may imply the essential difference in the way of colony growth. In fact, by microscope observation of growing zones (tips of outwardly growing branches or growing fronts) of bacterial colonies, we recognized two distinct types of growing processes. One is accompanied with no active movement of individual organisms (cells) when the growth was occurring on the surface of high C_a (hard agar) plates (regions A and B). For this type we can say that the growth is relatively "static": Outermost part of a colony grows by cell division, and population mass increases. It takes more than a few days to develop a specific pattern. Bacterial cells in the inner part change to spores and enter into rest phase or hibernation. They never grow afterwards. This means that we can observe the so-called "active zone"[5] experimentally in these regions.

The other type is accompanied with the active movement of individual cells for intermediate and low C_a (soft agar; regions C, D and E). The colony growth of this type is driven by the movement of bacterial cells and is remarkably "dynamic": The outermost growing front is enveloped by a thin layer (thickness ∿5 μm) of bacterial cells whose movement is dull. But cells inside the layer move around actively. Sometimes they collide with the layer, break through it and rush out. But the rushed-out cells become immediately dull. As a consequence, the layer or frontline expands or grows a little. Although the expansion of the frontline is not so speedy as the movement of bacterial cells inside it, the growth rate of a colony as a whole is still much higher than that of the first type. Deep inside the growth front the cell movement is again inactive for the colonies grown in the regions C and E.

It should be noted here that a thick broken line in Fig. 4 which indicates the beginning of the bacterial cell movement described above coincides with almost vertical phase boundaries of morphologies between regions A and E, and B and C in the figure. The verticality is understandable, because the cell movement is mainly influenced by the softness of agar plates. The bending from the verticality in the lowermost part (boundary between regions A and E) may be attributed to the inactivity of cells due to starvation coming from poor nutrient.

Since bacterial cells can move around by using flagella and the movement seems to induce the colony morphology change, we carried out the same experiments described so far but by using the immotile mutant with no flagella. We obtained surprisingly simple phase diagram, as shown in Fig. 9. Now the morphological change observed was only the gradual crossover from DLA- to Eden-like patterns in all the range of the agar concentration examined. We did not observe DBM-like nor homogeneously spreading patterns. In other words, the regions A and B in Fig. 4 expand into the entire region studied in the present experiments, and the regions C, D and E disappear. The bacterial cell movement seen for the wild type really triggers the morphological change.

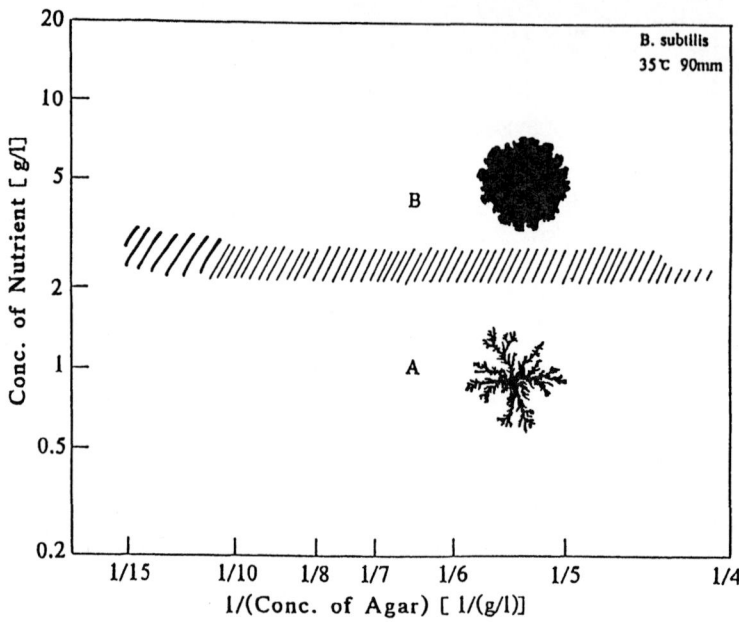

Fig. 9. Phase diagram of pattern change in colonies of immotile mutant of
B. *subtilis*.

SUMMARY AND DISCUSSION

We have inoculated bacteria called B. *subtilis* on the surface of agar
plates in which nutrient and agar concentrations were varied to investigate
the colony morphology change. Colony patterns were in fact found to change
drastically through the variation of the environmental conditions, and we
obtained the phase diagram of bacterial colony morphologies. Thus, we have
observed DLA, Eden-like, DBM-like and simply spreading patterns just by
varying environmental conditions for a single species of bacterium. Such
patterns have also been observed in nature and various physico-chemical
systems.[3-5,10,11,13-15] We believe, therefore, that these patterns are
universal in nature.

Usual bacteria have a tendency to swarm to form colonies on the surface
of solid medium such as agar plates used in the present experiments. This
implies the effective surface tension of the growth front of colonies.
Branches of DLA-like colonies observed in the region A look rather thick in
comparison with computer-simulated DLA clusters, as seen in Figs. 1 and 3.
This may be attributed to the effective surface tension. Finite nutrient
concentrations may also contribute to the branch thickness. Since B.
subtilis has simple rod-shaped cell form, it may not have any singular features in cell division (local growth process) that emerge on the appearence
of colony patterns except the branch thickness. In fact, we have also confirmed the DLA-like colony morphology for common rod-shaped bacterial species, E. *coli* and *Salmonella typhimurium*.[16]

Conversely, suppose we observe colony patterns which are very different
from DLA, in spite of having grown under the condition of diffusion-limited
growth. This may be regarded as the influence of some singular features in
the cell division on the colony morphology. The DLA growth can, therefore,
serve us as the basis for studying more complicated morphologies inherent in
biology.

In order to confirm real Eden-like colonies, we have to show that the growing surface is self-affine with particular scaling exponents.[13] Recently, Vicsek et al.[17] investigated the self-affine surface of colonies of bacterial species, *E. coli* and *B. subtilis*. We think that more extensive studies are clearly needed as for this, though.

Up to the present a model that describes unambiguously the DBM formation has not been established yet, although it has been observed in various systems.[6,7,14,15] However, we believe that there may be some clue to find the DBM mechanism in the characteristic behavior of colony growth.

ACKNOWLEDGEMENTS

The authors express their sincere thank to Dr. H. Fujikawa for providing bacterial strain (wild type of *B. subtilis*). This work was supported in part by the Grant-in-Aid for Scientific Research from the Ministry of Education, Science and Culture of Japan (Nos. 0280826 and 02804019).

REFERENCES

1. P. Singleton and D. Sainsbury, "Introduction to Bacteria", Wiley, Chichester (1981).
2. T. Matsuyama, M. Sogawa and Y. Nakagawa, FEMS Microbiol. Lett. 61, 243 (1989).
3. See, e.g., articles by M. Matsushita, G. Daccord and H. Van Damme, in: "The Fractal Approach to Heterogeneous Chemistry", D. Avnir, ed., Wiley, Chichester (1989).
4. J. Feder, "Fractals", Plenum, New York (1988).
5. T. Vicsek, "Fractal Growth Phenomena", World Scientific, Singapore (1989).
6. H. Fujikawa and M. Matsushita, J. Phys. Soc. Jpn. 58, 3875 (1989).
7. H. Fujikawa and M. Matsushita, J. Phys. Soc. Jpn. 60, 88 (1991).
8. M. Matsushita and H. Fujikawa, Physica A 168, 498 (1990).
9. T. A. Witten and L. M. Sander, Phys. Rev. Lett. 47, 1400 (1981).
10. P. Meakin, in: "Phase Transitions and Critical Phenomena", Vol. 12, C. Domb and J. L. Lebowitz, ed., Academic Press, New York (1988), p. 335.
11. M. Matsushita, in: "Formation, Dynamics and Statistics of Patterns", Vol. 1, K. Kawasaki, M. Suzuki and A. Onuki, ed., World Scientific, Singapore (1990), p. 158.
12. P. Meakin, J. Theor. Biol. 118, 101 (1986).
13. F. Family and T. Vicsek, ed., "Dynamics of Fractal Surfaces", World Scientific, Singapore (1991).
14. Y. Sawada, A. Dougherty and J. Gollub, Phys. Rev. Lett. 56, 1260 (1986).
15. D. Grier, E. Ben-Jacob, R. Clarke and L. M. Sander, Phys. Rev. Lett. 56, 1264 (1986).
16. T. Matsuyama and M. Matsushita, unpublished.
17. T. Vicsek, M. Cserzö and V. K. Horvath, Physica A 167, 315 (1990).

INTERFACIAL PATTERN FORMATION IN BIOLOGICAL SYSTEMS:

PRELIMINARY OBSERVATIONS DURING GROWTH OF BACTERIAL COLONIES

E. Ben–Jacob, H. Shmueli, O. Shochet and D. Weiss

School of Physics and Astronomy
Raymond & Beverly Sackler Faculty of Exact Sciences
Tel-Aviv University, Tel-Aviv 69978, Israel

INTRODUCTION

The pioneering studies of Fujikawa and Matsushita [1, 2] have demonstrated that the development of bacterial colonies grown on thin agar plates is an example of a quasi-two-dimensional biological growth process which is diffusion-controlled (diffusion of nutrients towards the colony). The patterns they have observed (fractals – DLA like [3, 4, 5, 6, 7], tip-splitting – DBM like [8, 9] and compact – Eden like [10]) are very similar to patterns which were predicted and observed in diffusion limited growth of azoic (non-living) systems. Moreover, they have shown that the various morphologies (morphology here refers to the global structure of the colony) can be organized in a morphology diagram indicating the existence of a morphology selection principle, again as was proposed for azoic system [9, 11, 12]. We have repeated these studies, focusing on the morphology transitions, and on comparison to inorganic systems that show similar patterning. We were seeking to reveal the common principles and to quantify the bacteria-bacteria interaction in terms of surface tension and surface kinetics [13]. Our main present motivation is two-fold: 1. To identify the additional principles that are distinct to living systems. 2. To perform experiments that will bring us closer to answer questions regarding adaptation and selection.

For spontaneous appearance of patterns, the system has to be driven out of equilibrium. It is now understood that the patterns result from competition between the dynamics on different levels [9, 14, 15]. The interplay between the microscopic level and the macroscopic level is self evident in the growth of snowflakes. The sixfold symmetry of the underlying ice crystal lattice is manifested in the six fold dendritic branches of the flake. How is the macroscopic level influenced by microscopic effects in this way? One can obtain qualitative understanding by considering the dynamics of non-equilibrium growth. In the formation of snowflakes, the stable solid phase propagates into the unstable supersaturated water-vapor phase. The rate of growth of the stable phase is limited by a diffusion process – the diffusion of water molecules from the gas phase towards the crystal. In the process, the kinetics of diffusion tend to drive the system towards the formation of "decorated" and irregular shapes. Diffusion kinetics determines the macroscopic approach towards equilibrium, and acts to intermingle the two phases on many length scales. By contrast, the microscopic dynamics occurring at the interface (determined by surface tension, surface kinetics and anisotropy) is associated with microscopic length scales and with organizational symmetries characteristic of the material on the micro-level. The basic diffusion field instability is channeled by the micro-level effects. Thus, in the example of the snowflake, the microscopic 6-fold symmetry channels the branching mechanism to form six main branches (dendrites). The ultimate interfacial pattern is then selected as a result of a two-way transfer of information between the microscopic and macroscopic levels.

Just as the microscopic dynamics affects the macroscopic morphology, so can macroscopic dynamics determine the microstructure of the system. If the system is sufficiently far from equilibrium, the interface may advance so rapidly that the newly formed phase does not have time to reach its lowest energy state on the microscopic level. The result may be a metastable state on the microstructural level. The growth of quasicrystals is an example of just such a process. Thus, the kinetics of the diffusion process, on the macroscopic scale, can reach down to the microdynamics and change the microscopic structure.

Mathematically, the micro-level dynamics act as singular perturbations in the dynamical equations for the macro-level, with an effect which is amplified the further we drive the system out of equilibrium. As we change the relative effects of the different dynamics, different morphologies are selected. We now understand why, on one hand, different systems share a common set of patterns, and on the other hand, the same system can show different shapes as we vary the control parameters.

In figure 1 we show three basic morphologies observed in Hele-Shaw experiment [9] and during electrochemical deposition [16, 9, 6]. They are: 1. Fractal growth, which is the characteristic shape when the diffusion instability and noise dominate the growth. 2. Dense branching morphology, which is characteristic of tip-splitting dynamics. It reflects the combination of the diffusion field instability and the stabilizing effects of surface tension and surface kinetics [9]. 3. Dendritic growth, which is observed when the crystalline anisotropy dominates the dynamics [9, 14, 15].

Ben-Jacob *et al.* [9] proposed the existence of a morphology selection principle ("The fastest growing morphology" principle). Such principle implies the existence of morphology diagrams in analogy with phase diagrams in equilibrium. For phases in equilibrium, for a given set of conditions the phase that minimizes the free energy is the selected one, irrespective of the prior history of the system. The concepts of a selection principle and a phase diagram go hand in hand. By contrast, non-equilibrium growth processes are time dependent, so it is not clear *a priori* that a morphology diagram should exist (that is, that the shapes will depend only upon the growth conditions and not on the history). However, if it does exist a selection principle must exist if a given morphology is reproducible for a given set of growth conditions. Given such a morphology selection principle, it is possible to generate a map of what shapes should be observed for what growth conditions. The existence of a morphology diagram has been confirmed experimentally in various systems, suggesting that a selection principle must exist. The existence of a morphology selection principle further implies the existence of morphology transitions. That is a sharp transition between morphologies as function of the controlled parameters.

In living systems (i.e. - a bacterial colony) there is an inherent additional level of complexity, as the building blocks themselves are living systems (the individual bacteria) that can go through major changes during the growth in response to the controlled parameters. It has been shown by Shapiro *et al.* [17, 18, 19] that macro-level changes (the morphology of the whole colony) go hand in hand with micro-level (the bacteria level) changes. We have classified three types of changes in the macro-level shape of the growing colony: 1. Sharp and extended (all around the interface) transition. These are similar to morphology transitions in nonliving systems. 2. Sharp and localized transitions (not observed in nonliving systems). 3. Gradual change over a fraction of the interface. The first type is a phenotypic adaptation, hence will be referred to as phenotypic morphology transition. We expect the second transitions (which will be referred to as a burst of new morphology) to arise from a mutation which improves the adaptation and become the dominant genotype (irreversible adaptation). At present, we are not certain about the micro-level origin of the third type. Our preliminary observations suggest that it might be related to reversible mutation (phase variation [20, 21]).

EXPERIMENTAL METHODS

The experimental methods that we have used are very similar to those used by Fujikawa and Matsushita [1]. The bacteria were grown on agar substrate in a standard plastic petri dishes with a diameter of 88mm. The medium consisted of 5 g of sodium chloride, 5 g of tri-Kaliumphosphat-3-hydrat ($K_3PO_4 \cdot 3H_2O$) and an amount of peptone (Bacto-peptone; Difco Laboratories) ranging from 0.3 g (poor nutrient) to 10 g (rich nutrient) and 15 g of agar (Bacto-agar; Difco) in one litre of distilled water at pH 7.0 ± 0.2. The mixture was autoclaved for 30 min. The agar is poured into the petri dishes (about 20 ml per plate). The plates were dried at room temperature ($23\text{-}25°C$) for 24 to 48 hrs. The amount of water evaporated is about 22% per day of drying determined by weight measurement. The bacteria we used were *Bacillus subtilis* strain 168. Inoculation was on the surface of the agar, usually at the center of the petri dish (unless more than one colony was grown to study competition). The colonies were incubated in a closed incubator at a temperature of $37 \pm 1.5°C$.

For photography, the colonies were stained with 0.1 % Coomassie Brilliant Blue R (Sigma chemical Co.) in methanol:acetic acid:distilled water (50:10:40). The strain is then washed using the same mixture

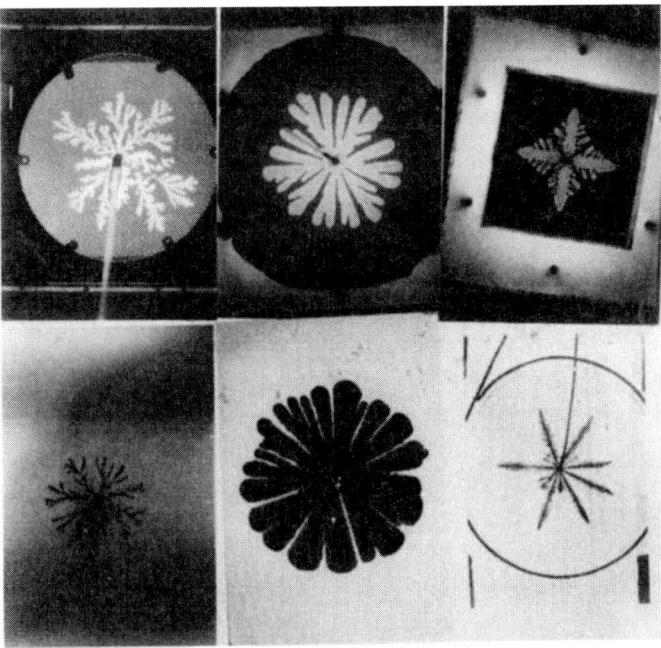

Figure 1. Three basic morphologies observed in Hele-shaw cell (upper row) and during electrochemical deposition (bottom row). Fractal growth (left), tip-splitting (middle) and dendritic (right).

without the stain. Microscopic observations were performed using an optical microscope (Olympus BH2-UMA) with ultra-long range objectives of up to 500× magnification. For higher magnifications we used a scanning electron microscope, utilizing the method developed by Shapiro [17].

MORPHOLOGY DIAGRAM AND TRANSITIONS

In fig. 2, we show three of the basic morphologies observed as we vary the nutrient concentration and humidity: These range from a compact (Eden like) growth, via tip-splitting growth (DBM), to a fractal (DLA like) growth. The observations can be summerized in a morphology diagram, as was found by Fujikawa and Matsushita [1]. Here, we are mainly interested in the range of parameters where the growth dynamics is via a cascade of tip-splitting leading to a dense-branching morphology (DBM). As was mentioned earlier, for this range of parameters the growth is diffusion controlled. Generally speaking, the reproduction rate which determines the overall growth rate is limited by the level of nutrient concentration c. The latter is limited by the diffusion of nutrient towards the colony. Hence, the process is similar to a diffusion limited growth in other azoic systems such as solidification from supersaturated solution, solidification from undercooled liquid, growth in Hele-Shaw cell, electro-chemical deposition etc. However, there are at least two essential differences [13]. First, the concentration field c extends into the colony. Second, the bacteria may secrete substances whose concentrations u could affect the dynamics of the bacteria including, the formation of a well defined envelope. In a forthcoming publication, we will describe in detail the model we propose for the growth of the bacterial colony and the predictions that follow [13]. The special nature of bacterial growth is reflected in the additional observed morphologies such as convective-like growth [22] (fig. 4b in ref. 2) and chiral growth described bellow.

Microscopic studies reveal the microscopic dynamics of the bacteria and its structure. Inside each of the colony tips the motion of the bacteria looks as a random motion of fluid molecules. Close to the boundaries, the motion is more organized. Further down the "finger" (older parts) the density of bacteria is higher, the movement is slower and the mean free path is much shorter. The individual bacterium has a rod like shape (vegetative mode) of about $0.6\mu m$ width and about $2 - 3\mu m$ in length. The length is sensitive to the level of nutrient concentration (longer for low concentration) and the age of the bacteria (short for "old" bacteria). The motion of longer bacteria shows higher degree of internal correlation.

13

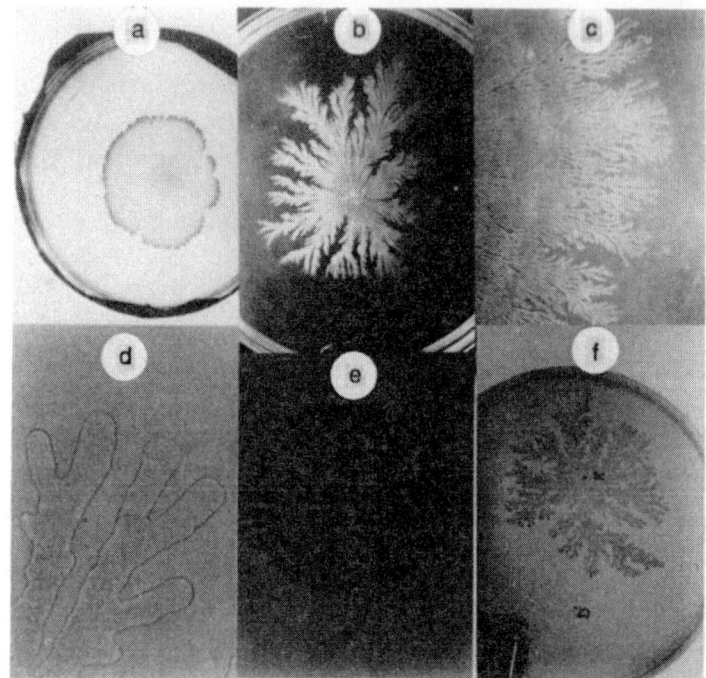

Figure 2. a. Compact growth (Eden-like) b-e. Dense-branching growth. f. DLA-like growth.

The last-decade developments in our understanding of interfacial pattern formation, predicts the existence of a morphology selection principle, the existence of a morphology diagram and the existence of a sharp morphology transitions [9]. In fig. 3 we show three transitions: fractal to DBM, compact to DBM and convective to DBM. For all three cases, the transition is very sharp and occurs simultaneously all around the interface. Reinoculation of either morphologies on the same plate lead to the same morphology irrespective of the origin. Hence, we identify the transition as a phenotypic morphology transition. We induce the transition by varying either the temperature or the humidity. In many cases, spontaneous transitions were also observed, resulting from changes in the humidity and nutrient concentration during growth. These transitions are reminiscent of the Hecker transitions [23, 24] observed in electro-chemical deposition experiments.

DENDRITIC MORPHOLOGY IN THE PRESENCE OF ANISOTROPY

It is now well understood that anisotropy is needed in order to have dendritic growth (one backbone with parabolic tip and side-branches) rather then a cascade of tip splitting. Hence, an additional test of the similarity of bacterial growth to interfacial pattern formation would be to impose anisotropy on the growth of the bacteria and to look for dendritic growth.

Anisotropy was imposed (however, not in a controlled manner) via ripples on the agar surface. While the agar is dried rapidly, wrinkles and ripples are formed as shown in fig. 4. When they are pronounced, they channel the growth of the colony (fig. 4) and provide anisotropy sufficient for dendritic growth (fig. 5).

Microscopic studies during dendritic growth show that the microscopic dynamics is very similar to that of the case of tip-splitting. It seems that the motion is more coordinated but quantitative measurements are still required.

BURSTS OF NEW MORPHOLOGY

In fig 6 we show a novel phenomenon (not observed in nonliving systems) in which a new morphology bursts at a localized point along the interface. The new morphology has higher growth velocity

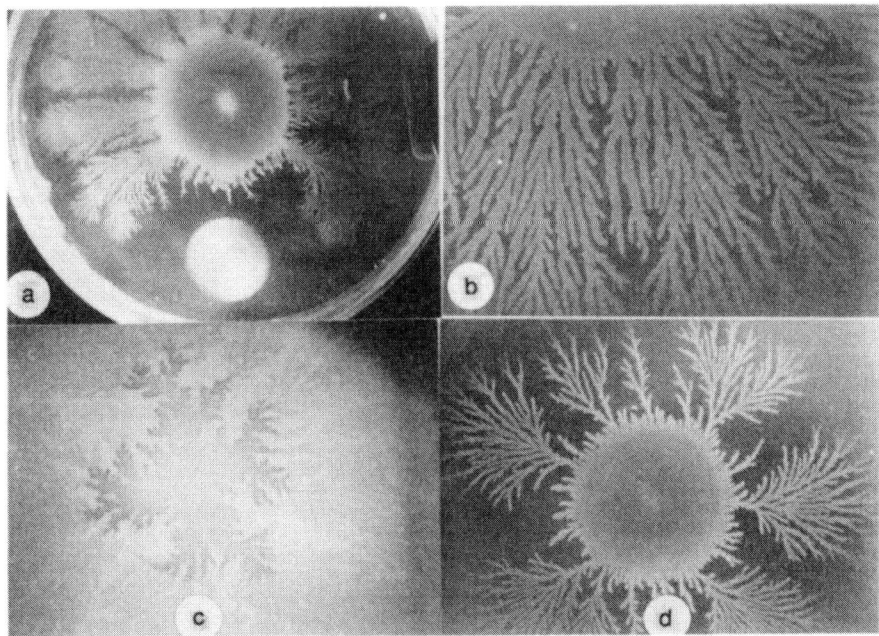

Figure 3. Morphology transitions during growth: a. compact to DBM. b. Closer look at the transition. c. fractal to DBM. d. convective to DBM. (The white circle at the bottom of fig. a is a fungus)

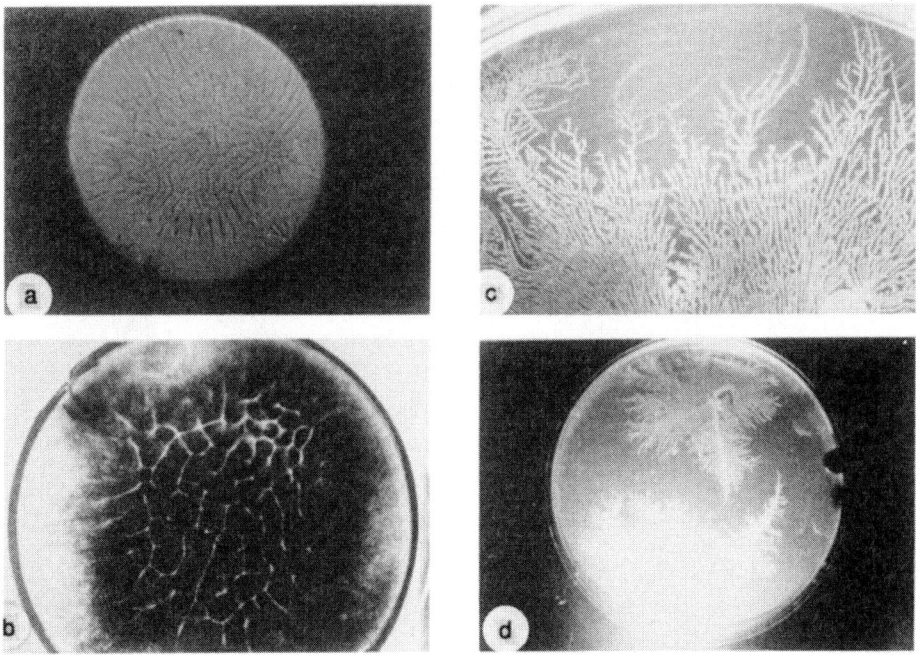

Figure 4. The effect of wrinkles and ripples on the agar. a,b. Pictures of the agar using diffraction of light to show the Deformations. c,d. Picture of the overall colonies which are affected by the deformations of the agar.

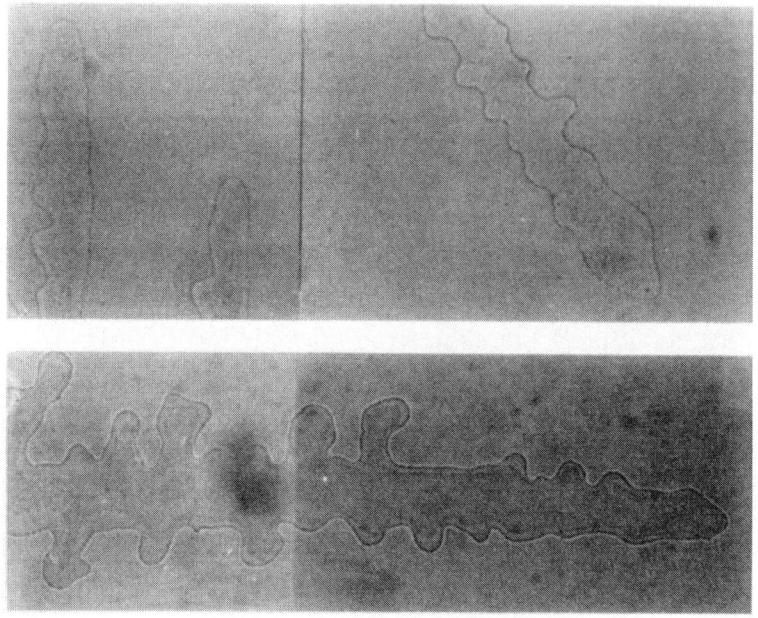

Figure 5. Close-up pictures of dendritic growth resulted from the deformations on the agar. The growth has clear main trunk with parabolic tip and emitted side branches.

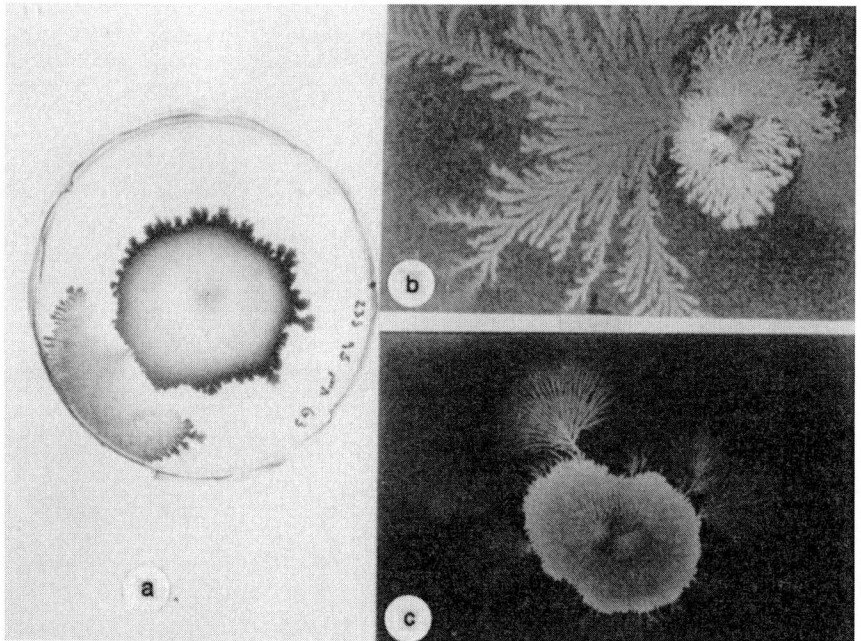

Figure 6. Bursts of new morphologies. a. A DBM burst from compact growth at a single localized point. b. A cascade of morphology bursts. Each new morphology starts at a localized point. c. Simultaneous bursts of a number of morphologies, some similar and some different

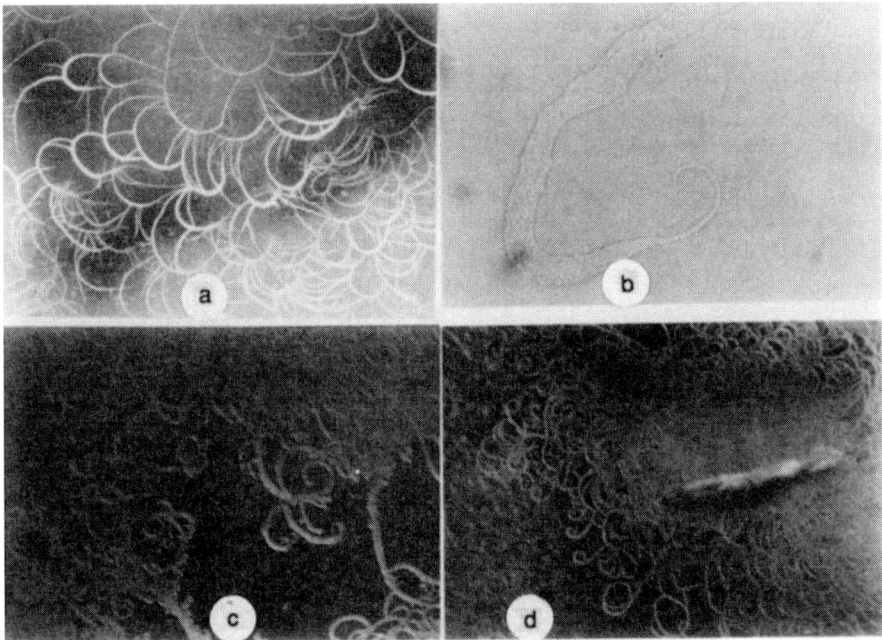

Figure 7: Chiral growth. a-c have the usual handedness-that we observe. d. show that the opposite handedness can also occur (but rarely under the conditions that we use). b is a closeup on a single chiral branch.

and outgrows the original morphology. We think that the burst of the new morphology results from a mutation of a new strain whose colony can adapt better to the growth conditions. In the few cases that we studied, reinoculations of colonies whose ancestors came from the original colony and from the burst, led to a different growth under the same growth conditions. Hence, while additional studies are needed, it is suggestive that the burst is due to a genotypic change.

CHIRAL MORPHOLOGY

In fig. 7 we show another interesting morphology observed during growth of bacterial colonies. That is a chiral growth in which the colony consists of chiral branches *all having the same handedness*. Similar patterning is also observed in organic (but non-living) systems: 2D solidification of phospholipids consisting of single handed izomers. Chiral growth is also observed during solidification of certain salts [25, 26]. In both cases, the mechanism in which the micro-level handedness is transferred to the macro-level is not yet understood.

Optical microscope observations indicate that, during chiral growth, the colony consists of a mixture of the ordinary rod-shaped bacteria and much longer, string-like, ones. The long bacteria look like typical *Bacillus subtilis* grown under conditions of low nutrient concentration, and have a chiral character as well. Mendelson [27] showed that long cells of *Bacillus subtilis* can grow in helices in which the bacteria are twisted around each other. He found left handedness below 39°, and right handedness above this temperature. However, this handedness is not the one that directly leads to the chiral growth that we have observed, as we didn't notice any helices of the kind reported by Mendelson.

The micro-level dynamics consists of slow swarming of the long bacteria and faster movement of the shorter ones. Unlike the case of tip-splitting, the motion is coordinated. In some cases, a circular stable motion inside wide branches is observed. The chiral behavior is not unique to the *Bacillus subtilis* Such dynamics is also observed in other bacteria (such as *Proteus* [28]).

Next, we studied the dependence of the chiral morphology on the growth conditions. The motivation was to find whether it has a specific range of parameters on the morphology diagram. We found that

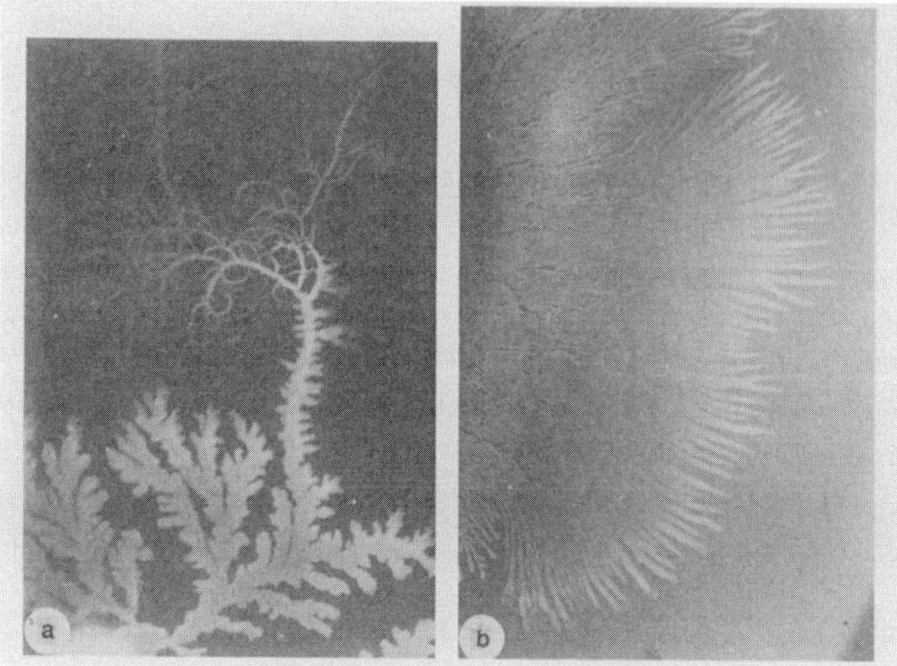

Figure 8. a. Spontaneous transition from tip-splitting to chiral growth. The transition was via intermediate stage of dendritic growth. Hence, we think that is was triggered by local deformation in the agar. b. Transition from DMB to convective growth, We expect this transition to be also reversible adaptation.

this is not the case. Instead it turns out that the occurrence is strongly dependent on history. We inoculated two colonies on the same agar plate, one from tip-splitting ancestors and one form chiral "parents". Both colonies grow as their ancestors, and keep doing it down the generations. These observations, suggest that the transition is not simply due to a phenotypic change. However, the fact that the transition occur simultaneously over a finite fraction of the interface might indicate that it is also not a simple mutation. Moreover, it seems to be possible to switch on and off between tip-splitting and the chiral growth. Hence, it appears to behave as a reversible mutation between two connected genotype in a manner that might be related to phase variation [20, 21]).

COMMENT ON FUTURE DIRECTIONS

Our preliminary observations present new phenomena which require additional detailed study before a solid interpretation can be derived. We have to study the chiral–tip-splitting transition as a function of growth conditions. In particular its temperature dependence, including the dependence of the handedness on temperature. A controlled method for imposing anisotropy has to be developed. Once this goal is achieved, it will be interesting to study the dendritic–tip-splitting transitions. Finally, our studies have to be completed with genetic studies. This is especially important in the case of bursts of new morphologies, to reveal if indeed it is related to mutations. Fig. 8c demonstrates some of the difficulties in interpretation as it show different transitions at the same time.

ACKNOWLEDGEMENTS

We are most greatful to Prof. M. J. Matsushita for many useful discussions and communicating with us his results prior to publications. We thank E. Z. Ron and J. A. Shapiro for introducing us to the "world" of bacterial growth and for their encouragment, many discussions and comments on our manuscript. We are thankful to G. Segal and O. Shmueli for their technical assistance. We have also benefited from conversations with Y. Aharonov, M.Y. Azbel, Y. Elkana, M. Kaizer, R. Kupferman, R. .G. Mintz and M. Sternberg.

This research was supported in part by a grant from the G.I.F., the German-Israeli Foundation for Scientific Research and Development, and by the Program for Alternative Thinking at Tel-Aviv university.

REFERENCES

[1] H. Fujikawa and M. Matsushita, J. Phys. Soc. Jap., Vol.58 No.11 (1989) 3875.

[2] H. Fujikawa and M. Matsushita, Phys. Soc. Jap., Vol.60 No.1 (1991) 88.

[3] T. A. Witten and L. M. Sander, Phys. Rev. Lett. 47 (1981) 1400.

[4] T. Vicsek, *Fractal Growth Phenomena* (World Scientific, New York, 1989).

[5] J. Feder Fractals (Plenum, New-York, 1988).

[6] L. M. Sander, nature 322 (1986) 789.

[7] P. Meakin, Phys. Rev. A 33 (1986) 3371.

[8] E. Ben-Jacob, G. Deutscher, P. Garik, N. D. Goldenfeld and Y. Lareah, Phys. Rev. Lett. 57 (1986) 1903.

[9] E. Ben-Jacob and P. Garik, Nature 343 (1990) 523.

[10] M. Eden, in: Proc. 4th Berkeley symp. on Mathematical Statistics and Probability, vol 4, F. Neyman, ed. (University of california Press, Berkeley, 1961) 223.

[11] E. Ben-Jacob, P Garik, T. Muller and D. Grier, Phys. Rev A 38 (1988) 1370.

[12] E. Ben-Jacob and P. Garik, Physica D 38 (1989) 16.

[13] E. Ben-Jacob, R. Kupferman, R.G. Mintz and O. Shochet, in preparation.

[14] J. S. Langer, Science 243 (1989) 1150.

[15] D. A. Kessler, J. Koplik and H Levine, Adv. Phys. 37 (1988) 255.

[16] M. Matsushita, M. Sano, Y. Hayakawa, H. Honjo and Y. Sawada, Phys. Rev. Lett. 53 (1984) 286.

[17] J. A. Shapiro, J. Bact. 169 (1987) 142.

[18] J. A. Shapiro and C. Hsu, J. Bact. 171 (1989) 5963.

[19] J. A. Shapiro and D. Trubatch, Physica D 49 (1991) 214.

[20] X. He, R Shen and Z. Seng, Act. Gen. Sin. 17 (1990) 216.

[21] G.D. Christensen, L.M. Baddour, M.B. Madison, J.T. Parisi and S.N. Abrahams, J. Inf. Dis. 161 (1990) 1153.

[22] P. Garik, J. Hetrick, B. Orr, D. Barkey and E. Ben–Jacob, Phys. Rev. Lett. 66 (1991) 1606.

[23] N. Hecker, Senior Thesis, University of Michigan, (1988).

[24] P. Garik, D. Barkey, E. Ben–Jacob, E. Bochner, N. Broxholm, B. Miller, B. Orr and R. Zamir, Phys. Rev. Lett. 62 (1989) 2703.

[25] H.E. Gaub, V.T. Moy and H.M. McConnell, J. Phys. Chem. 90 1721.

[26] R.M. Weis and H.M. McConnel, Nature 310 (1984) 47.

[27] N. H. Mendelson, Proc. Natl. Acad. Sci. USA, Vol 75, No 5, (1978) 2478.

[28] J. A. Shapiro, private communications.

AMOEBAE AGGREGATION IN DICTYOSELIUM DISCOIDEUM

Herbert Levine and William Reynolds

Department of Physics and Institute for Nonlinear Science
University of California, San Diego
La Jolla, California 92093

Abstract

Patterns formed during the aggregation phase of D. Discoideum have long been studied as a model system for biological organization . In particular, individual amoebae aggregate upon starvation and form a rudimentary multi-cellular slug as a precursor to stalk/spore differentiation and sporulation. Here, we discuss this system using recently developed ideas concerning waves (and their stability) in excitable media. We show that the onset of streaming can be attributed to an instability due to density fluctuations. Finally, we comment on possibilities for simulations and for comparisons to new experimental findings.

1 Introduction

The amoeba of the cellular slime mold Dictyoselium discoideum is a participant in one of the most striking examples of biological self-organization [1]. Upon starvation, 10^4 to 10^5 individual amoeba aggregate to a common site and form a rudimentary multicellular organism known as a slug or grex. This is accomplished by an elaborate chemical signalling system which causes the formation of outgoing (from the aggregation site) waves of cyclic adenine monophosphate (cAMP); these waves induce chemotactic motion of the cells during the aggregation phase. There is also evidence that cAMP signalling is important for slug motion and integrity [2], but that is not our focus here.

The cAMP signalling system has been studied extensively for many years. The necessary ingredients for a model of this system include mechanisms for detection of cAMP concentration by the cell, signal amplification via cAMP synthesis and a method for degradation of excess cAMP concentration. One compelling approach which has been put forth by Goldbeter and Martiel [3] makes use of a "receptor box" mechanism. Roughly, each cell has 10^6 receptors, each of which can be in four states, bound or unbound, actuated or deactivated. The activation/deactivation step is believed to be due to phosphoryllization. Upon bonding to cAMP, an autocatalytic production of cAMP inside the cell commences; the chemical is then released from the cell and, via diffusion, can activate other cells. Meanwhile, the local increase in cAMP tends to de-activate the receptor and causes the process to switch off at the same time as the cAMP is destroyed by phosphodisterase. In the next section, we will discuss more explicitly the "excitable" reaction-diffusion equations which result from this approach.

There are, of course, other proposed kinetics for the signalling system [4]. Although our calculations to date have all been done within the Martiel-Goldbeter (MG) approach, most of our results would remain qualitatively unchanged with our approaches to this "excitable" signalling medium.

Once the cAMP kinetic equations have been formulated, one can apply standard ideas from the theory of excitable systems [5] to find the form of the chemical waves seen in the aggregation. This will be reviewed in Section II, with particular emphasis on planar travelling waves. As far as these structures are concerned, this biological system is analogous to, say, CO catalysis on a metallic substrate [6] - this latter system is also describable by excitable kinetics and in both cases, the "species" responsible for the slow reaction does not diffuse. For our case, this species is the activated cell receptor and hence in the absence of chemotactic effects this density does not spread.

Of course, neglecting chemotaxis is only an approximation which enables us to make a preliminary estimate of the more rapid time evolution of the cAMP signal. Chemotaxis [7] must be included for two reasons, one practical and one fundamental. From the practical perspective, most measurements of the aggregation process actually image the moving cells, since these elongate as they move and offer a preferred axis for polarized light microscopy [8]. More fundamentally, we have shown that inclusion of chemotaxis leads to an instability in which density fluctuations are amplified. These issues will be discussed in Section III.

Finally, Section IV presents a perspective on our current efforts to formulate a fully nonlinear simulation of the coupled signalling-chemotaxis system. It is worthwhile mentioning here the underlying philosophy of this research. In general, biological systems are extremely complex and cannot be modeled with full faithfulness to all the biochemical underpinning of the "macroscopic" equations of interest on the multi-cellular scale. Instead, we hope to come up with phenemenological models that answer questions of principle - e.g., can such and such a pattern be explained merely on the basis of signalling and chemotaxis, or are cell contact forces necessary; can we understand the minimal amount of information from genes that could conceivably build a particular spatial structure, etc. The only way we know of doing this is to work out all the consequences of an assumed model, compare to experiments, add complexity to the model and so on.

2 The cAMP Signalling System

In this section, we review the model equations of Martiel and Goldbeter for the signalling system. We show how waves in this system can be studied using standard ideas from the theory of excitable media. This will set the stage for our discussion of chemotaxis in the next section.

In the Goldbeter-Martiel model, the complicated kinetics of cAMP reception, generation and release is reduced to the two equation system

$$
\begin{aligned}
\dot{\psi} &= \epsilon \nabla^2 \psi + \frac{1}{\epsilon} f(\psi, r) \\
\dot{r} &= g(\psi, r)
\end{aligned}
$$

where ψ is the (extracellular) cAMP concentration and r is the fraction of activated receptors. The kinetic laws are given by $f = s\Phi - \psi$ with

$$
\Phi(\psi, r) = \frac{\lambda_1 + \Lambda^2(\psi, r)}{\lambda_2 + \Lambda^2(\psi, r)} \quad \Lambda = \frac{r\psi}{1 + \psi}
$$

and $g = -r f_1(\psi) + f_2(\psi)(1 - r)$ with

$$
f_1 = \frac{1 + \sigma \psi}{1 + \psi} \quad f_2 = \frac{L_1 + \sigma L_2 h \psi}{1 + h \psi}
$$

The various parameters typically have values listed in Table I; it is important to remember that all of these depend on the cell density.

This set embodies the complex receptor box mechanism described earlier and can be understood in detail by studying the original Goldbeter-Martiel paper. For our purposes, we would like

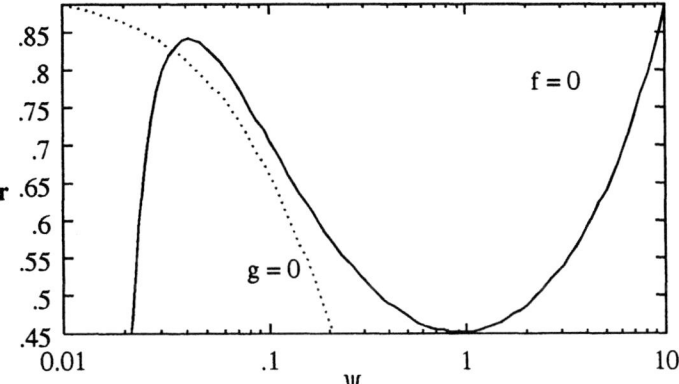

Figure 1. A Plot of the nullclines of the Martiel-Goldbeter Kinetics.

to point out that f represents both creation of cAMP by an auto-catalytic process dependent on the product of cAMP concentration ψ and activated, bound receptor $r/1 + \psi$ and also depletion via breakdown; similarly, g embodies the activation of de-activated receptors (the 1-r term) as well as the de-activation of activated receptors.

In Figure 1, we plot the nullclines of f and g with the assumed parameters. Note that at fixed r, there are three roots of the equation $f = 0$. Now in the kinetic equation the parameter ϵ represents the ratio of the g reaction rate to the f one. Since ϵ is small, the system will very quickly relax to one of the two stable branches of this nullcline and slowly evolve along them via the g dynamics. Since the $g = 0$ nullcline intersects the $f = 0$ one along a stable branch, the dynamical system exhibits a stable fixed point. However, if a large enough perturbation is applied, the system may jump to the large ψ branch (the excited one) and only recover slowly. This feature of the kinetic laws defines this system as an excitable medium.

Excitable systems support various types of waves including planar fronts, target patterns (caused by a point pacemaker) and rotating spirals [9]. An approximate procedure for finding these solutions proceeds as follows: [10] at fixed r, we can solve the ψ equation for a front which connects the excited phase $\psi = \psi_+(r)$ as $x \to -\infty$ to the quiescent phase $\psi = \psi(r)$, $x \to +\infty$. This front will move at a fixed velocity $c(r)$ and crosses zero at some unique value of r which we call the stall concentration r^*. Waves of relatively small velocities will have r field values which never get to be much different than r^*. We can then expand the kinetic equations to get the much simpler system

$$\dot{r} = a_\pm \ , \ a_\pm \equiv g(\psi_\pm(r^*), \ r^*)$$

where the subscript denotes in which phase a given point lies; this must be supplemented by the continuity of r and the boundary condition

$$c_n = c(r) - \epsilon \kappa$$

where c_n is the (normal) velocity of the wavefront, and κ is the wavefront curvature.

Let us illustrate the planar wave solution using this framework. Here, the entire pattern is assumed to move with velocity c_0; this requires the r values of the wavefront (transition from excited to quiescent as x increases) and waveback (quiescent to excited) to equal r_+, r_- respectively with

$$c_0 = \pm \ c(r_\pm).$$

Using $\dot{r} = -c_0 \frac{dr}{dx}$, the field solution is

$$r(x) \ = \ \begin{matrix} r_- - \frac{a_+ x}{c_0} & 0 < x < \lambda_+ \\ r_+ - \frac{a_-}{c_0}(x - \lambda_+) & \lambda_+ < x < \lambda_+ + \lambda_- \end{matrix} \qquad (1)$$

Requiring continuity of r at $x = \lambda_+$ and at $x = 0$ and $x = \lambda_+ + \lambda_-$, we obtain

$$\lambda_+ = \frac{c_0(r_- - r_+)}{a_+} \qquad \lambda_- = \frac{-c_0(r_- - r_+)}{a_-}$$

for the respective widths of the excited and quiescent phases.

One can develop a similar approach to target patterns and, at least outside the core, to rotating spirals [11]. Also, a straightforward extension of this method allows us to show that these waves are all linearly stable [12]. This will change dramatically when we introduce a coupling to cell chemotaxis.

3   Chemotaxis and the Streaming Instability

At the level of the Goldbeter-Martiel model, the cAMP signalling system is no different than a purely chemical excitable media. However, as we probe more deeply, we see that the cell density, assumed constant in the signalling treatment, is actually a dynamical variable which evolves via chemotaxis. As we shall see, the chemotactic response is the cause of an instability in the chemical wave system which ultimately leads to the formation of dense streams of cells from the initially homogeneous distribution.

Cell chemotaxis refers to the ability of a cell to detect a chemical gradient and move in the direction of increasing concentration. This can be accomplished in a number of ways. Bacteria appear to determine gradients by comparing concentrations at the beginning and end of a random tumbling motion [13]. Biased orientation of the next motion can lead to chemotaxis as a statistical bias in an otherwise random diffusive process. In the case of Dictyoselium amoeba, the method is more direct as it appears to involve gradient detection across a single non-moving cell by differences in receptor filling on opposite sides of the cell membrane [14]. Because of this, the chemotactic response is more pronounced - cells tend to move in the gradient direction with limited diffusive spreading [7]. This will affect our modeling approach to this phenomenon.

A detailed biochemical approach to chemotaxis in the amoeba is neither necessary nor even possible at this stage of knowledge. Instead, we will use some basic phenomenological facts, gleaned from actual experiments, to postulate a simple model [16]. First, the aforementioned direct response suggests that the gradient directly determines a cell velocity. If we combine this with the fact that a cell typically moves for some time after the gradient ends, we arrive at

$$\frac{d\vec{v}}{dt} = -\Gamma \, \vec{v} + k(\psi) \, \vec{\nabla} \, \psi$$

with Γ equal to the relaxation time. Now, the cells are much more responsive to gradients in the low cAMP state than in the high; this could simply be due to a saturation of the receptors. Therefore, the rate $k(\psi)$ is a (rapidly) increasing function of the cAMP concentration ψ. Finally, the density is merely advected along by the above velocity

$$\frac{\partial \rho}{dt} + \vec{\nabla} \cdot (\rho \, \vec{v}) = 0$$

The plane wave solution as determined in section II by solving the cell kinetic equations is in fact modified by the chemotactic coupling. We will use the idea that the largest gradient driving occurs as the wavefront passes the cell. During the wavefront passage, $\vec{\nabla} \psi$ becomes of order $1/\epsilon$, the inverse of the f reaction zone width. A similarly large gradient occurs during waveback passage; however, the value of k is much smaller in the excited phase (large ψ) than in the quiescent phase and hence this negative gradient <u>does</u> <u>not</u> elicit a large chemotactic response. In the small ϵ limit, then, we can use the simpler model

$$\frac{d\vec{v}_{\pm}}{dt} = -\Gamma \, \vec{v}_{\pm}$$

$$\vec{v}_{+} - \vec{v}_{-} \, \Big|_{\text{wavefront}} = \frac{k_{-}}{c_{0}} \, (\psi_{+} - \psi_{-})$$

Now, it is clear that the cell density will respond to the velocity kick and hence become non-uniform. In principle, then, we need to go back to the kinetic equations and put in a spatially varying set of parameters and solve a rather complex set of equations. Luckily, this is not necessary. The typical variation in density, as determined from the conservation law, will be of order v/c_{0}. Since cell speeds are typically only a few percent of signal speeds (tenths of a μm/sec as compared to 5-10 μm/sec), this variation can be safely neglected when deriving the wave dispersion formulas.

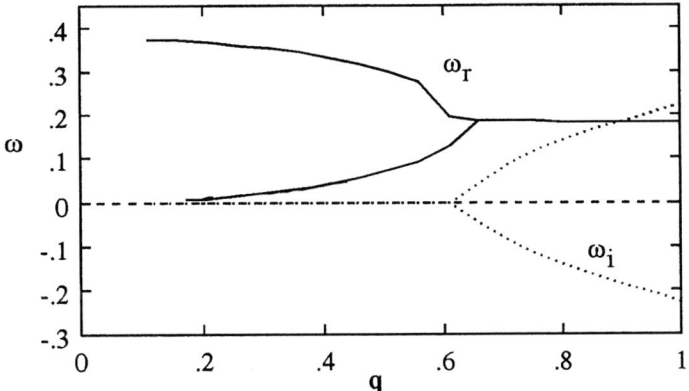

Figure 2. Real, ω_r, and imaginary, ω_i, parts of the growth rate of a perturbation as a function of wavevector, q.

As far as the plane wave is concerned, the chemotactic model is useful inasmuch as it predicts the waveform of cell density and concomitant cell velocity. As already mentioned, many measurements of this system make use of the fact that moving cells elongate in the direction of motion and hence can be imaged by polarization techniques. One could convert our results to a predicted waveform, given a relationship between shape asymmetry and velocity. This has not yet been tried.

More fundamentally, chemotaxis causes an instability in the entire signalling system which we believe is the cause of the observed breakup of the axisymmetric density field (and axisymmetric targets or spirals) to high density streams. The detailed derivation of this instability has been presented elsewhere and is rather tedious [15]. Roughly, we linearize around the planar wave solution, carefully including the perturbed density in the kinetic equations. This leads to a ten dimensional linear system (δr, $\delta \vec{v}$, $\delta \rho$ and the reaction zone location δx, for each phase) which must be solved to find the perturbation growth rate, ω. The result of a typical calculation is presented in Figure 2, clearly showing the instability with respect to a finite range of transverse wavevectors q.

It is important to understand the case of this instability and why it is a generic feature of this biophysical system. Imagine a local increase in cell density. This leads, via the density dependence of the signalling system parameters to an increased local emission of cAMP which spreads from this region. This increases the cAMP gradient, causing nearby cells to move towards this site and thereby increasing the density. This positive feedback loop will fail only at large enough q where the curvature of the wavefront will lead to reduced signal velocity and re-stabilize the problem at short scales. Any reasonable dynamics which involves coupling of the cell density to the cAMP field will invariably have a streaming instability.

If all such models predict the breakup of the initial axisymmetric aggregation, it is important to understand why we nonetheless observe this pattern over several hours. Within our model, reasonable choices for the various parameters lead to growth rates of order $1 - 2(hr)^{-1}$. This means that the cell density, which starts out being fairly uniform, takes a fairly long time (20 -100 signalling periods) to exhibit the effects of the unstable growth of local perturbations. Further study will be necessary to determine if this preliminary result remains valid when further experimental information is used to obtain more reliable values for the constants in the chemotactic response equations.

4 Discussion

As is evident, we have just begun to investigate the extent to which our model describes the actual processes governing the aggregation phase of the cellular slime molds. We are hopeful that this system, which has been the object of much biological interest as an accessible analog of many multi-cellular pattern formation examples, will prove amenable to the physicist's toolbox of models, dynamical techniques and simulation studies.

The most immediate direction which needs to be pursued is that of understanding the non-linear restabilization into high density streams[19]. The next essential step is to take our model systems and perform numerical simulations via a variety of techniques [17]. We already know that the simplest transcription of the model as it stands is not sufficient; the density becomes arbitrarily high and the system does not restabilize. We do not as yet know if this merely represents the insufficiency of the Goldbeter-Martiel model at large ρ or is more fundamental. If the latter is the case, introducing cell contact forces may be necessary. That this may not be so unreasonable can be seen from the fact that cells are almost abutting one another in typical videos of the streaming state.

We are greatly encouraged by the fact that several new experimental techniques are being applied to this problem and will greatly increase our knowledge of cellular motion in a variety of life-cycle phases. We have not yet made any serious effort to make quantitative predictions for these studies, and this too is a high priority for future work.

Finally, we return to issues regarding the perspective that working on this problem has given us regarding attempting to deal with biological systems. There is a fine line between reasonable simplifications (which could in principle be relaxed) and totally ad-hoc approaches which could not in principle be related to the actual system. Physicists know this difference well and will subdivide models into "toys" upon which one can sharpen mathematical methods but cannot predict anything physical and "models" which may be crude but are physically sound. Unfortunately, most attempts to discuss biological systems introduce "toy" models and then are left with interesting mathematical exercises with limited relevance [18]. The greatest challenge in this subject is to solve problems with experimental relevance and usher in the as yet unknown field of "physical biology". We hope that our work, as well as other similar efforts reported in these proceedings, provide some small steps in this direction.

Table 1. parameters for the Martiel-Goldbeter kinetic equations.

L_1	10
L_2	0.005
h	18.5
σ	10
λ_1	10^{-3}
λ_2	2.4
s	47
ϵ	0.01
Time Scale	28 minutes
Space Scale	8.2 mm

References

[1] Darmon, Michel and Brachet, Phillipe, in *Taxis and Behavior*, Hazelbauer, G. L. ed. London, 1978, 101-139; Devreotes, Peter N. and Zigmond, Sally H., *Ann. Rev. Cell Biol.*, 4, (1988) 649-686.

[2] Siegert, F. and Weijer C. J., *Physica D*, 49, (1991), 224-232.

[3] Martiel, Jean-Louis, and Goldbeter, Albert, *Biophys. J.*, 52, (1987) 807-828.

[4] Monk, P. B. and Othmer, H. G., Proc. Roy. Soc. B, 240, (1990), 555-589.

[5] Tyson, John J., and Keener, James P., *Physica D*, 32, (1988) 327-361.

[6] Levine, H. and Zou X., *J. Chem. Phys.*, 95, (1991), 3815-3825.

[7] Cellular chemotaxis is a vast field. Our simple model has been particularly motivated by: Alcantra F. and Monk, M., *J. Gen. Microbiol.*, 85, (1974) 321-334; Fisher, P.R., Merkl, R. and Gerisch, G., *J. Cell Bio.*, 108, (1989) 973-984; van Haastert, P. J. M., *J. Cell. Biol.*, 96, (1983) 1559-1565.

[8] Siegert F. and Weijer, C. J., *J. Cell Sci.*, 93, 325-335 (1989).

[9] Winfree, A. T., *When Time Breaks Down*, Princeton Univ. Press, Princeton, 1984.

[10] See, for example, Dockery, J. D., Keener, J.P. and Tyson, J. J., *Physica D*, **30**, 117-191, (1988) or Kessler D. and Levine, H., *Physica D*, **39**, 1-14 (1989).

[11] Karma, A., "The Scaling Regime of Spiral Propagation in Single-Diffusive Media", to appear in *Phys. Rev. Lett.*.

[12] This can be done using the methods described in Kessler, D. A. and Levine, H., *Phys. Rev. A*, **41**, 5418-5430, (1990).

[13] Berg, H. C. and Turner, L. *Biophys. J.*, **58**, 919-930, (1990).

[14] See Devreotes and Zigmond, *op. cit.*.

[15] Levine, H. and Reynolds, W. *Phys. Rev. Lett.*, **66**, 2400-2403 (1991).

[16] Earlier attempts at modelling chemotaxis include Keller, Evelyn F. and Segel, Lee A., *J. theor. Biol.*, **26**, (1970) 399-415; Grindrod, P., Murray, J.D. and Sinha, S. *IMA Journal of Mathematics Applied in Medicine & Biology*, **6**, (1989) 69-79.

[17] For simple approaches to simulation, see Parnas, Hanna and Segel, Lee A., *J. Cell Sci.*, **25**, (1977) 191-204; Parnas, Hanna and Segel, Lee A., *J. theor. Biol.*, **71**, (1978), 185-207; Mackay, Steven A. *J. Cell Sci.*, **33**, (1978) 1-16. Vasieva, O. O., Vasiev, B. N., Karpov, V. A., Zaikin, A. N., "A Computer Simulation of Autowave Aggregation of Dictyostelium Amoebae", preprint.

[18] Murray, J. D. *Nonlinear Science Today*, **1**, No. 3, 1-5 (1991).

[19] Newell, P. C. in *Biology of the Chemotactic Response*, Lackie, J. M., ed., Cambridge University Press, (1981), 89-114.

THE FRACTAL NATURE OF COMMON PATTERNS

Tamás Vicsek

Department of Atomic Physics
Eötvös University
Budapest, Puskin u 5-7
1088 Hungary

1. INTRODUCTION

The idea that fractal geometry is an important aspect of a wide range of phenomena was first extensively demonstrated using mathematical models imitating various objects in nature (Mandelbrot 1982). Later it has been shown that fractals have numerous applications in several branches of sciences, including specific examples from physics (Meakin 1987, Vicsek 1989).

Interestingly, less effort has been devoted to the analysis of the fractal aspects of samples obtained by directly collecting them from our everyday environment (what we shall call "nature"). Although not everyone agrees (Family 1991), it seems that learning about the details of the fractal scaling of patterns spontaneously developing in nature should be useful from the point of getting an insight into the applicability of the most general ideas on fractal behavior to the common particular examples. In this paper results concerning the fractal dimension and roughness exponent of a few natural patterns will be given. The corresponding numerical values have been calculated using simple techniques (Vicsek 1989) applied to the digitized images of the samples.

First, the fractal dimension of *self-similar* mineral dendrites of various origin will be given (section 2). In section 3 *self-affine* structures observed in simple experiments and collected from nature will be examined. The conclusions will be given in the last section.

2. THE FRACTAL DIMENSION OF MINERAL DENDRITES

In quarries it is no rarity to find black or reddish brown designs of tree-like structures on the surface of limestones. These patterns are known as *mineral den-*

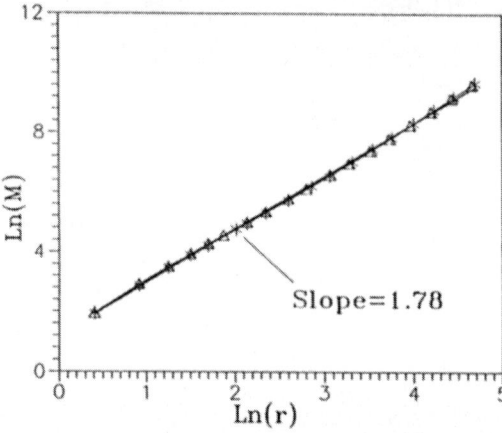

Figure 1. Manganese oxide deposit formed as a result of a reaction- diffusion process on the surface of a limestone found in Bavaria. Bottom: the fractal analysis this mineral dendrite is shown indicating that it has a fractal dimension close to 1.78 (Chopard et al 1991).

drites. In any Glossary of Geology they are defined, for instance, as: "Plant-shaped surface deposits of hydrons iron or manganese oxides found along joint, bedding and cleavage planes in rocks". An example is shown in Fig. 1a. What we know about their origin is that they are chemical deposits (oxides) that formed when at some point in the geological past the limestone was penetrated by a supersaturated solution of manganese or iron ions. The question of the actual mechanism of fractal pattern formation in such phenomena is still open (see also Ruiz 1991). For example, one would like to know how the shape and density of the deposites depends on the external parameters like concentration gradient or reaction rate.

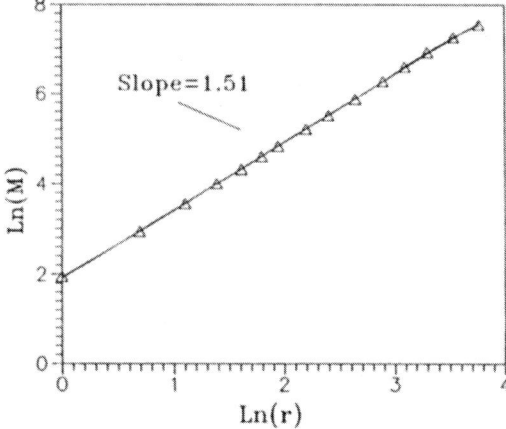

Figure 2. Manganese oxide deposits formed within the plane of a crack inside a quartz crystal. In the bottom part the fractal analysis this mineral dendrite is shown indicating that its fractal dimension is about 1.5 (Chopard et al 1991).

Here we present the scaling analysis of the shape of two rather different samples and mention that the development of these dendrites has been interpreted by a numerical model based on a simple reaction-diffusion lattice-gas model (Chopard et al 1991). The dimension of the dendrites can be determined by digitizing the pictures and using the sand box method (Vicsek 1989). This method involves the following procedure: i) a large number of "centers" is placed randomly on the digitized image, ii) the number $M(R)$ of "particles" (pixels belonging to the image) is counted in squares of increasing linear size R centered at the selected points, iii) an average is made over the centers. Then a log-log plot of $M(R)$ versus R is made in which the fractal dimension D is given by the slope fitted to the data since $M(R) \sim R^D$. The result ($D \simeq 1.78$) for the dendrite displayed in Fig. 1a is shown in Fig 1b. A similar analysis has been carried out for a set of pictures taken from a dendritic pattern which grew within the plane of a crack inside a quartz crystal originated from Brasil.

In this case the estimated dimension is about 1.5 as is demonstrated in Fig 2.

3. SELF-AFFINE SCALING OF SIMPLE WETTING FRONTS AND THE SURFACE OF LICHENS

If the conditions of the growth process are such that the development of the interface is marginally stable and the fluctuations are relevant, the resulting structure is a rough surface and can be well described in terms of self-affine fractals (Mandelbrot 1982, Vicsek 1989, Family and Vicsek 1991). Marginal stability means that the interface is neither stable nor unstable. In other words, the fluctuations (which die out quickly for a stable surface and grow exponentially for an unstable one) survive without drastic changes for a long time in the case of marginally stable growth.

Consider the time evolution of a rough interface in a d-dimensional space starting from an initially flat surface at time $t = 0$. Let us concentrate on a part of the surface having an extent L in $d - 1$ dimensions perpendicular to the growth direction. The surface typically can be described by a single-valued function $h(\vec{r}, t)$ which gives the height (distance) of the interface at position $\vec{r}$ at time t measured from the original $d - 1$ dimensional flat surface. During growth the interface heights fluctuate about their average value $\bar{h}(t) \simeq t$ and the extent of these fluctuations characterizes the width or the the thickness of the surface. The root mean-square of the height fluctuations $w(L, t)$ is a quantitative measure of the surface width and is defined by

$$w(L, t) = [\langle h^2(\mathbf{r}, t) \rangle_r - \langle h(\mathbf{r}, t) \rangle_r^2]^{1/2}, \tag{1}$$

According to scaling considerations w depends on L and t as (Family and Vicsek 1985, Kardar et al 1986)

$$w(L, t) = L^\alpha f(t/L^{\alpha/\beta}), \tag{2}$$

where $\alpha/\beta = z$ is the dynamic scaling exponent. An alternative approach to the characterization of self-affine surfaces changing in time is the determination of various *correlation functions*. The most convenient quantity is the so-called height-height correlation function $c(r, t)$ defined as $c(r, t) = \langle |h(\vec{r'}, t') - h(\vec{r'} + \vec{r}, t' + t)| \rangle_{\vec{r'}, t'}$, which is the average height difference measured for a time difference t at two points whose coordinates on the substrate are separated by $\vec{r}$. For long times and for $r << L$ the correlation function $c(r, 0)$ behaves as

$$c(r, 0) \sim r^\alpha. \tag{3}$$

This and expression (2) can be used to determine the value of the static exponent α from the digitized images of patterns obtained experimentally.

Now we turn to the analysis of two common processes leading to self-affine surfaces. The penetration of a wetting fluid into a porous medium is a typical everyday phenomenon in our immediate environment. One example is the kind of spots produced on the ceiling by a leaking roof. Another one is shown in Fig. 3, where the regions in a paper towel penetrated by coffee are shown. These type of processes are called two-phase viscous flows in porous media (one of the phases is air in the above examples).

In such processes the motion of the fluid-air interface is influenced by several factors. In the experiments with *wetting* fluids it can be achieved that at slow flow rates the stabilizing effect of the pressure distribution in the invading fluid (coffee in

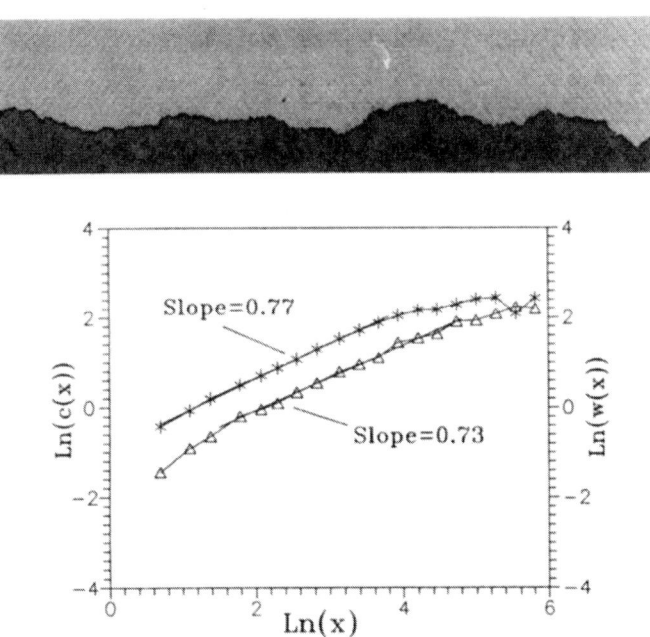

Figure 3. The top pattern was obtained by immersing a paper towel into coffee kept in a dish. Bottom: Fractal analysis of the pattern shown in the top part of this figure. The upper set of data corresponds to the dependence of the width on the linear size on which it is calculated, while the lower curve was obtained by calculating the height correlations. The two approaches give somewhat different estimates for the roughness exponent α which equals to the slope. However, both values are definitely different from the universal value 1/2.

this case) and the instabilities caused by capillary effects can be neglected (see e.g. Horváth et al 1991). This is why the experiment can be considered as a realization of marginally stable growth. The fluctuations are presented by the random distribution of free spaces of voids among the network of fibers from which papers are made of.

Fig. 4 shows the same phenomenon in the circular geometry, when a coffee drop penetrates into a sheet of paper radially. In this case the proper quantity to calculate

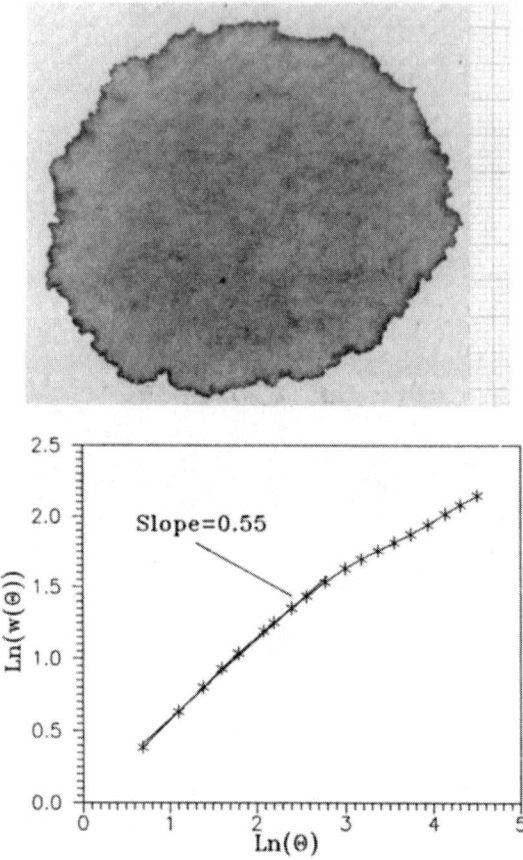

Figure 4. This pattern was obtained by dropping coffee onto a paper towel. Bottom: The dependence of the correlation function of eq. (4) on the angle θ. Interestingly, in this case the slope is close to the universal exponent 1/2.

is the correlation function

$$c(\theta) = \langle |R(\theta') - R(\theta' + \theta)| \rangle_{\theta'}, \qquad (4)$$

which is the average difference of the radii R corresponding to two surface points separated by an angle θ, where the radii and the angle is measured from the centre of mass of the structure.

Very recently there has been a growing interest in the fractal aspects of the living matter (Fujikawa and Matsushita 1989). In particular, many structures of biological origin display the kind of behavior shown in Fig.4. Fig. 5a. shows the picture of a lichen which has a shape very much reminding Eden clusters or coffee drops. Lichens are composed of a fungus whose mycelium forms a matrix for algae living in a symbiotic relation with the fungus. They can be found under a wide variety of climates growing on the surface of rocks and trunks. The results for $c(\theta)$ (Vicsek and Wolf 1991) are shown in Fig. 5b.

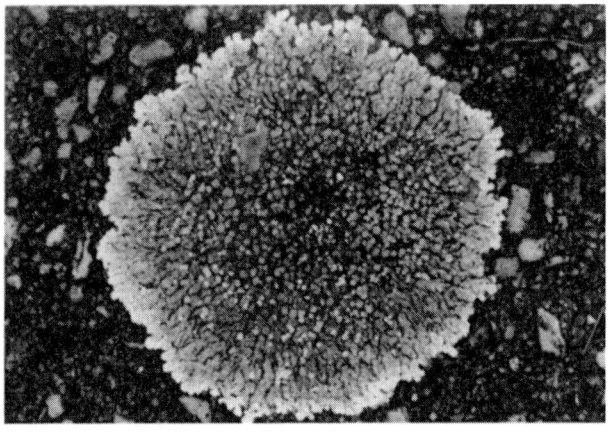

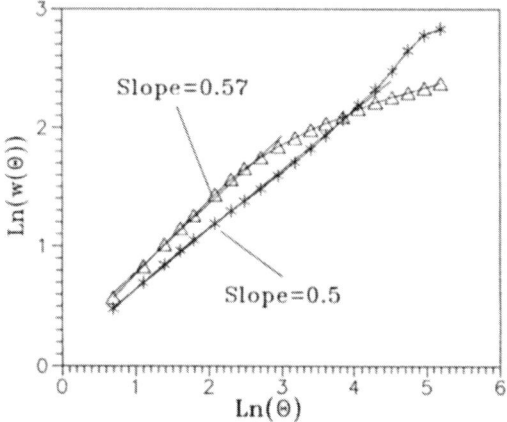

Figure 5. Picture of a plant called lichen. Bottom: For this pattern the correlation function $c(\theta)$ scales as a function of θ with an exponent about 0.5 (Vicsek and Wolf 1991), while the same quantity for another lichen is 0.57.

4. CONCLUSIONS

The idea of this work was to test the degree of applicability of the simplest models for fractal growth to the description of the common examples in our environment. Three types of phenomena have been considered, and the resulting patterns have been analyzed for the actual value of the exponents corresponding to fractal scaling.

Our main finding is that in many typical cases the particular values for the dimensions and the roughness exponents violate the assumption of a single universal exponent for a given kind of phenomenon. In addition, the range over which self-affine scaling could be observed was limited. It is expected, however, that the appropriate modifications of the existing simple models will account for the observed discrepancies.

Acknowledgements: The author thanks A. Jakó and D. E. Wolf for their help in carrying out different parts of this study. Inspiring discussions with S. Buldyrev, S. Havlin and H. E. Stanley during the author's visit to Boston University are also acknowledged. The present research was financially supported by the Hungarian Scientific Research Foundation Grant No. 693.

REFERENCES

Chopard, B., Herrmann, H. J. and Vicsek, T., 1991 to appear in *Nature*

Family, F. and Vicsek, T., 1985 *J. Phys.* **A18**, L75

Family, F. and Vicsek, T., 1991 *Dynamics of Fractal Surfaces*, (World Scientific, Singapore)

Family, F., 1991 private communication

Fujikawa, H. and Matsushita, M., 1989 *J. Phys. Soc. Jpn.* **58**, 387

Horváth, V. K., Family, F. and Vicsek, T., 1991a *J. Phys.* **A24**, L25

Kardar, M., Parisi, G. and Zhang, Y.-C., 1986 *Phys. Rev. Lett.* **56**, 889

Mandelbrot, B. B., 1982 *The Fractal Geometry of Nature* (Freeman, San Francisco)

Meakin, P., 1987 in *Phase Transitions and Critical Phenomena* Vol. 12 edited by C. Domb and J. L. Lebowitz (Academic Press, New York)

J. M. Ruiz, 1991 this volume

Stanley H E, Ostrowsky N, 1988 Editors *Random Fluctuations and Pattern Growth* (Kluwer, Dordrecht)

Vicsek T, 1989 *Fractal Growth Phenomena* (World Scientific, Singapore)

Vicsek, T, and Wolf, D. E., 1991 to be published

Witten T A and Sander L M, 1981 *Phys. Rev. Lett* **47**, 1400

STUDY OF SELF-AFFINE FRACTAL SURFACES WITH STM

J.M. Gómez-Rodríguez, A. Asenjo and A.M. Baró

Departamento de Física de la Materia Condensada,C-III
Universidad Autónoma de Madrid
E-28049 Madrid, Spain

INTRODUCTION

The measurement of the microtopography of surfaces is a difficult task. In the recent times, several kinds of Scanning Probe Microscopes (SPM) have appeared, where a probe tip is scanned at a very close distance from the sample surface (0.1 to 10 nm). The most old and popular of these instruments is the Scanning Tunneling Microscope (STM) invented by Binnig and Rohrer[1]. The tip-to-sample distance, s, is so small (0.5 nm) that a tunneling current flows through the vacuum between the two electrodes. Topographical images are obtained by keeping constant the distance s when scanning the probe tip over the sample. These SPM microscopes have several important properties: (i) they give three-dimensional images in real space, (ii) they give horizontal and vertical atomic resolution, (iii) since there is no mechanical contact between tip and surface and since the electron energy lies in the meV to low eV values, the method is non destructive, (iv) they are able to operate at several ambiances like air at atmospheric pressure, liquids, high pressure gases...

Although in some cases surfaces of materials are made of atomically smooth planes, in most cases and particulary those connected with natural and industrial processes, the surfaces of materials are extremely rough. In fact, roughness is perhaps the most striking property of solid surfaces. A quantitative description of roughness is normally done by using several parameters based on the calculation of averages of elevation and depression heights with respect to a reference surface. More powerful is to use the concepts of Mandelbrot's[2] fractal geometry since then we can have more elaborated quantities and expressions to characterize the roughness. Even more we can correlate the surface data with physical models. These models are used to make computer simulations. For example, simple models have been developed to study non equilibrium growth processes, such as those occurring in thin film deposition methods.

Material surfaces exhibit often different scaling behaviour in different directions, as a consequence of the preferred direction defined by the surface growth. In this case the description is in terms of self-affine rather than self-similar fractals.

In the last two years several methods[3,4,5] have been proposed for measuring fractal behaviours from STM micrographs. After Denley's[3] first attempts which did not fit properly to a fractal surface, mainly two methods have been established. One of them[4] is based on Fourier analysis of STM images. The other one, first proposed and fully developed by our

Fig. 1. (a) STM top view of a gold electrodeposit. (b) Perspective view showing artificially generated lakes. (c) Area-perimeter analysis for the lake pattern shown in (b).

group[5], is based on area-perimeter determination of lakes or islands artificially generated on STM digitized images, and it has been applied to obtain the fractal behaviour of gold[5] and platinum[6] electrodeposits as well as vacuum evaporated gold deposits[7].

As already reported[5] the **area**-perimeter method yields better results than Fourier analysis for typical file sizes used in STM (256×256 pixels[2]). This is why we will concentrate in the following section on the description of the area-perimeter method.

THE AREA-PERIMETER METHOD

In a similar way to that reported by Mandelbrot for the earth coastlines[2], lake or island patterns can be generated by "filling with water" STM 3-dimensional images. Based on the fact that the intersection of a self-affine surface and a plane generates self-similar lakes or islands, we analyze the perimeter (L) versus area (A) relationship of these islands or lakes. L and A for self-similar objects are related[2,8] by

$$L = cA^{D'/2} \qquad (1)$$

where c is a constant and D' is the fractal dimension of the coastlines. The area is defined as the number of pixels of a given object and the perimeter as the number of pixels of that object having neighbours not belonging to it. All the islands or lakes connected to the image borders are discarded as ill-defined. Very small islands ($A < 30b^2$, where b is the yardstick length, i.e. the size of a pixel in the grid) are also discarded according to some considerations on the yardstick length discussed in reference 5. All available heights in the digitized images are used in this process and the islands of each height are all included in the final plot.

Fig. 1 illustrates this method for a gold electrodeposit surface[5]. Just for clarity, only the lakes corresponding to one "water" height are plotted in fig. 1(c). The fractal dimension was found to be D'=1.5.

RESULTS ON VACUUM EVAPORATED GOLD SAMPLES AND DISCUSSION

The above mentioned area-perimeter method was used to test the fractal character of gold samples evaporated in vacuum at high rates[7].

Fig. 2(a) is a 3-dimensional STM shaded view of a 350×350 nm^2 vacuum evaporated gold sample. Fig. 2(b) shows the fractal analysis of all the islands generated by the "filling with water" simulation method. All available heights in the image (about 1000 in a 10 nm range) were included in this plot.

The most interesting feature in fig. 2(b) is the coexistence of two linear regions with different slopes. This seems to suggest that the gold surface is formed by two kinds of aggregates: The smaller ones are almost rounded in shape and lead to a very low fractal dimension (very close to the euclidean limit D'=1), while the larger ones are really ramified and have a very high D' value (1.72). The transition region between these two types of aggregates is around A=630 nm^2. This area corresponds to a lower cut-off length in the range of 20-30 nm.

In order to confirm the high fractal dimension at length scales bigger than 20-30 nm, larger areas of the sample were measured. Fig. 3(a) shows a 900×900 nm^2 3-dimensional image of the same sample. The area-perimeter analysis is shown in fig. 3(b). Only one linear region is present in this plot since the yardstick length is much larger than that of fig.2 and this does not allow the smaller clusters to be visible. The D' value obtained from the slope is quite close to the one calculated in fig. 2(b).

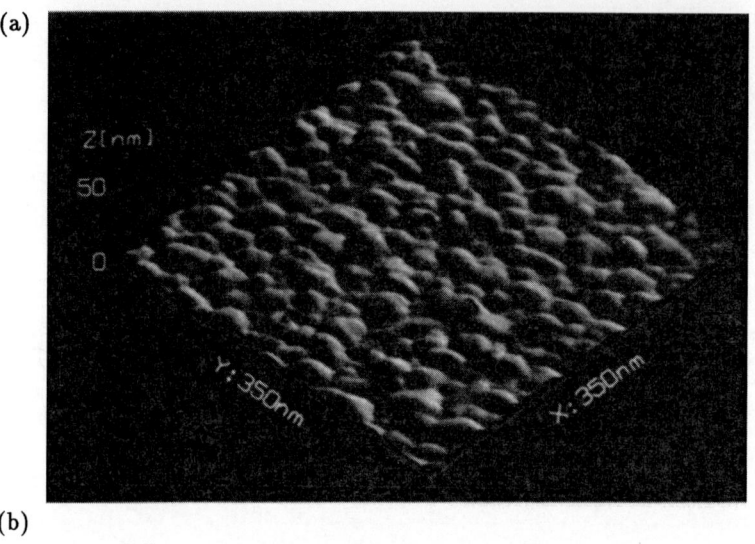

(a)

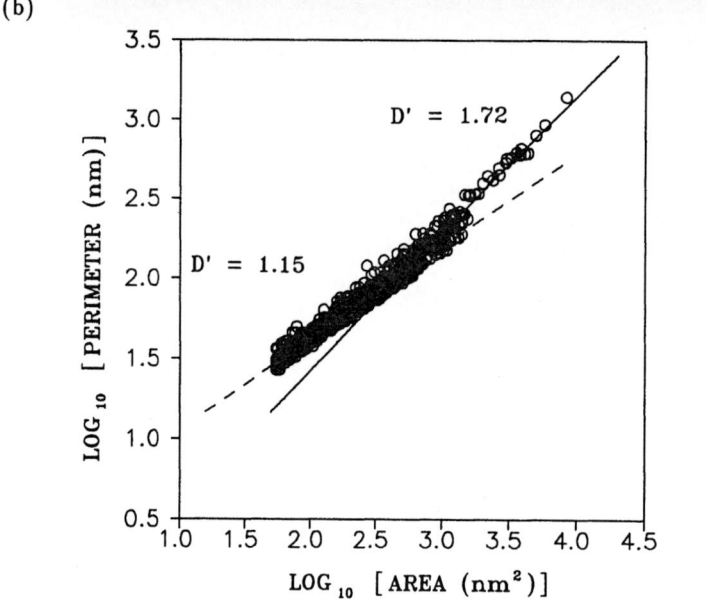

(b)

Fig. 2. (a) 3-dimensional shaded view of a 350x350 nm^2 STM measurement on vacuum evaporated gold. (b) Log L vs. log A plot.

Several more images were measured in order to get an accurate value of the high D' which was not affected by the probing tip influence or sample inhomogeneities. By averaging the results of more than 10 images, we got a final value D'= 1.72 ± 0.05.

In some previous studies of rough interfaces observed in experiments on displacement of viscous fluids in porous media[9,10], the dynamic scaling approach[11,12] has been succesfully used. A rough surface may be characterized by its interface width w(L), defined as the rms value of the height fluctuations, h($\vec{r}$), over a length scale L, i.e.,

$$w(L) = < [h(\vec{r}) - \overline{h_L}]^2 >^{\frac{1}{2}}$$ (2)

where $\overline{h_L}$ is the average height over a square of area LxL and the brackets stand for averaging

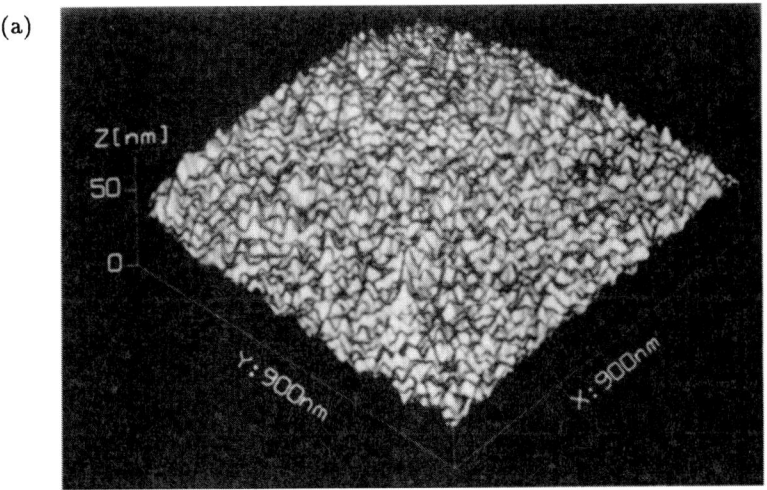

(a)

(b)

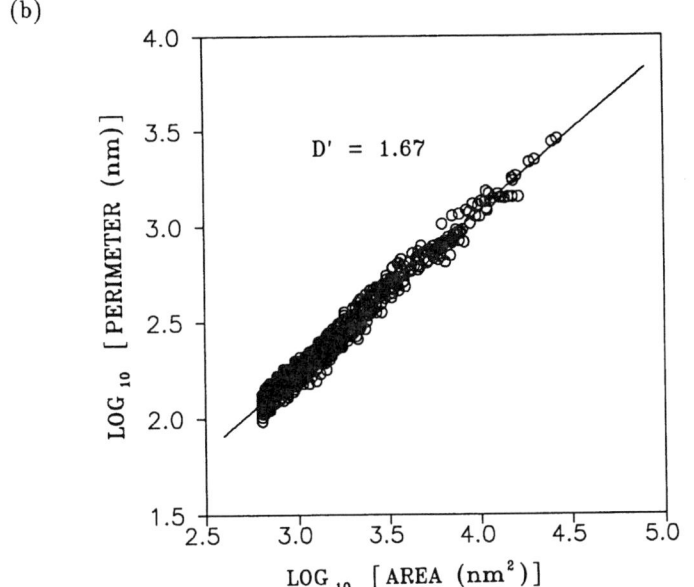

$D' = 1.67$

Fig. 3. (a) 3-dimensional shaded view of a 900x900 nm^2 STM image on vacuum evaporated gold. (b) Area-Perimeter analysis.

over all squares of area LxL.

According to the scaling approach, if a growing surface is a self-affine fractal in the steady-state its saturation width has a power law dependence on L:

$$w(L) \sim L^{\alpha} \tag{3}$$

where α is the so-called *roughness exponent*.

An alternative way to characterize a rough surface is by means of the height correlation function defined as

$$G_1(\vec{r}) = < [h(\vec{r_0}) - h(\vec{r_0} + \vec{r})]^2 > \tag{4}$$

where the brackets denote an average over all possible $\vec{r_0}$. It is usually more convenient to average $G_1(\vec{r})$ for all possible angles defining a new function only dependent on the distance r:

$$G(r) = < G_1(\vec{r}) >_{all\ angles} \tag{5}$$

The scaling form of this quantity is

$$G(r) \sim r^{2\alpha} \tag{6}$$

The relationship between the roughness exponent, α, and the fractal dimension, D', of a section of the surface and a plane is given by[13]

$$\alpha = 2 - D' \tag{7}$$

According to (7) the roughness exponent of our vacuum evaporated gold deposits would be $\alpha = 2 - 1.72 = 0.28$.

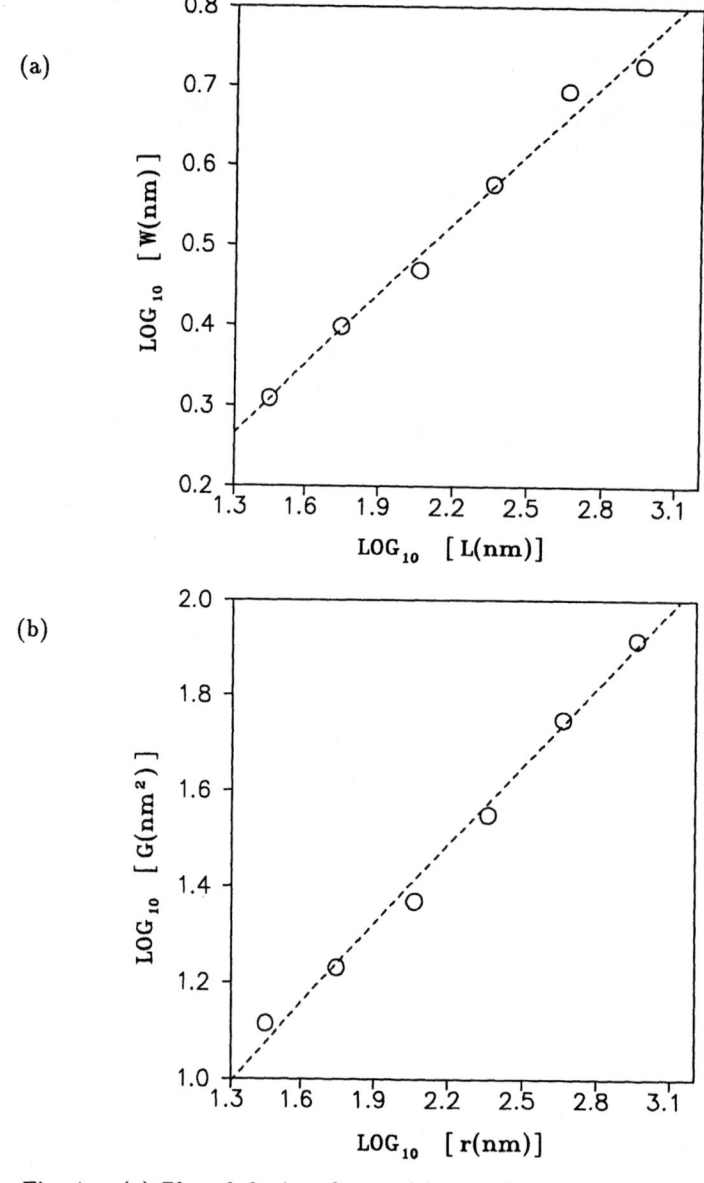

Fig. 4. (a) Plot of the interface width as a function of L for the STM image in fig. 3. The slope of the straight line is $\alpha = 0.29$. (b) Correlation function for the same image. The slope is $2\alpha = 0.55$.

A direct determination of α was done by using equation (3) and (6) on the data of fig. 3(a). Fig. 4(a) shows the interface width as a function of L for length scales larger than the above mentioned lower cut-off. The points reasonably fit to a straight line with slope $\alpha=0.29$. In fig. 4(b) the correlation G as a function of r is plotted only for length scales larger than the cut-off length. Again a straight line is observed for more than one decade, with slope equal to 0.55, i.e. $\alpha = 0.27$.

The roughness exponent calculated by three independent methods results to be approximately the same with a value around 0.3. This value is in very good agreement with that obtained from local growth models like ballistic deposition in (2+1)-dimensions ($\alpha = 0.33^{14} - 0.35^{12}$), suggesting that the evaporation of metals at very high rates produces deposits that grow far from equilibrium forming self-affine fractal surfaces in the steady-state.

CONCLUSIONS

We have shown the ability of Scanning Tunneling Microscopy to study the fractal character of growing surfaces at the nanometer level. The area-perimeter method, as well as direct determination of the roughness exponent have been used to prove the scaling behaviour of vacuum evaporated gold samples. For length scales larger than 20-30 nm, we have found an α value very close to the one obtained from ballistic deposition simulations.

ACKNOWLEDGEMENTS

We gratefully acknowledge P. Herrasti and P. Ocón for providing us with the gold evaporated samples. We are also indebted to L. Vázquez for measuring some of the STM images and to R.C. Salvarezza for fruitful discussions. This work has been financially supported by CICYT project number MAT90-0352-E.

REFERENCES

1. G. Binnig, H. Rohrer, Ch. Gerber and E. Weibel, Surface Studies by Scanning Tunneling Microscopy, Phys. Rev. Lett., 49:57 (1982).
2. B.B. Mandelbrot, "The Fractal Geometry of Nature", Freeman, New York (1983).
3. D.R. Denley, Scanning Tunneling Microscopy of Rough Surfaces, J. Vac. Sci. Technol. A, 8:603 (1990).
4. M.W. Mitchell and D.A. Bonnell, Quantitative Topographic Analysis of Fractal Surfaces by Scanning Tunneling Microscopy, J. Mater. Res., 5:2244 (1990).
5. J.M. Gómez-Rodríguez, R.C. Salvarezza and A.M. Baró, Fractal Characterization of Gold Deposits by Scanning Tunneling Microscopy, J. Vac. Sci. Technol.B, 9:495 (1991).
6. J.M. Gómez-Rodríguez, A.M. Baró, L. Vázquez, R.C. Salvarezza, J.M. Vara and A.J. Arvia, Fractal Surfaces of Gold and Platinum Electrodeposits. Dimensionality Determination by Scanning Tunneling Microscopy, J. Phys. Chem., in press (January 1992).
7. J.M. Gómez-Rodríguez, A. Asenjo,R.C. Salvarezza and A.M. Baró, Measuring the Fractal Dimension with STM: Application to Vacuum Evaporated Gold, Ultramicroscopy, in press (1992).
8. J. Feder, "Fractals", Plenum, New York (1988).
9. M.A. Rubio, C.A. Edwards, A. Dougherty and J.P. Gollub, Self-Affine Fractal Interfaces from Immiscible Displacement in Porous Media, Phys. Rev. Lett., 63:1685 (1989).
10. V.K. Horváth, F. Family and T. Vicsek, Dynamic Scaling of the Interface in Two-Phase Viscous Flows in Porous Media,J. Phys. A, 24:L25 (1991).

11. M. Kardar, G. Parisi and Y.C. Zhang, Dynamic Scaling of Growing Interfaces, Phys. Rev. Lett., 56:889 (1986).

12. F. Family, Dynamic Scaling and Phase Transitions in Interface Growth, Physica A, 168:561 (1990).

13. T. Vicsek, "Fractal Growth Phenomena", World Scientific, Singapore (1989).

14. P. Meakin, P. Ramanlal, L.M. Sander and R.C. Ball, Ballistic Deposition on Surfaces, Phys. Rev. A, 34:5091 (1986).

DYNAMIC SCALING IN SURFACE GROWTH PHENOMENA

Fereydoon Family

Department of Physics
Emory University
Atlanta, GA 30322, U.S.A.

INTRODUCTION

The most characteristic feature of many of the growth patterns that are formed in physical, chemical and biological processes is the existence of an evolving interface. Thus, in order to develop a better understanding of pattern formation in growth phenomena we need to study the structure and the dynamics of growing surfaces and interfaces. Recently considerable advances have been made [1] in this direction using the concept of dynamic scaling [2] which is based on the fact that growing surfaces exhibit non-trivial scaling behavior and naturally evolve to a steady-state without a characteristic time or spatial scale. Due to its generality, the dynamic scaling approach has become a standard language in the study of growing surfaces and interfaces [1]. In this review I will discuss some recent advances in theoretical, simulational and experimental studies of surface and interface growth phenomena using this approach.

DYNAMIC SCALING

If growth conditions are such that an evolving interface is neither stable nor unstable against perturbations, the resulting interface is a self-affine fractal which can be characterized by a single-valued functions $h(\mathbf{r}, t)$. The function $h(\mathbf{r}, t)$ gives the height of the interface at position $\mathbf{r}$ at time t measured from an initially flat $d - 1$ dimensional surface at time $t = 0$. For multi-valued surfaces, $h(\mathbf{r}, t)$ is the maximum height of the surface at $\mathbf{r}$. We concentrate on a section of the surface having an extent L in $d - 1$ dimensions perpendicular to the growth direction. As the surface evolves, the roughness or the width of the surface increases with time. The root mean-square of the height fluctuations $w(L, t)$ gives a quantitative measure of the surface width and is defined by

$$w(L, t) = [\langle h^2(\mathbf{r}, t) \rangle_r - \langle h(\mathbf{r}, t) \rangle_r^2]^{1/2}, \tag{1}$$

where $\langle \ldots \rangle_r$ denotes an average over $\mathbf{r}$. In the absence of a characteristic time in the growth process $w(L, t)$ increases with some power of time [2],

$$w(L, t) \sim t^\beta. \tag{2}$$

Within a region of length L the surface fluctuations cannot increase indefinitely but reach

Growth Patterns in Physical Sciences and Biology, Edited
by J. M. Garcia-Ruiz *et al.*, Plenum Press, New York, 1993

a steady-state with a constant value of the width which depends on L. These steady-state values, $w(L, t \to \infty)$, have a power law dependence on L and can be characterized by the exponent α [2],

$$w(L, t \to \infty) \sim L^\alpha. \tag{3}$$

The exponents α and β defined in (2) and (3) characterize the structure and time evolution of a growing surface, respectively. The dependence of $w(L, t)$ on t and L can be combined into the dynamic scaling form [2],

$$w(L, t) = L^\alpha f(t/L^z). \tag{4}$$

where the dynamic exponent z is defined by $z = \alpha/\beta$, and the scaling function $f(x)$ is a constant for $x \gg 1$, and for $x \ll 1$ it has to be of the form $f(x) \sim x^\beta$. Similar scaling behavior can be observed in the height difference correlation function $c(r, t)$, defined by,

$$c(r, t) = \langle [\tilde{h}(r', t') - \tilde{h}(r + r', t' + t)]^2 \rangle_{r', t'}, \tag{5}$$

where $\tilde{h} = h - \langle h \rangle$ and $t' \gg L^z$. Dynamic scaling implies that,

$$c(r, 0) \sim r^{2\alpha}, \quad \text{for} \quad r \ll L, \tag{6}$$

and for fixed r and short times, $c(0, t)$ scales as,

$$c(0, t) \sim t^{2\beta}, \quad \text{for} \quad t \ll L^z. \tag{7}$$

The correlation functions $c(r, 0)$ and $c(0, t)$ reach a constant steady-state value in the limits $r \gg L$ and $t \gg L^z$, respectively.

SURFACE GROWTH MODELS

In recent years a number of models have been proposed for describing growing surfaces [1]. Here, I describe two computer simulation models, the ballistic deposition [2, 4, 5] and Eden model [6-10], and the Kardar, Parisi and Zhang (KPZ) model [11,12] which is a nonlinear Langevin equation.

The ballistic deposition model has been used extensively to simulate surface growth by random deposition of molecules on cold substrates, which is a common experimental arrangement particularly in thin film growth. In this model particles follow a straight-line trajectory until they encounter a particle on the surface, or a particle in one of the nearest-neighbor columns. The surface of the ballistic deposition is a self-affine fractal and its evolution has been shown to be described by the dynamic scaling approach [2,4,5].

The Eden model was originally proposed [6] as a model of cell growth in biological systems. In the Eden model [6] every surface site, which are the nearest-neighbor perimeter sites of the cluster, can grow with equal probability. Thus, in each step of the simulation a randomly chosen perimeter site is occupied. In this way a compact pattern is formed which has a fluctuating interface. Surface properties of the Eden model have been extensively investigated using the dynamic scaling approach [7-10].

A phenomenological model based on a nonlinear Langevin type equation has been proposed by Kardar, Parisi and Zhang [11] for describing the evolution of growing surfaces and interfaces. Kardar *et al* have argued that the evolution of an interface can be described by the equation

$$\frac{\partial h}{\partial t} = \nu \nabla^2 h + (\lambda/2)|\nabla h|^2 + \eta(\mathbf{r}, t) \tag{8}$$

where ν is the surface tension, the parameter λ is the growth velocity perpendicular to the interface, and the noise term $\eta(\mathbf{r}, t)$ is assumed to be Gaussian with delta-function correlation:

$$\langle \eta(\mathbf{r}, t)\eta(\mathbf{r}', t')\rangle = 2D\delta(\mathbf{r} - \mathbf{r}', t - t'). \tag{9}$$

The KPZ equation has been analyzed [11,12] by renormalization group methods in $d = 2$, i.e. for a one dimensional interface, and the results are $\alpha = 1/2$, $\beta = 1/3$, and $z = 3/2$. In addition, on the basis of general arguments [4] it has been shown that,

$$\alpha + \alpha/\beta = 2. \tag{10}$$

Simulations of various surface growth models, such as the ballistic deposition [2,4,5] and the Eden model [7-10], are in excellent agreement with the exact results $\alpha = 1/2$, $\beta = 1/3$, and $z = 3/2$ in two dimensions. Similar exact results are not available for $d > 2$ from the KPZ equation, and the numerical simulation results for various models are somewhat controversial [1,5,13,14]. On the basis of numerical results several conjectures have been made for α and β in $d > 2$ [9,13,14]. In particular, the most recent calculations [10,15] appear to be in much better agreement with the Kim-Kosterlitz conjecture [14], $\alpha = 2/(d+2)$ and $\beta = 1/(d+1)$, than with the other conjectures [9,13].

NUMERICAL SOLUTION OF THE KPZ EQUATION

Since the values of α and β for $d \geq 3$ are controversial [1] we have carried out [15] a direct solution of the KPZ equation in $d = 3$. We first perform a change of scale $h = y\sqrt{2D/\nu}$ and $t = \tau/\nu$ and write (8) in the form,

$$\frac{\partial y}{\partial \tau} = \nabla^2 y + \epsilon|\nabla y|^2 + \xi(\mathbf{r}, \tau) \tag{11}$$

where the non-linearity parameter has been written as $\epsilon = \lambda^2 D/2\nu^3$, and the noise correlations as, $\langle \xi(\mathbf{r}, \tau)\xi(\mathbf{r}', \tau')\rangle = \delta(\mathbf{r} - \mathbf{r}', \tau - \tau')$. We have integrated (11) on a discrete grid in $d = 2$, and 3 for $\epsilon = 1, 2, 5, 10,$ and 25 using finite-difference methods. As a test of our simulation method, we duplicated the known exact results, $\alpha = 1/2$ and $\beta = 1/3$ in $d = 2$. In addition, for the linear model ($\epsilon = 0$) we recovered the exact solution in $d = 2$ and 3.

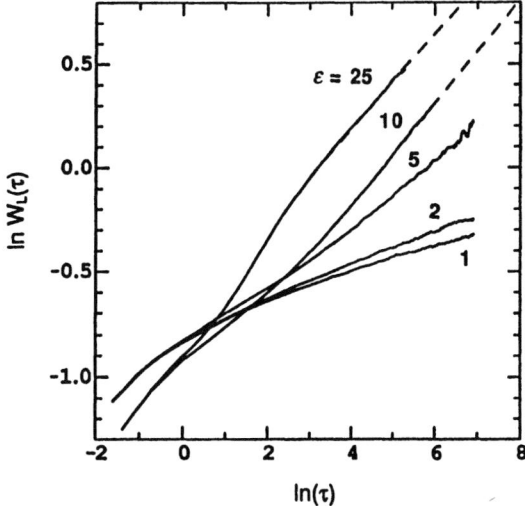

Fig. 1. Log-log plots of width $w_L(\tau)$ for the KPZ equation in $d = 3$ for $\epsilon = 1$, 2, 5, 10 and 25.

Figure 1 shows our results [15] for the early-time growth behavior from our numerical solution of the non-linear equation in $d = 3$. For large ϵ ($\epsilon = 10$ and 25), β appears to be close to the value of $\beta = 1/4$ conjectured by Kim and Kosterlitz [14]. Similarly for the surface roughness exponent α we find 0.39 ± 0.01 for $\epsilon = 25$, which is also close to the Kim-Kosterlitz value $\alpha = 0.4$. For $\epsilon = 10$, the slope 0.37 ± 0.02 is somewhat below this value but appears to be increasing with increasing L.

Our results for the KPZ exponents in $d = 3$ are significantly closer to the conjecture of Kim and Kosterlitz [14] than to any of the other conjectures [9,13]. In this connection, recent simulations of the Eden model [10], ballistic deposition model [5] and directed polymers [16] in $d = 3$ appear to show an increase towards the Kim-Kosterlitz values as well. In addition, we see no evidence of a phase transition in $d = 3$ to logarithmic behavior for $\epsilon \geq 1$. While not completely ruling out the possibility of a transition [17, 18] , our results seem to support Tang, Nattermann and Forrest's idea [19] of a very long crossover region for small ϵ.

A NEW APPROACH TO SCALING IN SURFACE GROWTH

As indicated above, one of the most challenging questions in surface growth phenomena is an analytic determination of the exponents for the KPZ equation in arbitrary dimensions. Here we present one approach [20] which is not exact, but does enable us to determine the surface exponents for the KPZ equation [11]. This approach is quite general and we have shown that in addition to surface growth phenomena it can be applied to a wide variety of other nonequilibrium systems.

The key element in our approach [20] is the analogy between the Langevin-type equations in growth phenomena and the forced Navier-Stokes equation. Let us begin by noting that if all the transport coefficients such as ν, λ, D, in the KPZ equation depend on purely microscopic length-scales a; then on scales $l \gg a$ this equation describes the macroscopic behavior in the same manner as the Navier-Stokes equation describes turbulent flow. This suggests that modified and generalized versions of the type of scaling arguments introduced by Kolmogorov [21] in the context of turbulence might be useful as a method to identify the different scaling regimes observable in surface growth phenomena and other extended open dissipative systems.

Basically for any Langevin-type equation such as the KPZ equation to show scaling each separate term (including the noise), when coarse-grained over length scales l, must be of the same order of magnitude or negligible. Only under these circumstances can scaling behavior arise. To apply this concept to the KPZ equation, we assume that at long times $t \gg t_l$, the typical magnitude of the fluctuations in the interfacial height averaged over a length scale l scale as $\langle (h(\mathbf{r}+\mathbf{l},t) - h(\mathbf{r},t))^{1/2} \rangle_l \sim h_l$. This implies that $h_l \sim l^\alpha$. Then, apart from the noise, for times $t \gg t_l$ and averaged over scales l, the various terms in the KPZ equation may be estimated as $\langle |\partial h/\partial t| \rangle_l \sim h_l/t_l$, $\nu \langle |\nabla^2 h| \rangle_l \sim \nu h_l/l^2$, and $\lambda/2 \langle (\nabla h)^2 \rangle_l \sim \lambda h_l^2/l^2$.

To proceed further we need to estimate the average noise on these length and time scales. For white noise we estimate [20] its mean square fluctuations on length scales l and time scales t_l as $\eta_l \sim \sqrt{D/(S_l t_l)}$ where S_l is the average surface area of the interface on length scales l. This is a simple consequence of adding uncorrelated random variables. We estimate the surface area of the growth on length scales l as $S_l \sim (h_l^2 + l^2)^{(d-1)/2}$. In the limit $l >> h_l$ we have $\eta_l \sim \sqrt{D/l^{(d-1)}t_l}$. We have analyzed a variety of growth models using our scaling arguments and this form of the coarse-grained noise and have found the exact results in all cases.

In order to obtain the Kim-Kosterlitz conjecture for the KPZ exponents [14], however, we must assume [20] that the coarse-grained noise has the form $\eta_l \sim (D/h_l^{(d-1)}t_l)^{1/2}$. In addition, we assume that at sufficiently large length scales the nonlinear term in the KPZ equation will dominate the surface diffusion. Equating the $\partial h/\partial t$ term with the nonlinear term implies that a typical fluctuation lasts for time $t_l \sim l^2/\lambda h_l$. The scaling behavior of these two terms implies $\alpha + z = 2$. Equating our estimate for the noise fluctuation to the inertial term then yields

$$h_l \sim (D/\lambda)^{1/(d+2)} l^{2/(d+2)}, \tag{12}$$

and consequently $\alpha = 2/(d+2)$. The result $\beta = 1/(d+1)$ immediately follows from the scaling relation $\alpha + \alpha/\beta = 2$.

Thus we have derived theoretically the expressions conjectured by Kim and Kosterlitz for α and β. However, this result is clearly based on the specific form we used for the noise term [20]. As already mentioned, we have also used this approach [20] to study a number of other models and obtained the exact results in all cases. This approach is quite similar to those used in Flory theory for equilibrium systems and it should be useful in the study of a wide variety of non-equilibrium problems.

INTERFACE DYNAMICS IN FLUID FLOW IN POROUS MEDIA

We have experimentally studied [22] the evolution of the interface in quasi-two-dimensional displacement of viscous fluids in inhomogeneous media in order to test the dynamic scaling idea and to determine the exponents. The experimental setup [22] was a linear Hele-Shaw cell made of parallel plexiglass plates of size 24 cm×100 cm. To produce a porous medium we packed 220 μm diameter glass beads between the plates. The beads were spread randomly and homogeneously in one layer and glued to the lower plate.

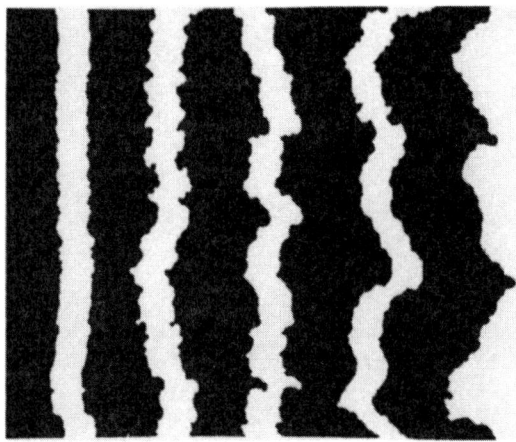

Fig. 2. The evolution of the fluid-air interface, moving from left to right, is shown by plotting the digitized image of the meniscus at different times.

The upper plate was placed directly on the beads and iron rods and clamps were used to prevent the lifting of the plates. Colored glycerine with 4 vol% of water was injected at a fixed flow rate into air between the plates along a line at a shorter sidewall. The viscosity of the glycerine was $\sim$180 cP and the air-liquid surface tension was about $\sim$65 dyn/cm.

The evolution of the interface was recorded on a videotape and digitized with 768×620 spatial resolution. Each digitized image was saved as an array of height values and this process was continued until all images in a run were recorded. We used the average height

$\bar{h} = \langle h(x,t) \rangle_t$ as the time scale in the calculations, because $\bar{h}$ is proportional to the time due to the constant flow velocity and the absence of holes in the region occupied by the more viscous fluid. Fig. 2 shows the development of the fluid-air interface at different times. Clearly the initially smooth interface roughens as it moves into the porous medium.

The dynamic scaling of the growing interface is demonstrated by the data [22] displayed in Fig. 3, where the logarithm of the square root of the height-height correlation function for $x = 0$ is plotted versus $\ln t$. The slope gives $\beta \simeq 0.65$. We have also studied [22] the spatial scaling of the correlations by plotting the logarithm of the square root of the correlation function against $\ln x$. The slope of the straight line fitting the initial part of the data gave [22] $\alpha \simeq 0.81$. These values are different from $\alpha = 1/2$ and $\beta = 1/3$ obtained in various models of interface growth [11]. Our values of α and β are consistent with the scaling relation $\alpha + \alpha/\beta = 2$.

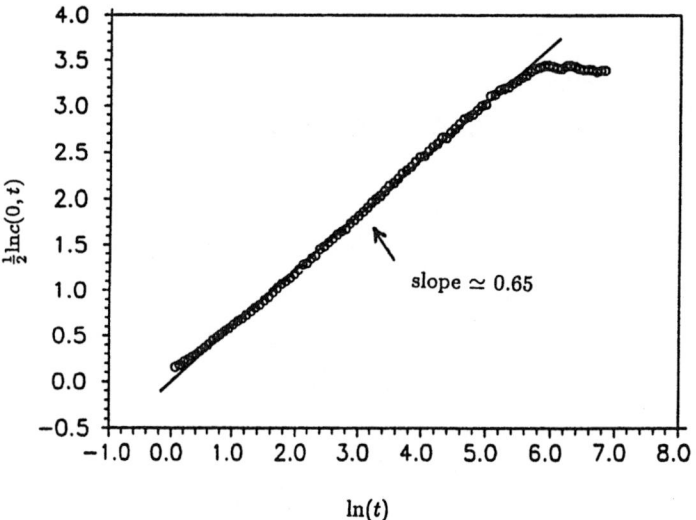

Fig. 3. The logarithm of the square root of the height-height correlation function for $x = 0$ is plotted versus $\ln t$. The slope gives $\beta \simeq 0.65$.

The non-universality of our results is most likely due to the fact that the fluctuations of the interface are not governed by a Gaussian noise [12]. One possibility is that there exists temporal or spatial correlation in the noise. This has been shown to lead to non-universality by Medina et al [12] and Amar, Lam and Family [23], but the results indicate that $\alpha > 2/3$ and $\beta > 1/2$ is impossible for translationally invariant spatially correlated noise. In order to account for the anomalous exponents, Zhang [24] has proposed that the amplitude of the noise is distributed according to a power law instead of a Gaussian distribution. In the next section we discuss the results of the Zhang model and we will compare them with the experimental findings.

SURFACE GROWTH WITH POWER-LAW NOISE

Zhang [24] has suggested that the anomalous exponents arise from the fact that the amplitude of the random noise in the experiments [22] has a non-Gaussian, power law distribution of the form

$$P(\eta) \sim \frac{1}{\eta^{1+\mu}} \quad for \quad \eta > 1 \ ; P(\eta) = 0 \ otherwise, \tag{13}$$

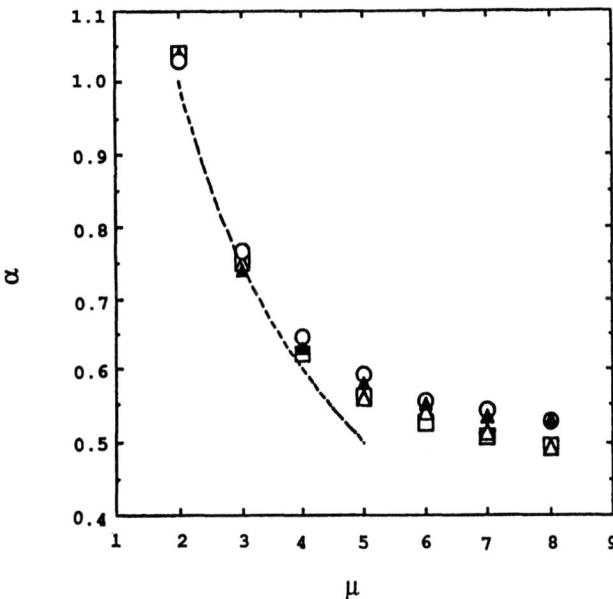

Fig. 4. Variation of the exponent α with μ in the Zhang model and variants in two dimension. Dashed line is prediction of the Flory formula.

where η is the delta-correlated noise. We have carried out [25,26] extensive simulations of two different variants of the Zhang model as well as a ballistic deposition model with power-law noise in order to study the effects of power law noise in surface growth.

Figure 4 summarizes our results [25] for the exponent α for all three models, for $2 \leq \mu < 7$. We also find [25] that the relation $\alpha + z = 2$ is approximately satisfied by all three models and therefore it is reasonable to assume that all 3 models belong to the KPZ universality class. These results indicate that for $\mu < 7$, the exponents α and β for all three models remain clearly above the Gaussian values $\alpha = 1/2$, $\beta = 1/3$. However at $\mu = 8$, for both versions of the Zhang model [24], we obtained results [25,26] which are essentially indistinguishable within error bars from the Gaussian values. We note that our results for α and β for all three models are somewhat higher than the Flory-type formula [27] $\alpha = (1 + d)/(\mu + 1)$ which predicts $\mu_c = 5$ in $d = 2$. We note that when a cutoff is introduced in the power law distribution, the exponents are observed to crossover to the known Gaussian values [26]. We also find [25] power-law distribution of the height fluctuations which should be useful in testing the existence of power-law noise in experiments.

NOISE DISTRIBUTION IN FLUID FLOW IN POROUS MEDIA

We have directly determined [28] the distribution of the noise in experiment on two-phase flow in porous media in order to determine whether the Zhang model [24] does or does not apply to the experiments. In all our calculations we used the digitized images and additional averaging was made over independent experiments with the same system. We applied several different definitions of noise [28] in order to extract the noise distribution from the digitized interfaces. The basic approach was to determine the distribution of amplitudes between two successive profiles separated by a short time interval.

The resulting noise distribution [28] averaged over the experiments is shown in Fig. 5 with circles. Our experimental data points [28] can be fitted by a straight line on log-log

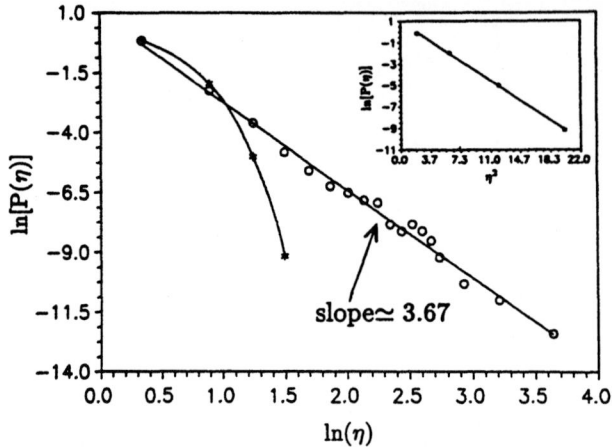

Fig. 5. The logarithm of the noise distribution $P(\eta)$ in the experiment is plotted against the logarithm of the noise amplitude η. In the inset the logarithm of $P(\eta)$ is plotted against η^2 for a computer model, demonstrating that the noise distribution in the simulations is Gaussian.

plot indicating the algebraic decay of noise amplitudes. The exponent of the corresponding power-law behavior $P(\eta) = c\eta^{-(1+\mu)}$ is $\mu = 2.67 \pm 0.19$. From figure 4 we see that for $\mu = 2.67 \pm 0.19$ the value of $\alpha = 0.81 \pm 0.02$ in close agreement with the experimental result [22] of $\alpha \simeq 0.81$. This agreement suggests that in fact the anomalously large values of the exponents found in various experiments could be due to the presence of a power-law noise distribution of the form suggested by Zhang [24]. Clearly this possibility should be tested for other experiments.

UNIVERSAL SCALING IN SURFACE GROWTH

Despite the great activity [1] in determining the exponents from discrete models and continuum equations, there has been no rigorous demonstration of the relation between the KPZ equation and these models. Thus, a more detailed study of the scaling behavior of the KPZ equation would be helpful in establishing a connection between the discrete models and the continuum description.

We have derived [29] expressions for the scaling behavior of the asymptotic coefficients C_t ($C_t = w(\infty, t)/t^{1/3}$) and C_L ($C_L = w(L, \infty)/L^{1/2}$) as a function of the hydrodynamic parameters λ, D, and ν from the scaling properties of the KPZ equation in $d = 2$. Our scaling analysis predicts the existence of a universal scaling function as well as universal amplitude ratios in $d = 2$. We have tested these predictions by simulations of three different surface growth models and by a mode-coupling calculation [29, 30]. Similar results have been derived for the correlation function $c(r, t)$ in the steady-state limit [30].

The scaling behavior of C_t and C_L as a function of D, ν, and λ may be derived as follows. We perform a scale change to a parameterless equation with *dimensionless* variables, $h' = (\lambda/2\nu) h$, $\tau = (\lambda^4 D^2/4\nu^5)t$, $x' = (\lambda^2 D/2\nu^3)x$, and $\xi = (2\nu^4/\lambda^3 D^2)\eta$, so that (8) may be rewritten in $d = 2$ as,

$$\partial h'/\partial\tau = \nabla'^2 h' + |\nabla' h'|^2 + \xi(x', \tau) \tag{14}$$

where $< \xi(x_1', \tau_1)\xi(x_2', \tau_2) > = \delta(x_1' - x_2')\delta(\tau_1 - \tau_2)$. If we write $w'(L', \tau) = g(L', \tau)$, then transforming back to x, h and t we obtain a universal scaling form for the surface width,

$$w(L,t) = L^{1/2} C_L \ F(|\lambda| C_L \ t/L^{3/2}) \tag{15}$$

where $F(x)$ is a universal scaling function. Similarly, we can define [29] a universal amplitude ratio R as,

$$R = C_t/(|\lambda| C_L^4)^{1/3} \tag{16}$$

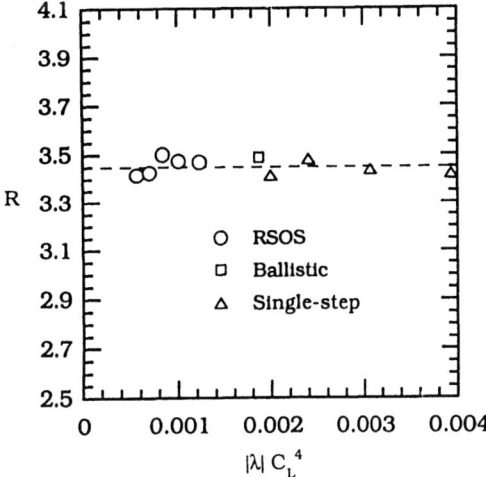

Fig. 6. Universal amplitude ratio R as a function of the scaling parameter λC_L^4 is shown for three different growth models. Average value (dashed line) is 3.45 ± 0.05.

In order to verify our predictions we simulated three different growth models in the KPZ universality class in $d = 2$. Fig. 6 shows a summary of our results [29, 30] for the amplitude ratio R plotted as a function of the scaling parameter λC_L^4 as well as the driving force f for each model. Average value (dashed line) is 3.45 ± 0.05. We also numerically solved the KPZ equation in $d = 2$ from which we obtained $R \simeq 3.2$ in approximate agreement with this value.

In order to test the universality of the scaling function in (15) we also determined $w(L,t)$ for two models. Fig. 7 shows a scaling plot of our results [29, 30]. The asymptotic scaling functions for both models are essentially identical when scaled in the form of (15) as predicted.

The universal amplitude ratio R and the scaling function $F(x)$ were also calculated using a mode-coupling approximation [30]. The solid line in figure 7 shows the mode-coupling results obtained for the scaling function, which is in good agreement with our simulation results.

We note that the same scaling analysis that we have applied to the KPZ equation may be applied to a variety of other models of surface growth at and below their critical dimension d_c, including models with power-law noise. We are currently carrying out simulations on different models with power-law noise in order to test our predictions.

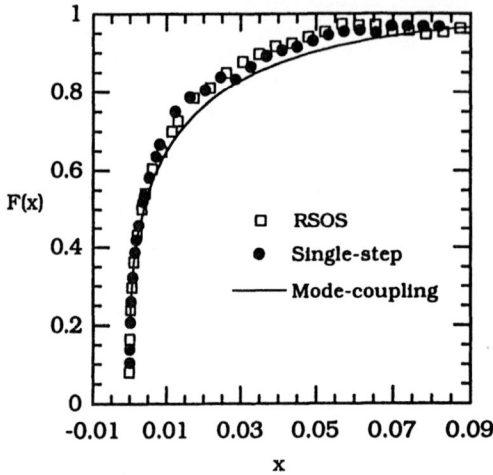

Fig. 7. Scaling function $F(x)$ defined in (16) for two different discrete models. The solid line is result of mode-coupling calculation.

CONCLUSIONS

Studies of the scaling properties of rough surfaces and interfaces provide an effective method for describing the evolution and the structure of growth patterns in physical, chemical and biological processes. The dynamic scaling approach has become a standard tool in analyzing the data from simulations, analytical theories and experiments on growing surfaces. Although considerable progress has been made in recent years in understanding surface growth phenomena, many unresolved questions still remain. For example, only a single experiment has been carried out that has directly investigated the dynamic scaling behavior of a two-dimensional interface. The idea of universal scaling has only recently been developed and could provide an important tool fin understanding the relation between experiments and microscopic models. Finally, both the results of the experiments and discrete models indicate that in order to fully understand surface growth phenomena a much clearer understanding of the nature of the noise must be developed.

ACKNOWLEDGEMENTS

I would like to thank Jacques Amar, Viktor Horváth, Tamás Vicsek and Pui-Man Lam for their collaborations on some of the works discussed here. I would also like to thank Jacques Amar for his helpful comments on the manuscript. This research was supported by the Office of Naval Research and the Petroleum Research Fund administered by the American Chemical Society.

REFERENCES

1. F. Family and T. Vicsek, eds., *Dynamics of Fractal Surfaces* (World-Scientific, Singapore, 1990).
2. F. Family and T. Vicsek, *J. Phys. A* **18**, L75 (1985).
3. B. B. Mandelbrot, *The Fractal Geometry of Nature* (Freeman, San Francisco, 1982).

4. P. Meakin, P. Ramanlal, L. M. Sander and R. C. Ball, *Phys. Rev. A* **34**, 5091 (1986).

5. F. Family, *Physica A* **168**, 561 (1990).

6. M. Eden, in *Proceedings of the Fourth Berkeley Symposium on Mathematical Statistics and Probability, Volume IV: Biology and Problems of Health*, edited by J. Neyman (Univ. of Calif. Press, Berkeley, 1961), pp. 223-239.

7. R. Jullien and R. Botet, *J. Phys. A* **18**, 2279-2287 (1985).

8. D. Stauffer, and J. G. Zabolitzky, *Phys. Rev. Lett.* **57**, 1809 (1986); *Phys. Rev. A* **34**, 1523 (1986).

9. J. Kertész and D. E. Wolf, *J. Phys. A* **21**, 747 (1988); *Europhys. Lett.* **4**, 651 (1989).

10. P. Devillard and H. E. Stanley, *Physica A* **160**, 298 (1989)

11. M. Kardar, G. Parisi and Y.-C. Zhang, *Phys. Rev. Lett.* **56**, 889 (1986).

12. E. Medina, T. Hwa, M. Kardar, and Y.-C. Zhang, *Phys. Rev. A* **39**, 3053 (1989).

13. T. Halpin-Healy, *Phys. Rev. A* **42**, 711 (1990).

14. J. M. Kim and J. M. Kosterlitz, *Phys. Rev. Lett.* **62**, 2289 (1989).

15. J. G. Amar and F. Family, *Phys. Rev. A* **41**, 3399 (1990).

16. J. M. Kim, M. A. Moore and A. J. Bray, *Phys. Rev. A* **44**, 2345 (1991).

17. H. Yan, D. Kessler and L. M. Sander (1990), *Phys. Rev. Lett.* **64**, 926 (1990).

18. Y. P. Pellegrini and R. Jullien, *Phys. Rev. Lett.* **64**, 1745 (1990).

19. L.-H. Tang, T. Nattermann and B. M. Forrest, *Phys. Rev. Lett.* **65**, 2422 (1990); J. M. Kim, A. J. Bray and M. A. Moore, , *Phys. Rev. A* **44** R4782 (1991).

20. H. G. E. Hentschel and F. Family, *Phys. Rev. Lett.* **66**, 1982 (1991).

21. A. N. Kolmogorov, A. N., *Dokl. Akad. Nauk. SSSR* **30**, 299 (1941).

22. V. K. Horváth, F. Family and T. Vicsek *J. Phys. A* **24**, L25 (1991).

23. J. G. Amar, P.-M. Lam and F. Family, *Phys. Rev. A* **43**, 4548 (1991).

24. Y. C. Zhang, *J. de Physique* **51**, 2129 (1990).

25. J. G. Amar, and F. Family, *J. Phys. A* **24**, L79 (1991).

26. J. G. Amar, J. G. and F. Family, *J. de Physique I* **1**, 175 (1991).

27. Y. C. Zhang, Y.-C., *Physica A* **170**, 1 (1990).

28. V. K. Horváth, F. Family and T. Vicsek *Phys. Rev. Lett.* **67**, 3207 (1991).

29. J. G. Amar and F. Family, *Phys. Rev. A* **45** (Rapid Communications), *in press* (1992).

30. J. G. Amar and F. Family, *Phys. Rev. A* **45** , *in press* (1992).

MBE GROWTH AND SURFACE DIFFUSION

David A. Kessler*, Herbert Levine+, and Leonard M. Sander*

*Physics Department
University of Michigan
Ann Arbor, MI 48109-1140 USA

+Department of Physics
and
Institute for Nonlinear Science
University of California, San Diego
La Jolla, CA 92093

There has been much recent interest in the statistical properties of nonequilibrium surfaces. In particular, much attention has been focussed on theoretical models to describe thin film growth by Molecular Beam Epitaxy. A number of groups [1,2,3] have proposed that the statistical properties of MBE growth are given by the fourth order continuum equation [1]

$$\dot{h} = -D_4\nabla^4 h + \lambda_4\nabla^2(\nabla h)^2 + \eta \tag{1}$$

where $h(\vec{x}_\parallel, t)$ is the height of the surface and η is a noise source with correlations

$$\langle \eta(\vec{x}_\parallel, t) \rangle = 0$$

$$\langle \eta(\vec{x}_\parallel, t)\eta(\vec{x'}_\parallel, t') \rangle = S\delta^d(\vec{x}_\parallel - \vec{x'}_\parallel)\delta(t - t') \tag{2}$$

The above equation, which attempts to model surface diffusion (SD) incorporates a conservation law wherein the total volume of the film is conserved on average. As is well known, this is true only if both desorption and overhangs are ignored. If either of these are significant, then the scaling properties of the interface are expected to be those of the the Kardar-Parisi-Zhang (KPZ) equation [4]

$$\dot{h} = D_2\nabla^2 h + \lambda_2(\nabla h)^2 + \eta \tag{3}$$

What has not yet been adequately addressed is the question of *when* overhangs and desorption are important. Of these two, it is our belief that overhangs are potentially

Growth Patterns in Physical Sciences and Biology, Edited
by J. M. Garcia-Ruiz *et al.*, Plenum Press, New York, 1993

most relevant for MBE growth, due to the relatively large energy barrier for desorption (compared to that for surface diffusion). This is especially so at lower temperatures, which as we shall see is where overhangs are most important. In this paper we present a simple model which captures the essence of the physics of deposition and surface diffusion (in the absence of desorption) so that the crossover between the SD and the KPZ regimes can be elucidated. We shall ignore the possibility of desorption in the following, saving a detailed discussion for future work.

As a preface, it is useful to review the physical motivation of Eq. (1), which lies in the fact that the essential mechanism of surface relaxation is surface diffusion. Thus, the growth of the surface should be described by [5]

$$v_n = D_4 \nabla^2_{\mathrm{L-B}} \kappa + \mathcal{F} \tag{4}$$

where v_n is the normal velocity of the interface, κ the curvature, $\nabla^2_{\mathrm{L-B}}$ the Laplace-Beltrami (intrinsic Laplacian) operator on the surface, and $\mathcal{F}$ the (fluctuating) flux. Assuming the flux is normally incident so that growth is primarily in the transverse $(\hat{x}_t)$ direction and expanding about a flat interface yields

$$\dot{h} = -D_4 \nabla^4 h + \mathcal{F} + \dots \tag{5}$$

This is similar to Eq. (1) above, albeit with some important differences. First, there is *no* nonlinear term of the form $\nabla^2 (\nabla h)^2$. The physical basis of this fact is that the chemical potential is independent of the orientation of the surface. This symmetry argument would seem to imply that such a term cannot be generated by coarse-graining, and so λ_4 should be zero to all orders in a renormalization-group expansion. This is consistent with the fact that λ_4 is not renormalized, [3] so that if the bare theory has λ_4 equal to zero, it cannot be induced by other nonlinear terms. Furthermore, in the absence of the λ_4 term, the nonlinearities implicit in Eq. (4) are all relevant in $d = 1$ and marginal in $d = 2$ so that the scaling behavior of surface diffusion dominated regime is in fact not obvious.

The second important difference between Eq. (4) is that the flux $\mathcal{F}$ has not been specified. The crossover to KPZ behavior is, as we shall see, due to the fact that, on sufficiently large length and time scales, the flux $\mathcal{F}$ *cannot* be modelled by a noise η with correlations as in Eq. (2), as a result of the physical process of "sticking on the sides" of surface features.

In discussing models, let us start with that of Das Sarma and Tamborenea[7] and Wolf and Villain,[2] which we shall refer to as the DT/WV model. In this model particles are dropped at a random value of $\vec{x}_\parallel$ onto the top of that column so that the height at $\vec{x}_\parallel$ is increased by 1. At this point the just-dropped particle is relaxed, moving to the top of a neighboring column if it increases the particle's coordination. This model was found to produce a surface which scales according to the linear version of Eq. (1), i.e. $\lambda_4 = 0$. In $d = 1$ the width of the surface grew as t^β, $\beta = 3/8$ for $t \ll L^z$, $z = 4$ and saturated at a value proportional to L^α, $\alpha = 3/2$, for long times, $t \gg L^z$. As a model for MBE growth, (as opposed to a model which simply reproduces the scaling behavior of Eq. 1) this model suffers from two defects which render it *unphysical*. First, the particles drop directly onto the top of a column, bypassing an arbitrary number of particles in neighboring taller columns on the way down. This "drop-through" rule is in contrast to the drop rule of the ballistic aggregation model, [8] where the dropped particle sticks as soon as it encounters any particle of the aggregate. We emphasize that this "sticking on the side" is physically correct and would be present in any experiment or realistic (Molecular-Dynamics, say) simulation [9] with finite-range forces. The second defect is that relaxation is allowed to occur between nearest-neighbor *columns* regardless of the difference in heights involved. This corresponds to an extremely anisotropic and unphysical diffusion, where diffusion is finite in the directions parallel to the substrate but infinite in the transverse direction. Note that both these defects arise from the solid-on-solid (SOS) restriction which is at the heart of the DT/WV Model. Whereas the SOS constraint is appropriate in some contexts, including some models of KPZ-type growth,[10] in the present case it is inappropriate since it enforces a symmetry (particle number conservation, on average) not present in the physical situation.

We now turn to a model of ideal MBE growth in which both surface diffusion and deposition are treated physically. In this model, the "drop-through" rule is replaced by that of ballistic aggregation, where the dropped particle sticks upon first encountering the aggregate, either below or to one side. We also adopt a more physical surface diffusion rule, wherein a randomly chosen surface (not fully coordinated) particle moves to a new position chosen from among the best-coordinated sites in a hypercubical box with sides of length $2L_D + 1$ (in *all* $d + 1$ dimensions) centered on the particle's current position. The diffusivity is governed by L_D and D, the number of diffusion steps a surface particle performs on average in the the time it takes to lay down a monolayer, and scales for small D as $L_D D^{1/2}$. This surface diffusion is not to be confused with height diffusion, which has also been studied[11] in the context of ballistic aggregation drop rules). We claim that this new model, ballistic aggregation with surface diffusion (BASD), is a physically correct analogue of ideal MBE growth and should correctly describe the statistical properties of surfaces formed in this manner. Furthermore, while other more realistic models involving activated hopping [6] could in principle elucidate the issues posed, their computational complexity makes them unsuited for a detailed studied of scaling properties.

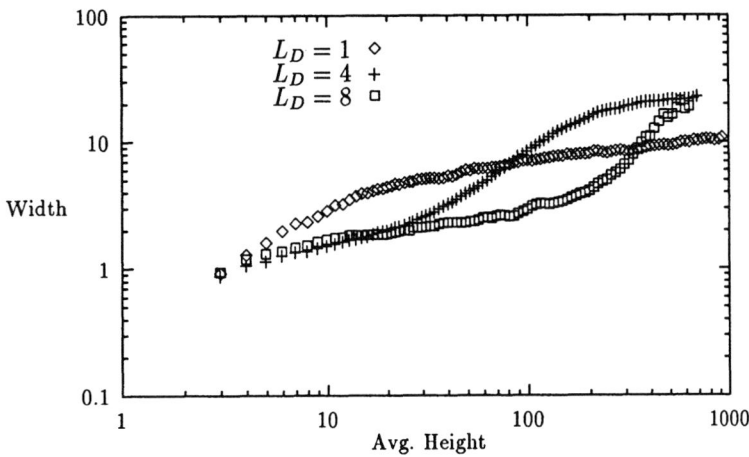

Figure 1. Width vs. average height for the BASD Model with varying L_D=1,4,8. Other parameters are $L = 500$, $D = 5$.

Results of simulation of our BASD Model in 1+1 dimension for varying L_D are presented in Fig. 1. After a short transient, there is a regime, lengthening with L_D, where w grows with time, apparently as t^β, $\beta \approx .25$. Then then follows a period of rapid growth of the width, following which power-law again sets in. This latter regime then terminates in a saturation regime. Pictures of the aggregate in the first power-law and final saturated regimes are shown in Fig. 2a and 2b respectively. Fig. 2a is almost perfectly dense, with essentially no defects or overhangs. Fig. 2b, on the other

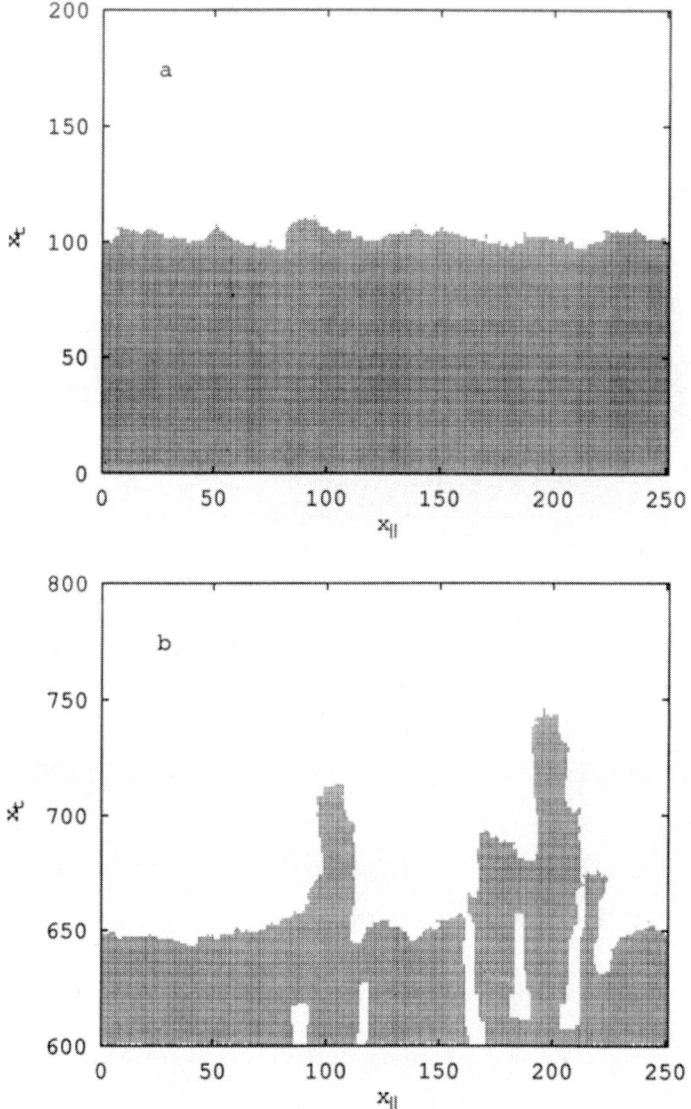

Figure 2. Snapshots of the (top of the) aggregate in the BASD Model with $L = 250$, $L_D = 6$, $D = 5$: a) after 100 nominal layers; b) after 600 nominal layers.

hand, resembles a coarsened version of a standard ballistic aggregate. We thus interpret the first power-law regime as representing surface diffusion scaling, whereas the second power-law regime and subsequent saturation are described by the KPZ equation. One interesting and unexpected feature to note is the sharp rise in width separating the two regimes. This is in contrast to the results of the exactly solvable linear model with both $\nabla^2 h$ and $\nabla^4 h$ terms, where the crossover is much smoother.

It is interesting to consider what value of β should characterize the surface-diffusion regime. While we do not expect $\beta = 1/3$, since λ_4 is zero, it would appear that we are not seeing the $\beta = 3/8$ characteristic of the linear model either. This point clearly requires further study.

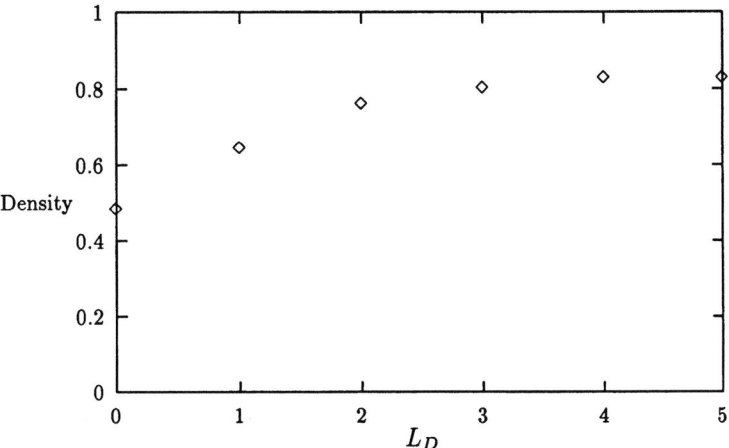

Figure 3. Asymptotic Density vs. L_D for the BASD Model.

The short distance coarsening in the KPZ regime reflects the smoothing due to surface diffusion. This smoothing leads to the surprising result, evident in Fig 1, that the saturated width of our BASD model *increases* with L_D. Essentially, the BASD Model produces ballistic aggregates made with blocks whose size L_{block} increases with L_D. The effective length of the system is thus L/L_{block} so that w is reduced by a factor $L_{\text{block}}^{-\alpha}$, where α is the long-time exponent of BA. However, excursions in width are now measured in units of L_{block} so that the measured width in the original scale has actually increased, by a factor $L_{\text{block}}^{1-\alpha}$.

Another consequence of this is that the average density of the aggregate does not approach 1 at large L_D. If the above argument that the aggregates at large L_D were regular ballistic aggregates grown with larger blocks were exactly correct, the average density of the aggregate would in fact be *independent* of L_D. The measured density as a function of L_D is displayed in Fig. 3. While the average density does increase with L_D for small L_D, it appears to be saturating at a value of approximately .83, so that our argument is asymptotically correct.

The above implies that Eq. (1) is not an appropriate description of the physics, on length scales larger than a diffusion length. Instead, the physics is that of the KPZ equation. This is so despite the absence of desorption. It is important to understand the origin of the $\nabla^2 h$ term in the KPZ description of our BASD Model. It does not arise from any explicit gravitational force or tendency for aggregated particles to move downhill. Rather, it arises from the lateral spreading of columns (which leads to an effective diffusion of h). Equivalently, "sticking on the sides" leads to an $\dot{h}$ which depends on the local slope, since the larger the slope, the higher up the next particle will stick. This produces the nonlinear $(\nabla h)^2$ term in Eq. (3). Upon coarse-graining this term gives rise to a $\nabla^2 h$ term, as shown by renormalization group studies. [4] Furthermore, and most importantly, the magnitude of the $\nabla^2 h$ coupling, D_2, is inevitably of order 1, in units of (lattice spacing)**2 per monolayer deposition time.

Given the above we can estimate the crossover time by recognizing that the most general continuum description of our growth model should contain both the D_4 term of Eq. 1 and the D_2 term of the KPZ equation, with the relative strength $D_4/D_2 \approx L_D^2$, where lengths are measured in units of a lattice spacing. Therefore, on sufficiently large length scales, the lower order relaxation should dominate the growth process as soon as $(D_2 t)^{1/z_2} \geq (D_4 t)^{1/z_4}$ or equivalently

$$t \geq L_D^{2z_2/(z_4 - z_2)} \tag{6}$$

where z_2, z_4 are the dynamical exponents for the KPZ equation and the appropriate surface diffusion model, respectively. Similarly, the crossover length is given by

$$L^* = L_D^{2/(z_4 - z_2)} \tag{7}$$

Since L_D can be varied over a wide range by altering the substrate temperature during deposition, this prediction should be amenable to experimental test.

A final point is the porosity of the films produced by the BASD model, which is not typical of real MBE films. It is important to note that growth conditions in the laboratory are carefully chosen to produce "good" films. For example, films are typically grown at high temperatures where diffusion lengths are of order .1 microns. Low temperature films do become porous if grown too thick.[12] This is consistent with the initial compact, SD regime seen in Fig. 2a. Also, interrupted growth is frequently employed to improve the film quality. Moreover, any desorption and imperfect sticking will compactify the film; it would be useful to quantify these effects so as to understand the variation of D_2 with depositional parameters. It is an important challenge to identify the operative mechanism(s) responsible for compact films in the parameter ranges appropriate to experimental conditions.

In conclusion, we have seen how a physically reasonable model for ideal MBE growth leads to a crossover from SD to KPZ scaling on a time scale which grows with diffusion length. We maintain that similar results would obtain from a study of any physically correct model, including more realistic Monte-Carlo or Molecular- Dynamics simulations. Thus we would expect KPZ scaling to be realizable in experimental film growth. We believe further study of our BASD Model, especially of the 2+1 dimensional case, will prove fruitful.

ACKNOWLEDGEMENTS

We acknowledge useful conversations with Y. Tu, B. Orr, and H. Yan, and the insightful comments of J. Villain. The work of DAK is supported by the U.S. DOE under grant DE-FG-0285ER54189 and that of LMS by the U.S. NSF under grant DMR91-17249.

REFERENCES

[1] J. Villain, J. Physique I **1**, 19 (1991).

[2] D. E. Wolf and J. Villain, Europhysics Lett. **13**, 389 (1990).

[3] Z. W. Lai and S. Das Sarma, Phys. Rev. Lett. **66**, 2348 (1991).

[4] M. Kardar, G. Parisi, and Y.-C. Zhang, Phys. Rev. Lett. **56**, 889 (1986).

[5] W. W. Mullins, J. Appl. Phys. **28**, 333 (1957).

[6] S. Das Sarma, J. Vacuum Sci. Tech. **A8**, 2714 (1990); I. K. Marmorkos and S. Das Sarma, unpublished.

[7] S. Das Sarma and P. Tamborenea, Phys. Rev. Lett. **66**, 325 (1991).

[8] M. J. Vold, J. Colloid Interface Sci. **14**, 168 (1959); H. J. Leamy, G. H. Gilmer, and A. G. Dirks, in *Current Topics in Materials Science*, edited by E. Kaldis, (North-Holland, Amsterdam, 1980) Vol. 6; for a recent review, see L. Sander, in *Solids Far From Equilibrium: Growth, Morphology and Defects*, edited by C. Godreche (Cambridge, 1991).

[9] H. J. Leamy, et. al., Ref. [8].

[10] P. Meakin, P. Ramanlal, L. M. Sander, and R. C. Ball, Phys. Rev **A34**, 5091 (1986).

[11] H. Yan, D. A. Kessler, and L. M. Sander, Phys. Rev. Lett. **64**, 926 (1990).

[12] D. D. Perovic, G. C. Weatherly, J.-P. Noël, and D. C. Houghton, J. Vac. Sci. Technol. **B9**, 2034 (1991).

GROWTH IN SYSTEMS WITH QUENCHED DISORDER

Mark O. Robbins, Marek Cieplak,* Hong Ji,† Belita Koiller,‡
and Nicos Martys**

Department of Physics and Astronomy
The Johns Hopkins University
Baltimore, MD 21218

INTRODUCTION

In this paper we consider the effect of quenched disorder on growth. Two specific examples are considered to illustrate the general nature of the changes induced by disorder: magnetic domain growth[1-3] and immiscible fluid invasion.[4-8] In each case there are two domains which have different spin orientations or fluid composition. An applied force, magnetic field or pressure, favors growth of one domain.

Several universal features are found in all systems studied.[1-8] The first is that disorder pins the interface between domains. The interface can only advance continuously when a critical force, f_c, is exceeded. Three types of morphology are found for the marginally stable interfaces at f_c: faceted, self-affine and self-similar. Faceted interfaces are found in the limit of weak disorder where lattice anisotropy becomes important. When the degree of disorder is large, the interface is a self-similar fractal characteristic of percolation.[9] Self-affine interfaces are found at intermediated degrees of disorder in some models. The transitions between these growth morphologies are critical phenomena characterized by diverging coherence lengths.

The onset of interface motion is also a critical phenomena in the self-similar and self-affine regimes. Increasing the force towards f_c produces larger and larger incremental advances of the interface. Their mean size diverges at f_c and they follow a power-law distribution. The exponents describing these and other quantities can be related through scaling laws.[6] Only the fractal dimensions and a single additional exponent are needed to determine all other exponents.

We begin by describing the growth rules used in our studies. Then results are presented with an emphasis on universal trends. These are primarily illustrated for the magnetic system which is conceptually simpler. The final section of the paper presents a summary and conclusions.

GROWTH MODELS

The Hamiltonian and rules for domain wall motion are described in detail in reference.[1] Ising spins $s_i = \pm 1$ are placed on each site i of a 2 or 3D array. Their Hamiltonian, in

Growth Patterns in Physical Sciences and Biology, Edited
by J. M. Garcia-Ruiz *et al.*, Plenum Press, New York, 1993

units of the exchange coupling, is

$$\mathcal{H} = - \sum_{<i,j>} s_i s_j - \sum_i (h_i + H) s_i \ . \tag{1}$$

The first term is a nearest-neighbor spin-spin exchange interaction. The second term describes the interaction of each spin with the sum of the local random field h_i and a uniform external magnetic field H. Values of h_i are generated randomly following some probability distribution function $P(h)$. Several forms for P have been studied,[2] but the behavior is mainly dominated by the bounds of $P(h)$. In this paper we restrict our attention to uniform distributions: $P(h) = (2\Delta)^{-1}$ for $|h| < \Delta$. The value of Δ characterizes the degree of disorder. It represents the ratio of the size of the random fields to the exchange coupling which favors aligned spin states.

Spins at the bottom edge of the simulation cell are originally "flipped" ($s = +1$), while all other spins are "unflipped" ($s = -1$). To mimic the process of fluid invasion, growth occurs at zero temperature and *only spins at the interface are allowed to flip*. An interface spin is flipped when this lowers the total energy of the system. If there is more than one unstable spin on the interface, the most unstable spin is flipped first. Each spin-flip alters the exchange interaction on neighboring interface spins or adds new spins to the interface. Thus a single spin-flip may produce a chain reaction. Once a spin is flipped, it does not return to the unflipped state.

The overall advance of the interface is controlled by the value of the external field H, which is varied quasistatically. It is initially taken to be the smallest value, H_0, that causes a single spin on the interface to flip. This single change in the spin array may cause neighboring spins to flip, which may induce subsequent spin flips. The external field value is kept equal to H_0 until a stable interface is attained. Then H is increased so that a single spin flip occurs on the new interface, and the procedure is repeated until the interface reaches the top of the system. The value of H required for the pattern of flipped spins to span the system rapidly approaches a limiting critical field H_c as the number of spins along an edge of the cell, L, increases.

It is convenient at this point to identify quantities which describe the importance of the local environment in producing spin-flips.[2] These will be useful in the discussion of the critical transitions below. Whether a spin s_i at the interface flips depends on the value of the external field H, the value of the local random field h_i, and the state of the z nearest-neighbor spins. The fraction of spins with n flipped neighbors which will flip at the critical field is

$$f_n = \int_{-\infty}^{H_c} P(z - 2n - H) dH = (H_c - z + 2n + \Delta)/2\Delta \ , \tag{2}$$

where the last equality holds only for uniform distributions and for $|H_c - z + 2n| < \Delta$.

Fluid invasion has been studied in two different types of model 2D porous media.[4-8] The first consists of ducts of random width placed along the bonds of a square network.[8] Such models have been studied in experiments by several groups.[10-13] The second class of porous media was constructed[4-6] by placing disks of random radii on the sites of a square or hexagonal lattice of lattice constant a. This model is closer to a cross-section through a glass bead pack – another common model system.[14-19]

The effective degree of disorder in these porous media depends on two factors. The first factor is the degree of geometrical disorder in the pore space. In the case of duct networks this is characterized by $R = d_{max}/d_{min}$, the ratio between the maximum and minimum duct widths. The ratio between the maximum and minimum widths of the

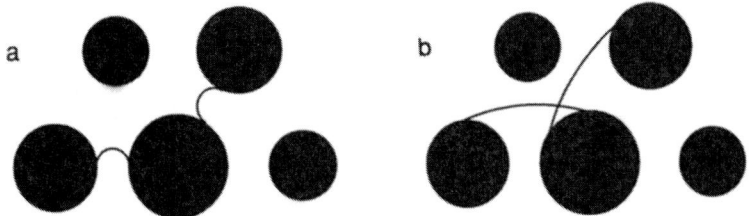

Figure 1. Arcs connecting successive beads along the interface for (a) $\theta = 180°$ and (b) $\theta = 30°$. The invading fluid is advancing from below, and the bond angle between three successive disks along the interface is $\alpha = 120°$.

throats between disks provides a similar measure in the disk model. The second important factor is the relative wetting tendency of the invading fluid. This can be quantified by the contact angle θ (measured through the invading fluid) at which the fluid interface intersects the solid.

Figure 1 illustrates why θ controls the effective interaction between neighboring pores. The invading fluid is driven forward by an external pressure P. In quasi-static invasion this pressure drop appears everywhere along the interface. In 2D, the interface consists of circular arcs with curvature P/γ, where γ is the surface tension and the boundary of the invading fluid is convex for positive pressures. These arcs must also intersect the disks or duct walls at θ.

When the invading fluid is non-wetting, the narrow ducts or throats between disks are the hardest to penetrate because the required pressure is largest (Fig. 1(a)). Arcs are confined within their respective throats, and neighboring arcs advance independently. The situation is much like that in a magnetic system with no exchange coupling between spins. When the invading fluid is more wetting, arcs are sucked rapidly through the narrow throats and the pore spaces are the hardest regions to pass. Since arcs in neighboring throats share a common pore space, interactions between them are enhanced. In Fig. 1(b) each arc would be stable individually. However, when both are present in the pore they *overlap* and the kink in their combined interface is unstable. The probability of an overlap depends on the bond angle α between successive disks along the interface. In the figure $\alpha = 120°$. If the interface connected three beads in a line ($\alpha = 180°$) the arcs would no longer overlap for this pressure, contact angle, and geometry. Thus overlaps preferentially remove sharp bends in the interface (small α). They act much like an exchange coupling in maintaining a coherent domain. However, in the magnetic case only the number of surroundings spins of each type is important. In the fluid case, the relative position of the throats determines α and thus the condition for stability. Since overlaps become more likely as θ decreases, the effective degree of coupling increases and the relative degree of disorder decreases.

Growth was initiated from a flat interface or central ring. The algorithm is much like that for magnetic domain growth, with pressure P replacing the external field H. The initial value of P was set to P_0, the pressure at which the first segment of the interface became unstable. This segment of the interface was advanced. Any resulting instabilities were advanced in turn until a new stable interface was reached. The value of P was then increased until a segment of this interface became unstable, and the process was repeated. For each θ we identified a critical pressure P_c at which the invading fluid would first span an infinite system.

The major difference between our magnetic and fluid studies is in the treatment of "trapped regions".[11,20] In some cases, advancing a segment causes the interface to completely surround a region of the defending fluid. This trapped region can not shrink further if the fluid is incompressible. We implemented this rule in studies of the random disk model, but not in studies of the duct network or magnetic domain models. Trapping may change the fractal dimension slightly[20] in 2D, but does not change the transitions in growth morphology discussed below.[1-8]

The above growth algorithms are useful for studies of the approach to the critical force from the pinned regime. Different algorithms are needed to study steady-state motion at forces greater than the critical force. Two rules for advancing unstable segments of the interface were used. One was an Eden-like model[21] in which the growth site was chosen randomly from the set of unstable segments. The second was a synchronous update rule where all unstable segments at a given time step were advanced before any subsequent instabilities were considered. These algorithms produce very different large-scale structure as shown below.

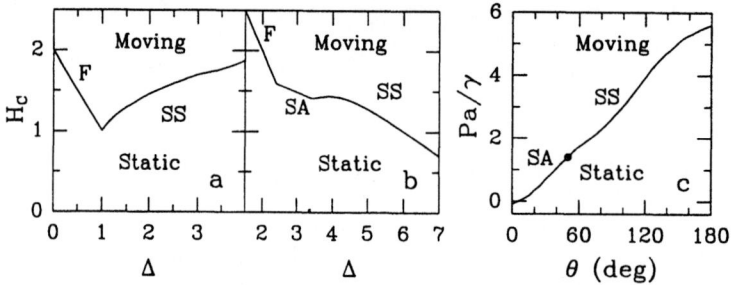

Figure 2. Phase diagrams for magnetic domain growth on the (a) square and (b) cubic lattices, and for fluid invasion of (c) a hexagonal array of disks. A curve in each panel indicates the critical force separating static and moving states of the interface. The large scale structure of the interface at the critical force is indicated by SS for self-similar interfaces, SA for self-affine interfaces and F for faceted interfaces. In (a) there is a direct transition from self-similar to faceted growth. In (b) there is an intermediate self-affine regime. A dot indicates the transition to self-affine growth in (c). The faceted regime is suppressed in this highly disordered system.

PHASE DIAGRAMS FOR GROWTH

Fig. 2 shows phase diagrams for growth[1-6] as a function of the driving force (H or P) and effective degree of disorder (Δ or θ). Two types of transitions are evident in all phase diagrams. The first is the onset of steady growth at H_c or P_c. The second involves changes in the growth morphology. The large-scale structure at the critical force changes from self-similar to faceted as the effective degree of disorder decreases. In some cases there is an intervening regime where the interface is a self-affine fractal. We will begin by discussing growth in the limit of large disorder, and describe each transition in detail. To simplify the discussion we focus on magnetic domain growth and then comment briefly on differences in the case of fluid invasion.

When the disorder is very large, the exchange coupling can be neglected. The probability that a spin will flip becomes independent of its surroundings (ie., n). The problem reduces to site percolation since all spins with $H + h_i > 0$ will flip and all others will not. The fraction of spins which flip at H_c must equal the critical probability for normal site-percolation: $p_c = 0.5927$ and 0.3117 for square and cubic lattices, respectively.[9] Figure 3 shows that this condition is satisfied. All probabilities f_n merge to p_c as $\Delta \to \infty$.

As Δ decreases, the local environment becomes increasingly important in initiating spin-flips. Spins with larger n are more and more likely to flip before other spins (Fig. 3). This leads to cooperative invasion of neighboring regions, and the pattern of flipped spins becomes smoother (Fig. 4). A measure of this change is given by the average finger width, w, calculated as the mean width of segments of adjacent flipped spins.[4] Fig. 5 shows w versus Δ for the square and cubic lattices. In all systems studied, w rises rapidly near the transition to a new type of growth morphology and then saturates at a value

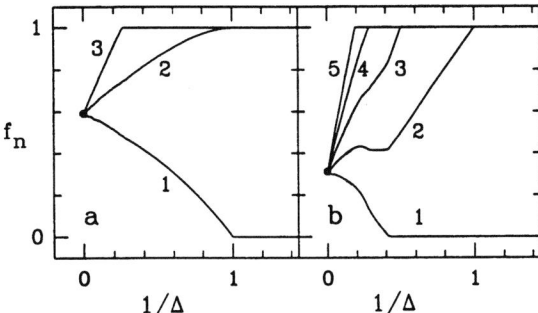

Figure 3. Probability f_n that a spin surrounded by n flipped spins will flip at H_c as a function of $1/\Delta$ on the (a) square and (b) cubic lattices. Curves are labeled by n.

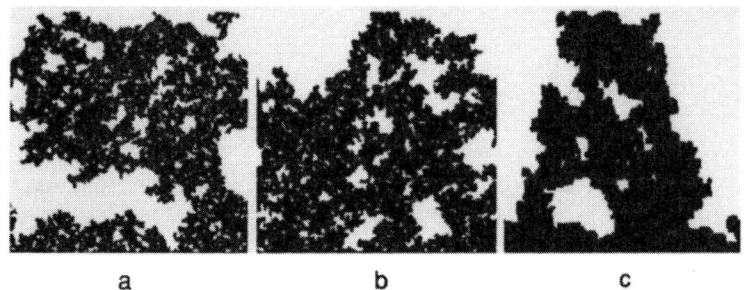

Figure 4. Invaded domains in a duct network model with $L = 512$ and $R = 10$ as θ decreases towards the critical angle $\theta_c = 48°$. The values of θ and w are: (a) $\theta = 170°$, $w = 6.8a$; (b) $\theta = 100°$, $w = 17a$; and (b) $\theta = 60°$, $w = 80a$. Equivalent changes occur in all other models as the effective degree of disorder decreases.

close to the system size.[1-8] From the behavior of the probabilities f_n and from finite-size scaling fits we have determined that $w \propto (1 - \Delta_c/\Delta)^{-\nu'}$ with $\Delta_c = 1.$, $\nu' = 1.9 \pm 0.1$ on the square lattice,[2] and $\Delta_c = 3.4$, $\nu' = 2.5 \pm 0.2$ on the cubic lattice. The exponent ν' is not universal and depends, for example, on the distribution of random fields.[2] However, one can identify universal exponents[2] relating w to the growth probabilities f_n.

Another type of universality applies to growth at all $\Delta > \Delta_C$. Despite the large change with Δ in the domains of Fig. 4, the structure at scales greater than w is always self-similar. Moreover, the fractal dimension remains equal to that for normal site percolation.[9] The box-counting method gives $D_f = 1.89 \pm 0.01$ in 2D and 2.5 ± 0.3 in

Table 1. General scaling laws obeyed by critical exponents for invasion, and specialized forms for compact growth.[6] Here D_f and D_e are the fractal dimensions of the bulk and external hull. For compact growth D_f equals the spatial dimension and $D_e = D_f - 1$.

General	Compact Growth
$\psi = \nu(D_f + 1 - d)$	$\psi = \nu$
$\omega = \nu(D_e + 1 - d)$	$\omega = 0$
$\phi = \nu B = \nu(D_f - D_e) + 1$	$\phi = 1 + \nu$
$\tau' D_f = D_f + D_e - 1/\nu$	$\tau' d = 2d - 1 - 1/\nu$

3D, where the largest systems had $L = 4000$ and 300, respectively.[1-8]

The exponents describing the approach to the critical force can also be related to exponents for normal percolation. The correlation length, ξ, corresponds to the forward advance of the interface. As H increases to H_c, ξ, the total domain volume V, the total interface area S, and the incremental increases dV after changes in H all diverge:

$$\xi \propto (H_c - H)^{-\nu}, \quad V \propto (H_c - H)^{-\psi}, \quad S \propto (H_c - H)^{-\omega}, \quad \langle dV \rangle \propto (H_c - H)^{-\phi}. \tag{3}$$

At H_c there is a power law distribution of incremental advances[22]

$$P(dV) \propto dV^{-\tau'}. \tag{4}$$

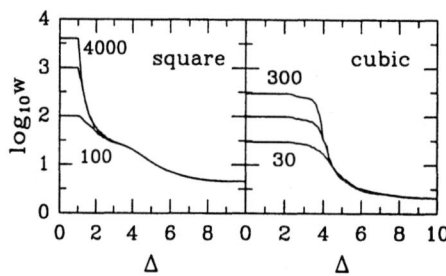

Figure 5. Variation of w with Δ on the square lattice with $L = 100$, 1000 and 4000 and the cubic lattice with $L = 30$, 100 and 300. Bends in the curves occur at points where one of the f_n becomes equal to 1 or 0.

Table 1 gives scaling laws relating these exponents to the fractal dimension and coherence length exponent ν for normal percolation. Table 2 shows that numerical values from finite-size scaling are consistent with normal percolation exponents. Note that 2D values were obtained for the disk model which includes trapping. It is known that trapping may decrease D_f in a non-universal way,[20] but our results indicate that any decrease is less than 0.05 in our model.

In Fig. 2(a) the interface becomes faceted for $\Delta < \Delta_C$. There is also a direct transition from self-similar to faceted growth in the duct network model.[8] These transitions may be understood by considering the effect of local environment on growth in the square lattice. The starting interface for our simulations was chosen to be along a line connecting nearest-neighbor spins. Unflipped spins on the interface have n=1 flipped spins and $z - 1 = 3$ unflipped spins as neighbors. The initial value of the external field in an infinite system is $H_0 = 2 - \Delta$, the lowest value at which a spin with $n = 1$ will flip. Once the first spin above the original flat interface flips, two of its neighbors have

Table 2. Critical exponents from finite-size scaling fits for wetting and non-wetting invasion of the 2D random disk model and for domain wall motion at intermediate and large disorder on the cubic lattice. Error bars in the last significant digit are indicated in parentheses. Within these uncertainties, all exponents are consistent with the scaling laws of Table 1. Within our accuracy, exponents for non-wetting invasion and domain growth at large disorder equal those for ordinary percolation.[9,22] The exponents for self-affine growth are also universal.

	$\theta = 25°$ $P_c = .4935(5)$	$\theta = 179°$ $P_c = 5.62(2)$	2D Perc.	$\Delta = 3$ $H_c = 1.480(1)$	$\Delta = 6$ $H_c = 1.011(1)$	3D Perc.
D_f	2	1.88(4)	1.89...	3	2.5(1)	2.5(1)
D_e	1	1.32(2)	1.33...	2	2.5(1)	2.5(1)
α	0.81(4)			0.67(3)		
ν	1.30(5)	1.32(7)	1.33...	0.8(1)	0.9(1)	0.9(1)
τ'	1.125(25)	1.30(5)	1.31...	1.3(1)	1.6(1)	1.6(1)

equal numbers ($z/2 = 2$) of up and down spins as neighbors. For $\Delta < 1$ these spins will also flip since the maximum field required to flip a spin with $n = 2$ is $H_2^* = \Delta < H_0$. This changes the environment of two further spins to $n = 2$, and the process continues until all spins on the initial interface have flipped. Each row of spins then flips in turn until the interface reaches the top of the system. As a consequence, $H_c = H_0 = 2 - \Delta$ for $\Delta \leq \Delta_C$ and all spins in the system are flipped. In Fig. 3 the value of f_1 goes to 0 at Δ_C and $f_2 \to 1$.

The key feature in the above argument is that there is a gap between the highest force needed to flip one spin configuration and the lowest force needed to flip spins with one fewer flipped neighbor. In such cases there is no critical divergence in the invaded area as H increases to H_c. Instead, the interface is confined to a given facet over the range of forces in the gap between H_2^* and H_c. Then there is a sudden onset of growth at H_c. The moving state may exhibit critical phenomena as H is *decreased* to H_c because the fraction of sites which can flip with fewer neighbors will vanish as H decreases. The typical length scale between such sites may act like a mean-field coherence length.

Self-affine growth was only found for magnetic domain growth in 3D and highly disordered random disk models in 2D. Random disk models with small variations in throat width exhibited a direct transition to faceted growth like that in Fig. 2(a). As in the case of self-similar growth, the self-affine regime shows universal large-scale structure and critical exponents. The critical exponents satisfy the scaling relations of Table 1 and values determined from finite-size scaling[3,6] are quoted in Table 2. The calculated roughness exponent is 0.81 ± 0.04 for the disk model[5] (Fig. 6) and 0.67 ± 0.03 for the 3D Ising model.[3] The latter value is consistent with the prediction[23] of scaling arguments: $\alpha = 2/3$. The result for the random disk model is not yet fully understood, but agrees well with experimental values for invasion of glass bead packs by a wetting fluid.[17-19] We discuss the comparison between theory and experiment further in the following section.

LARGE-SCALE STRUCTURE FOR H ≠ H$_c$

In the preceding section we discussed the structure of pinned domains at H_c. Self-similar, self-affine or faceted structures were found depending upon the degree of disorder (Fig. 2). For $H < H_c$ the interface is pinned, but it advances a finite distance of order ξ in response to increases in H. The value of ξ diverges as $H \to H_c$ in the self-similar and self-affine growth regimes. On length scales smaller than ξ the interface exhibits the scaling behavior found at the critical force. At larger length scales the structure reflects that of the starting interface. This was completely flat in most of our simulations.

Figure 6 shows that the same type of crossover occurs for forces greater than the critical force. Each panel shows the scaling of the rms width of the interface h over intervals of width ℓ. Results were typically averaged over at least ten different configurations of disks or random fields and at several different times. For a self-affine curve $h \propto \ell^\alpha$ where α is the roughness exponent.[10]

Panel (a) shows the scaling for fluid invasion at and above $P_c = 0.4935\gamma/a$ for $\theta = 25°$. Crosses indicate results for static interfaces obtained at the highest pressure before the interface reached the top of the system. They fall on a reasonably straight line with slope $\alpha = 0.81$. Other symbols depict results at successively higher P. Note that all exhibit roughly the same slope at intermediate scales before rounding over to a slope of about 0.5 at large scales. The crossover occurs at a dynamic coherence length which decreases as P increases. Preliminary results[26] indicate that it scales with the same exponent ν that is found below H_c (Table 2).

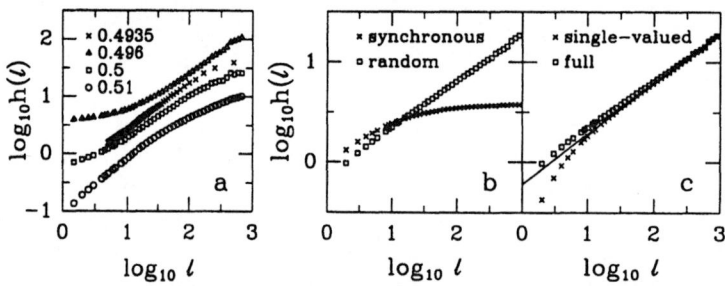

Figure 6. Fluctuation in the interface height $h(\ell)$ as a function of interval width ℓ for (a) fluid invasion of the disk model at $\theta = 25°$ and (b,c) domain growth in 2D for $\Delta = 6$ and $H = 4$. Results for the indicated pressures are compared in (a). Panel (b) compares results for synchronous and random growth algorithms and (c) compares results for the full interface and a single-valued interface consisting of the highest point at each position.

For P close to P_c moving interfaces may actually be rougher than static interfaces, especially at small ℓ. This reflects the influence of overhangs which develop on the growing interface. The value of h has little to do with ℓ if the surface is not a single-valued function of position along the bottom of the system. Instead h reflects the typical vertical distance between the highest and lowest points above a single position, and the frequency of overhangs. Overhangs are an essential feature of wetting invasion. The disks impede growth more in some areas than others. However, even areas that are impassable do not pin the entire interface. The interface is free to surround the bad spots and either leave small trapped regions behind or invade them from an easier direction. Overhangs form in the process of surrounding such regions, and are more prevalent in snapshots of slowly moving interfaces than in static ones. At still higher pressures the role of unfavorable regions is suppressed and there are few overhangs.

Experiments on fluid invasion have revealed similar changes in the scaling behavior with pressure.[17-19] At low velocities there is self-affine scaling with α near 0.8. At higher velocities a crossover to a lower α is observed.[18] As in Fig. 6(a) the interface appears to be self-affine with α near 0.5 at the largest scales. However, neither our simulations nor the experiments have gone to large enough scales to confirm that 0.5 is the asymptotic value of the roughness exponent. There has also been some variation in the reported values of α at small scales.[17-19] To address these issues we have studied several models of 2D magnetic domain growth.

The structure of 2D magnetic domains at H_c is either faceted or self-similar.[1,2] As in fluid invasion, this structure persists at small length scales for $H > H_c$. New scaling behavior appears at large scales. The correlation length where the crossover occurs decreases as H increases. Figure 6(b) shows results for the self-similar regime with $\Delta = 6$, $L = 8000$ and $H = 4$. Since H is much larger than $H_c = 2.4$, the crossover occurs at $\ell \approx 10$ and the large scale behavior is revealed. When all unstable spins are flipped synchronously, the large scale structure is flat. When they are chosen randomly, $\alpha = 1/2$. These results are exactly what would be obtained in the absence of disorder where the random model becomes identical to the Eden model.[21] Thus there seems to be a common trend in the scaling of growing interfaces. They scale as if they were at the critical point up to a length $\xi \propto (f - f_c)^{-\nu}$, where f is either P or H. At larger scales, the interface behaves as if there was no disorder: Synchronous update algorithms produce flat interfaces and Eden-like algorithms produce self-affine interfaces with $\alpha = 1/2$ in 2D.

One may construct a single-valued interface from the actual interface by taking the highest value at each point. This will not change the large-scale structure of the interface if it is self-affine. However, as shown in Fig. 6(c), the behavior at small scales may be very different. The difference becomes most pronounced as H decreases towards H_c where overhangs produce a plateau in $h(\ell)$ at small ℓ for the real interface (Fig. 6(a)). The structure is self-similar at these scales and it is wrong to identify a slope in this regime with a roughness exponent. The safest way to ensure that this region of the data is not included in a fit for α is to compare the results for h with and without overhangs and exclude regions where the results differ.[26]

Overhangs may explain the difference between reported results for the roughness exponent in fluid invasion experiments. Rubio et al.[17] analyzed the full interface and found $\alpha = 0.73 \pm 0.03$, while Horvath et al.[18] analyzed the single-valued interface and found a larger value $\alpha = 0.81$. Overhangs would tend to make the former value smaller than the actual value, and the latter value larger. Analysis of both interfaces would provide an estimate of one source of systematic error. There may also be a difference in the importance of overhangs in the two experiments since they used different fluids. We find that the size and frequency of overhangs depend on θ within our model.

The observation of interfaces with $\alpha > 0.5$ was unexpected and sparked substantial interest. Most growth models map into the continuum growth equation of Kardar, Parisi and Zhang.[24] This yields $\alpha = 0.5$ in 2D and $\alpha < 0.5$ in 3D for uncorrelated noise. Larger values of α can only be produced by assuming power-law correlated noise.[25] It is not clear what sort of physical mechanism would lead to annealed power-law noise in fluid invasion. However, the power-law distribution of growths at f_c might play an analogous role.[5] The crossover to a smaller value of α at large f would then be associated with a cutoff in the distribution of growths at ξ^{D_I} and in the corresponding effective noise. A detailed relation between the disorder induced distribution of growths and the effective noise has yet to be established.

Two other explanations for the anomalously large value of α have been proposed. The most recent[28,29] are based on directed percolation models and give values of α which are slightly smaller than experiment: $\alpha \approx 2/3$. A previous calculation included quenched disorder in a continuum model of domain growth, where the interface was forced to be single-valued.[27] At large scales α approached $1/2$, but there was an apparent value of $\alpha = 0.75$ over a range of ℓ. While this result is suggestive, it is not likely to be relevant to the experiments. The calculation was done in the self-similar regime of the phase diagram where the single-valued approximation is invalid. In a more complete calculation the domain structure at small scales would be consistent with percolation. There is no evidence of such structure in the experiments. Moreover, the region where $\alpha = 0.75$ would coincide with the region where overhangs affect the scaling of $h(\ell)$. Rubio et al.'s analysis included overhangs and would yield a plateau in this regime.[17]

SUMMARY AND CONCLUSIONS

We have illustrated the rich variety of behavior which is found when quenched disorder influences growth. At small driving forces growth is blocked by the disorder. The large-scale structure of the interface reflects the initial conditions. At large driving forces quenched disorder is unimportant. The large-scale structure is consistent with models with annealed disorder. A new large-scale structure occurs at the critical force where motion starts. For f near f_c this structure persists up to a correlation length ξ which diverges at f_c.

When the degree of disorder is large the growing domain has the form of a percolation cluster at f_c. In the extreme limit each spin or interface segment acts independently, as assumed in previous percolation models.[10,11,20] As the degree of disorder decreases, larger clusters of spins act coherently. The domains have a characteristic finger width which diverges at a transition to self-affine or faceted growth.

The transition to self-affine growth is analogous to the onset of magnetic order at an equilibrium phase transition. One can define a surface-normal correlation function: $S(l) = \langle \hat{n}(l'+l) \cdot \hat{n}(l') \rangle$, where l is the arc length along the interface and $\hat{n}$ the local surface normal. In the self-affine regime the long-distance limit of this correlation function is a constant, $S(\infty)$. The value of $S(\infty)$ rises continuously from zero at the critical transition and becomes unity in the faceted regime.[7] One may also define an effective surface tension which becomes non-zero in the self-affine regime.[5,7] Recent work[2] shows that the magnetic domain model is equivalent to a continuous set of generalized bootstrap percolation models.[30] It may be possible to obtain analytic results using this analogy.

The critical behavior which occurs as f increases to f_c is fairly well understood. Scaling relations have been derived and checked in simulations.[3,6] Studies of the behavior above f_c are in progress. The exponent ν appears to be the same on both sides of f_c. Other exponents, such as the one relating the velocity to $f - f_c$, have not yet been determined.

ACKNOWLEDGEMENTS

Support from the National Science Foundation through Grant DMR-9110004 and the Donors of the Petroleum Research Foundation, administered by the American Chemical Society is gratefully acknowledged. M.O.R. also thanks the members of the Theoretical Physics Institute at the University of Minnesota for their hospitality during the period when this work was completed.

NOTES

*Permanent address: Institute of Physics, Polish Academy of Sciences, 02-668 Warsaw, Poland.

†Current address: Physique de la Matière Condensée, College de France, 11, Place Marcelin-Berthelot, 75231 Paris Cedex 05, France.

‡Permanent address: Departamento de Física, Pontifícia Universidade Católica do Rio de Janeiro, CEP 22453, Rio de Janeiro, Brazil.

**Current address: National Institute of Standards and Technology, 226/B348, Gaithersburg, MD 20899.

REFERENCES

1. H. Ji and Mark O. Robbins, Phys. Rev. A44, 2538 (1991).

2. B. Koiller, H. Ji and M. O. Robbins, submitted to Phys. Rev. B.

3. H. Ji and Mark O. Robbins, to be published.

4. M. Cieplak and M. O. Robbins, Phys. Rev. Lett. 60, 2042 (1988); Phys. Rev. B41, 11508 (1990).

5. N. Martys, M. Cieplak, and M. O. Robbins, Phys. Rev. Lett. 66, 1058 (1991).

6. N. Martys, M. O. Robbins and M. Cieplak, Phys. Rev. B44, 12294 (1991).

7. N. Martys, PhD. thesis, (1990); N. Martys, M. O. Robbins and M. Cieplak, in *Scaling in Disordered Materials: Fractal Structure and Dynamics*, edited by J. P. Stokes, M. O. Robbins and T. A. Witten (Materials Research Society, Pittsburgh, 1990), p. 67.

8. B. Koiller, H. Ji and M. O. Robbins, Phys. Rev. B, in press.

9. D. Stauffer, *Introduction to Percolation Theory* (Taylor and Francis, London, 1985).

10. J. Feder, *Fractals* (Plenum Press, New York, 1988).

11. R. Lenormand and S. Bories, C. R. Acad. Sci. Ser. B291, 279 (1980). R. Chandler, J. Koplik, K. Lerman and J. F. Willemsen, J. Fluid Mech. 119, 249 (1982).

12. M. M. Dias and A. C. Payatakes, J. Fluid Mech. 164, 305 (1986).

13. R. Lenormand, J. Phys.: Condens. Matter 2, SA79 (1990). R. Lenormand and C. Zarcone, Phys. Rev. Lett. 54, 2226 (1985).

14. J. P. Stokes, D. A. Weitz, J. P. Gollub, A. Dougherty, M. O. Robbins, P. M. Chaikin, and H. M. Lindsay, Phys. Rev. Lett. 57, 1718 (1986).

15. A. P. Kushnick, J. P. Stokes and M. O. Robbins, *Fractal Aspects of Materials: Disordered Systems*, (Materials Research Society, Pittsburgh, 1988) Ed. by D. A. Weitz, L. M. Sander and B. B. Mandelbrot, p. 87, and to be published.

16. J. P. Stokes, A. P. Kushnick and M. O. Robbins, Phys. Rev. Lett. 60, 1386 (1988).

17. M. A. Rubio, C. Edwards, A. Dougherty and J. P. Gollub, Phys. Rev. Lett. 63, 1685 (1989); *ibid.* 65, 1339 (1990).

18. V. K. Horváth, F. Family, and T. Vicsek, Phys. Rev. Lett. 65, 1388 (1990); J. Phys. A24, L25 (1991).

19. P. Z. Wong, private communication.

20. D. Wilkenson and J. F. Willemsen, J. Phys. A16, 3365 (1983).

21. See for example, T. Vicsek, *Fractal Growth Phenomena* (World Scientific, Singapore, 1989).

22. The exponent τ' is related to, but different from, the exponent τ describing the cluster distribution in ordinary percolation.

23. G. Grinstein and S. Ma, Phys. Rev. B28, 2588 (1983).

24. M. Kardar, G. Parisi, and Y. Zhang, Phys. Rev. Lett. 64, 543 (1990).

25. E. Medina, T. Hwa, M. Kardar and Y.-C. Zhang, Phys. Rev. A39, 3053 (1989). J. Amar and F. Family, J. Phys. A24, L79 (1991).

26. B. Koiller, N. Martys and M. O. Robbins, to be published.

27. D. A. Kessler, H. Levine and Y. Tu, Phys. Rev. A43, 4551 (1991).

28. L. H. Tang and H. Leschhorn, to be published.

29. A.-L. Barabási, S. V. Buldyrev, F. Caserta, S. Havlin, H. E. Stanley and T. Vicsek, to be published.

30. J. Chalupa, P. L. Leath and G. R. Reich, J. Phys. C. 12, L31 (1981): P. M. Kogut and P. L. Leath, J. Phys. C 14, 3187 (1981); J. Adler and J. Aharony, J. Phys. A 21, 1387 (1988).

KINETIC ROUGHENING WITH ALGEBRAICALLY DISTRIBUTED NOISE AMPLITUDES OR WAITING TIMES

János Kertész

Institute for Teichnical Physics
Hungarian Academy of Sciences
Budapest, P.O.Box 76
H-1325, Hungary

1. INTRODUCTION

In this paper we report on numerical simulation results on Zhang's prediction[1] about the relevance of the distribution of noise amplitudes η in kinetic surface roughening[2]. For power law distributions $P(\eta) \sim \eta^{-(1+\mu)}$ the roughening exponent depends on μ in both $1 + 1$ and $2 + 1$ dimenions[1,3-6]. A new characteristic length occurs in this problem which separates a regime with multiscaling behavior from a regime where conventional scaling is found[7]. Anomalous exponents characterize the surface also in the case when waiting times on the growth sites are distributed according to a power law for "negative λ models", i.e. if the growth velocity decreases with tilting the substrate[8].

It is widely accepted that kinetic surface roughening in a number of computer models is adequately described by the Kardar-Parisi-Zhang (KPZ) theory (for references see the reprint volume Ref. 2). The KPZ equation of motion of the height variable $h(x, t)$ above a d-dimensional substrate (i.e. for the "$d+1$-dimensional problem") is

$$\frac{\partial h}{\partial t} = \nu \nabla^2 h + \lambda (\nabla h)^2 + \eta \qquad (1)$$

where ν is an effective surface tension and λ is the coupling constant of the nonlinear term due to lateral growth. λ is positiv (negativ) if the velocity increases (decreases) with a tilt of the surface. The term η is usually assumed to be uncorrelated and bounded (e.g. Gaussian) noise.

The surface can be characterized by the roughening exponent α, and the dynamic exponent z which occur in the scaling form of the correlation function or of the surface width $w = \sqrt{< (h- <h>)^2 >}$:

$$w \sim L^\alpha f(t/L^z) \qquad (2)$$

where L is the linear size of the substrate and for the scaling function $f(y \to \infty) =$ const. and $f(y \to 0) \sim y^\beta$ with $\beta = \alpha/z$. Concerning the exponents, good agreement between analytic results, numerical solution of (1) and simulation of diverse computer models have been obtained.

According to the renormalization group calculations the values of the exponents are determined by a single coupling constant $g \sim \lambda^2$. For a given dimension a critical value g_c separates a strong coupling phase $g > g_c$ from a week coupling one. (For $d \leq 2$ the weak coupling phase shrinks to the point $g = 0$.) The weak coupling exponents are given by the linear theory: $\alpha = (2-d)/2$ and $z = 2$. The strong coupling exponents satisfy the equality

$$\alpha + z = 2 \qquad (3)$$

and their exact values are known for $d = 1$ ($\alpha = 1/2$ and $z = 2/3$).

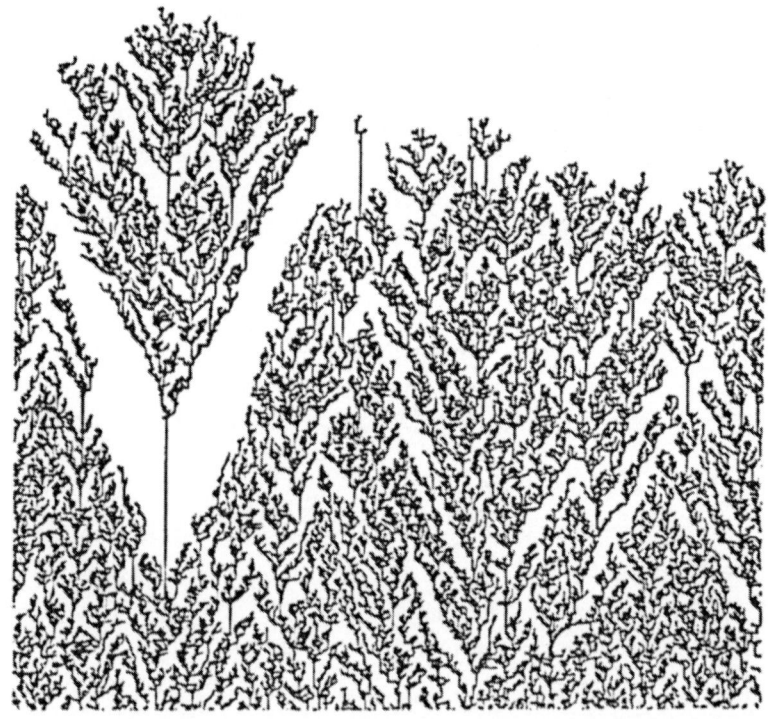

Figure 1. Ballistic deposition of rods with lengths distributed according to a power law (Ref. 3); $\mu = 3$.

2. ANOMALOUS EXPONENTS DUE TO POWER LOW NOISE

Recent experiments on quasi-two-dimensional two-phase viscous flows[9,10] and bacterial colony growth[11] have lead to exponents different from the above mentioned ones: The results for α are in the range $0.75 - 0.85$ instead of $1/2$. As a possible resolution of this problem (the appearance of new universality classes) it has been suggested[1] that a power law distribution of the noise amplitudes

$$P(\eta) = \frac{\mu}{\eta^{1+\mu}} \quad \text{for} \quad \eta \geq 1 \quad \text{and} \quad P(\eta) = 0 \quad \text{otherwise} \tag{4}$$

with $\mu > d + 1$ could lead to new exponents (roughening dominated by rare events).

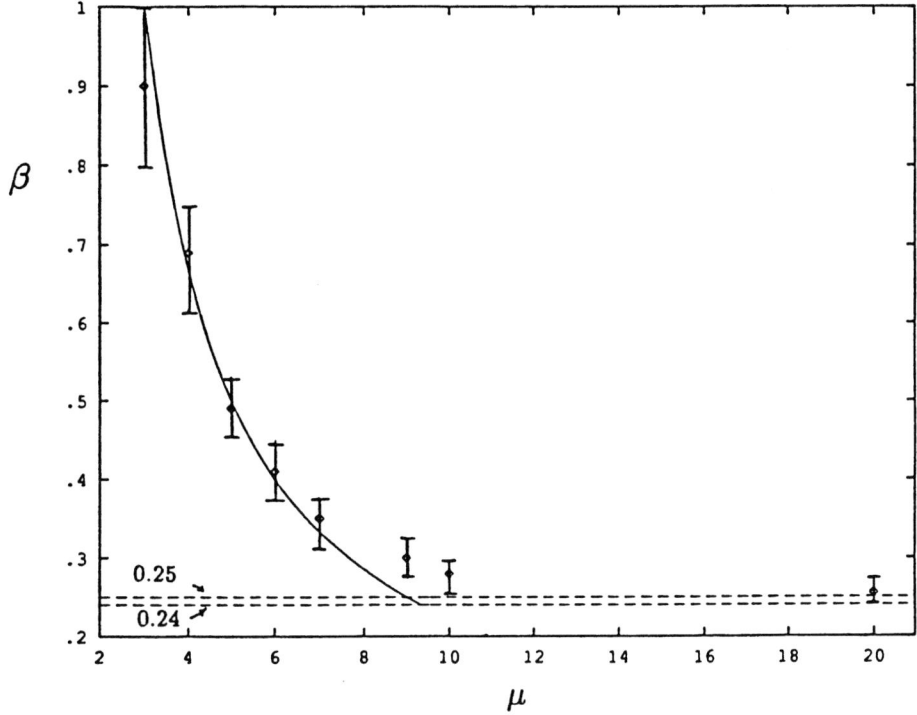

Figure 2.   The exponent β for rare-events governed growth in 2+1 dimensions (Ref. 5), as calculated from (6) (solid line) and from simulation of model (5).

In fact, the μ-dependence of the exponents α and β has been observed in computer simulations1,3−6. Fig. 1 shows a snapshot of a generalized ballistic deposition model in 1+1 dimesions[3]. In this model vertical rods of different lengths distributed according to (4) fall down at random positions and get sticked to the aggregate whenever they get into touch with it. The original version of ballistic deposition is obtained for $\mu \to \infty$ i.e. $P(\eta) = \delta(1 - \eta)$. In this case the model is known to belong to the KPZ strong coupling universality class. Numerical evidence shows that the exponents depend in fact on μ.

In 2+1 dimensions we investigated the following model[5]:

$$h(\mathbf{x}, t+1) = \min[h(\mathbf{x}', t) + \eta(\mathbf{x}', t)] \tag{5}$$

where the min has to be taken over the nearest neighbors of $\mathbf{x}$ and over $\mathbf{x}$ itself. The exponents agree within the numerical accuracy with the usual KPZ strong coupling ones if η is a bounded white noise. Fig. 2 shows the dependence of β on μ if the noise amplitudes are distributed according to (4).

A simple mean field type reasoning[12,13] suggests

$$\alpha(\mu) = \frac{d+2}{\mu+1} \qquad \text{for} \qquad d+1 < \mu < \mu_g \tag{6}$$

and $\alpha(\mu) = \alpha(\infty)$ for $\mu > \mu_g$ where the value of μ_g is determined by equating the expression on the r.h.s. of (6) with $\alpha(\infty)$. The other exponent can be calculated by using (3). The full line in Fig. 2 is the calculated β exponent corresponding to (5). Presently, it is not clear whether there is in fact a finite value of μ above which the KPZ ($\mu = \infty$) exponents can be observed, as suggested by (6).

3. MULTIAFFINITY IN SURFACE GROWTH

In analogy with the case of self-similar fractals, the concept of multifractality can be useful for rough surfaces as well. In this context, it has been suggested recently[14] that for a class of surfaces the qth order height-height correlation functions should be studied which, for fixed time t in 1+1 dimensions, are expected to exhibit the following scaling:

$$c_q(x, t) = \frac{1}{L} \sum_{i=1}^{L} |h(x_i, t) - h(x_i + x, t)|^q \sim x^{qH_q},$$

where H_q is an exponent continuously changing with q. Surfaces with height correlations showing this behavior are called "multiaffine" and using the multifractal formalism relating the local singularities of the surface to the H_q spectrum they can be described in terms of multifractality.

We found[7] numerical evidence of multiaffine scaling in the model of rare events dominated roughening. This multifractality is specific to this type of surface growth; models obeying the KPZ equation with bounded (e.g. Gaussian) noise result in a constant H_q.

We have simulated the 1+1 dimensional version of the model given by (5). The noise amplitude η was taken from a power-law distribution (4). We have chosen $\mu = 3$ for which the change in the roughening exponent is known to be significant (α is in the range 0.7-0.8 instead of 1/2).

Figure 3. presents data for $L = 65536$ at time $t = 602890$, i.e., deep in the saturation regime; an average over five runs was taken. This figure suggests that: i) The initial part of the data sets for each q exhibits scaling behaviour with a unique slope depending on q i.e. multifractal scaling is present, ii) this kind of scaling crosses over into the uniform scaling behaviour for x exceeding some characteristic crossover length $x_\times$.

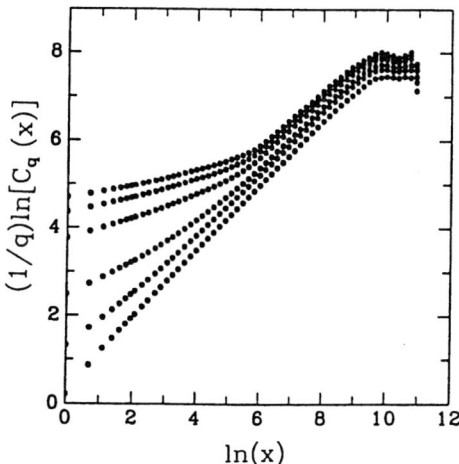

Figure 3. The q-th order correlation functions for rare-events governed growth with $\mu = 3$ and for $q = $1, 2, 3, 5, 7, 9. An average over 7 runs was taken (Ref. 7).

Feature ii) means that in addition to the characteristic length $\sim t^{\beta/\alpha}$ always present in kinetic roughening until the system size L is reached at saturation, a new characteristic length $x_\times$ occurs in surface growth dominated by rare events. For $x < x_\times$ the q–th order correlations show multifractal scaling behavior. For $x_\times < x < t^{\beta/\alpha}$ conventional scaling sets in while no correlations are present for $x > t^{\beta/\alpha}$.

The new characteristic length $x_\times$ is an immediate consequence of the rare events. A perturbation due to a large jump in the surface propagates in the lateral direction linearly until the rest of the surface catches up (see Figure 1). In the mean field approach[12,13] the surface width at saturation is identified with the characteristic large jumps, therefore the size of such jumps scales with the system size as $\sim L^\alpha$. Due to the linear propagation of the perturbation a characteristic length in the substrate direction occurs which is of the same order as the width. For distances smaller than this length the correlations are dominated by the large jumps[15] while on larger lengths the self-affine structure of the whole surface becomes apparent. This consideration implies[16] $x_\times \sim L^\alpha$ and this means that the relative size of the region over which the multifractal scaling behavior can be observed on a logarithmic scale larger becomes dominating in the large L limit.

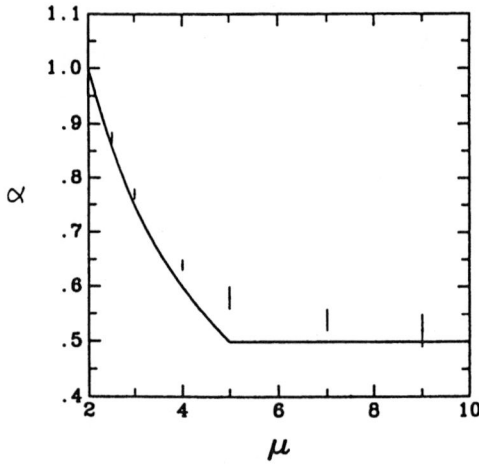

Figure 4. The exponent α as a function of μ in a 1+1 dimensional model with waiting times distributed according to a power law (7) (Ref. 8). The solid line represents the mean field result (6).

It is natural to assume that the behavior for the diverging and non-diverging moments is qualitatively different; this is expected to lead to a change in the H_q spectrum often described as a "phase transition".

4. WAITING TIMES WITH POWER LAW DISTRIBUTION

In model (5) the rare but dramatic increments of the surface result in new morphologies characterized by exponents different from the ususal KPZ ones. The exponents are sensitive to the power low distribution of noise because for model (5) the average velocity increases with tilting the substrate, i.e., it is a "positive λ model" (see the Introduction). For a "negative λ model" a possibility study the effects of rare eventg is to introduce waiting times at the growth sites[8].

The model we studied numerically is the generalization of the so called single step growth model, defined in 1+1 dimensions on the square lattice tilted by 45°. Growth sites are those unoccupied sites which have two neighbors below them occupied and they get occupied if the waiting time expires (see Fig. 1. in Lei-Han Tang's article of this volume, where a detailed discussion of the waiting time formulation of surface growth can be found). The resulting configuration has at most height differences of unity.

We took waiting times distributed according to the following law

$$P(\tau) = (\tau + \frac{1}{2})^{-\mu} - (\tau + \frac{3}{2})^{-\mu}, \qquad \tau = \frac{1}{2}, \frac{3}{2}, ... \tag{7}$$

which corresponds asymptotically to $\mu\tau^{-\mu-1}$. (A similar model was independently introduced by Jensen and Procaccia[17].) In this case the rare events block the growth for long times. Since the single step model is a "negative λ model", we expect that the power law distribution (7) influences the exponerits. Figure (4) shows that this is in fact the case. Again, the mean field exponents are indicated by a solid line.

5. CONCLUSION

It is now clear that rare events dominated surface growth leads to anomalous exponents. For "positive λ models" the incremental noise amplitudes are distributed according to a power law while a similar distribution of the waiting timnes leads to the desired effect for the "negative λ models". For a given μ universality seems to be valid (within the numerical accuracy). However, there are significant differences in the surfaces: For the positive λ case the higher order correlation functions diverge in the thermodynamic limit while they remain finite for the waiting time models. It should be clarified whether the observed multifractal behavior is related to the divergence of the higher order correlation function.

Finally we mention that recent evaluation of experimental data on two fluid displacement in quasi two-dimensional random medium provided evidence for power law output noise[18]. The mechanism by which the power law distribution could occur is presently unclear; possibly it is related to crossover phenomena at the pinning transition point characteristic for kinetic roughening with quenched randomness[19-21].

Acknowledgements: The author thanks A.L. Barabási, R. Bourbonnais, S. Buldyrev, S. Havlin, M.H. Jensen, H.E. Stanley, L.H. Tang, T. Vicsek, D.E. Wolf and Y.C. Zhang for cooperation. This research was financially supported by the Hungarian Scientific Research Foundation, by SFB-341, by a HAS-NSF cooperation grant and by the Humboldt Foundation.

REFERENCES

1 Y.-C. Zhang *J.Physique* **51**, 2129 (1990).

2 F. Family and T. Vicsek, *Dynamics of Fractal Surfaces*, (World Scientific, Singapore 1991)

3 S.V. Buldyrev, S. Havlin, J. Kertész, H.E. Stanley and T. Vicsek, *Phys. Rev.* **A43**, 7113 (1991)

4 J.G. Amar and F. Family *J. Phys. A : Math. Gen.* **24**, L-79 (1991)

5 R. Bourbonnais, J. Kertész and D. E. Wolf *J. Physique II* **1** 493 (1991)

6 R. Bourbonnais, H. J. Herrmann and T. Vicsek, *Intl. J. Mod. Phys. C*

7 A.L. Barabási, R. Bourbonnais, M.H. Jensen, J. Kertész, T.Vicsek and Y.C. Zhang, preprint for Phys. Rev. Lett.

8 L.H. Tang, J.Kertész and D.E. Wolf, *J. Phys. A* **24**, L1193 (1991)

9 M. A. Rubio, C. A. Edwards. A. Dougherty, and J. P. Gollub, *Phys. Rev. Lett.* **63**, 1685 (1989).

10 V. K. Horváth, F. Family and T. Vicsek, *J. Phys. A*, **24**, L25 (1991)

11 T. Vicsek, M. Cserző, and V. Horváth, *Physica A*, **167**, 315 (1990)

12 Y.-C. Zhang *Physica A* **170**, 1 (1990)

13 J. Krug *J. Physique I.* **1**, 9 (1991)

14 A.-L. Barabási and T. Vicsek, *Phys. Rev.* **A44**, 2730 (1991)

15 A.-L. Barabási *J. Phys. A* **24**, L1013 (1991)

16 D.E. Wolf and J. Kertész unpublished

17 M.H. Jensen and I. Procaccia, J. Physique I, **1**, 1139 (1991)

18 V.K. Horváth, F. Family and T. Vicsek, *Phys. Rev. Lett.* (December 2. 1991)

19 D. Kessler, H. Levine and Y. Tu, *Phys Rev. A* **43**, 4548 (1991)

20 M. Leschhorn and L.H. Tang, (Cologne preprint 1991)

21 A.L. Barabási, S. Buldyrev, F. Caserta S. Havlin, H. E. Stanley and T. Vicsek (preprint 1991)

ANOMALOUS SURFACE ROUGHENING: EXPERIMENT AND MODELS

S. Havlin,[1,2] A.-L. Barabási,[1] S. V. Buldyrev,[1] C. K. Peng,[1]
M. Schwartz,[1,3] H. E. Stanley,[1] and T. Vicsek[4]

[1]Center for Polymer Studies and Department of Physics
Boston University, Boston, MA 02215   United States

[2]Department of Physics, Bar-Ilan University
Ramat-Gan   Israel

[3]Department of Physics and Astronomy
Tel-Aviv University, Ramat-Aviv, 69978 Tel-Aviv   Israel:

[4]Department of Atomic Physics, Eötvös University
Budapest, H-1445   Hungary

ABSTRACT

We review briefly recent studies based on power law distribution of noise to explain the anomalous surface roughening found in several experiments. We study the probability distribution of the height fluctuations in $d = 1 + 1$ by mapping the surface to a Lévy walk. We also review numerical studies for the effect of long-range correlated noise on (i) the KPZ equation and the related directed-polymer (DP) problem and (ii) the ballistic deposition (BD) model. We describe measurements of the interface formed when a wet front propagates in paper with anomalous roughening exponent $\alpha = 0.63 \pm 0.04$. We suggest a model based on propagation and pinning of a self-affine interface in the presence of quenched disorder, with erosion of overhangs. By mapping our model to directed percolation, we find $\alpha \simeq 0.63$.

I. INTRODUCTION

Recently considerable efforts have been made in understanding the dynamics of non-equilibrium interface growth in the context of a variety of models, analytical theories and experiments.[1-4] Many recent investigations have concentrated on the dynamic scaling properties of interfaces obtained in experiments and in various surface growth models. Particular attention has been devoted to the scaling properties of the rms interface width

$$w(\ell, t) \equiv \langle (h(x,t) - <h(x,t)>)^2 \rangle^{1/2} \sim \ell^\alpha f(t/\ell^{\alpha/\beta}). \tag{1}$$

Here $h(x,t)$ is the surface height at time t, the angular brackets denote the average over x belonging to an interval of size ℓ; also, $f(u) \sim u^\beta$ for $u \ll 1$ and $f(u) \to Const$ for $u \gg 1$.

Growth Patterns in Physical Sciences and Biology, Edited
by J. M. Garcia-Ruiz *et al.*, Plenum Press, New York, 1993

It has been widely believed that many such problems belong to the same universality class as the the Kardar-Parisi-Zhang (KPZ) equation,[5]

$$\frac{\partial h(x,t)}{\partial t} = \nabla^2 h + \frac{\lambda}{2}(\nabla h)^2 + \eta(x,t).$$ (2)

Here $\eta(x,t)$ is a random noise term. One such model is ballistic deposition,[6] for which the surface width exponents α and β satisfy the general scaling relation[7,8]

$$\alpha + \frac{\alpha}{\beta} = 2$$ (3a)

and can be calculated exactly in the case of normally distributed uncorrelated noise $\eta(x,t)$ in one dimensions ($d = 1+1$):

$$\alpha = \frac{1}{2}, \qquad \beta = \frac{1}{3}.$$ (3b)

There have recently appeared several experiments on surface growth which yield exponents quite different from those of (3b). For example, in some experiments on immiscible displacement of viscous fluids in porous media it was found that $\alpha = 0.73 \pm 0.003$ [Ref.9], $\alpha \approx 0.81$, $\beta = 0.65$ [Ref. 10] and, in recent experiments on the growth of bacteria colonies $\alpha = 0.78 \pm 0.06$ [Ref.11].

II. ROUGHENING WITH POWER LAW DISTRIBUTED NOISE

Recently Zhang[12] suggested that the anomalous roughening found in experiments can be explained by an uncorrelated "noise" $\eta(x,t)$ obeying a power-law distribution, $p(\eta) \sim \eta^{-\mu-1}$ where $\eta \geq 1$. This anomalous noise can be simulated in a deposition context by depositing rods of size ℓ sampled from a power law probability distribution,

$$p(\ell) \sim \ell^{-\mu-1}.$$ (4)

The growth rule is similar to the conventional ballistic deposition rule, i.e., a deposited rod is attached to the highest nearest neighbor surface site. The site at which deposition next occurs can be chosen either *deterministically*[12,13] or *randomly*.[14]

The Zhang idea[12] has received recent support from both mean field theory[15] and numerical simulations,[12-14] both of which suggest that the exponents α and β are anomalously large, and depend continuously on the parameter μ (at least for $\mu < \mu_c \approx 5$).[13-15]

Figure 1 shows the comparison of our results[14] for different values of μ with the Zhang-Krug prediction,

$$\alpha = \frac{3}{\mu+1}, \qquad \beta = \frac{3}{2\mu-1}, \qquad [\mu \geq 2].$$ (5)

It can be seen that for $\mu \geq \mu_c = 4.5 \pm 0.5$ both α and β are almost independent of μ and are very close to the classical values $\alpha = 1/2$, $\beta = 1/3$. For $\mu < \mu_c$ both exponents deviate from their "classical" values and approach the limiting values $\alpha = 1$, $\beta = 1$ predicted by the Zhang-Krug relation for $\mu = 2$. As a final consistency check, we found excellent agreement with the scaling relation $\alpha + \alpha/\beta = 2$ in the entire range of studied values of $\mu \geq 2$. For $\mu < 2$, see our discussion below.

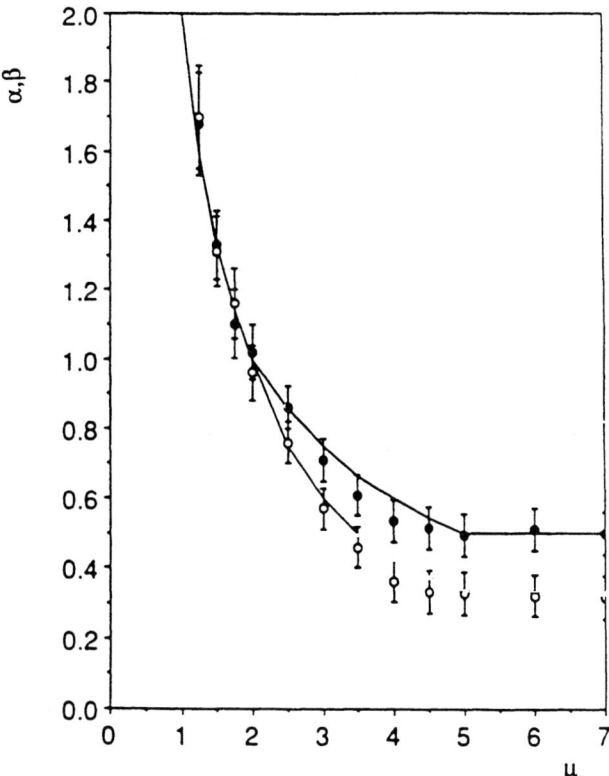

Fig. 1. Comparison of numerical results for exponents α ($\bullet$) and β ($\circ$) with theoretical predictions given for $\mu \geq 2$ by Eqs. (5) (solid line), and for $\mu \leq 2$ by Eq. (10) (solid line). After Ref. 16.

In a recent work[16] a closed-form expression for the probability distribution for the fluctuations $\delta h(x,t) \equiv h(x,t) - \langle h(t) \rangle$ in surface height $h(x,t)$ was suggested based on a formal analogy between anomalous roughening and the statistics of a Lévy walk.

By a Lévy walk we mean that each unit of time a random walker steps a unit length. The walker moves ℓ successive steps in the same direction before randomly changing direction, and ℓ is taken from a Lévy distribution[17] $p(\ell) \sim \ell^{-\mu-1}$. The probability density $P(x,n)$ that the walker is at position x after n steps has a tail distribution ($x \gg n^{1/\mu}$) of the form[18]

$$P(x,n) \sim n/x^{\mu+1} = \frac{1}{n^{1/\mu}} \left(\frac{x}{n^{1/\mu}} \right)^{-\mu-1}. \qquad (6)$$

By mapping the surface growth to Lévy walk we obtain[16] that for $t \gg t_\times \sim L^z$ the distribution of $h(x,t)$ has the asymptotic form

$$P(\delta h, L, t \gg t_\times) \sim \frac{1}{L^{(3-\alpha)/\mu}} \left(\frac{\delta h}{L^{(3-\alpha)/\mu}} \right)^{-\mu-1} \qquad [\delta h \gg L^{(3-\alpha)/\mu}]. \qquad (7)$$

Since for $t \gg t_\times$, $\delta h \sim w \sim L^\alpha$, we find a self consistent equation for α, $\alpha = (3-\alpha)/\mu$, or $\alpha = 3/(\mu+1)$, the same as Eq. (5). Note that α assumes its classical value $\alpha = 1/2$

at $\mu = \mu_c = 5$. Substituting $\alpha = 3/(\mu + 1)$ into (7), we obtain the predicted scaling form

$$P(\delta h, L, t \gg t_\times) \sim \frac{1}{L^{3/(\mu+1)}} \left(\frac{\delta h}{L^{3/(\mu+1)}} \right)^{-\mu-1} \quad (8)$$

To test (8), we performed simulations of ballistic deposition using Eq. (4) for several values of μ and for a sequence of values of L.[16] Figure 2 supports the predicted data collapse of Eq. (8) for $\mu = 1.5$ and 3.

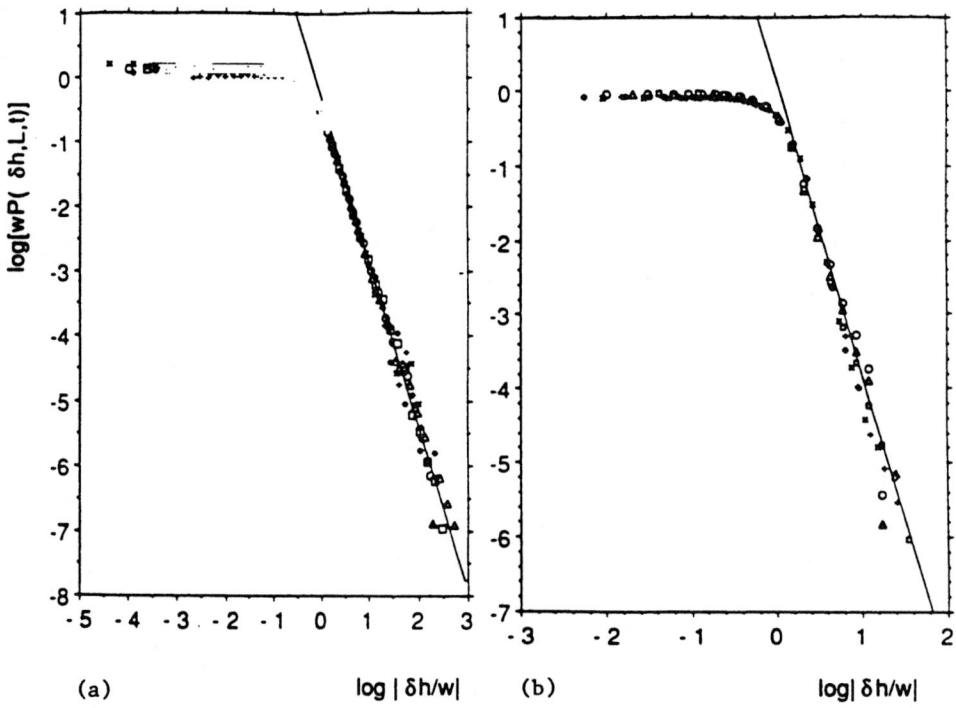

(a)   log | δh/w| (b)   log| δh/w|

Fig. 2. Log-log scaling plot of $wP(\delta h, L, t)$ against $|\delta h|/w$. Here w is the *first* moment of $P(\delta h, L, t)$ for (a), $\mu = 1.5$ and for (b), $\mu = 3$. Symbols in (a) are: $+$ ($L = 1028, t = 64$); $\Diamond$ ($L = 2048, t = 256$); $\times$ ($L = 2048, t = 1024$); $\triangle$ ($L = 256, t > 4096$); $\square$ ($L = 512, t > 4096$); o ($L = 1024, t > 4096$). Symbols in (b) are: $\square$ ($t = 32, L = 2048$); $\triangle$ ($t = 128, L = 2048$); o ($t = 512, L = 2048$); $+$ ($t > 16384, L = 512$); $\times$ ($t > 16384, L = 1024$); $\Diamond$ ($t > 32768, L = 2048$). The straight lines have slopes of (a) 2.5 and (b) 4, as predicted by Eq. (7). After Ref. 16.

To obtain the early time ($1 \ll t \ll t_\times$) dependent probability we use again the time-space scaling relation $t \equiv L^{\alpha/\beta}$ of (1) and L in (7) should be replaced by $tL = t^{1+1/z}$ yielding

$$P(\delta h, t) \sim \frac{1}{t^{(z+1)/\mu z}} \left(\frac{\delta h}{t^{(z+1)/\mu z}} \right)^{-\mu-1}. \quad (9)$$

Since $\delta h \sim t^\beta$ we obtain $\beta = (z+1)/\mu z = 3/(2\mu - 1)$, the same as in (5). Figure 2 shows also data for the time-dependent probability density supporting (9).

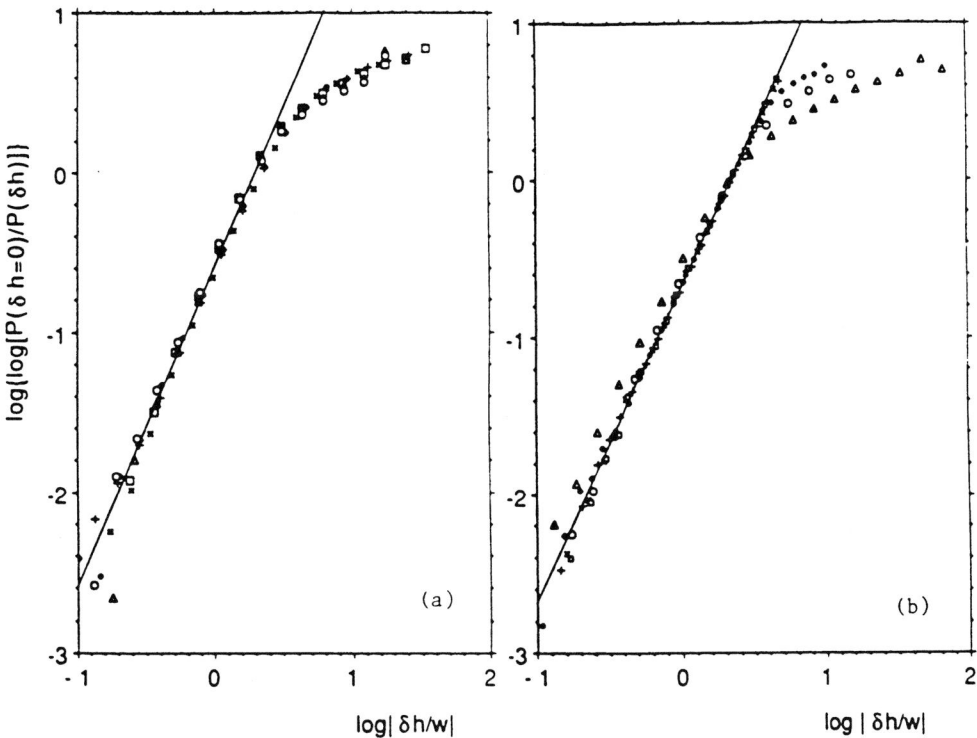

Fig. 3: (a) Log-log scaling plots of $\log[P(\delta h = 0, L, t)/P(\delta h, L, t)]$ versus $|\delta h|/w$ for $\mu = 3$ for different system sizes L and times t. The straight line has the slope 2, as should be for Gaussian distribution. The symbols are the same as those used in Fig. 2(b). (b) Log-log plot of $\log[P(\delta h = 0, L, t)/P(\delta h, L, t)]$ as a function of $|\delta h|/w$ for different values of μ (the data for each value of μ obtained for $L = 1024$ and large t, greater than $t_\times$): $\square$ ($\mu = \infty$); $\times$ ($\mu = 6$); $+$ ($\mu = 5$); $\Diamond$ ($\mu = 4$); o ($\mu = 3$); $\triangle$ ($\mu = 2$). The crossover value of $\delta h/w$ at which the behavior changes from Gaussian (straight line with slope 2) to power law increases gradually with the value of μ. After Ref. 16.

The above considerations, Eqs. (7)-(9), are valid for $\mu > 2$. For $\mu \leq 2$, the relation $t_\times L \sim L^{z+1}$ fails since $t_\times$ is bounded from below by L. Thus one must repeat the arguments leading to Eqs. (7)-(9) using $t_\times L = L^2$ from which follows[16]

$$\alpha = \beta = \frac{2}{\mu}, \qquad [\mu < 2]. \tag{10}$$

Note that (10) complements the Zhang-Krug prediction (5) for values of μ below 2. Numerical data supporting (10) are shown in Fig. 1. The analogy to Lévy walks predicts not only the tails of the probability densities but also their behavior in the range of small fluctuations. In this range and for $\mu \geq 2$ the distribution of Lévy walks is known to be Gaussian,[18] predicting that for $\delta h < w$ the probability density of surface heights is also Gaussian. Indeed, plotting the data in Fig. 3 as $\log\{\log[P(\delta h, L)/P(0, L)]\}$ versus $\log \delta h$ for several values of L shows a clear range of slope 2 supporting a Gaussian form. The crossover from Gaussian to a power-law occurs at a value of $y = \delta h/w$ which increases as μ increases as expected from the analysis of theory in Ref. 18.

As seen from Fig. 2, both time and size dependence have the *same* scaling relation. Indeed, Eqs. (8) and (9) and the Gaussian form found for small fluctutions can be combined for $\mu \geq 2$ to a single scaling relation

$$P(\delta h, L, t) \sim \frac{1}{w} F\left(\frac{\delta h}{w}\right),$$ (11a)

where $w = w(L, t)$ is given by Eq. (1), and

$$F(y) \sim \begin{cases} \exp(-ay^2) & [y < y_c] \\ y^{-\mu-1} & [y > y_c]. \end{cases}$$ (11b)

The data collapse shown in Fig. 2 supports (11). Similar numerical results, supportng (11) for $\mu = 3$ were presented in Ref. 13.

From the analogy to Lévy walks we expect that in the case of conventional ballistic deposition ($\mu = \infty$) the distribution will be Gaussian:

$$P(\delta h, t) \sim \frac{1}{t^{1/3}} e^{-a(\delta h)^2/t^{2/3}}$$ (12a)

for $t \ll t_\times$, and

$$P(\delta h, L) \sim \frac{1}{L^{1/2}} e^{-a(\delta h)^2/L}$$ (12b)

for $t \gg t_\times \sim L^{1/z}$. In Fig. 4 we show numerical data supporting (12).

The Zhang[12] model, which is based on the assumption that the noise in the system has power law distributed amplitudes, may be the explanation for the anomalous surface roughening in experiments but the origin of such a noise in real systems remains unclear.[19]

III. CORRELATED NOISE

Anomalous surface roughening can be also due to long-range correlated noise. Next we review recent studies on the effect of long-range correlated noise on surface growth models.

When the noise itself is the result of another stochastic process, then the noise *cannot be treated as random*—the noise is correlated in space and/or time.[20] In this case, the exponents depend on the strength of the correlation. Medina et al[5] used dynamical renormalization-group analysis to study the KPZ equation with long range correlated noise. The noise they studied has the correlation

$$\langle \eta(\mathbf{x}, t)\, \eta(\mathbf{x}', t') \rangle \sim |\mathbf{x} - \mathbf{x}'|^{2\rho-(d-1)} \, |t - t'|^{2\theta-1},$$ (13a)

where d is the overall dimension of the system ($d-1$ is the dimension of the surface). If the noise has no temporal correlation, i.e.,

$$\langle \eta(\mathbf{x}, t)\, \eta(\mathbf{x}', t') \rangle \sim |\mathbf{x} - \mathbf{x}'|^{2\rho-(d-1)} \, \delta(t - t'),$$ (13b)

the exponents obey the relation $\alpha + z = 2$. Since then there is only one independent scaling exponent, it is sufficient to give β; for $d = 1 + 1$

$$\beta = \begin{cases} 1/3 & 0 < \rho \leq \frac{1}{4} \\ (1+2\rho)/(5-2\rho) & \frac{1}{4} < \rho \leq 1. \end{cases}$$ (14a)

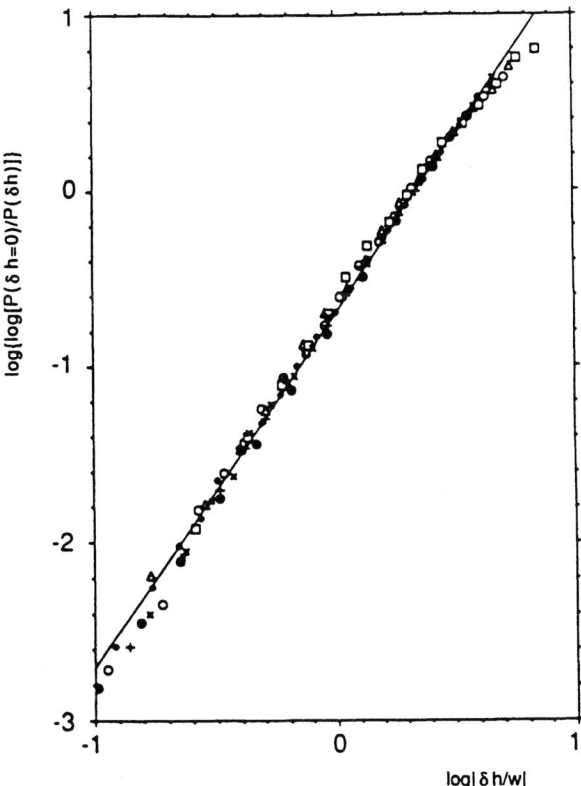

Fig. 4. Log-log plot of $\log[P(\delta h = 0, L, t)/P(\delta h, L, t)]$ as a function of $|\delta h|/w$ for $\mu = \infty$, which is the case of *conventional* ballistic deposition. Here w scales according to Eq. (1) with $\alpha = 1/2$, $\beta = 1/3$. The Gaussian behavior is found in the entire range of δh: $\square$ ($t = 64, L = 4096$); $\triangle$ ($t = 256, L = 4096$); o ($t = 1024, L = 4096$); $+$ ($L = 512, t > 16384$); $\times$ ($L = 1024, t > 16384$); $\Diamond$ ($L = 2048, t > 32768$); $\bullet$ ($L = 4096, t > 32768$). After Ref. 16.

The other feature of the KPZ equation is that it can be mapped to the directed polymer (DP) problem[21]. The noise plays the role of a time-dependent random potential. Thus, the results of Ref. 5 can also apply to the DP problem in a correlated potential field.

Zhang[22] used a replica method to study the DP problem with correlated noise η given by Eq. (13b). Due to the analogy between the DP problem and the KPZ equation, Zhang predicts for $d = 1 + 1$

$$\beta = \begin{cases} (1+2\rho)/(3+2\rho) & 0 < \rho \le \frac{1}{2} \\ (1+2\rho)/(5-2\rho) & \frac{1}{2} < \rho \le 1. \end{cases} \tag{14b}$$

Hentschel and Family[23] studied the scaling behavior for dissipative dynamical systems and proposed a new relation:

$$\beta = \frac{1}{3 - 2\rho}, \qquad 0 \le \rho \le \frac{1}{2}. \tag{14c}$$

Note that the three predictions [Eqs. (14a), (14b), and (14c)] differ for $0 < \rho < 1/2$.

There have been several prior attempts to verify the analytical results with correlated noise.[24] This work relies on numerical methods that probably generate undesired correlations in the noise. Here we review a recent work[25] where we generate algebraically-correlated noise,[26] integrate numerically the KPZ equation, and also simulate the DP growth in a correlated potential field. The results of Peng et al.[25] for both KPZ and DP agree with each other, and qualitataively agree somewhat better with (14c) than with (14a) or (14b). Finally we implement correlated noise into the BD model, and were surprised to find surface roughening exponents that differ from *both* the KPZ equation *and* the DP problem.

To construct the algebraically-correlated noise, we first generate a representation of random Gaussian uncorrelated noise $\eta_o(\mathbf{x}, t)$, then Fourier transform it to obtain $\eta_o(\mathbf{q}, \omega)$. We define

$$\eta(\mathbf{q}, \omega) \equiv |\mathbf{q}|^{-\rho} |\omega|^{-\theta} \eta_o(\mathbf{q}, \omega). \tag{15}$$

The noise $\eta(\mathbf{x}, t)$ is obtained by Fourier transforming $\eta(\mathbf{q}, \omega)$ back into the space and time domain. It is straightforward to verify that $\eta(\mathbf{x}, t)$ obtained in this way has the correct correlations (13a). We restrict ourselves to the $d = 1 + 1$ case and the noise has only spatial correlation ($\theta = 0$) as in Eq. (13b).

(i) Consider first the KPZ equation with noise η described by (13b). For a one-dimensional surface, the discrete form of Eq. (2) is

$$h_{t+\Delta t}(i) = h_t(i) + \Delta t[h_t(i + 1) + h_t(i - 1) - 2h_t(i)]$$
$$+ \frac{\lambda \Delta t}{2} \left[\frac{h_t(i + 1) - h_t(i - 1)}{2} \right]^2 + \sqrt{\Delta t} \eta_t(i). \tag{16}$$

Small Δt is needed to obtain good convergence, and we choose the appropriate time step by verifying that smaller time steps do not change our results. We obtain the exponent β from $w(\ell, t)$ defined in Eq. (1), since $w \sim t^\beta$ for $\Delta t \ll t \ll t_\times$.

We start with the case $\lambda = 0$ (no non-linearity) for which z and β can be found exactly from dimensional analysis[2,27]: a change of scale $x \to bx$ and $t \to b^z t$ implies $h \to b^\alpha h$ and

$$\eta(x, t) \to b^{\rho - 1/2 - z/2} \eta(x, t)$$

[from Eq. (13b)]. Equation (2) is scale invariant for the choice

$$z_0 = 2 \quad ; \quad \beta_0 = \frac{1}{4} + \frac{\rho}{2}. \tag{17}$$

Our numerical simulation for $\lambda = 0$ confirms (17).

When $\lambda \neq 0$ the exponents change. We find that the exponent β approaches the same value for non-zero λ. Since changing λ should not change the universality class, we carry out our simulation for that value of λ which gives the fastest convergence to the correct value of β; then we vary the parameter ρ. The results are shown in Fig. 5. The solid, dashed, and dotted lines are the predictions from three theories [Eqs.(14a), (14b), and (14c)], respectively.

To check our results, we also study the DP growth. By a simple transformation $W(\mathbf{x}, t) \equiv \exp[(\lambda/2)h(\mathbf{x}, t)]$, we obtain from (2)

$$\frac{\partial W}{\partial t} = \nabla^2 W + \frac{\lambda}{2} \eta(\mathbf{x}, t) W. \tag{18}$$

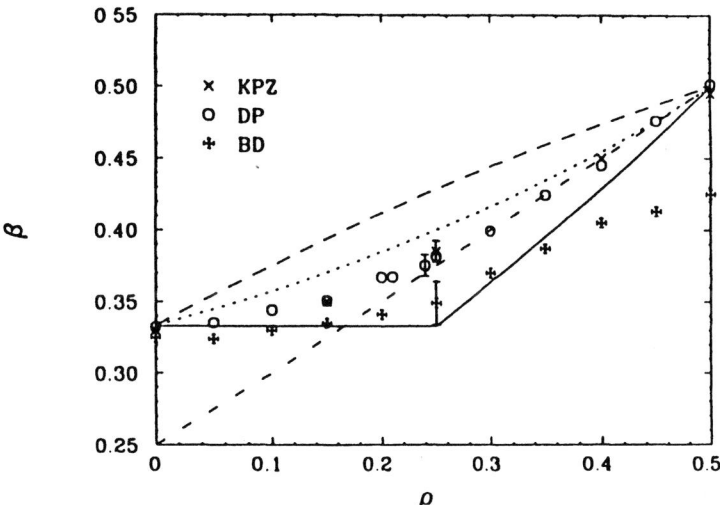

Fig. 5. Comparison of our numerical results and theoretical predictions of (14a), (14b), and (14c) (solid, dashed, and dotted lines respectively). Typical error bars are shown for each of the three models treated. The dot-dashed line, Eq. (17), is obtained by neglecting the non-linear term in Eq. (2). After Peng et al., Ref. 25.

Here W is the sum of Boltzmann weights for all configurations of a DP connecting $(0,0)$ and $(\mathbf{x},t)$, and $\eta(\mathbf{x},t)$ is the potential field. The Boltzmann weight for all paths joining the points $(0,0)$ and (x,t) is

$$W(x,t) \equiv \sum_c \exp[-E_c/kT]. \tag{19}$$

Here E_c is the sum of the potential field η on configuration c, and the sum is over all configurations joining the two end points.

The typical transverse fluctuation scales with the length of the polymer t as $\langle x^2(t)\rangle^{1/2} \sim t^\nu$. At zero temperature, only the optimal path (configuration with minimum energy) makes a contribution. Since the optimal path still dominates at finite but low temperature, we choose $T = 0$ to simplify our numerical task. We generate a representation of $\eta(x,t)$ [obeying Eq. (13b)], and record the end point of the optimal path $x(t)$. We average over many realizations (typically 10^5 of $\eta(x,t)$). The exponent ν is related to the dynamic exponent $z = \alpha/\beta = 2-\alpha$ of KPZ equation via $\nu = 1/z$. Hence to compare with the KPZ results, we define $\beta_{DP} \equiv 2\nu - 1$ and show the results in Fig. 5. The agreement with our numerical results for the KPZ equation provides an excellent consistency check on our numerical methods.

(ii) Next we study the BD model with algebraically spatial correlated noise. For *uncorrelated* BD,[1-4] particles rain down vertically onto the substrate until they reach one of the growth sites. A growth site is defined as the highest site on each column that belongs to the nearest neighbors of the deposition surface. Once the particle reach the growth site it stops and become a part of the deposit. Note that the deposition rule defined above allows lateral growth, which is believed to be described by the non-linear term $[(\nabla h)^2]$ in Eq. (2).

We introduce correlated noise according to (13b). As seen from Fig. 5, we find significant differences between exponents obtained from the BD model and the DP growth (or the KPZ model).

IV. EXPERIMENT AND A DIRECTED PERCOLATION MODEL

Next we present experiments in which ink, coffee and other suspensions are absorbed by a hanging paper, forming a rough interface between wet and dry regions. We analyze this morphology and measure its roughness exponent α, Eq. (1). Based on the experiment we propose a new model for interface roughening. Both the model

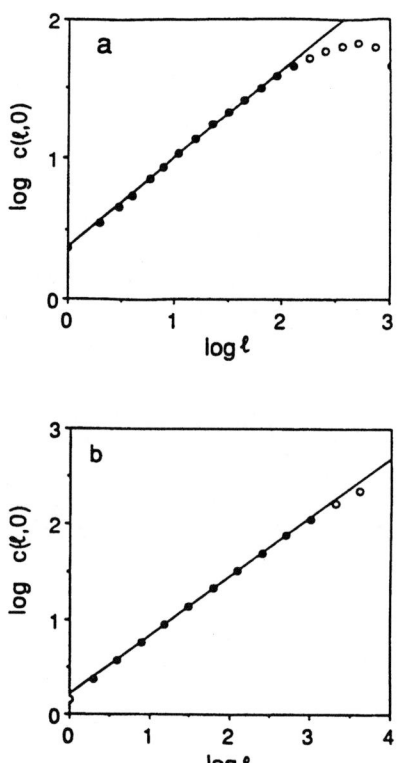

Fig. 6. Log-log plots showing the dependence on length scale ℓ of the height-height correlation function $c(\ell, 0) \sim w(\ell)$ for (a) the experimental data (averaging over 10 different experiments), and (b) the numerical results (averaging over 1000 different realizations for system size $L = 16,384$ and for $p = 0.4675$, very close to p_c for the infinite system). The slope for the set of experimental points indicated by solid circles (two decades) is 0.63 ± 0.04, while the slope for the simulation point indicated by solid circles (three decades) is 0.63 ± 0.02. After Barabási et al. Ref. 28.

and the experiment produce interfaces with an anomalously large value of α.[28]

(a) Experiment. The experiment was performed by clipping paper to a ring stand, and allowing it to dip into a basin filled with suspensions of ink or coffee. The suspension was absorbed into the paper, forming a rough interface between the wet and the dry regions. We allow the interface to rise until it stops and no change in

either height or shape of the interface is observed. The stopping can be attributed to the evaporation of the fluid in the wet regions. After drying, we digitize this rough interface. We then calculate the height-height correlation function[10] $c(\ell, 0) \equiv w(\ell)$ on different length scales ℓ, averaging over ten different interfaces. Figure 6a shows the data, which support a scaling of the form $c(\ell, 0) \equiv w(\ell) \sim \ell^{\alpha}$ with $\alpha = 0.63 \pm 0.04$.

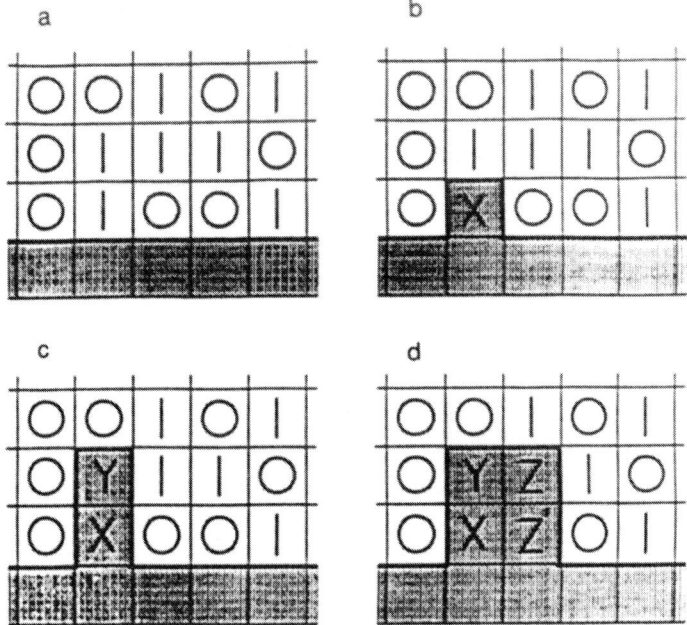

Fig. 7. Explanation of the model for interface growth with erosion of overhangs. Wet cells are indicated by shaded cells. Dry cells are randomly blocked with probability p (indicated by 0 or unblocked with probability $1 - p$ (indicated by 1). The interface between wet and dry cells are shown by a heavy line. (a) $t = 0$, (b) $t = 1$, (c) $t = 2$, and (d) $t = 3$. After Barabási et al., Ref. 28.

(b) Model. The model we propose is defined as follows: on a square lattice of edge L (with periodic boundary conditions) we block a fraction p of the cells to correspond to the inhomogeneous nature of the paper towel. At $t = 0$, we regard the "interface" to be the bold horizontal line shown in Fig. 7a. At $t = 1$ we randomly choose a cell (labeled X in Fig. 7b) which is one of the unblocked dry cells that are nearest neighbors to the interface. We wet cell X and *any cells that are below it in the same column.* This process is then iterated. For example, Fig. 7c shows that at $t = 2$ we choose cell Y a second unblocked cell to wet, while Fig. 7d shows that at $t = 3$ we wet cell Z *and also cell Z' below it.*[29]

We find that for p below a critical threshold[30] $p_c = p_c(L)$ the interface propagates without stopping, while for p above p_c the interface is pinned. Figure 6b displays the scaling behavior of the model at criticality, and we find that $\alpha = 0.63 \pm 0.02$, a value identical to the experimental value of Fig. 6a.

Next we argue that the model presented above is connected to directed percolation,[4,31] thereby providing a theoretical basis for the observed and calculated values of the anomalous roughening exponent α. The propagation of the interface will stop when it reaches for the first time a directed path of blocked cells leading from West to East—this path is such that one can walk on it from West to East without turning to the West. Such a 'directed path' is a path on the directed percolation cluster formed by the cells labeled 0. We assume that a single transverse length characterizes the directed percolation clusters so that the width w of this interface scales as the transverse correlation length $\xi_\perp$ of the directed percolation problem ($\xi_\perp$ is a rigorous upper bound). Thus we assume $w(\ell) \sim \xi_\perp$ and $\ell \sim \xi_\parallel$, where $\xi_\parallel$ is the longitudinal correlation length in the corresponding directed percolation problem. Since $\xi_\perp \sim \xi_\parallel^{\nu_\perp/\nu_\parallel}$ we identify[4,31,32] $\alpha = \nu_\parallel/\nu_\perp \simeq 0.63$.

To probe the *dynamics* of the growing interface in the model, we study the height-height correlation function $c(\ell, t)$. Our numerical results support an exponent $\beta = 0.64 \pm 0.04$. The usual exponent identity attributed to the Galilean invariance, which is known to be valid for the KPZ equation,[5] is violated; we find $\alpha + z \simeq 1.69$, smaller than two. This is a consequence of the strong anisotropy of the mechanism which excludes the overhangs: An infinitesimal tilting of the pinned interface will result in removing blocked cells, thus allowing the interface to propagate further.

Further support for the directed percolation model can be obtained if we consider a finite system at a fixed value $p_o < p_c$. If $\xi_\parallel(p_o)$ is larger than the system size L, the interface may be stopped by the directed percolation path. Thus we identify two regimes: Regime I where $\xi_\parallel > L$ and Regime II where $\xi_\parallel < L$. In Regime I, we observe only anomalous roughening ($\alpha \simeq 0.68$), while in Regime II we predict a crossover to behavior described by the KPZ exponent ($\alpha = 0.5$). A similar crossover[33] is observed, both in our calculations and even in some recent experiments (Fig. 3 of Ref. 10).

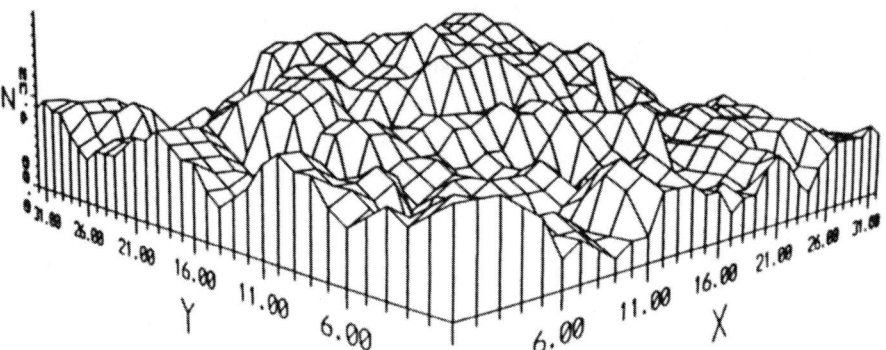

Fig. 8. A two-dimensional "directed surface" with erosion of overhangs using the rules described in Fig. 7. The critical probability of blocked sites is $p_c \simeq 0.74$.

We also studied several variants of the above model. One interesting variant arises if we replace the blocked-unblocked percolation substrate by one used in invasion percolation. At every cell of the lattice we put a random number between 0 and 1, and advance the interface to the nearest neighbor with the smallest random number. We use the same mechanism as in the previous model to erode the overhangs. In this model the interface never stops propagating unless we introduce a cutoff (a particular value of $p = p_1$; cells possessing a random number $p > p_1$ are blocked cells). Thus the interface that stops propagating is generated by the same mechanism as in the normal percolation model, i.e. they are in the same universality class. However, one significant difference is that the normal percolation interface is generated basically by a local growing rule, while in the invasion percolation the growing rule is global. Numerical studies on the *moving* interface—both of invasion and normal percolation substrate—give the same roughness exponent, $\alpha = 0.68 \pm 0.05$, slightly larger than for the pinned interface.

Barabási et al [28] also studied the above model in higher dimensions, for which there are no theoretical predictions on the values of the scaling exponents (see Fig. 8). Our results suggest $\alpha = 0.51 \pm 0.05$ for $2+1$ dimensions, which is larger than the most accurate available numerical result[34] for KPZ growth $\alpha = 0.4 \pm 0.01$. Using similar arguments as for $d = 1+1$, we obtain here a two-dimensional "directed surface" with correlation exponents $\nu_\parallel$ and $\nu_\perp$, which obey $\alpha = \nu_\parallel/\nu_\perp = 0.51$.

We wish to thank K. Shaknovich for technical assistance, M. Araujo, F. Caserta, M. Gyure, G. Huber, M. O. Robbins, S. Schwarzer, and G. H. Weiss for helpful discussions, F. Caserta for collaborating on some of the experiments, and the Hungary-USA exchange program of the Hungarian Academy of Sciences and the NSF for support.

REFERENCES

1. F. Family and T. Vicsek (eds.),*Dynamics of Fractal Surfaces* (World Scientific, Singapore, 1991).

2. For a recent review, see J. Krug and H. Spohn, in *Solids Far from Equilibrium: Growth, Morpholgy and Defects*, edited by C. Godréche (Cambridge Univ. Press, Cambridge, England, 1991).

3. T. Vicsek, *Fractal Growth Phenomena* (World Scientific, Singapore, 1989).

4. A. Bunde and S. Havlin (eds.), *Fractals and Disordered Systems* (Springer-Verlag, Berlin, 1991).

5. M. Kardar, G. Parisi, and Y.-C. Zhang, Phys. Rev. Lett. **56**, 889 (1986); E. Medina, T. Hwa, M. Kardar and Y.-C. Zhang, Phys. Rev. A **39**, 3053 (1989).

6. F. Family and T. Vicsek, J. Phys. A **18**, L75 (1985).

7. P. Meakin, P. Ramanlal, L. M. Sander, and R. C. Ball, Phys. Rev. A **34**, 5091 (1986).

8. J. Krug, Phys. Rev. A **36**, 5465 (1987).

9. M. A. Rubio, C. A. Edwards, A. Dougherty, and J. P. Gollub, Phys. Rev. Lett. **63**, 1685 (1989).

10. V. K. Horváth, F. Family, and T. Vicsek, J. Phys. A **24**, L25 (1991)

11. T. Vicsek, M. Cserzö, and V. K. Horváth, Physica A **167**, 315 (1990).

12. Y.-C. Zhang, J. de Physique **51**, 2113 (1990).

13. J. Amar and F. Family, J. Phys. A **24**, L79 (1991).

14. S. V. Buldyrev, S. Havlin, J. Kertész, H. E. Stanley, and T. Vicsek, Phys. Rev. A **43**, 7113 (1991).

15. Y.-C. Zhang, Physica **170**, 1 (1990); J. Krug, J. Physique I, **1**, 9 (1991).

16. S. Havlin, S. V. Buldyrev, H. E. Stanley and G. H. Weiss, J. Phys. A **24**, L925 (1991).

17. Since for $\mu > 1$, $\langle \ell \rangle$ is finite, the Lévy walk model will have the same distribution as the Lévy flight where the walker makes ℓ steps in one unit of time.

18. G. Zumofen, J. Klafter, and A. Blumen, Chemical Physics **146**, 433 (1990); M. Araujo, S. Havlin, G. H. Weiss and H. E. Stanley, Phys. Rev. A **43**, 5207 (1991).

19. V. K. Horváth, F. Family and T. Vicsek, Phys. Rev. Lett. **67**, 3207 (1991).

20. V. Yakhot and S. A. Orszag, Phys. Rev. Lett. **57**, 1722 (1986); V. Yakhot and S. A. Orszag, Phys. Rev. Lett. **60**, 1840 (1988); and references therein.

21. M. Kardar and Y.-C. Zhang, Phys. Rev. Lett **58**, 2087 (1987).

22. Y.-C. Zhang, Phys. Rev. B **42**, 4897 (1990).

23. H. G. E. Hentschel and F. Family, Phys. Rev. Lett. **66**, 1982 (1991).

24. P. Meakin and R. Jullien, Europhys. Lett. **9**, 71 (1989); P. Meakin and R. Jullien, Phys. Rev. A **41**, 983 (1990); A. Margolina and H. E. Warriner, J. Stat. Phys. **60**, 809 (1990).

25. C. K. Peng, S. Havlin, M. Schwartz, and H. E. Stanley, Phys. Rev. A **44** 2239 (1991); J. G. Amar, P.-M. Lam, and F. Family, Phys. Rev. A **43**, 4548 (1991) consider one of the models we treated (BD) and find agreement with (14a) over the range $0 \leq \rho \leq 0.43$.

26. S. Havlin, R. Selinger, M. Schwartz, H. E. Stanley, and A. Bunde, Phys. Rev. Lett. **61**, 1438 (1988); C. K. Peng, S. Havlin, M. Schwartz, H. E. Stanley, and G. H. Weiss, poster presentation at Cargèse NATO ASI (July 1990).

27. S. F. Edwards and D. R. Wilkinson, Proc. R. Soc. London, Sect. A **381**, 17 (1982).

28. A.-L. Barabási, S. Buldyrev, F. Caserta, S. Havlin, H. E. Stanley, and T. Vicsek (preprint); see also L. H. Tang and H. Leschhorn (preprint).

29. The rule corresponding to the erosion of cells is introduced to take into account that under the influence of the locally advanced interface the pinned parts of the interface are likely to be 'released' and consequently catch up in a short time. See the discussion in M. Cieplak and M. O. Robbins, Phys. Rev. **B41**, 11508 (1990) and M. O. Robbins (private communication).

30. D. Stauffer and A. Aharony, *Introduction to Percolation Theory*, 2nd edition (Taylor & Francis, London, 1992).

31. W. Kinzel, in *Percolation Structures and Processes*, edited by G. Deutscher, R. Zallen, and J. Adler (A. Hilger, Bristol, 1983).

32. B. Hede, J. Kertész, and T. Vicsek, J. Stat. Phys. **64**, 829 (1991).

33. The general scaling formalism extended to the case when there is a diverging length in the system ($\xi_{||}$) as the critical point p_c is approached has been proposed in J. Kertész and D. E. Wolf, Phys. Rev. Lett. **62**, 2517 (1989).

34. B. M. Forrest and L.-H. Tang, Phys. Rev. Lett. **64**, 1405 (1990).

WAITING-TIME FORMULATION OF SURFACE GROWTH AND

MAPPING TO DIRECTED POLYMERS IN A RANDOM MEDIUM

Lei-Han Tang

Institut für Theoretische Physik
Universität zu Köln
D-5000 Köln 41, Germany

INTRODUCTION

One class of growth patterns which have been observed in physical and biological systems is the development of a seed into a compact cluster with a rough surface or interface.[1] Examples of this type include vacuum-deposited thin films[2] at low surface mobility and bacteria colonies[3] on an agar plate. Recent theoretical studies[4] of the roughness of a moving surface have focused on the analysis and simulation of models with simple kinetic growth rules. Many of these models have been found to exhibit a universal dynamic scale invariance.[5] The origin of universality among these models has been elucidated by Kardar, Parisi, and Zhang[6] within a continuum theory.

The purpose of this paper is to show that there exists another theoretical framework where the concept of universality can be developed. Basic ideas of this approach were introduced in a recent article by Roux, Hansen, and Hinrichsen,[7] and in other earlier works.[8,9] Within this framework, different lattice models can be compared directly without going to the continuum limit. By mapping lattice growth models to the directed polymer problem[10] on a lattice, the role of randomness in the growth rules becomes immediately clear. Results and insights on the directed polymer problem can be translated immediately to the growth problem, and vice versa.

WAITING-TIME FORMULATION

The waiting time formulation can be applied to any cluster growth process that has the following characteristics. At a given instant of the growth process, all cells on a lattice can be divided into three groups, black, white, and active. The black cells are cells on the existing cluster (occupied cells). The active cells are those (or part of those) empty cells in the immediate neighborhood of the black cells. The white cells are the remaining ones that are not under consideration for occupation at this moment. An active cell $\mathbf{R}$ turns into a black cell after a certain waiting period $\tau(\mathbf{R})$ determined by the kinetics of cluster growth.

For some growth processes such as diffusion-limited-aggregation,[11] the waiting time $\tau(\mathbf{R})$ is a dynamical variable that depends on the evolution of the whole cluster, so that the determination of $\tau(\mathbf{R})$ from the growth kinetics is a highly nontrivial problem. (See other contributions in this volume.) In reaction limited processes, which is the subject of concern here, growth usually depends only on the local environment so that the task for determining $\tau(\mathbf{R})$ can be greatly simplified. The nontrivial problem is

Growth Patterns in Physical Sciences and Biology, Edited
by J. M. Garcia-Ruiz *et al.*, Plenum Press, New York, 1993

then to determine the overall morphology of the cluster (such as surface roughness) from a known waiting-time distribution.

A general mathematical formulation of waiting-time growth was presented by Richardson[8] some twenty years ago. The main result of his study is that, if the distribution of waiting times obeys certain conditions, the resulting cluster is compact. He showed that these conditions are satisfied by the growth rules of the Eden model and a number of other models.

In the case of Eden growth, a waiting-time description has been worked out in detail by Roux *et al.*[7] In the following we consider two other well-known examples, the single-step model[12] and ballistic deposition model,[13] whose growth rules are anisotropic. We shall see how the active cells are identified in the two cases, and how the distribution of waiting times is determined from the growth rules. For simplicity we shall focus on the (1+1)-dimensional case (one-dimensional surface in two dimensions) but generalization to higher dimensions is straightforward.

Single-step Model

The single-step model in (1+1) dimensions describes the growth of a crystal on a square lattice tilted by 45°, as illustrated in Fig. 1 following a similar representation given by Meakin *et al.*[12] The surface of the crystal is specified by a set of column heights $\{h_i\}, i = 1, \ldots, L$, with the "single-step" constraint $h_{i+1} - h_i = \pm 1$ and periodic boundary conditions $h_{i+L} = h_i$. An active cell, indicated by a cross in Fig. 1, is a cell which lies just above a local minimum of the surface. The activation of an empty cell $\mathbf{R}$ thus involves occupation of two of its neighbors $\mathbf{R_1}$ and $\mathbf{R_2}$ on the lower left and right. The active cells are then filled according to one of the following schemes. In *random sequential updating*, an active cell is picked up at random and filled. The list of active cells is then updated before another random selection is made, and the process continues. In *sublattice updating*, active cells on a given sublattice (e.g. cells in odd columns) are considered simultaneously and each has a probability p to become occupied. The list of active cells is then updated before one moves on to the next sublattice until all sublattices have been considered in rotation.

To give a waiting-time description of the growth process defined above, it is convenient to introduce the following definitions of time t.[14] For random sequential updating, we let t increase in units of $\Delta\tau = 1/L$. During each time interval $\Delta\tau$, one column is randomly selected and the active cell in this column is filled if there is one.

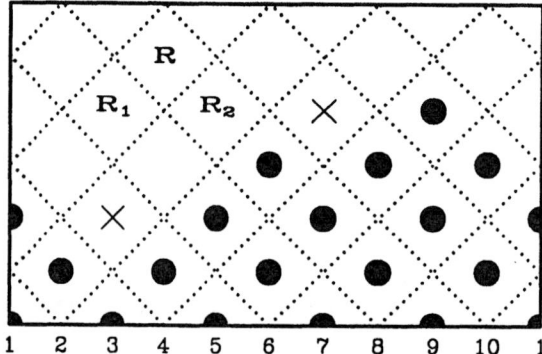

Fig. 1. The single-step model with periodic boundary conditions. Solid circle: occupied cell; Cross: active cell. A cell $\mathbf{R}$ is activated by the occupation of *both* neighbors $\mathbf{R_1}$ and $\mathbf{R_2}$ on the lower left and lower right.

The probability that a fresh activated cell is occupied after a waiting period $\tau = n\Delta\tau$ is given by

$$P(\tau)\Delta\tau = \left(1 - \frac{1}{L}\right)^n \frac{1}{L}, \tag{1}$$

which, in the limit $L \to \infty$, yields

$$P(\tau) = e^{-\tau} \qquad \text{(random sequential)}. \tag{2}$$

For sublattice updating, we let t increase in units of $\Delta\tau = 1/s$, where s is the number of sublattices. In the simplest case $s = 2$, which is the case we will focus on, an empty cell is activated by the occupation of active cells on the other sublattice, so that the waiting time τ takes only half-integer values. It is easy to show that τ has the following distribution,

$$P(\tau) = p(1 - p)^{\tau - 1/2}, \quad \tau = \frac{1}{2}, \frac{3}{2}, \dots \qquad \text{(two - sublattice)}. \tag{3}$$

For both updating schemes, the waiting time is a stochastic variable completely independent from cell to cell, and independent of the overall morphology of the surface. The time $t(\mathbf{R})$ that a cell $\mathbf{R}$ becomes occupied is related to the occupation time of its two lower neighbors $\mathbf{R}_1$ and $\mathbf{R}_2$ by

$$t(\mathbf{R}) = \tau(\mathbf{R}) + \max\{t(\mathbf{R}_1), t(\mathbf{R}_2)\}. \tag{4}$$

We say that $\mathbf{R}_1$ ($\mathbf{R}_2$) is the *parent* of $\mathbf{R}$ if $t(\mathbf{R}_1) > t(\mathbf{R}_2)$ $[t(\mathbf{R}_2) > t(\mathbf{R}_1)]$.

Ballistic Deposition

Let us assume the deposition to be normal to the substrate. Each time a particle is dropped from above at a randomly chosen column i, falls down along a vertical trajectory, and sticks at the first contact with the existing cluster, as shown in Fig. 2. The height of column i after the growth event is given by

$$h_i' = \max\{h_i + 1, h_{i-1}, h_{i+1}\}. \tag{5}$$

This equation serves as the definition of active cells $\mathbf{R} = (i, h_i')$. Adopting the definition of time as in the random sequential case of the single-step model, the waiting time for a fresh activated cell to become occupied follows the same distribution as given by Eq. (2). The occupation time of an active cell $\mathbf{R} = (i, h)$ is related to those of neighboring

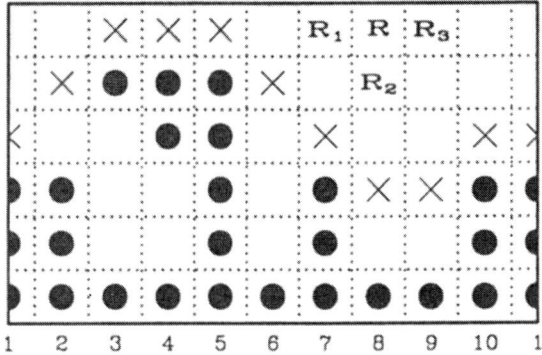

Fig. 2. The ballistic-deposition model at normal incidence. Solid circle: occupied cell; Cross: active cell. A cell $\mathbf{R}$ is activated by the occupation of *one* of the neighbors $\mathbf{R}_1$, $\mathbf{R}_2$, and $\mathbf{R}_3$.

cells $\mathbf{R}_1$, $\mathbf{R}_2$, and $\mathbf{R}_3$ (see Fig. 2) by

$$t(\mathbf{R}) = \tau(\mathbf{R}) + \min_{n=1,2,3}\{t(\mathbf{R}_n)\}. \tag{6}$$

In this case the neighbor that is occupied first is called the *parent* of $\mathbf{R}$, i.e., it is the cell which activated $\mathbf{R}$. Note, however, that the waiting period in ballistic deposition can be prolonged to infinity when a growth event at one of the neighboring columns takes place before $t(\mathbf{R})$ is reached.

MAPPING TO THE DIRECTED POLYMER PROBLEM

The directed polymer problem can be stated as follows. Consider a lattice with a distribution of random site-energies $e(\mathbf{R})$. Define a *directed path* (=directed polymer) S on the lattice as a sequence of sites $\mathbf{R}_i, i = 0,\ldots,l$, such that $\mathbf{R}_i$ and $\mathbf{R}_{i+1}$ are neighboring sites in some given sense, and that $\delta_i = \mathbf{R}_{i+1} - \mathbf{R}_i$ is a vector with (e.g.) a nonnegative component along a chosen direction. Here l is known as the *length* of the path. At zero temperature, we want to compute, for a *fixed* end point $\mathbf{R} = \mathbf{R}_l$ (but not necessarily a fixed l), the ground state among all possible directed paths satisfying a given constraint on the starting point $\mathbf{R}_0$. The energy of the ground state E obviously satisfies the following iterative relation,

$$E(\mathbf{R}) = e(\mathbf{R}) + \min_{\delta}\{E(\mathbf{R} - \delta)\}, \tag{7}$$

where the minimum is taken over all possible neighboring sites with δ having a nonnegative component in the chosen direction.

We now see an obvious analogy between the directed polymer problem as defined by Eq. (7) and the waiting-time growth problems as formulated by Eqs. (4) and (6). For the single-step model, Eq. (4) becomes identical to Eq. (7) if one performs the substitution[14]

$$\tau(\mathbf{R}) = -e(\mathbf{R}), \qquad t(\mathbf{R}) = -E(\mathbf{R}) \qquad (\text{single} - \text{step model}). \tag{8}$$

In contrast, for ballistic deposition, a cell is activated by the occupation of one of its neighbors below or on the side, so that the following identification is appropriate,

$$\tau(\mathbf{R}) = e(\mathbf{R}), \qquad t(\mathbf{R}) = E(\mathbf{R}) \qquad (\text{ballistic deposition}). \tag{9}$$

For a growth process initiated from some surface Σ (substrate) at $t = 0$, the corresponding condition on the starting point $\mathbf{R}_0$ of the directed paths is that $\mathbf{R}_0$ should always be in contact with Σ, but is otherwise free to move so as to minimize the total energy of the path. The minimum energy path from the substrate to a surface cell $\mathbf{R}$ is nothing but an ancestry table for $\mathbf{R}$ under the definition of the parental relationship given above.

Figures 3(a) and (b) show the surface configurations (thick lines on the top) and the corresponding minimum energy paths obtained from one realization of the single-step model at two different times. The sublattice updating scheme is used, with $p = 1/2$. All cells on the surface are connected to the substrate by the minimum energy directed paths which form a tree-like structure (ancestry tree). (Due to the discrete set of values taken by τ, there are loops on the tree reflecting the degeneracy of the paths involved.) For two surface cells on the same tree, their height is correlated due to the overlap in the corresponding minimum energy paths. The amount of correlation, of course, depends on how far back along the tree where the parentage becomes identical. The latter is obviously related to the distance between the two cells in the direction parallel to the surface. As time goes on, some trees flourish by taking more of the surface cells as their offsprings, while others die out.

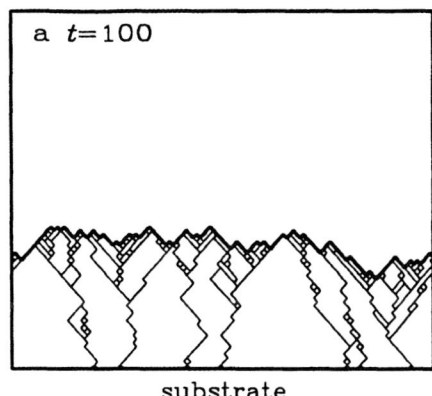 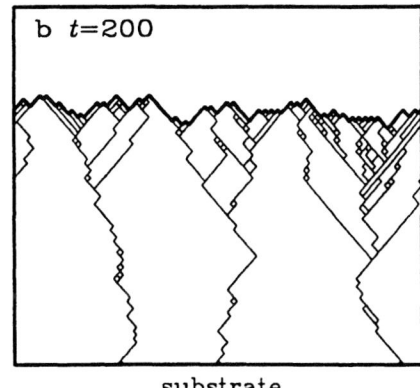

Fig. 3. Surface configurations (thick lines at the top) of the single-step
model during a single run at (a) $t = 100$ and (b) $t = 200$ starting
from a flat substrate. The corresponding minimum energy paths
(ancestry trees) are also shown. Note the decrease in the number of
trees that stem from the substrate as time increases. Here $L = 128$.

Using the above picture, it is easy to work out the relations between the scaling
behavior of flutuating quantities in the growth and directed polymer problems. For
example, let us consider the equal-time height-height correlation function of a growth
process initiated from a flat substrate at $t = 0$, which is expected to obey the following
scaling law in the limit $L \to \infty$,

$$C(r,t) = \langle[h(r + r_0, t) - h(r_0, t)]^2\rangle \simeq r^{2\zeta} F(t/r^z). \tag{10}$$

Here and elsewhere $\langle\cdot\rangle$ denotes average over the waiting time or random energy distribution. The scaling function $F(x) \sim x^{2\zeta/z}$ for $x \ll 1$ and approaches a constant in
the limit $x \to \infty$. Equation (10) can be understood in the directed polymer picture
as follows. Let us first rewrite $C(r,t)$ as

$$C(r,t) = \langle[h(r + r_0, t) - \langle h(t)\rangle]^2\rangle + \langle[h(r_0, t) - \langle h(t)\rangle]^2\rangle$$
$$- 2\langle[h(r + r_0, t) - \langle h(t)\rangle][h(r_0, t) - \langle h(t)\rangle]\rangle. \tag{11}$$

Here $\langle h(t)\rangle = \langle h(r_0, t)\rangle = \langle h(r + r_0, t)\rangle$ is the average height of the surface. Instead of
(11), one may consider the fluctuation of occupation times (or equivalently, ground-
state energies) among cells at a given height $h = \langle h(t)\rangle$,

$$(\Delta E)^2 = \langle E^2\rangle - \langle E\rangle^2 \simeq h^{2\omega}. \tag{12}$$

Provided that $\Delta E \ll \langle E\rangle \sim h$, the first two terms in (11), which are identical to each
other, are proportional[7] to $(\Delta E)^2$. The last (correlation) term in Eq. (11) becomes
negligible when r is much greater than the typical lateral span ξ of the trees. The
equivalence of (10) and (12) in this limit yields $\zeta/z = \omega$. * The crossover of $C(r,t)$
from t dependence to r dependence takes place when $r \simeq \xi \simeq t^{1/z}$, at which point
the paths of two surface cells separated by a distance r start to overlap. Assuming
that the typical lateral span of a tree is of the same order as the typical transverse

---

* A weak point in this argument is the assumption that all surface cells at a given
t have the same occupation time $t(\mathbf{R})$, which is not necessarily true. However, the
result is still valid if the fluctuation in $t(\mathbf{R})$ at a fixed t is of the order of ΔE at a fixed
h or less.

component of the end-to-end displacement of a minimum energy path, $1/z$ is identified as the wandering exponent ν of a minimum energy directed path.[14] The same picture can be applied to obtain finite-size scaling relations of the two problems. as was done by Roux *et al.*[7]

WAITING-TIME DISTRIBUTION AND SURFACE ROUGHNESS

In the waiting-time formulation, the basis for universality resides on the general nature of the mapping to the directed polymer problem. Different growth models (or different updating schemes) can be compared by examining the geometry of allowed paths and the random-energy distribution of the corresponding directed polymer problem. In the remaining part of the paper we shall consider two examples to see how the waiting-time (or equivalently the random-energy) distribution influences quantitatively or even qualitatively the roughness of the surface.

Noise-reduction

Our first example is the single-step model with "noise-reduction".[15,16] Here the growth rules are the same as we discussed earlier, except that a counter is attached to every active cell. This counter, initially set at zero, increases by 1 whenever an occupation event as defined previously takes place. The actual occupation occurs only when the counter reaches a given value M. In a parallel updating scheme, the waiting-time distribution for this problem is given by

$$P_M(\tau) = \frac{(\tau - \frac{1}{2})!}{(M-1)!(\tau - M + \frac{1}{2})!} p^M (1-p)^{\tau - M + 1/2}, \quad \tau = M - \frac{1}{2}, M + \frac{1}{2}, \dots \quad (13)$$

which, in the case $M = 1$, reduces to (3). The mean and standard deviation of this distribution are given by

$$\tau_0(M) = \sum_\tau \tau P_M(\tau) = \frac{M}{p} - \frac{1}{2}, \quad (14a)$$

$$\sigma^2(M) = \sum_\tau (\tau^2 - \tau_0^2) P_M(\tau) = M(1-p)/p^2, \quad (14b)$$

respectively.

Knowing the waiting-time distribution in this case actually allows one to make quantitative predictions about the scaling behavior of the surface at different M. Let us first note that, according to Eq. (4), the ground-state paths are unaltered by the substitution $\tau \to \lambda\tau, \lambda > 0$. In addition, *for the single-step model*, all possible directed paths from a flat substrate to a surface cell $\mathbf{R}$ have the same length, $l = h(\mathbf{R})$. Thus the substitution $\tau \to \tau + \tau_0$ has no influence on the ground-state configurations either. By invoking these symmetries, it is easy to see that, if the waiting time satisfies a gaussian distribution [or any other distribution parametrized as $P(\tau) = A(\sigma)\hat{P}((\tau - \tau_0)/\sigma)$] with a mean τ_0 and variance σ^2, the energy of a ground-state path of length h must be of the following form at large h,

$$E \simeq -[\tau_0 + \sigma e_0]h + \sigma e_1 h^\beta, \quad (15)$$

where e_0 is a numerical constant, e_1 is a random variable of order 1. and $\beta = \zeta/z$. To calculate $C(r, t)$, we keep $E = -t$ fixed and solve for $h \simeq \langle h(t) \rangle + \Delta h$ from Eq. (15). Equating the averaged and fluctuating parts separately, we obtain,

$$t \simeq (\tau_0 + \sigma e_0)\langle h(t) \rangle, \qquad \Delta h \simeq [\sigma e_1 / (\tau_0 + \sigma e_0)]\langle h(t) \rangle^\beta, \quad (16)$$

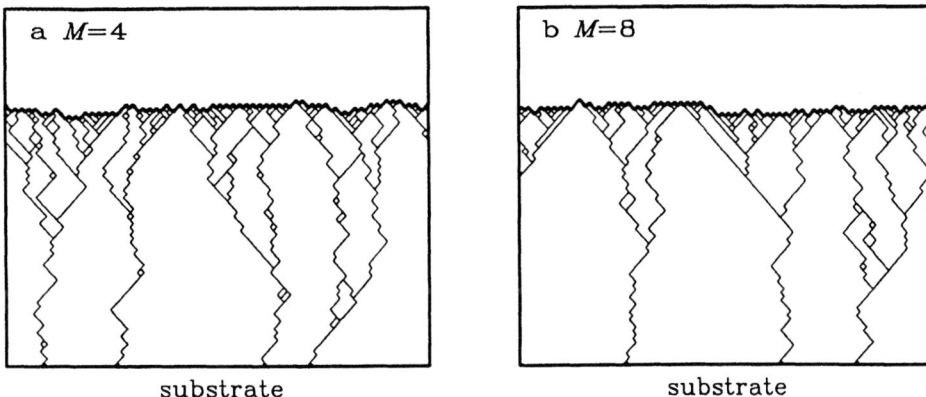

Fig. 4. Surface configurations (thick lines at the top) of the single-step model with noise-reduction. (a) $M = 4, t = 680$; (b) $M = 8, t = 1320$. The corresponding minimum energy paths (ancestry trees) are also shown. Here $L = 128$.

where we have assumed $\Delta h \ll \langle h(t) \rangle$. For $r \gg t^{1/z}$, Eq. (16) yields

$$C(r,t) \simeq 2\langle(\Delta h)^2\rangle \simeq \frac{2\sigma^2 \langle e_1^2 \rangle}{(\tau_0 + \sigma e_0)^2} \left(\frac{t}{\tau_0 + \sigma e_0}\right)^{2\beta}. \tag{17}$$

The crossover to a r-dependent behavior at $r \simeq \xi$ is determined by the geometry of the minimum energy paths which is not affected by the values of τ_0 and σ. Thus we have

$$\xi \simeq h^{1/z}, \tag{18}$$

where the numerical factor in front of the power-law (not written explicitly) does depend on the shape of the distribution. Using Eqs. (16)-(18), we arrive at a general scaling form

$$C(r,t) \simeq \frac{2\sigma^2}{(\tau_0 + \sigma e_0)^2} r^{2\zeta} \Phi\left(\frac{t}{(\tau_0 + \sigma e_0)r^z}\right), \tag{19}$$

where $\Phi(x)$ is a universal scaling function for a given class of distributions.

Coming back to the distribution (13), we see that, for $M \gg 1$, $P_M(\tau)$ approaches a gaussian distribution (central limit theorem) with $1 \ll \sigma(M) \ll \tau_0(M)$. Results of the above discussion are thus applicable in this limit. Figure 4 shows the surface configurations for two different values of the noise-reduction parameter M at roughly the same height and the ancestry trees. [Compare also with Fig. 3(b).] As expected, noise-reduction has no influence on the transverse fluctuations of the minimum energy paths, but does suppress the fluctuation of h on a "constant energy" surface.

Using Eqs. (14) and (19) we find, for $M \gg 1$,

$$C(r,t) \simeq \frac{1-p}{M} r \Phi\left(\frac{pt}{Mr^{3/2}}\right), \tag{20}$$

where we have set $\zeta = 1/2$ and $z = 3/2$. For $pt \ll Mr^{3/2}$, we have

$$C(r,t) \simeq (1-p)p^{2/3} t^{2/3} M^{-5/3}, \tag{21a}$$

while for $pt \gg Mr^{3/2}$ we have

$$C(r,t) \simeq (1-p)r/M. \tag{21b}$$

In both cases possible numerical prefactors are not included. I have performed simulations of the single-step model for $M = 2$ to 12 and $L = 15000$ or 25000. Figure 5 shows the data for the mean-square width of the surface $w^2(t)$ plotted against $t^{2/3}M^{-5/3}$ for $t \ll L^z$. [For $r \gg t^{1/z}$, the relation $C(r,t) = 2w^2(t)$ holds.] The data collapse is in good agreement with Eq. (21a). A somewhat surprising feature of Fig. 5 is that curves at different M, when extrapolated to $t = 0$ using the linear part at large t, intersect the vertical axis at about the same value 0.35. This correction to scaling term may have its origin in the discreteness of the height h in our model. Since the amplitude of the scaling part decreases rapidly with increasing M, plotting $w^2(t)$ directly against t on a log-log scale would give rise to a crossover regime with a lower exponent β, as observed previously by Kim and Kosterlitz.[16] However, this type of crossover introduced by a noise-reduction algorithm can be easily removed when schemes which eliminate the constant from $w^2(t)$ are used in analyzing the data.

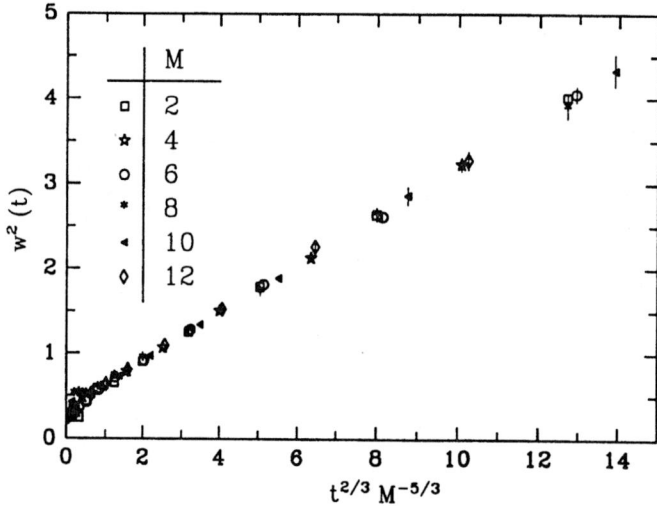

Fig. 5. Mean-square surface width $w^2(t)$ against the scaled variable $t^{2/3}M^{-5/3}$ for six different values of M, as given in the legend.

It is worth emphasizing that Eq. (19) was derived under the condition that *all* allowed paths connecting the substrate to a surface cell **R** have the same length. This condition is not satisfied by the Eden model or ballistic deposition model. Under such a situation, increasing the mean waiting time τ_0 makes side walks (wandering) less favorable, resulting in an increase of the stiffness of minimum energy paths. The lateral correlation length ξ at a given film thickness (height) is actually *reduced* by noise-reduction in these cases.

Power-law Distributed Waiting-times

The noise-reduction algorithm discussed above changes the scaling amplitudes but not the exponents. Here we consider an example where, by assigning a power-law tail to the waiting-time distribution in the single-step model, the exponents also vary.

The specific waiting-time distribution we shall discuss is given by[14]

$$P_\mu(\tau) = (\tau + \frac{1}{2})^{-\mu} - (\tau + \frac{3}{2})^{-\mu}, \qquad \tau = \frac{1}{2}, \frac{3}{2}, \ldots, \tag{22}$$

which, in the limit of large τ, falls off as $\mu\tau^{-\mu-1}$. *

The reason why a different scaling law may emerge under (22) comes from the mapping (8) for the single-step model. Due to the slow power-law falloff of the distribution at large τ, there is a non-negligible probability for having a cell of a large negative energy (known as a rare event) which may distort the minimum energy paths in a significant way. Whether rare events actually alter the scaling behavior of the paths depends, of course, on how frequent they appear, which in turn depends on μ. The following mean-field type argument, first given by Zhang[18] and by Krug[19], is in order. A directed path of length l will typically sample $N \simeq l^{d+(1/z)}$ cells, where d is the dimensionality of the surface. The largest value τ_{max} of a set of N independent random variables $\tau_1, \ldots, \tau_N$ with a distribution $P(\tau)$ has the distribution

$$\hat{P}(\tau_{\text{max}}) = NP(\tau_{\text{max}})\Big[\sum_{\tau \leq \tau_{\text{max}}} P(\tau) \Big]^{N-1}. \tag{23}$$

Using (22) and (23), one finds that the root-mean-square fluctuation of τ_{max} grows as $N^{1/\mu}$ at large N. If one identifies this fluctuation with the fluctuation of the ground state energy, l^β, one finds,

$$d + \frac{1}{z} = \mu\beta. \tag{24}$$

If in addition one assumes **

$$\zeta + z = 2, \tag{25}$$

then

$$\zeta = \frac{d+2}{\mu+1}, \qquad \beta = \frac{d+2}{2\mu-d}. \tag{26}$$

For $d = 1$, Eq. (26) yields $\zeta > 1/2$ for $\mu < 5$, in which case the rare-events are expected to be relevant.[18,19] Our simulations[14] of the model suggest that ζ may become bigger than $\frac{1}{2}$ even at $\mu > 5$. Other simulations[20] of related models gave conflicting results. For details on rare-event dominated growth the reader is referred to the article by J. Kertész in this volumn.

SUMMARY AND CONCLUSIONS

In this paper we considered in some detail the waiting-time formulation of surface growth models. We have shown that, through a suitable definition of time, the commonly used schemes for simulating stochastic growth can be readily transcribed into waiting-time growth problems. The temporal ordering of growth events allows us to introduce a parental relationship on a local level. For the single-step model, the mathematical formulation of waiting-time growth is exactly the same as that of the zero-temperature directed polymer problem. The directness of paths in this case comes from the spatial ordering of neighboring growth events. In other cases such as the Eden model, the normal of the surface provides a natural direction for the preferred orientation of the ancestry line so that a mapping to a directed polymer problem can also be made.[7] The ancestry of surface cells is represented by the tree-like structure

---

* A similar model has been considered by Jensen and Procaccia in Ref. 17.

** In the directed polymer problem, Eq. (25) is obtained if the average energy of a minimum energy path directed along a line at an angle θ from the vertical axis varies quadratically with θ. See Refs. 10 and 18.

of minimum energy paths which tells us about the amount of correlation in the height of the surface cells.

Although the connection between surface growth models and the directed polymer problem has been known for some time,[6] its full meaning is revealed only in the waiting-time formulation. The direct mapping offers a framework for studying the universal scaling behavior of surface roughness as well as other quantitative issues such as the effect of noise-reduction on scaling amplitudes. It is our hope that this approach will bring us not only a better intuitive understanding of both growth and directed polymer problems but also much needed quantitative results on the roughening behavior of various surface models.

ACKNOWLEDGEMENTS

Many of the ideas presented here grew out of a collaboration with János Kertész and Dietrich E. Wolf. I wish to thank the Institut für Festkörperforschung des Forschungszentrums Jülich for hospitality, where part of the work was carried out. Financial support from the workshop organizers is also gratefully acknowledged.

REFERENCES

1. T. Vicsek, *Fractal growth phenomena*, 2nd Edition, (World Scientific, Singapore, 1991); *Kinetics of Ordering and Growth at Surfaces*, edited by M. Lagally, (Plenum, New York, 1990), and references therein.
2. R. Messier and J. E. Yehoda, J. Appl. Phys. **58**, 3739 (1985); J. Chevrier, V. Le Thanh, R. Buys, and J. Derrien, Europhys. Lett. **16**, 737 (1991); R. Chiarello, V. Panella, J. Krim, and C. Thompson, Phys. Rev. Lett. **67**, 3408 (1991).
3. T. Vicsek, M. Cserzö, and V. K. Horváth, Physica A **167**, 315 (1990).
4. *Dynamics of Fractal Surfaces*, edited by F. Family and T. Viscek, (World Scientific, Singapore, 1991).
5. F. Family and T. Vicsek, J. Phys. A **18**, L75 (1985); R. Jullien and R. Botet, J. Phys. A **18**, 2279 (1985).
6. M. Kardar, G. Parisi, and Y.-C. Zhang, Phys. Rev. Lett. **56**, 889 (1986).
7. S. Roux, A. Hansen, and E. L. Hinrichsen, J. Phys. A **24**, L295 (1991).
8. D. Richardson, Proc. Camb. Phil. Soc. **74**, 515 (1973).
9. B. Derrida, lecture at the HLRZ workshop on "Kinetic Roughening and Self-organized Criticality", Feb. 4-6, 1991, Jülich, Germany.
10. B. Derrida and H. Spohn, J. Stat. Phys. **51**, 817 (1988); T. Halpin-Healy, Phys. Rev. A **42**, 711 (1990); D. S. Fisher and D. A. Huse, Phys. Rev. B **43**, 10728 (1991), and references therein.
11. T. A. Witten and L. M. Sander, Phys. Rev. Lett. **47**, 1400 (1982).
12. P. Meakin, P. Ramanlal, L. M. Sander, and R. C. Ball, Phys. Rev. A **34**, 5091 (1986); M. Plischke, Z. Rácz, and D. Liu, Phys. Rev. B **35**, 3485 (1987).
13. M. J. Vold, J. Colloid Sci. **14**, 168 (1959).
14. L.-H. Tang, J. Kertész, and D. E. Wolf, J. Phys. A **24**, L1193 (1991).
15. D. E. Wolf and J. Kertész, Europhys. Lett. **4**, 561 (1987); Phys. Rev. Lett. **63**, 1191 (1989); J. Kertész and D. E. Wolf, J. Phys. A **21**, 747 (1988).
16. J. M. Kim and J. M. Kosterlitz, Phys. Rev. Lett. **62**, 2289 (1989).
17. M. H. Jensen and I. Procaccia, J. Physique II **1**, 1139 (1991).
18. Y.-C. Zhang, Physica A **170**, 1 (1990).
19. J. Krug, J. Physique I **1**, 9 (1991).
20. J. G. Amar and F. Family, J. Phys. A **24**, L79 (1991); S. V. Buldyrev, S. Havlin, J. Kertész, H. E. Stanley, and T. Vicsek, Phys. Rev. A **43**, 7113.

SCALING FAR FROM THERMAL EQUILIBRIUM

H. G. E. Hentschel

Emory University
Department of Physics
Atlanta, GA 30322, USA

INTRODUCTION

Patterns and flows far from thermal equilibrium such as interfacial growth[1,2] and fluctuations in sandpiles[3] have been shown to exhibit scaling in both their spatial and temporal behaviour. Similar scaling behaviour has been known to exist in fully developed turbulent flows in both the inertial and dissipative regimes for well over half a century[4]. These diverse phenomena are united by the fact that they are all describable by a special class of non-linear partial differial equations.

These equations are dissipative and contain stabilizing diffusive terms (which depending on the physical process under examination may be due to the viscosity v or surface tension σ, as well as to diffusion D) which make themselves felt at small scales and result in these equations exhibiting only stable fixed point solutions when not driven. These solutions are temporally quiescent and spatially translationally invariant. In the presence of forcing (on the large stirring length scale in the case of turbulence, and on the microscopic scale in the case of interfacial growth and sandpile models), however, the interaction of the nonlinearities and dissipative terms with the forcing can lead to intricate patterns of spatiotemporal fluctuations described by a variety of anomalous exponents. Such complex behaviour in infinite degree of freedom systems should no longer surprise us, as even a simple one degree of freedom non-linear oscillator in the presence of dissipation and forcing such as the Duffing oscillator can exhibit stochastic behaviour as the forcing strength and frequency is varied[5].

The oldest of these phenomena which we shall discuss, and also the most studied from both the experimental and theoretical viewpoints is fully developed turbulence. We know from experience that forcefully stirring a fluid with velocity fluctuations v_0 at a large 'stirring' length scale l_0 leads to chaotic mixing at smaller scales due to the stretching and folding of material lines and surfaces by the turbulent motion. This motion exhibits scaling behaviour over a region of length scales $l_d \ll l \ll l_0$ between the stirring length scale and some inner length scale l_d. This scaling range exists whether we study velocity fluctuations $v_l \sim <|v(x+l, t)-v(x,t)|> \sim l^{1/3}$, the energy spectrum $E(k) \sim k^{-5/3}$, or the diffusion of passive scalars in the velocity field $<R(t)^2> \sim t^3$. By increasing the stirring it is possible to control the range of this the inertial regime (so called because of the large Reynolds numbers $Re = v_0 l_0/v$ involved which imply that the viscous term can be neglected compared to the non-linear inertial term in the Navier-Stokes equation), and in the atmosphere often a range of $\log_{10}(l_0/l_d)\sim 8$ orders of magnitude can be achieved when the stirring is ultimately due to the sun.

In 1941 Kolmogorov[6] introduced a very powerful scaling argument to account for the observed exponents. The argument has been by far the most successful theoretical approach in describing the host of experimentally measurable exponents as well as the magnitude of fluctuations observed, with the caveat that the argument must represent some type of spatial mean -field average, as it known that clear deviations from
Kolmogorov scaling appear when high moments of the velocity differences $<|v(x+r,t)-v(x,t)|^q>$ are measured[7], and these are now known to be due to the multifractal character[8]

of turbulent flow. Kolmogorov's argument is essentially based on dimensional analysis using physically relevant parameters, but it is possible to reinterpret his arguments in terms of the strength of an effective time dependent noise driving the small scale motion and use it to rederive Kolmogorov's results. We shall show this in section II.

The point about such an identification is that it can very easily be generalized to numerous other forced partial differential equations[9] as we show in Section III — provided the the magnitude and spatial dependence of the effective force can be measured. We first consider two models for which our scaling arguments agree explicitly with Dynamic Renormalization Group calculations : i) Interfacial growth with a consevation law[10], and ii) Self Organized Criticality[11]. Then we consider the analogous case of interfacial growth without a conserved order parameter governed by the Kardar-Parisi-Zhang equation[12] which is closely related to the Burger's equation[13] and show that while direct application of our scaling arguments do not appear to apply to this case, it is possible to derive the Kim-Kosterlitz (KK) exponents[14] for this model provided the effective noise is renormalized to that expected for a wrinkled surface.

Finally in section IV we discuss the structure of our scaling scheme which can be regarded both in terms of simplicity of implementation, and use of relevant length scales (though not in terms of the existence of a free energy functional that has to be minimised) as a non-equilibrium extension of Flory theory.

SCALING IN FLUID TURBULENCE

The equations governing fully developed isotropic turbulence in an incompressible fluid are the forced dissipative Navier-Stokes equations

$$\partial v/\partial t + v.\nabla v = -1/\rho \; \nabla p + v\nabla^2 v + f \qquad (1)$$

complemented by the null divergence condition $\nabla.v = 0$. The fluid is forced randomly and isotropically on macroscopic length scales l_0 — the stirring length scale — with an energy input per unit mass per unit time ε leading to typical velocity fluctuations $v_0\sim(\varepsilon l_0)^{1/3}$ on the stirring length scale and therefore

$$<f(r,t).f(r',t')> = \varepsilon \; Q(|r-r'|/l_0, (t-t')/t_0) \qquad (2)$$

where the stirring time scale $t_0 \sim l_0/v_0 \sim \varepsilon^{-1/3}l_0^{2/3}$.

On the stirring length scale the Reynolds number $Re = v_0 l_0/v >> 1$ and the viscous dissipative term can be neglected when compared with the inertial term in the Navier-Stokes equation, until an inner length scale $l_d\approx(v^3/\varepsilon)^{1/4}$ is reached where $Re_d = v_d l_d/v \approx 1$ and viscosity becomes relevant. It is worth noting that $(l_0/l_d)\sim Re^{3/4}$, and therefore the range of scales over which viscosity can be neglected — the inertial range — can be made arbitrarily large by increasing the magnitude of the stirring. The inertial range $l_d \ll l \ll l_0$ can can be considered as consisting of a statistically stationary energy cascade whereby the energy input at the stirring length scale is passed to smaller scales through the mode coupling caused the non-linear inertial term $v \cdot \nabla v$ until the viscous term becomes large enough at l_d for the kinetic energy in the velocity fluctuations to dissipate into heat. Kolmogorov pointed out[6] that this cascade process would imply that all information about the stirring length scale would soon be lost and as viscosity was insignificant all correlation or structure functions could only depend on the local length scale l, and ε, which using simple dimensional analysis fixes both the magnitude and scaling exponents for any variable of interest, for example $<(v(r+l) - v(r))2> \sim \varepsilon^{2/3} l^{2/3}$.

From Kolmogorov's argument we can estimate the magnitude of the fluctuations over a length scale l of each term in the Navier-Stokes equation $<| \partial v/\partial t |>_l \sim v_l/t_l \sim \varepsilon^{2/3} l^{-1/3}$, $<| v \cdot \nabla v |>_l \sim v_l^2/l \sim \varepsilon^{2/3} l^{-1/3}$, $<| 1/\rho \, \nabla p |>_l \sim 1/\rho \, p_l/l \sim \varepsilon^{2/3} l^{-1/3}$, $<| v \nabla^2 v |>_l \sim 0$. If we examine these expressions two points stand out. First, we see that each term is of the same order of magnitude or negligible. This is a consequence of the existence of scale invariance —if this were not the case scaling solutions could not exist. But second, Kolmogorov goes further and gives us explicit estimates for both the magnitude of these fluctuations. Is it possible to extract a general method for estimating their size which goes beyond the case of fully developed turbulence and yet retains the simplicity of the original arguments?

We note that each of the terms estimated above has the dimensions of an effective forcing term of magnitude $\eta_l \sim \varepsilon^{2/3} l^{-1/3}$. This term can be rewritten $\eta_l \sim \sqrt{(\varepsilon/t_l)}$, and in this form the forcing term is very suggestive, because it is just the estimate one would make for the size of the noise flucuations with correlation

$$<\eta(r,t) \cdot \eta(r',t')> = \varepsilon \, \delta(t-t') \qquad (3)$$

averaged over a time scale t_l. Note there is no spatial delta correlation in Eq.(3), and we interpret this as a consequence of the fact that the stirring occurs on scales $l_0 \gg l$ that we are interested in. Thus if we equate this estimate for the forcing term on length scales l to each of the non-zero estimates for localized fluctuations of the various terms in the Navier-Stokes equation, identical results to the Kolmogorov ansatz will be achieved by construction. It appears that one interpretation of the existence of Kolmogorov scaling is that at small scales $l \ll l_0$ the inertial non-linearity effectively renormalizes the forcing on the stirring length scale into a white noise with correlation function given by Eq.(3). This noise is then continually creating localized fluctuations of the velocity by driving the Navier-Stokes equation.

This appearance of white noise in chaotic non-linear dynamical systems is ubiquitous. For example, the Henon map $x_{n+1} = y_n + 1.0 - a x_n^2$, $y_{n+1} = b x_n$ can be regarded as the one of the simplest dynamical systems related to turbulent behaviour[15] in the sense that it was created to mimic a Poincare map with constant negative Jacobian -b of a Lorenz-type set of differential equations with constant negative divergence; while the Lorenz equations[16] in turn were designed to represent simplified forced dissipative hydrodynamic flow in the weather system. And the important point here is that even in this simple system an effective internally generated noise appears which results in a smooth ergodic measure in regions of parameter space: Thus apart from the well known strange attractor with the longitudinal structure of a simple curve and the transverse structure of a Cantor set which appears as the asymptotic attractor above $a \approx 1.06$ for b=.3, due to the stretching and folding induced by the map, below $a \approx 1.06$, a sequence of period-doubling bifurcations can be observed, which lead to chaotic bands with a smooth clustered measure in phase space. This measure can be regarded as due to the internal generation of white noise by the dynamical system[17].

There is nothing unique in this viewpoint as regards the Navier-Stokes equations and this therefore this "effective noise" scaling ansatz should be extendedable to any forced dissipative partial differential equation in several directions: (i) To the dimensional case; (ii) To forcing noises of different types; (iii) To scaling of spatiotemporal patterns and fluctuations in open dissipative systems obeying different symmetries and conservation laws from the Navier-Stokes equation. Thus the program is clear: 1) If scaling solutions are to exist then each term in the partial differential equation when coarse-grained over length scale l must be of the same order of magnitude or negligible; 2) The fluctuations of the effective forcing will control scaling in the dissipative dynamical system — all we need do is find the size of the effective forcing. Let us consider some examples of our approach.

APPLICATIONS TO NON-EQUILIBRIUM GROWTH AND PATTERNS

Scaling in far from equilibrium systems has been observed in both temporal and spatial behaviour, a situation strongly reminiscent of turbulence. For example, scaling has been observed for the height fluctuations $h(r,t)$ in driven surface growth[1,2] and for the fluctuations in flowing sand piles[3], and a comparison of such fluctuations with the intermittent velocity fluctuations observed in boundary layer turbulence close to a wall[18], will strike one with their similarity. These and related models have been the subject of extensive investigations in recent years and have resulted in the recognition[19] that in several models the width fluctuations $W(L,t) = <|h(r+L,t) - h(r,t)|>$ on interfaces of dimensions L^{d-1} all obey a special form of self-affine dynamic scaling: Initially the fluctuations increase with time $W(L,t) \sim t^\beta$ if $t << t_L$ and then saturate to a length scale dependent final value $W(L,t) \sim L^\alpha$ at long times $t >> t_L$. The time scale $t_L \sim L^z$ also scales with an exponent $z = \alpha/\beta$. This last identity can be derived from the existence of a scaling form. To estimate fluctuations on length scales $l << L$, the system size, we can of course extend these scaling forms, and if we examine a small window of length scale l, we can again assume that at long times $t >> t_l$ the typical magnitude of the fluctuations in the interfacial height scale as $< |h(r+l,t) - h(r,t)|> \sim h_l \sim l^\alpha$ and that these fluctuations last for times of the order $t_l \sim l^z$.

The simplicity of the noise driven scaling arguments examined above suggests immediate application to a variety of such non-equilibrium phenomena, and comparison with renormalization group approaches where they exist.

Surface Growth with Conservation Law

Consider a growing interface which can change its shape by diffusive processes as well as by non-linear growth with velocity λ transverse to the instantaneous surface — but only in such a manner that in the absence of external forcing the total volume of material in the interface is conserved. In order to study this form of interfacial growth Sun, Guo, and Grant[10] (SSG) used the dynamic renormalization group to study the non-linear Langevin equation

$$\partial h/\partial t = -\nabla^2[\nu \nabla^2 h + \lambda/2(\nabla h)^2] + \eta(r,t),$$
$$(4)$$

where $<\eta(r,t)\eta(r',t')> = -2D\nabla^2\delta(r-r')\delta(t-t')$. If we neglect the diffusion term as small at large enough length scales in the for the same reason that the viscous contribution to the Navier-Stokes equation was neglected — namely the diffuse term becomes negligible compared to the non-linear growth term — and equate our estimate for the time variation in the height fluctuations $< |\partial h/\partial t |>_l \sim h_l/t_l$ to our estimate for the non-linear term $\lambda/2 < \nabla^2[(\nabla h)^2]>_l \sim \lambda h_l^2/l^4$ in Eq.(4), we find the identity $\alpha + z = 4$, which is an exponent equality known to be obeyed by conserved surface growth. To actually get the magnitude and form of the exponents we need estimates for the forcing on length scale l and the simple summation of spatiotemporal random variables on this length scale and associated time scale t_l yield the estimate for the noise $\eta_l \sim \sqrt{(D/l^{d+1} t_l)}$. This is the essential core of the physics. Once this estimate for the noise fluctuations have been made, equating with the other non zero terms yields $h_l \sim (D/\lambda)^{1/3} l^{(3-d)/3}$, and consequently $\alpha = (3-d)/3$ with $d_c=3$; and $t_l \sim (D/\lambda^2)^{-1/3} l^{(9+d)/3}$, and consequently $z = (9+d)/3$ in this regime; from scaling $\beta = \alpha/z = (3-d)/(9+d)$. These results are in agreement with SGG. But could this scaling ansatz hold up in a more complex situation? A good ground for testing these ideas is self organized criticality.

Self-Organized Criticality

Self-organized criticality[3,20] is the process whereby a non-equilibrium system self-organizes itself into a marginally stable macroscopic state having no intrinsic length scale and showing self-similar critical fluctuations on all length scales. Such self-organiztion is argued to be robust as the critical point is an attractor of a dynamical system with a large basin of attraction. Typical examples of such systems include avalanches, earthquakes, and the structure of macroscopic terrains. Models for such phenomena include the Bak, Tang, and Weisenfeld discrete sandpile model[3] and the asymptotic configurations adopted by coupled sets of non-linear dissipative oscillators.

The example we shall consider are the the current fluctuations and avalanches in a flowing sandpile. Hwa and Kardar[11] introduced a driven Langevin equation incorporating the symmetries and conservation laws of the original Bak, Tang, and Weisenfeld discrete sandpile model for self-organized criticality, though the Langevin equation itself is derived in the limit of small fluctuations and therefore may not incorporate the full richness of the discrete models. Assume the sand has a macroscopic flow direction and consider small fluctuations $h(\mathbf{r},t)$ transverse to the macroscopic slope. The resulting equation for these spatiotemporal fluctuations

$$\partial h/\partial t = \nu_\parallel \partial_\parallel^2 h + \nu_\perp \nabla_\perp^2 h - \lambda/2\partial_\parallel h^2 + \eta(\mathbf{r},t), \tag{5}$$

is anisotropic due to the existence of a macroscopic flow direction; and the noise due to the added sand grains is taken to be white $< \eta(\mathbf{r},t)\eta(\mathbf{r}',t')> = 2D\delta(\mathbf{r} - \mathbf{r}')\delta(t-t')$.

Again in this model the dynamic exponent z, and the roughening exponent α can be found. But in this case because the problem is both self-affine and anisotropic, there are three length scales rather than two involved in the analysis. It is this aspect of the scaling that represents a stringent test of our scaling approach. Thus if l is a typical scale to be studied along the flow direction, then associated with this parallel length scale is a transverse length scale $l_\perp \sim l^\zeta$ where ζ is the spatial anisotropy exponent, and again we call a fluctuation in the sandpile height on these scales h_l. Using these three length scales the various terms in Eq.(5) can be estimated as $<|\partial h/\partial t|>_l \sim h_l/t_l$, $<|\nu_\parallel \partial_\parallel^2 h|>_l \sim \nu_\parallel h_l/l^2$, $<|\nu_\perp \nabla_\perp^2 h|>_l \sim \nu_\perp h_l/l_\perp^2$, and $<|\lambda/2\partial_\parallel h^2|>_l \sim \lambda h_l^2/l$. We now again require an estimate of the noise fluctuations on length scale l. Estimating the area of a correlated region of the sandpile surface on length scales l to be $S_l \sim l l_\perp^{d-2}$ gives a noise estimate $h_l \sim \sqrt{(D/S_l t_l)} \sim \sqrt{(D/(l l_\perp^{d-2} t_l))}$. Neglecting at large length scales the parallel component of sandpile relaxation through surface tension compared to the parallel non-linear transport term then yields the exponent equalities $\alpha - z = \alpha - 2\zeta = 2\alpha - 1 = -(z+1 + (d-2)\zeta)/2$; which can be reexpressed as $z = 6/(8-d)$, $\alpha = (2-d)/(8-d)$, and $\zeta = 3/(8-d)$ in agreement with the dynamic renormalization approach of Hwa and Kardar. More fully we find $t_l \sim (\nu_\perp^{d-2}/D^2\lambda^4)^{1/(8-d)} l^{6/(8-d)}$, $h_l \sim (D^2\lambda^{d-4}/\nu_\perp^{d-2})^{1/(8-d)} l^{(2-d)/(8-d)}$, while $l_\perp \sim (\nu_\perp^3/D\lambda^2)^{1/(8-d)} l^{3/(8-d)}$.

It appears that this scaling approach is most satisfactory. All the cases we have considered thus far have, however, involved conserved variables. Let us now therefore examine the situation for a typical non-linear fluctuation involving non-conserved variables.

Surface Growth without a Conservation Law

The rough interfaces observed experimentally in such diverse phenomena as flow of a wetting fluid through porous media[1], and the surface of bacterial colonies[21], has prompted many investigators to examine the underlying cause of this apparent universality. Microscopic models of restricted driven growth as well as Langevin equations which may describe the large length scale fluctuations have been introduced. Typical of such equations is the Kardar, Parisi, Zhang[12] (KPZ) equation for the height fluctuations $h(r,t)$ in an interface growing with a velocity λ normal to the interface

$$\partial h/\partial t = \nu \nabla^2 h + \lambda/2(\nabla h)^2 + \eta(r,t), \tag{6}$$

where $<\eta(r,t)\eta(r',t')> = D\delta(r-r')\delta(t-t')$. The most important difference between the KPZ equation, and all the other non-linear equations we have considered so far is that Eq.(6) does not conserve interfacial volume in the absense of forcing.

The results of dynamic RG[22] and numerical integrations[23] are consistent with each other in d=2 giving, $\alpha = 1/2$, and $\beta = 1/3$. There are no exact results in d>2, but on the basis of numerical evidence alone, Kim and Kosterlitz[14] (KK) have suggested $\alpha = 2/(d+2)$ and $\beta = 1/(d+1)$. In d=3 the Kim-Kosterlitz exponents are close to the values obtained by numerical solution of the KPZ equation.

Then, apart from the noise, and averaged over scales l, the various terms in the KPZ equation may be estimated as $<|\partial h/\partial t|>_l \sim h_l/t_l$, $<|\nu\nabla^2 h|>_l \sim \nu h_l/l^2$, and $\lambda/2<(\nabla h)^2>_l \sim \lambda h_l^2/l^2$. Neglecting the surface tension contribution at long length scales and equating the time dependent term to the non-linear growth term $h_l/t_l \sim \lambda h_l^2/l^2$ yields the well known exponent relationship $\alpha + z = 2$. But as before to proceed further we need to estimate the average noise on these length and time scales. For white noise we naturally estimate its mean square fluctuations on length scales l and time scales t_l as $\eta_l \sim \sqrt{(D/S_l t_l)}$ $\sim \sqrt{(D/(l^{d-1} t_l))}$ where S_l is the average surface area of the interface on length scales l. As before this estimate is a simple consequence of adding uncorrelated random variables. We then find that $h_l \sim (D/\lambda)^{1/3} l^{(3-d)/3}$, and consequently $\alpha = (3-d)/3$ with $d_c=3$ exactly the same exponents as for conserved growth; but $t_l \sim D^{-1/3} l^{(3+d)/3}$, and consequently $z = (3+d)/3$. The exponent $\beta = \alpha/z = (3-d)/(3+d)$. The only problem is the disagreement between these exponents and experiment. Why are cases of conserved and non-conserved variables apparently so different, and where has the argument above broken down? Clearly the mode coupling renormalises the effective noise in a manner which lies beyond the scope of our scaling argument, but may be understandable in a full dynamic renormalization group treatment. Is it possible, nevertheless, to argue the direction in which a solution lies? Consider for the moment the case of a mathematically perfect self-affine surface. In this case as l —> 0 we know that $h_l \gg l$, and the proper estimate of the surface area of the highly wrinkled interface would be $S_l \sim (h_l^2+l^2)^{(d-1)/2} \longrightarrow h_l^{d-1}$. As a consequence the noise fluctuations would scale as $\eta_l \sim \sqrt{(D/S_l t_l)} \sim \sqrt{(D/(h_l^{d-1} t_l))}$ and as we shall show in a moment using this estimate for the noise we can derive the Kim-Kosterlitz exponents. It is, however, in the limit l —> ∞ that the Kim-Kosterlitz exponents are known to apply and in this limit $h_l \ll l$. It is still possible to use the argument above for large l provided only that $h_l \gg l$, and derive an intermediate range of length scales, which diverge as the dimensionless constant $R = \lambda^{d-1}D/\nu^d$ diverges, in which the KK exponents are still valid. This dimensionless constant is very reminiscent of the Reynolds number and the KK range derived using these assumptions goes between an inner length scale $l_{in} \sim (\nu^{d+2}/D\lambda^{d+1})$ and an outer length scale $l_{out} \sim (D/l)^{1/d}$. The whole approach shows strong similarities to the inertial range in turbulence which lasts between the inner Kolmogorov length scale l_d and the stirring length scale l_0. Unfortunately this approach does not save the situation because there no sign experimentally of such a range and indeed in most computer simulations $h_l \ll l$ is observed. The only alternative therefore is to simply make the observation that the relevant length scale describing the magnitude of the noise fluctuations in the KPZ equation appears to be h_l and examine the consequences.

Using this ansatz we equate our estimate for the noise fluctuation in a rough interface $\eta_l \sim \sqrt{(D/(h_l^{d-1} t_l))}$ to the inertial term. This then yields $h_l \sim (D/\lambda)^{1/(d+2)} l^{2/(d+2)}$, and therefore $\alpha = 2/(d+2)$ while $t_l \sim (\lambda^{d+1}D)^{-1/(d+2)} l^{2(d+1)/(d+2)}$, and consequently $z =$

2(d+1)/(d+2). We can find the scaling behavior of h_t with time at short times by reexpressing h_l in terms of t_l and assuming scaling is valid, with the result $h_t \sim (Dt)^{1/(d+1)}$, and therefore $\beta = 1/(d+1)$. Thus we have derived theoretically the expressions conjectured by Kim and Kosterlitz α and β on the basis of numerical results, as well as their fluctuation amplitudes provided the ansatz that h_l is the relevant length scale controlling the driving is accepted. Clearly open questions remain, and it would be very useful to have a clear answer to this apparent dichotomy observed between conserved and non-conserved order parameters in far from equilibrium growth.

DISCUSSION

We have tried to present a new approach to the study of the scaling behavior in non-equilibrium systems naturally described in terms of non-linear dissipative Langevin equations in the same spirit that Flory theory can be applied to problems naturally described in terms of the minimization of a free energy. The approaches have a number of similarities, namely: both are simple scaling arguments based on the identification of relevant variables in a problem. In the case of Flory theory this involves identifying the physically competing entropy and energy terms, in our case the systems are dissipative and presumed quiescent in the absense of forcing. The physics in this case involves an estimation of the forcing term with length scale. This forms the essence of each approach, and once these terms have been identified the rest of the calculation can be done on the back of an envelope. The problem with the two methods is also identical—if the physical intuition is incorrect then all else fails, and even if the physical intuition is correct, the complete justification for the argument lies outside the scope of the theory. The advantages are also similar — namely a great deal of physical insight at very little cost. These approaches can be compared with dynamic renormalization group approaches which theoretically are exact or least any desired accuracy can be achieved — but at a great mathematical expense and often only in a limited range of dimensions. A final advantage of our approach lies in the possibility that the fields of fluid mechanics and non-equilibrium statistical physics will find a larger common language which will enhance investigations in both fields.

We have illustrated the method by applying it to a number of systems that have been investigated recently such as surface growth and fluctuations in self-organized criticality. In particular, we showed how the expressions conjectured by Kim and Kosterlitz for the scaling exponents in the KPZ equation may be derived using this approach. In addition to providing a method for determining the scaling exponents of complex non-linear equations, this approach provides insight into the crossover between scaling regimes that can be observed in the microscopic parameter space of different systems. The different scaling regimes manifest themselves in regions where a particular term in the equation becomes relevant. The type of arguments used in this approach are quite similar to those used in Flory theory for equilibrium systems. Therefore, due to lack of standard methods for studying non-equilibrium phenomena, this approach will be useful in the study of a wide variety of related problems.

For instance these scaling arguments are not only useful to estimate the magnitude, and exponents of correlation functions — a task that is of equal ease in any dimension — but they can be used to study other consequences of these fluctuations. Among the most ubiquitous and puzzling is the apparent spatial dependence of the measured transport coefficients. This is not only a phenomenon associated with flows in the parameter space of complex and often physically inpenetrable dynamic renormalization group equations, but can appear under quite simple guises in fluid mechanics. Consider, for example the effective turbulent kinematic viscosity $v_{turb}(y)$, at a distance y from the wall in a turbulent boundary layer. A constant momentum flux τ can be assumed to be transferred from the distant fast flowing fluid to the wall, and typical velocity fluctuations of size $\sqrt{(\tau/\rho)}$ can therefore be expected in the turbulent boundary layer. The effective turbulent viscosity can in consequence be expected to scale with distance y from the wall as $v_{turb}(y) \sim \sqrt{(\tau/\rho)}y$ provided molecular viscosity can be neglected — a clear spatial dependence has appeared in the transport coefficient.

Very similarly the spatial dependence of transport coefficients in non-equilibrium phenomena such as growth can be easily derived from a direct consideration of how these exponents are measured in simulations, and can be related to the bare coefficients of the underlying partial differential equation. Thus if we wished to measure spatial of the effective surface tension in the KPZ equation we would expect that $\nu_l \sim <|\partial h/\partial t|>_l /<|\nabla^2 h|>_l \sim l^2/t_l \sim C_\nu l^{2-z} \sim (\lambda D^{d+1})^{1/(d+2)} l^{2/(d+2)}$. Thus it can be seen that the same physical quantity can be expressed either in terms of bare microscopic coefficients, or in terms of renormalized spatially dependent coefficients, and both simply represent two different descriptions of the same physical phenomena, they both contain the same information.

We have seen, however, when we apply this program that that a fundamental difference appears in our scaling approach when dealing with conserved and non-conserved order parameters which lies outside the provenance of our approach, but is surely explicable in a full dynamical renormalization group treatment of the noise fluctuations, namely: For conserved variables the unrenormalized length scales can be found self-consistently from the magnitude of neglected terms compared with those kept; while for non-conserved variables the nonlinearity dynamically renormalizes the relevant length scale controlling the asymptotic scaling and this length scale has to be used to find the asymptotic exponents.

Finally it should be noted that both Kolmogorov scaling and our arguments represent non-equilibrium mean-field type arguments and not exact exponents except in special cases. For example it is known that Kolmogorov scaling is an approximation which assumes that fully developed turbulence is space filling. But the velocity and dissipation fluctuations in the case of intermittent turbulence actually have multifractal measures associated with their spatial distribution and it would be interesting to consider how such variations can be accounted for by the forcing.

REFERENCES

1. V.K. Horváth, F. Family and T. Vicsek, Dynamic Scaling of the Interface in Two-Phase Viscous Flows in Porous Media, *J.Phys. A:Math.Gen* 24:L25 (1991).
2. F. Family and T. Vicsek, "Dynamics of Fractal Surfaces," World Scientific, Singapore (1991).
3 P. Bak, C.Tang, and K. Weisenfeld, Self-Organized Criticality: An Explanation of 1/f Noise, *Phys. Rev. Lett.* 59:381 (1987).
4. See for instance L.D. Landau and E.M. Lifshitz, "Fluid Mechanics, Course of Theoretical Physics Vol.6," Pergamon Press, Oxford (1979).
5. J.P. Crutchfield and B.A. Huberman, Fluctuations and the Onset of Chaos, *Phys. Lett.* 72A:407 (1980).
6. A. N. Kolmogorov, The Local Structure of Turbulence in Incompressible Viscous Fluid for Very Large Reynolds Numbers, *C.R. (Dokl.) Acad. Sci. URSS* 30:301, 538 (1941).
7. C. Meneveau and K.R. Sreenivasan,, in "The Physics of Chaos and Systems far from equilibrium",M-D Van ed., North-Holland, Amsterdam (1987).
8. H.G.E. Hentschel and I. Procaccia, The Infinite Number of Generalized Dimensions of Fractals and Strange Attractors., *Physica* 8D:435 (1983).
9. H.G.E. Hentschel and F. Family, Scaling in Open Dissipative Systems, *Phys. Rev. Lett.* 66:1982 (1991).
10. T. Sun, H. Guo, and M. Grant,Dynamics of driven interfaces with a conservation law, *Phys. Rev.* A40:6763 (1989).
11. T. Hwa and M. Kardar, Dissipative Transport in Open Systems: An Investigation of Self-Organised Criticality , *Phys. Rev. Lett.* 62:1813 (1989).
12. M. Kardar, G.Parisi, and Y.-C. Zhang, Dynamic Scaling of Growing Interfaces, *Phys. Rev. Lett.* 56:889 (1986).
13 J. M. Burgers, "The Nonlinear Diffusion Equation," Riedel, Boston, (1974).
14. J. M. Kim and J. M. Kosterlitz, Growth in a Restricted Solid-on-Solid Model , *Phys. Rev. Lett.* 62:2289 (1989).
15. M.Henon, A Two-Dimensional Mapping with a Strange Attractor, *Comm. Math. Phys.* 50:69 (1976).

16. E.N.Lorenz, Deterministic Nonperiodic Flow, *J. Atmospheric Sci.* 20:130 (1963).

17. A. Wolf and J. Swift, Universal Power Spectra for the Reverse Bifurcation Sequence,
Phys. Lett. 83A:184 (1981).

18. See for instance M. Van Dyke, "An Album of Fluid Motion," Parabolic Press, Stanford California, (1982).

19. F. Family and T. Vicsek, Scaling of the Active Zone in the Eden Process on Percolation Networks and the Ballistic Deposition Model, *J. Phys.* A 18: L75 (1985).

20. C. Tang and P. Bak, Critical Exponents and Scaling Relations for Self-Organized Critical Phenomena, Phys. Rev. Lett. 60:2347 (1988).

21.T. Vicsek, M. Cserzö and V.K. Horváth, Self-Affine Growth of Bacterial Colonies, *Physica A* 167:315 (1990).

22 E. Medina, T. Hwa, M. Kardar, and Y. Zhang, Burgers equation with correlated noise: Renormalization-group analysis and applications to directed polymers and interface growth, *Phys. Rev.* A39:3053 (1989).

23. J.G. Amar and F. Family, Numerical solution of a continuum equation for interface growth in 2+1 dimensions, *Phys. Rev.* A41:3399 (1990).

DISCRETE POTENTIAL FLOW SIMULATION OF A PREMIXED FLAME FRONT

J.C. Antoranz, A. López-Martín, J. L. Castillo and P.L. García-Ybarra

Departamento de Física Fundamental, U.N.E.D.
Apartado Correos 60.141, 28080-Madrid, Spain

INTRODUCTION

The propagation of a premixed flame front in a gaseous medium provides an example of non-Laplacian growth, contrary to cases as viscous fingering, non-equilibrium solidification, electrochemical deposition, etc., where interfacial pattern formation is controlled by Laplace equation.[1] In the flame front problem Laplace equation for the velocity potential holds only in the upstream fresh gas zone (when the flow is assumed potential at infinity) where the temperature is constant and low enough to keep the chemical reaction frozen. In fact, through the flame front, the temperature rises exponentially, up to the combustion temperature, due to the diffusion of heat released in a narrower reaction zone.[2] Then, inside the thermal flame thickness (typically of 0.1 mm width) gas thermal expansion occurs and the continuity equation, written in terms on a velocity potential, ϕ, takes the form of a Poisson equation with a non-homogeneous term equal to the relative rate of gas volume increase

$$\Delta\phi = -\frac{d\,\ln\rho}{dt} \tag{1}$$

where ρ and t are the gas density and time, respectively. Thus, the gas velocity field induced by the gas expansion results from the addition of a set of distributed sources located along the flame front, whose strength is fixed by the local flux of fresh gas through the flame and by the mass conservation requirement, i.e., by the normal burning velocity of the flame front and by the expansion rate. The flame velocity is a characteristic of the thermodynamical properties of the reactive gas mixture and depends on the local flow configuration: front curvature and stretch, heat loses, etc.[3]

In the downstream burned gas zone the temperature is also homogeneous and equal to the combustion temperature, although the flow is no longer potential. In general, vorticity is generated across the flame front when the gradients of pressure and density are not collinear (only planar flame fronts produce no vorticity). When a fluid element is crossing through the flame thickness a gradient of density is established inside it and the action of the local pressure gradient produces an instantaneous torque that generates an angular velocity in the fluid element. From Euler equation, the vorticity production rate through the flame is given by

$$\frac{d\left(\nabla\times\mathbf{v}\right)}{dt} = -\left(\nabla\rho^{-1}\times\nabla p\right) \tag{2}$$

Kinematically the fluid velocity field can be decomposed in two contributions: a

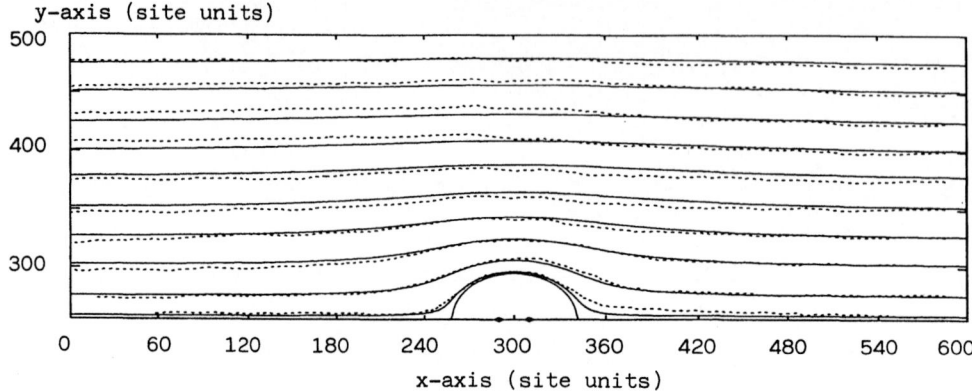

Fig. 1. Potential flow around a cylinder of elliptic cross section. Full lines correspond to the exact streamlines. Dashed lines have been calculated with the automaton in a 1000×600 site lattice by following the trajectories of 6×6 site boxes located initially on the right end wall where one thousand sources have been placed to generate the mainstream. The cylinder is simulated by a source-sink doublet located in the middle of the x-axis.

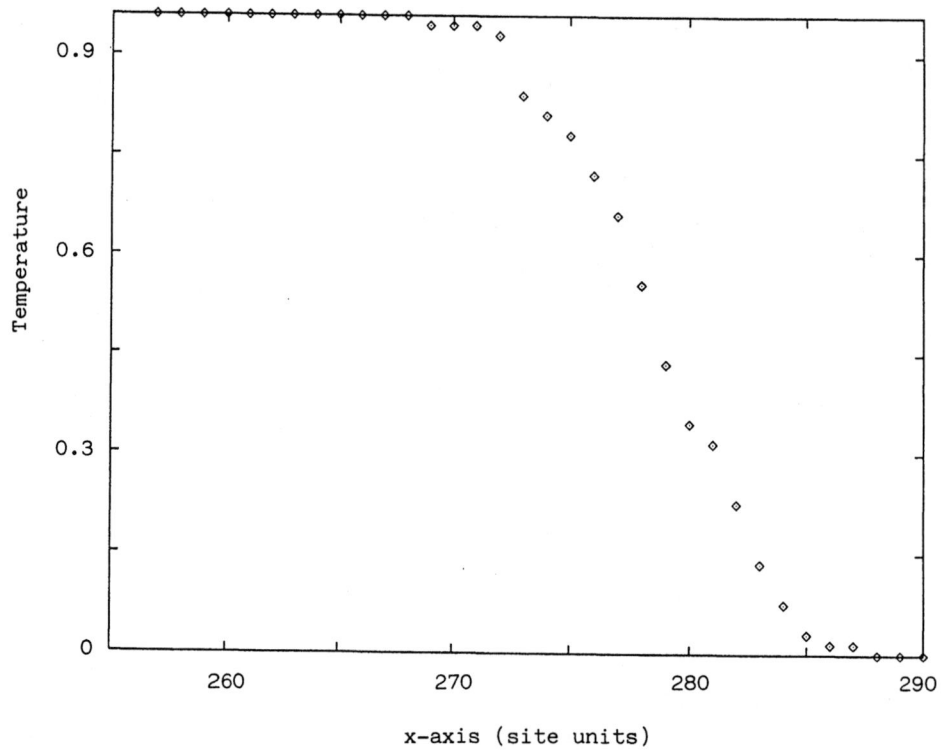

Fig. 2. Typical profile of *temperature* across the front. At each point the temperature has been taken as the relative number of burned (black) particles in a 8×8 site box centered at that point.

potential (rotational free) part, that can incorporate the effects of a density change in the fluid and a solenoidal (divergenceless) field accounting for the rotational contribution. In the present case, the physical phenomena described just above lead naturally to this mathematical result. Nevertheless, it is known that the intrinsic instability of a flame front (the Darrieus-Landau instability[4,5]) is completely due to the potential part of the flow induced by the thermal expansion, the rotational part of the flow playing only a weak stabilizing role.[6] Even more, in the limit of a small thermal expansion, it has been shown that a purely irrotational model leads qualitatively to the correct non-linear evolution equation for the flame front, up to second order in the density change parameter.[7] Thus, instead of dealing with the whole problem, a potential flow approach can be used to get valuable insights concerning the non-linear flame front dynamics with a considerably smaller effort.

In this paper a cellular automaton, previously introduced,[6,8] will be used to simulate a propagating flame front as well as the potential part of the self-generated flow. In fact, the automaton is useful to compute any incompressible potential flow generated by distributions of sources and sinks. The automaton simulates a discrete fluid on a square lattice where each site is occupied by a single fluid particle. No empty sites nor multi-occupancy are allowed. During the simulation, *new* fluid particles emanate from a distribution of site sources, according to their strength. Each new particle takes a site surrounding the source and pushes away the particle formerly placed there. This one pushes the neighbor in the same direction and so on. This site exchange take place up to the automaton boundaries where different boundary conditions can be imposed. Analogously, the action of a sink is to *suck up* neighboring particles at a given rate along some randomly chosen directions. To show the accuracy of this procedure, the two dimensional potential flow around a cylinder of elliptic cross section computed with the automaton is compared with the theoretical calculation in Figure 1.

In the next section the ability of the automaton to simulate the burning flame velocity is discussed. Then, a section is devoted to the front stability properties and pattern selection study. Finally the concluding remarks are collected in the conclusions.

FRONT ADVANCEMENT AND INTERNAL NOISE

The flame front surface is the boundary between burned and fresh gas and advances through the latter with a normal burning velocity dictated by the progress of the chemical reaction. In general, combustion phenomena are controlled by chemical reactions with high global activation energies, in such a way that the reaction takes place only when the local temperature reaches a value close to the combustion temperature. In the automaton output, different colors are assigned to fresh and burned particles (white and black, let say) and a local pseudo-temperature can be defined at each site as the relative number of black particles in a neighborhood of the site. Then, to simulate the chemical reaction, a rule for the color change of a particle *from white to black* is applied, based on the local temperature, that results in an effective front advancement relative to the fresh gas (the normal burning velocity of actual flames). Once such a color change occurs, the increase in particle size simulating the thermal expansion is supposed to take place during the same time step and each new *burned* particle acts like a source of new black particles (with a strength fixed by the expansion rate), as was explained in the introduction.

In actual flames, the balance between the processes of (heat and species) diffusion on the one side and the advection induced by the flame progress on the other side, leads to a steady thickness of the front due to the development of exponential (temperature and concentration) profiles. In the automaton, the discrete nature of the expansion procedure induces random displacements of the particles and provides an inter-mixing between the black and the white ones. According to our definition of temperature, we can exploit this noisy spreading of the front to simulate the heat diffusion and interpret it as the actual thermal flame thickness. This artifact links the thermal expansion to the heat diffusivity whose effective value is determined a posteriori once the expansion rate is fixed. At present, we can not yet conclude whether this particle dispersion corresponds to some stationary diffusivity or the front thickness growths in time with a power law. A typical front temperature profile is shown in Figure 2.

The procedure described in this section, when implemented on a computer, leads to front evolution pictures like those represented in the Figures 3 and 4. As expected, planar fronts exhibit Darrieus-Landau instability whose peculiarities are analyzed in the next section.

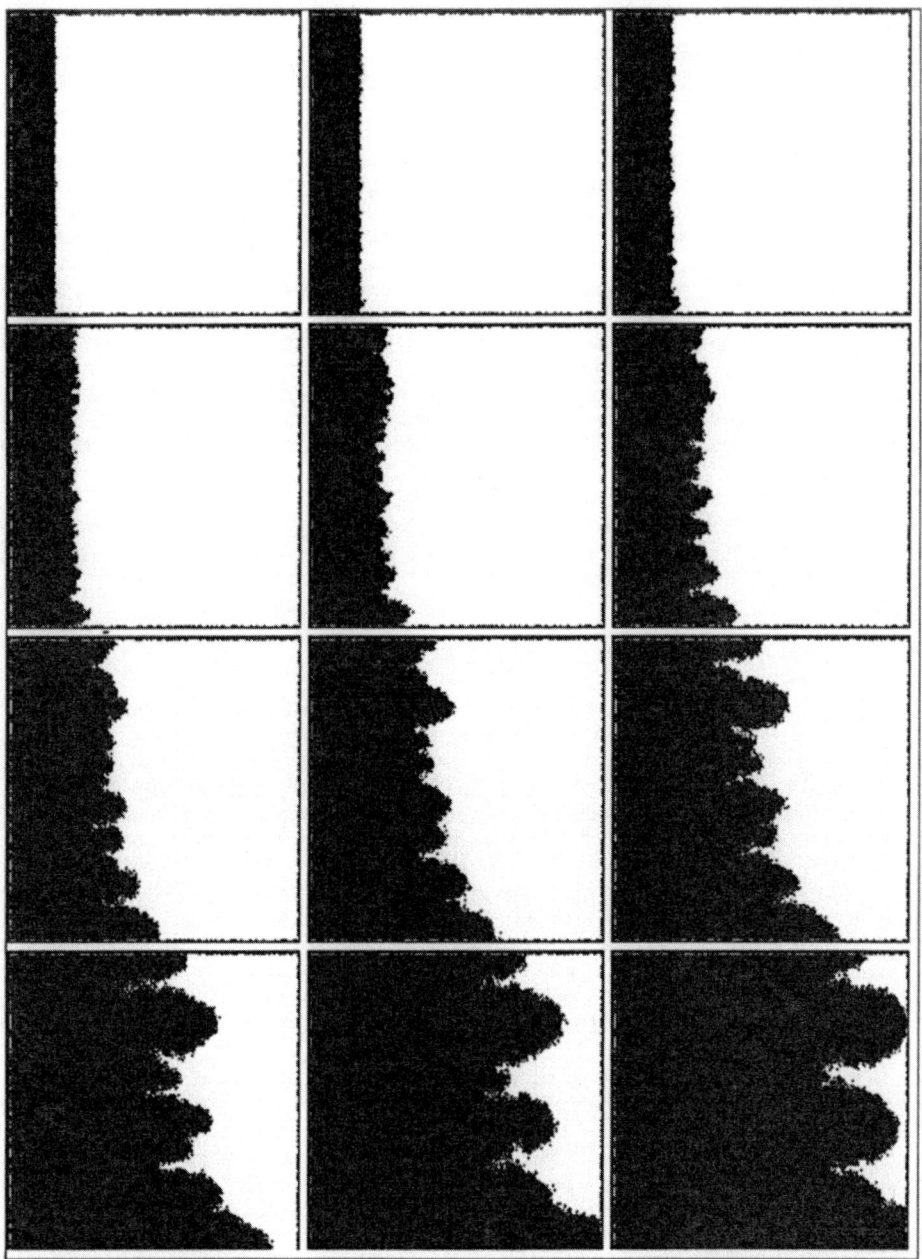

Fig. 3. Time evolution of an initially planar front in a lattice of 800×800 sites with reflecting boundary conditions on the upper and lower walls and open right and left walls. Time runs from left to right and then from top to bottom, starting at t=2 time-steps and with successive increments of 4 time-steps.

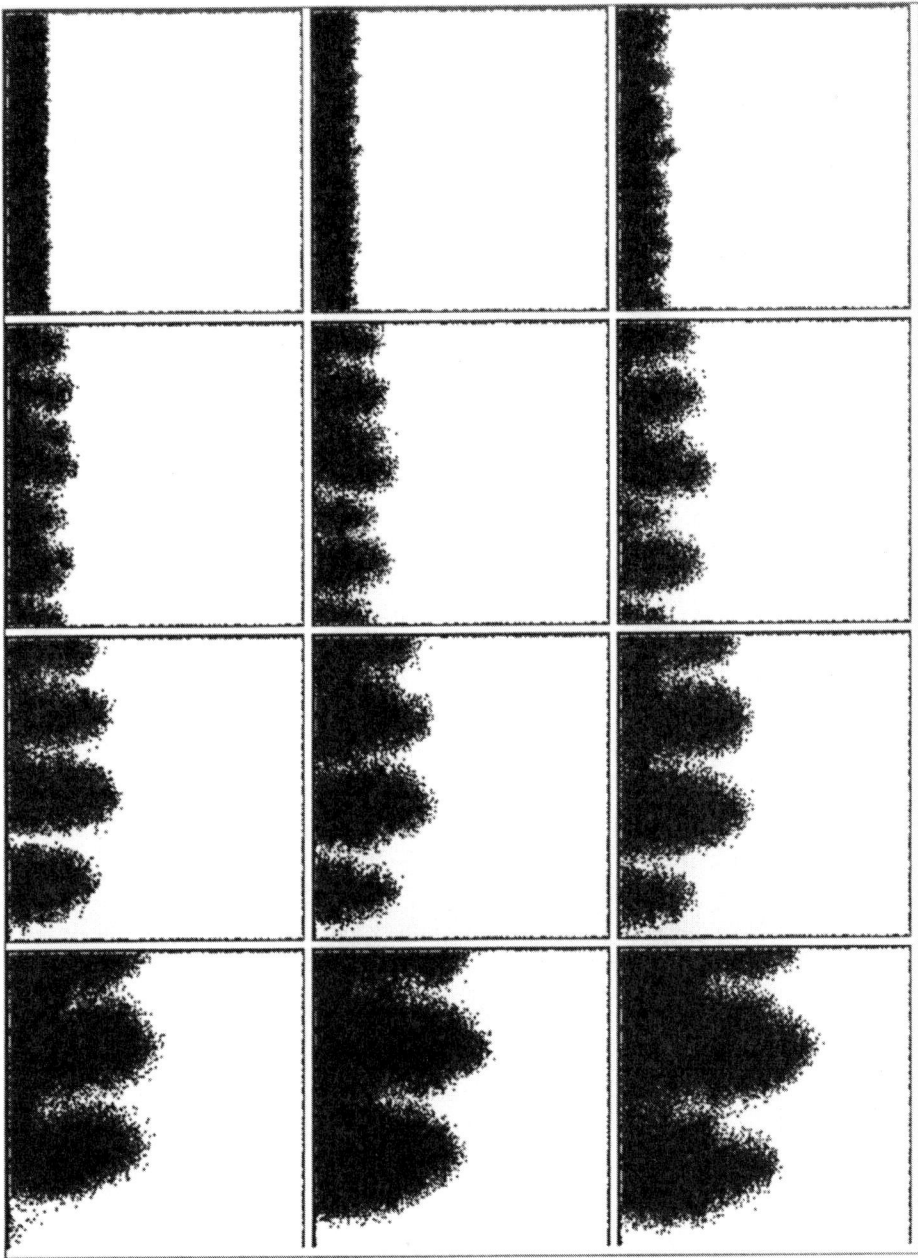

Fig. 4. Time evolution of an initially planar front in a lattice of 800×800 sites with reflecting boundary conditions on the walls except in the left wall which is assumed open. Time runs from left to right and then from top to bottom, starting at t=2 time-steps and with successive increments of 4 time-steps.

During the linear stage of the instability development, the amplitude A_k of each wave number k, in the Fourier transform of the distorted front position, can be assumed to growth exponentially as $\exp(\sigma t)$. Then, in the limit of an infinitely thin front and assuming potential flow everywhere, the linear dispersion relation provides the growing rate $\sigma(k)$ as a linearly increasing function of the front corrugation wavenumber k,[6]

$$\sigma = \frac{1-\rho}{2\rho} u\, k$$

(3)

where u is the flame burning velocity, with respect to the fresh gas, and $\rho \equiv \rho_2/\rho_1$, $(0 < \rho \leq 1)$, with ρ_1 and ρ_2 being the unburnt and burnt gas densities, respectively. When the effects of finite front thickness are incorporated,[9-11] the diffusive and convective effects inside the flame width provide the necessary stabilizing cutoff for the large wavenumbers, through a term mainly proportional to $-k^2$. This result indicates that planar fronts (of infinite extent) are always unstable because a non-vanishing band of wavenumbers $0 < k < k^*$ have positive growing rates and will distort the front that will evolve towards non-planar shapes dictated by the non-linear mode interaction.

The time evolution of planar fronts simulated with the automaton shows this kind of strong instability as can be seen in the two cases displayed in the Figures 3 and 4, corresponding to propagation in two-dimensional channels with specular boundary conditions on the lateral walls and open and closed right end-walls, respectively. In both cases, cells of relatively small wavelength starts to develop on the initially planar front and their non-linear interaction leads to the appearance of larger and larger structures. However, the front is thinner and advances, with respect to the lattice, at a higher velocity in the open channel case (Figure 3) than in the closed channel (Figure 4). In the first case, the unburnt (white) fluid particles may escape through the right end-wall as they are pushed by the expansion procedure and the fresh gas moves with a non-vanishing average velocity towards the right. Whereas in the second case, specular conditions have been imposed not only on the lateral walls but also on the right end-wall. Then, a fluid particle arriving on any of these boundaries by the expansion procedure pushes another particle (which was originally allocated on that wall site) in the specular direction and the expansion progresses now along the new defined direction. Thus, when the right end-wall is reached, the expansion effect returns back to the front, crosses the front and continues up to the left end-wall. Therefore, the unburnt fluid is at rest in the average and, moreover, due to the back flow of particles the amount of scattering processes in the front doubles and the front thickens.

Figures 5 and 6 quantify the mode interaction behavior for the discrete modes $k_m = 2\pi m/L$, with L being the channel width. They depict the time evolution of the (logarithmic) power spectral density for the front Fourier transform (the front position was assumed to be the locus of the lattice points where the temperature reached the value 0.9). As can be seen, the instability provides a continuous input of energy into the modes up to a saturation value is achieved after which the mode energy content fluctuates randomly. The smaller the wavenumber the latter the saturation appears. Thus, for long times, the energy is being concentrated in the large wavelengths until, ultimately, a single cell is expected to fill the whole channel width. Longer runs would be needed to arrive to this latter state but this was effectively the behavior observed when the boundary condition on the lateral walls was changed to permit the particles escape freely through the walls.[8] In that case, the final single-cell pattern is quickly obtained after about twenty time steps. It is worthwhile to notice that this long time behavior coincides with recent numerical integrations of the Sivashinsky flame front evolution equation.[12]

CONCLUSIONS

A cellular automaton, running on a square lattice and simulating the action of discrete sources and sinks, has been implemented. The automaton describes accurately simple potential flows with reasonably good space and time resolutions and can be a promising tool to compute unsteady potential flows of any kind.

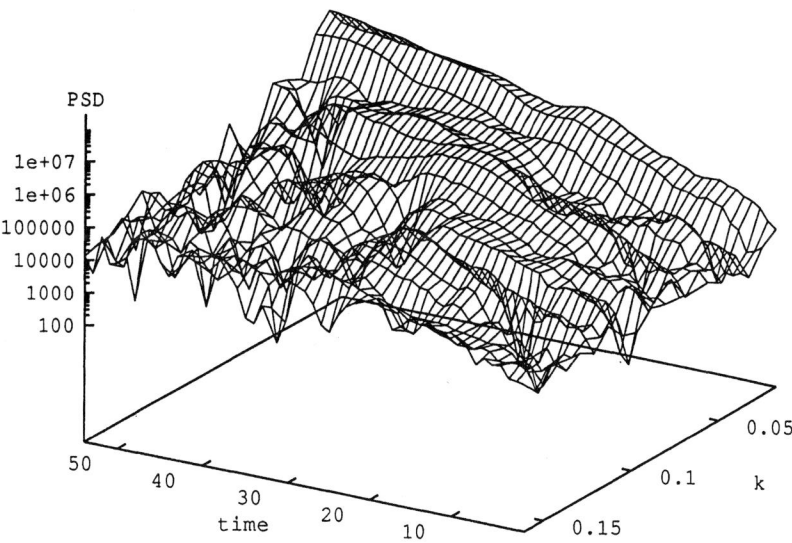

Fig. 5. Time evolution of the power spectrum density (PSD) associated to the Fourier transform of the front propagation depicted in Figure 3. Note the logarithmic scale in the vertical coordinate.

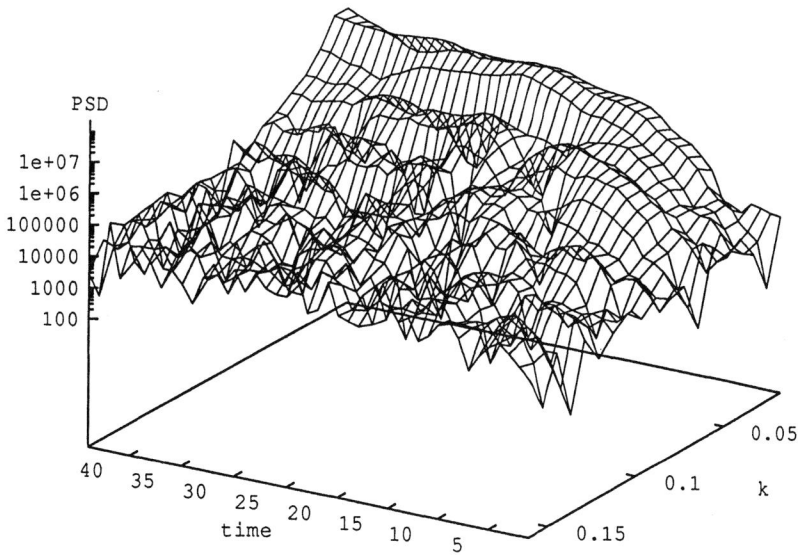

Fig. 6. Time evolution of the power spectrum density (PSD) associated to the Fourier transform of the front propagation depicted in Figure 4. Note the logarithmic scale in the vertical coordinate.

Also, an exothermic chemical reaction can be modelled by incorporating a simple rule for the color change of the fluid particles. The local density changes (due to the thermal expansion) at the sites where the chemical reaction is taking place is simulated by locating there site sources whose strength is determined by the specific volume increase. Thus, the automaton serves to approximately simulate an advancing flame front through a premixed mixture of reactive gases. The generation of vorticity by actual flames cannot be described by this automaton because the flow induced by any distribution of sources is always irrotational. However, the rotational part of the flame self-generated flow is expected to play a minor role in the flame front stability and further long time evolution.

The cellular automaton simulations agree well with the results of Sivashinsky's equation for the qualitative behavior of premixed flames propagating in ducts through quiescent gases. Both approaches predict the formation of a single cell with a wavelength limited by the channel width. The transition from the early stages of planar front corrugation towards the final single-cell pattern *via* the non linear mode interaction has been characterized by the time evolution of the front Fourier transform power spectrum. After the linear stage of exponential growth, the energy content in each mode saturates in a characteristic time which increases continuously with the mode wavelength. In real space, this inverse energy cascade, corresponds to the growth of the small disturbances to larger size structures.

Furthermore, the reconstruction of the linear dispersion relation from the Fourier transform of the front position will permit to have an indirect measurement of the effective Markstein length, scale below which the disturbances are stabilized by the diffusive (heat and mass) transport inside the flame thickness.[3]

ACKNOWLEDGEMENTS

The authors acknowledge financial support by the JOULE program of the Commission of the European Communities under contract JOUE 42-C (EDB) and by Dirección General de Investigación Científica y Técnica (DGICYT, Spanish Ministry of Education and Science) under projects PB88-0159 and CE91-0005.

REFERENCES

1. T. Vicsek, "Fractal Growth Phenomena," World Scientific, Singapore (1989).
2. F. A. Williams, "Combustion Theory," The Benjamin/Cummings, Menlo Park (1985).
3. P. Clavin, Dynamic Behavior of Premixed Flame Fronts in Laminar and Turbulent Flows, *Prog. Energy Combust. Sci.* 11:1 (1985).
4. G. Darrieus, Propagation d'un front de flamme. Essai de théorie des vitesses anormales de déflagration par développement spontané de la turbulence. Presented at *La Technique Moderne* (1938) and at *Le Congrès de Mécanique Appliquée* (1945).
5. L. Landau, On the Theory of Slow Combustion, *Acta Physicochimica* U.S.S.R, Vol. XIX, N°1:77 (1944).
6. P. L. García-Ybarra, J. C. Antoranz and J. L. Castillo, Simulation of Flame Fronts by Sources of Fluid Volume, *in*: "Nonlinear Phenomena Related to Growth and Form," M. Ben Amar, P. Pelcé and P. Tabeling, ed., Plenum, New York (In press).
7. G. I. Sivashinsky and P. Clavin, On the Nonlinear Theory of Hydrodynamic Instability in Flames, *J. Physique* 48:193 (1987).
8. J. C. Antoranz, P. L. García-Ybarra and J. L. Castillo, Premixed Turbulent Flame Simulation by a Cellular Automaton, *in* "Proceedings of the Second International Workshop on Turbulent Premixed Combustion," I. Gökalp, ed. (In press).
9. P. Pelcé and P. Clavin, Influence of Hydrodynamics and Diffusion upon the Stability Limits of Laminar Premixed Flame, *J. Fluid Mech.* 124:219 (1982).
10. P. Clavin and P. García-Ybarra, The Influence of the Temperature Dependence of Diffusivities on the Dynamics of Flame Fronts, *J. Méc. Théor. Appli.* 2:245 (1983).
11. P. L. García-Ybarra, Flame Front Stability with General Intermolecular Interaction Potential, *Prog. Astro. Aero.* 95:115 (1984).
12. S. Gutman and G.I. Sivashinsky, The Cellular Nature of Hydrodynamic Flame Instability, *Physica D* 43:129 (1990).

FRACTAL LANDSCAPES IN PHYSICS AND BIOLOGY

H.E. Stanley[1], S.V. Buldyrev[1], F. Caserta[1], G. Daccord[2], W. Eldred[1],
A. Goldberger[3], R.E. Hausman[1], S. Havlin[1,4], H. Larralde[1], J. Nittmann[1],
C.K. Peng[1], F. Sciortino[1], M. Simons[5], P. Trunfio[1], and G.H. Weiss[4]

[1]Boston University, Boston, MA 02215, USA

[2]Dowell Schlumberger, 42003 St. Etienne, France

[3]Harvard Medical School, Boston, MA 02215, USA

[4]Physical Sciences Laboratory, NIH, Bethesda, MD 20892, USA

[5]Biology Department, MIT, Cambridge, MA 02139, USA

The purpose of this talk is to describe some recent progress in three aspects of statistical physics that overlap biology. The first of these represents a discovery about the real world for which we presently have no theoretical understanding.[1] The second concerns the solution of a rich model of many body physics which should apply to a wide range of biological situations but presently applies primarily to mathematical ecology.[2] The third concerns what on the one hand might be a mere numerical coincidence or, on the other hand, might represent evidence that diffusion limited aggregation describes the physical mechanism underlying the growth of nerve cells—at least the quasi-two-dimensional nerve cells that grow in the retina.[3]

I. LONG-RANGE CORRELATIONS IN NUCLEOTIDE SEQUENCES

DNA nucleotide sequences have been analyzed using models, such as an n-step Markov chain, which incorporate the possibility of *short-range* nucleotide correlations[4]. We discovered in the nucleotide sequence a remarkably *long-range* power law correlation that is significant because it implies a new scale invariant property of DNA. We found such long-range correlations in intron-containing genes and in non-transcribed regulatory DNA sequences, but not in cDNA sequences or intronless genes.

We are using a novel method for studying the stochastic properties of nucleotide sequences by constructing a 1:1 map of the nucleotide sequence onto a walk—which we term a DNA walk; it is this mapping that we use to provide a quantitative measure of the correlation between nucleotides over long distances along the DNA chain. For the conventional one-dimensional random walk model, a walker moves either up $[u(i) = +1]$ or down $[u(i) = -1]$ one unit length for each step i of the walk.[5] The DNA walk is defined by the rule that the walker steps up $[u(i) = +1]$ if a pyrimidine occurs at position a linear distance i along the DNA chain, while the walker steps down $[u(i) = -1]$ if a purine occurs at position i. The question we ask is whether such a walk displays only short-range correlations (as in an n-step Markov chain)

Growth Patterns in Physical Sciences and Biology, Edited
by J. M. Garcia-Ruiz *et al.*, Plenum Press, New York, 1993

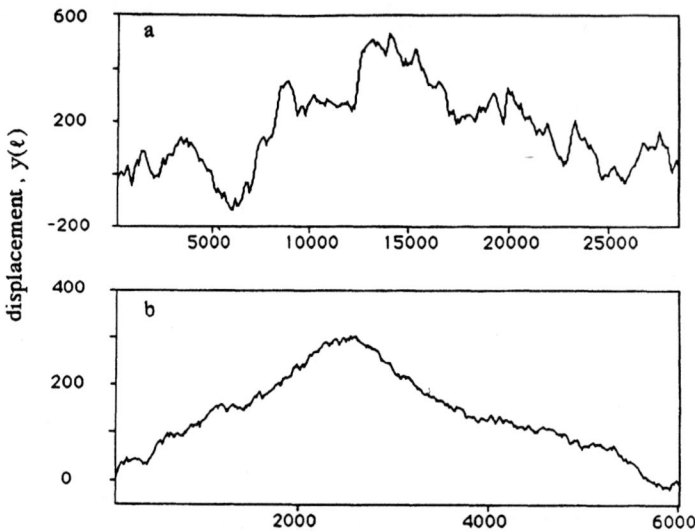

Fig. 1. The DNA walk representations of (a) intron-rich human β-cardiac myosin heavy chain gene sequence, (b) its cDNA. Note the more complex fluctuations observed for the intron-containing gene in (a) as compared with the intron-less sequences (b). The very jagged contour of the DNA walk in (a), characteristic of intron-containing genes, is associated with long-range correlations. After Ref. 1.

or long-range correlations (as in critical phenomena and other scale-free "fractal" phenomena).

This DNA walk provides a novel graphical representation for each gene and permits the degree of correlation in the nucleotide sequence to be directly visualized (Fig. 1). It naturally motivates a quantification of this correlation by calculating the "net displacement" $y(\ell)$ of the walker after ℓ steps, which is the sum of the unit steps $u(i)$ for each step i,

$$y(\ell) \equiv \sum_{i=1}^{\ell} u(i). \tag{1.1}$$

An important statistical quantity characterizing any walk is the root mean square fluctuation $F(\ell)$ about the average of the displacement. For the case of an uncorrelated walk, the direction of each step is independent of the previous steps. For the case of a correlated random walk, the direction of each step depends on the history ("memory") of the walker.

The calculation of $F(\ell)$ can distinguish three possible types of behavior. (i) If the nucleotide sequence were random, then $F(\ell) \sim \ell^{1/2}$ (as expected for a *normal* random walk). (ii) If there were a local correlation extending up to a characteristic range (such as in Markov chains), then *the asymptotic behavior $F(\ell) \sim \ell^{1/2}$ would be unchanged from the purely random case*. (iii) If there is no characteristic length (i.e., if the correlation were "infinite-range"), then the fluctuations will be described by a power law

$$F(\ell) \sim \ell^{\alpha}, \tag{1.2}$$

with $\alpha \neq 1/2$.

The power-law form of Eq. (1.2) implies a self-similar (fractal) property in the DNA walk representation. To visualize this finding, one can magnify a segment of the

DNA walk to see if it resembles (in a statistical sense) the overall pattern. Fig. 2(a) shows the DNA walk representation of a gene and Fig. 2(b) shows a magnification of the central portion. Note the similar fluctuation behavior on the two different length scales shown.

We calculated α from the slope of double logarithmic plots of the mean square fluctuation $F(\ell)$ as a function of the linear distance ℓ along the DNA chain for a broad range of representative genomic and cDNA sequences across the phylogenetic spectrum. In addition, we analyzed other sequences encoding a variety of other proteins as well as regulatory DNA sequences. We discovered that remarkably long-range correlations ($\alpha > 1/2$) are characteristic of intron-containing genes and non-transcribed genomic regulatory elements. In contrast, for cDNA sequences and genes without introns, we find that $\alpha \cong 1/2$ indicating no long-range correlation. Thus, the calculation of $F(\ell)$ for the DNA walk representation provides a new, *quantitative* method to distinguish genes with multiple introns from intron-less genes and cDNAs based

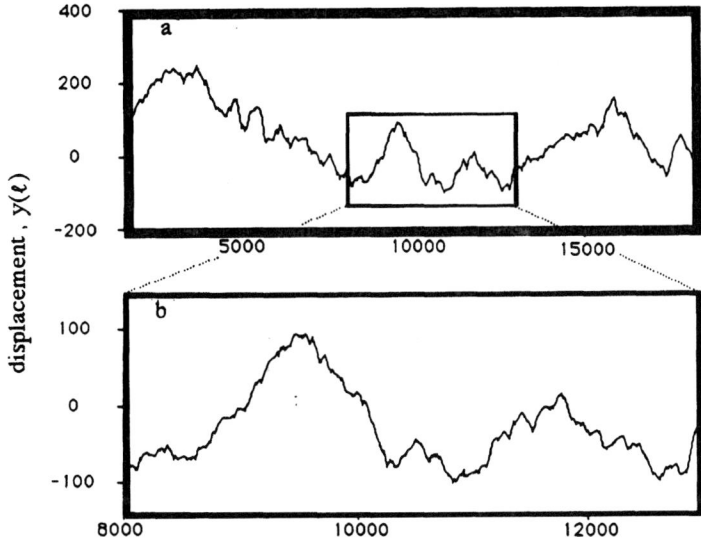

Fig. 2. The DNA walk representation for the rat embryonic skeletal myosin heavy chain gene ($\alpha = 0.63$). (a) The entire sequence. (b) The magnification of the solid box in (a). The statistical self-similarity of these plots is consistent with the existence of a scale-free or fractal phenomenon termed a fractal landscape. After Ref. 1.

solely on their statistical properties. The finding of long-range correlations in intron-containing genes appears to be independent of the particular gene or the encoded protein—it is observed in genes as disparate as myosin heavy chain, beta globin and adenovirus. The functional (and structural) role of introns remains uncertain, and although our discovery does not resolve the "intron-late" vs. "intron-early" controversy about gene evolution,[6] it does reveal intriguing fractal properties of genome organization that need to be accounted for by any such theory.

II. TERRITORY COVERED BY N DIFFUSING PARTICLES

Next, I wish to discuss an application of basic ideas of random walks to characterize the spread of a population. This concerns a hitherto unsolved problem of interest in physics, chemistry, metallurgy and, of course, ecology.

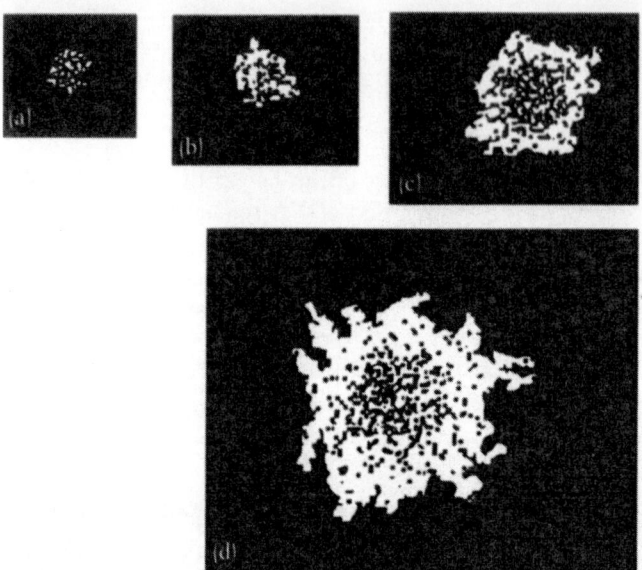

Fig. 3. Snapshots at successive times of the territory covered by N random walkers for the case $N = 1000$ for a sequence of times t_j: (a) $t_1 = 100$, (b) $t_2 = 1000$, (c) $t_3 = 10000$ and (d) $t_4 = 100000$. The roughening in parts (c) and (d) is characteristic of the experimental findings for the diffusive spread of a population.[11] After Ref. 2.

The mean number of distinct sites visited by a single random walker after a time t is a quantity of longstanding interest since it is a direct measure of the territory covered by a diffusing particle. This quantity enters into the description of many phenomena of interest in ecology,[7] metallurgy,[8] chemistry,[9] and physics.[10] Previous analysis[11] has been limited to the mean number of distinct sites visited by a *single* random walker, $\langle S_1(t) \rangle$. The nontrivial generalization to $\langle S_N(t) \rangle$, the mean number of distinct sites visited by N walkers, is particularly relevant to the classic problem in mathematical ecology of defining the territory covered by N members of a given species[7,12] and is also related to the study of the Smoluchowski model for the rate of chemical reactions of the form $A + B \rightarrow B$, taking into account the possibility of a number of B's rather than the single B envisaged by Smoluchowski.[9]

We have recently obtained an analytic solution[2] to the problem of calculating $\langle S_N(t) \rangle$ on a d-dimensional lattice, for $d = 1, 2, 3$. We have confirmed the analytic arguments by Monte Carlo and exact enumeration methods. We also found a remarkable transition in the actual geometry of the set of visited sites. This set initially grows with the shape of a disk with a relatively smooth surface until it reaches certain size, at which the surface becomes increasingly rough (see Fig. 3). This phenomenon may have been observed by Skellam,[12] who plotted contours delineating the advance of the muskrat population and noted that initially the contours were smooth but at later times they become rough (see Fig. 4).

Specifically, we find that the mean number of distinct sites, $\langle S_N(t) \rangle$, passes through several distinct growth regimes in time (Fig. 5). At very short times, we find the simple expression

$$\langle S_N(t) \rangle \sim At^d \qquad [t \ll t_\times], \qquad [\text{Regime I}] \qquad (2.1)$$

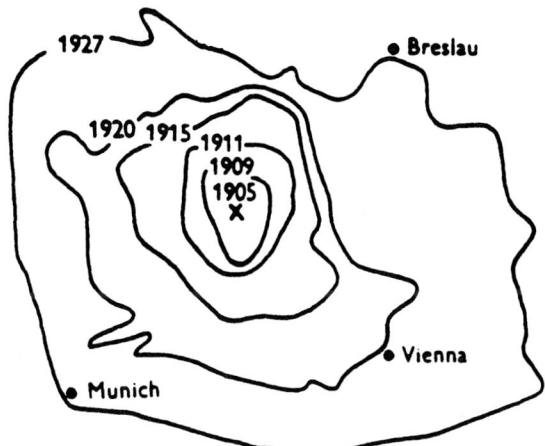

Fig. 4. Contours showing the spread of the muskrat population in Central Europe after the species first appeared in 1905. After Skellam (Ref. 12).

where A depends on the lattice. Eq. (2.1) simply states that every accessible site is occupied by a walker.

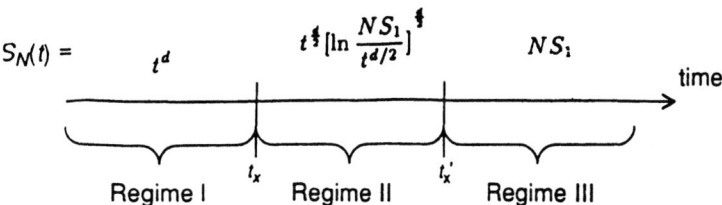

Fig. 5: Schematic illustration of the results for the number of distinct sites visited by N random walkers initially at the origin, indicating the fact that depending on the time t there are quite different behaviors. Here, $\langle S_1 \rangle$ is given by Eq. (2.7).

Regime I holds so long as there are many walkers at every accessible site–i.e., so long as $N P_{\min}(t) \gg 1$, where $P_{\min}$ is the smallest non-zero occupation probability on the lattice at time t. Then $P_{\min}(t) = z^{-t}$, where z is the number of nearest neighbors of a site, so Regime I must terminate at a crossover time $t_\times$ which scales logarithmically with N,

$$t_\times \sim \ln N. \qquad (2.2)$$

To discuss times greater than $t_\times$, we will calculate $\langle S_N(t) \rangle$ using generating function techniques. This analysis leads to a compact scaling expression for $\langle S_N(t) \rangle$.

$$\langle S_N(t) \rangle \sim t^{d/2} f(x) \qquad [t \gg t_\times], \qquad (2.3)$$

where the tilde denotes the fact that (2.3) holds for N and t both large. The scaled variable x is given by

$$x \equiv \begin{cases} N & [d=1] \\ N/\ln t & [d=2] \\ N/\sqrt{t} & [d=3], \end{cases} \qquad (2.4)$$

and the scaling function $f(x)$ by

$$f(x) = \begin{cases} (\ln x)^{d/2} & t_\times \ll t \ll t'_\times \quad \text{[Regime II]} \\ x & t \gg t'_\times. \quad\quad \text{[Regime III]} \end{cases} \tag{2.5}$$

Here the second crossover time $t'_\times$ is

$$t'_\times \sim \begin{cases} \infty & [d=1] \\ e^N & [d=2] \\ N^2 & [d=3]. \end{cases} \tag{2.6}$$

The appearance of Regime III (for $d \geq 2$) can be understood from the following heuristic argument. In Regime II, all but an exponentially small fraction of the walkers are contained within a d-dimensional sphere of radius $\xi \sim t^{1/2}$. Hence $\langle S_N(t) \rangle$ must be bounded from above by the volume of this sphere, $V(t) \sim t^{d/2}$. A second upper bound on $\langle S_N(t) \rangle$ is $N \langle S_1(t) \rangle$, where

$$\langle S_1(t) \rangle \sim \begin{cases} t^{1/2} & [d=1] \\ t/\ln t & [d=2] \\ t & [d=3] \end{cases} \tag{2.7}$$

is the number of distinct sites visited by one random walker. A crossover in $\langle S_N(t) \rangle$ will occur if the system passes from one constraint to the other. For $d = 1$, $V(t) < N \langle S_1(t) \rangle$ for all t, so no crossover occurs—Regime II holds for arbitrarily large t, confirming the result (2.6a) above. For $d = 2, 3$, we find $V(t) < N \langle S_1(t) \rangle$ initially, but for sufficiently large t, $V(t) > N \langle S_1(t) \rangle$. Thus $t'_\times$ is obtained from the condition

$$V(t'_\times) \sim N \langle S_1(t'_\times) \rangle. \tag{2.8}$$

For $d = 2$, (2.7) and (2.8) lead to $t'_\times \sim N t'_\times / \ln t'_\times$, so that $t'_\times \sim e^N$; this confirms the result (2.6b) above. Similarly, for $d = 3$, $(t'_\times)^{3/2} \sim N t'_\times$ implies $t'_\times \sim N^2$, confirming the result (2.6c). One can interpret $t'_\times$ as the time up to which the walkers visit the same places very frequently. For times longer than $t'_\times$, the walkers "almost" do not see each other, and can be treated independently. Thus one would expect that $\langle S_N(t) \rangle \sim N \langle S_1(t) \rangle$ under these conditions.

Following the same kind of reasoning, we can generalize the above argument to any spatial dimension d. The crossover time to the final regime will be given by

$$t'_\times \sim N^{2/(d-2)} \qquad [d > 2]. \tag{2.9}$$

This result is a consequence of the fact that $\langle S_1(t) \rangle \sim t$ for any dimension larger than 2. Equation (2.9) shows the effect of the space dimension on $t'_\times$; it shows that when the dimension increases, the walkers become "independent" at shorter times $t'_\times$.

The remarkable feature is the appearance of Regime II. The behavior in Regime I corresponds to the limit $[N \to \infty, t \text{ fixed}]$; the interface of the set of visited sites is smooth and $\langle S_N(t) \rangle$ is easy to understand ($\langle S_N \rangle \sim t^d$). The behavior in Regime III corresponds to the opposite limit $[t \to \infty \ N \text{ fixed}]$; the interface of the set of visited sites is extremely rough and $\langle S_N \rangle$ is also easy to understand ($\langle S_N \rangle \sim N \langle S_1 \rangle$). In Regime II, the function $\langle S_N \rangle$ takes on a unexpected and nontrivial form. The walkers are largely confined to a sphere of radius $\sqrt{t}$ (in contrast to Regime I, where they populate a sphere of radius t); the interface of the set of visited sites undergoes a progressive roughening, which is readily apparent on visual inspection of the set of visited sites (Fig. 3).

We also carried out numerical calculations for $\langle S_N(t) \rangle$ using both the methods of Monte Carlo and exact enumeration. In particular, we confirmed the scaling form (2.3).

III. PHYSICAL MECHANISM UNDERLYING NEURITE OUTGROWTH

Neurons in the central nervous system, and the retina in particular, have a characteristic morphology—a cell body from which radiate processes (neurites) termed the neuronal arborization.[13] The ability to identify a neuron based on the *qualitative* shape of its arborization has long been recognized. However, meaningful *quantitative* analyses remain elusive. The shape attained by a neuron is thought to result from *environmental* as well as genetic influences.[14] Many local environmental effects, such as growth factors and electrical charge, are known to influence these directional choices.[15,16]

The critical question is how these "local" effects result in the complex branching pattern of a neuron, in contrast to the behavior of, say, a phototropic plant like *Phycomyces* which adopts a branchless structure when grown in the presence of a point source of light.[17] Complex branching patterns in other kinds of growth[18,19] are based on diffusion-limited processes, and are quantitatively described using fractal analysis.[18-21] The DLA model[20] has recently been shown to model both outgrowth and aggregation processes.[21]

We digitized photographs of the neurons with a video camera, using a grid of $2^{16} = 65,536$ pixels. The fractal dimension d_f of the digitized patterns was determined as follows. First we compute the center of gravity and radius of gyration. Then we take as the origin one point within a square centered at the center of gravity and with a side equal to the radius of gyration. Every point on the structure within this square is chosen as a local origin and the cluster mass (number of occupied pixels) within a distance r of this local origin is calculated. Averaging over all possible choices of local origin, excluding the empty sites, we find the averaged cluster mass $M(r)$ scales with r as

$$M(r) \sim r^{d_f}. \tag{3.1}$$

Thus the slope of a double logarithmic plot of $M(r)$ against r gives a quantitative value of d_f. We also used the correlation method to calculate d_f and obtained similar values of d_f. Both the box counting and correlation protocols have been successfully applied to a wide range of fractal objects.[18,19] We have chosen the retina as a model system because it contains many neurons with unique dendritic arborizations that lie primarily in two dimensions (with an aspect ratio of approximately $10 : 1$). This both facilitates analysis (minimizing complicating effects arising from growth in a third dimension) and allows comparisons to culture conditions. Our analysis of retinal neurons *in vivo* in adult animals shows neurons with well developed axons and dendrites.

Figures 6a and 6b show β ganglion cells from the cat retina.[22]. Figures 6c and 6d are the corresponding double logarithmic plots of $M(r)$ against r. From the slopes of the linear portions we estimate $d_f = 1.71$ and 1.69 for Figs. 6c and 6d respectively. Averaging over all the patterns of 11 neurons *in vivo*, we find $d_f = 1.68 \pm 0.15$ using the box counting method and $d_f = 1.66 \pm 0.08$ using the correlation method.

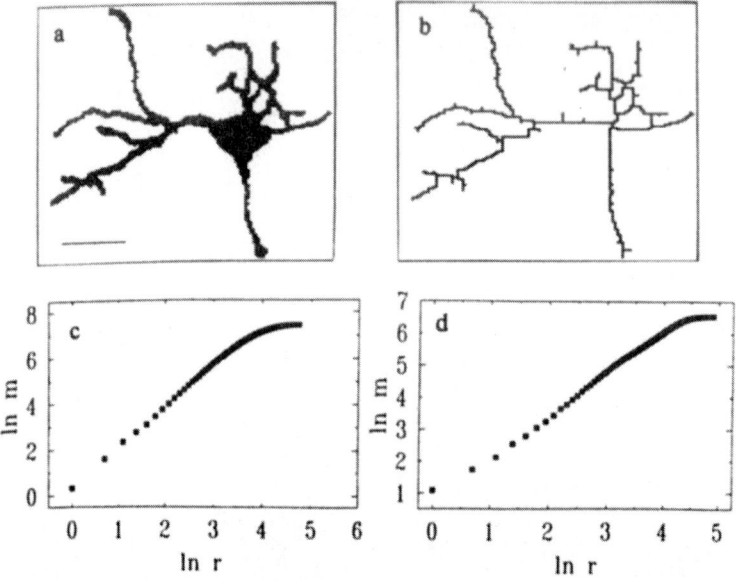

Fig. 6. (a), (b) Digitized images of β ganglion cells in a cat retina (Ref. 22), taken at early and later stages of development, respectively. The scale bar is 10 μm. (c), (d) The corresponding fractal analysis.

This value for the fractal dimension is the same as that found for the diffusion limited aggregation (DLA) model, a model that has been found to describe a vast range of "diffusion-limited" phenomena. The DLA model incorporates two features that might apply to neurite outgrowth at the molecular level: (i) the factors controlling the growth are those inherent in a diffusion equation, and (ii) the growth proceeds by *stochastic* growth rules. Concerning point (i), we note that at least three diffusion-limited physical processes which may result in neurons having fractal shapes are electrical fields, chemical gradients, and viscosity differences. Two of these, electrical fields and chemical gradients, are known to modify the shape of neurons.[15,16]

The key point is that the growth follows rules that faithfully represent the solution of the equations for a diffusion-limited process, *including the presence of stochastic noise*. It has recently become established that the resulting clusters accurately describe a class of growth phenomena in which the diffusion equation or Laplace equation $\nabla^2\phi = 0$ controls the essential physics. Thus growth phenomena governed by chemical gradients (in which case ϕ is the concentration), by electrical gradients (ϕ is an electrical potential), or by a difference in viscosity between the inside and outside of the pattern (ϕ is the pressure) all are believed to be described by the DLA model.[20,21] It is known that growing neurons respond to chemical gradients and electric fields, and it is also known that there is a difference in viscosity between the neuronal cytoplasm and the surrounding intercellular matrix. For this reason, it is not implausible that DLA might represent a zeroth order description of neuron growth.

In summary, many biological phenomena appear to be fractal,[23−24] including for example the chick embryo circulatory system,[25] structure of the bronchial tree,[26−28] and the human retinal circulatory system;[29] the present study is among the first to systematically partly characterize a biological *structure*, the fully developed retinal neuron *in vivo* using fractal mathematics. Moreover, the reproducibility of the numerical values for d_f of fully developed neurons *in vivo* suggests that d_f is *not* a parameter devoid of physical content; indeed, for non-biological objects, accurate measurements of d_f have led to subsequent physical understanding.[30,31]

ACKNOWLEDGEMENTS

We wish to thank M. Araujo, A. L. Barabasi, C. DeLisi, J. Hausdorff, H. J. Herrmann, G. Huber, J. Lee, L. Liebovitch, P.H. Poole, R. D. Rosenberg, M. Schwartz and R. Voss for help at various stages of this work, and AHA, CONACYT, NIH, NSF, ONR and the US-Israel Binational Foundation for support.

REFERENCES

1. C. K. Peng, S. Buldyrev, A. Goldberger, S. Havlin, F. Sciortino, M. Simons, and H. E. Stanley, Nature **355**, xxx (1992); preprint.

2. H. Larralde, P. Trunfio, S. Havlin, H. E. Stanley, and G. H. Weiss, Nature **355**, 423 (1992); Phys. Rev. A **45**, xxx (1992). This work is placed in the context of the great body of literature on random walks in M. F. Shlesinger, "New Paths for Random Walkers" Nature **355**, 396 (1992).

3. F. Caserta, H. E. Stanley, W. Eldred, G. Daccord, R. E. Hausman, and J. Nittmann, Phys. Rev. Lett. **64**, 95-98 (1990); F. Caserta, R. E. Hausman, W. D. Eldred, H. E. Stanley, and C. Kimmel, Neurosci. Letters **135**, xxx (1992).

4. S. Tavaré and B. W. Giddings, in *Mathematical Methods for DNA Sequences*, Eds. M. S. Waterman (CRC Press, Boca Raton, 1989), pp. 117-132.

5. E. W. Montroll and M. F. Shlesinger, "The Wonderful World of Random Walks," in *Nonequilibrium Phenomena II. From Stochastics to Hydrodynamics*, eds. J.L. Lebowitz and E. W. Montroll, pp. 1-121 (North-Holland, Amsterdam, 1984).

6. W. Gilbert, *Nature* **271**, 501 (1978); J. E. Darnell, Jr., *Science* **202**, 1257-1260 (1978); W. F. Doolittle, *Nature* **272**, 581-582 (1978).

7. E. C. Pielou, *An Introduction to Mathematical Ecology* (Wiley-Interscience, NY, 1969); L. Edelstein-Keshet, *Mathematical Models in Biology* (Random House, NY, 1988).

8. R. J. Beeler and J. A. Delaney, *Phys. Rev.A* **130**, 926 (1963); R. J. Beeler, *Phys. Rev. A* **134**, 1396 (1964); H. B. Rosenstock, *Phys. Rev. A* **187**, 1166 (1969).

9. M. v. Smoluchowski, *Z. Phys. Chem.* **29**, 129 (1917); S. A. Rice, *Diffusion-Controlled Reactions* (Elsevier, Amsterdam, 1985).

10. J. W. Haus and K. W. Kehr, *Physics Reports* **150**, 263-416 (1987); S. Havlin and D. Ben-Avraham, *Adv. Phys.* **36**, 695-798 (1987); J.-P. Bouchaud and A. Georges, *Physics Reports* **195**, 127-293 (1990).

11. M. N. Barber and B. W. Ninham, *Random & Restricted Walks* (Gordon & Breach, NY, 1970); H. C. Berg, *Random Walks in Biology* (Princeton University Press, Princeton, 1983).

12. J. G. Skellam, *Biometrika* **38**, 196-218 (1951).

13. S. Ramón y Cajal, *The Structure of the Retina*, Transl. 1972, (Charles Thomas, Springfield, IL).

14. R.O. Lockerbie, Neuroscience **20**, 719-729 (1987).

15. R. W. Gundersen and J. N. Barrett, Science **206**, 1079-1080 (1979).

16. L. F. Jaffe and M.-M. Poo, J. Exp. Zool. **209**, 155-128 (1979).

17. See, e.g., E. P. Fischer and C. L. Lipson, *Thinking About Science, Max Delbrück and the Origins of Molecular Biology* (W. W. Norton & Co., New York, 1988).

18. J. Nittmann, G. Daccord, and H. E. Stanley, Nature **314**, 141-144 (1985); G. Daccord, J. Nittmann, and H. E. Stanley, Phys. Rev. Lett. **56**, 336-339 (1986).

19. G. Daccord, Phys. Rev. Lett. **58**, 479-482 (1987); G. Daccord, and R. Lenormand, Nature **325**, 41-43 (1987); recent insight on chemical dissolution via renormalization group considerations can be found in T. Nagatani, J. Lee, and H. E. Stanley, Phys. Rev. Lett. **66**, 616 (1991); T. Nagatani, J. Lee, and H. E. Stanley, Phys. Rev. A **45**, 2471 (1992).

20. T. A. Witten and L. M. Sander, Phys. Rev. Lett. **47**, 1400 (1981); Phys. Rev. B **27**, 5686 (1983); L. M. Sander, Nature **332**, 789 (1986). Recent work on the dynamics of DLA growth is described in S. Schwarzer, J. Lee, S. Havlin, H. E. Stanley, and P. Meakin, Phys. Rev. A **43**, 1134-1137 (1991) and refs. therein. A "void-channel" model for DLA structure is described in J. Lee, S. Havlin and H. E. Stanley,' Phys. Rev. A **45**, 1035 (1992).

21. For applications of DLA, see J. Feder, *Fractals* (Plenum, New York 1988); H. E. Stanley, N. Ostrowsky (eds.), *Random Fluctuations and Pattern Growth: Experiments and Theory* (Proceedings 1988 Cargèse NATO ASI Series E: Applied Sciences, Vol. 157) Kluwer, Dordrecht 1988; T. Vicsek, *Fractal Growth Phenomena* (World Scientific, Singapore 1989); A. Bunde and S. Havlin (eds.), *Fractals and Disordered Systems* (Springer-Verlag, Berlin, 1991). Color photographs displaying DLA-based phenomena appear in D. Stauffer and H. E. Stanley, *From Newton to Mandelbrot: A Primer in Theoretical Physics* (Springer Verlag, Heidelberg & New York, 1990), and in E. Guyon and H. E. Stanley: *Les Formes Fractales* (Palais de la Decouverte, Paris, 1991) [English Translation: *Fractal Forms* (Elsevier, Amsterdam 1991).

22. J. Maslim, M. Webster and J. Stone, J. Comp. Neurol. **254**, 382-402 (1986).

23. L. Liebovitch, in *Advanced Methods of Physiological Systems Modeling*, Vol. II, ed V. A. Marmarelis (Plenum, NY, in press).

24. P. Meakin, J. Theo. Biol., **118**, 101 (1986).

25. A. A. Tsonis and P. A. Tsonis, Perspectives in Biology and Medicine, **30**, 355 (1987).

26. B. J. West, Bull. Am. Phys. Soc., **34**, 716 (1989).

27. B.J.West and A.L. Goldberger, J. Appl. Physiol., **60**, 189 (1986).

28. B.J. West and A.L. Goldberger, Am. Sci., **75**, 354 (1987).

29. F. Family, B.R. Masters, and D.E. Platt, Physica D **38**, 98 (1989).

30. Applications are described in many recent review articles. See, e.g., P. Meakin in *Phase Transitions and Critical Phenomena* (eds. C. Domb and J. L. Lebowitz), Vol. 12 (Academic, Orlando, 1988).

31. After completing Ref. 3, we learned that Smith et al [Smith, T. G., Marks, W. B., Lange, G. D. , Sheriff Jr., W. H., Neale, E. A., J. Neuroscience Methods **27**, 173-180 (1989)] also measured d_f for unspecified vertebrate central nervous system neurons in culture. Our work is complementary to that of Smith et al in that we suggest that a known growth process (DLA) may in part underlie this value of d_f. Thus our work suggests a possible physical basis, and should stimulate discussion of the underlying biophysical processes.

INTERFACE KINETICS AND OSCILLATORY GROWTH

IN DIRECTIONAL SOLIDIFICATION OF BINARY MIXTURES

B. Caroli*, C. Caroli

Groupe de Physique des Solides, associé au CNRS,
Universités Paris 7 et Paris 6
2, Place Jussieu, 75005 Paris
* Also : Faculté des Sciences Fondamentales et Appliquées,
Université de Picardie, 33, rue Saint Leu, 80039 - Amiens, France

When binary mixtures of species soluble at equilibrium are submitted to directional solidification (by pulling at a constant imposed velocity V in an imposed thermal gradient), the solids thus obtained are often found to exhibit quasi periodic composition modulations. Since we are considering here soluble species[#], these inhomogeneities are necessarily due to the dynamics of growth of the solid from its melt. We can distinguish, among these situations, two main cases :
(a) Composition related striations are, at least roughly, parallel to the growth velocity.
(b) The striations are essentially normal to V, i.e. parallel to the growth front.

Case (a) is typical of metallic alloys and organic plastic materials grown by slow directional solidification (V < a few 100 μ/sec.). The physical mechanism responsible for the striations in this situation is well understood : it is the diffusive Mullins-Sekera[1, 2, 3] instability. Let us recall briefly the corresponding model of solidification.

The liquid-solid first order transition produces latent heat and solute rejection which must be evacuated by <u>diffusion</u> away from the growth front in order for solidification to proceed. Solute diffusion being the slower process is the limiting one. It takes place in the liquid. This transport mechanism is generic - i.e. independent of the physico-chemical specificities of the mixture. In order to get a complete description of the dynamics, one must also specify the attachment kinetics on the front, i.e. describe at least phenomenologically how the atoms or molecules of the liquid get incorporated into the ordered solid lattice.

In the simplest version of the MS model, one assumes that the sticking process is so fast (on the scale of the time for growing one atomic layer of the solid) that there is local thermodynamic equilibrium between the two phases at each point on the front. Then, the temperature at a point r_{fr} is related to the local concentration $C(r_{fr})$ on the liquid side and to the local curvature $K(r_{fr})$ by the Gibbs-Thomson equation :

$$T(r_{fr}) = T_{eq}(C(r_{fr})) - T_0 d_0 K(r_{fr})$$

where T_0 is a reference temperature and d_0 a capillary length proportional to the L/S surface tension.

At very small V, the front is found to be isothermal, i.e. planar. C_{fr} is constant, the emerging solid alloy is homogeneous. Solute rejection induces, ahead of this stationary front, in the liquid, a concentration gradient of range the diffusion length D/V.

[#] - We do not consider the case of eutectic solidification - i.e. of species which have only limited solubility in the solid phase.

If a fluctuation creates at some time a bump on this front, the isoconcentration lines ahead of it get closer, so the diffusion current ($J = - D \nabla C$) increases, so, by mass conservation, the front accelerates in order to increase correspondingly solute rejection : diffusion destabilizes planar fronts.

This effect is counteracted by two stabilizing ones :
- the shorter the wavelength of the deformation, the higher its cost in surface energy : capillarity stabilizes the front against large wavevector distorsions.
- distorting the front means getting some solid (resp. liquid) to move into hotter (resp. colder) regions : the external thermal gradient is stabilizing in particular at long wavelengths.

The interplay of these effects leads to the MS cellular instability[1] : for $V > V_{MS}$ (of order, typically, a few μ/sec for dilute mixtures in gradients of order 100 K/cm), the front becomes unstable against quasi periodic deformations of finite wavelength. The corresponding modulation of concentration along the front is frozen in the emerging solid, leading to striations parallel to the growth velocity. When the velocity increases above V_{MS}, the cells become deeper, then develop dendritic side-branching, but, at a given growth velocity, the primary spacing of these structures remains roughly periodic and stationary[4]

The MS model also predicts that, due to capillarity, the planar front should restabilize at large velocities, when $V > V_m = D/d_o$.

In the isotropic-nematic transition, where V_m is in the 100 μ/sec range, this is indeed observed[5]. In metallic alloys (where D is much larger), restabilization is expected to occur only under rapid solidification conditions, with velocities in the m/sec range[3].

However, what is observed in many Al-based alloys[6] and in Ag-Cu[7] is that there is a velocity range in which the solids obtained are non-homogeneous but <u>with striations parallel to the growth front</u>, called banded structures.

Such "banding" is also observed in a variety of other cases, for example growth of some doped semiconductors from an unstirred melt, and electrodeposition of some metals under well specified electrolytic conditions[8]. There also exists a class of naturally grown rocks, plagioclases, with a composition oscillating in the direction where they have grown[9].

There is a priori not much in common between all these materials nor between their growth conditions but, as we will now discuss, for one element : interface kinetics cannot be neglected when describing how they are grown.

This is clear when comparing the restabilization phenomena described above : the primary difference between the liquid crystal and metallic cases is the order of magnitude of V_m. For $V \simeq 1$m/sec, the time for growing a new solid layer is of order 10^{-10} sec, i.e. becomes comparable with typical times for atomic motion : attachment kinetics can no longer be considered instantaneous.

The processes by which atoms or molecules attach to the solid lattice may be viewed differently depending on whether the S/L interface is facetted or rough on the microscopic scale.

(i) If the surface is a perfect facet, since a step on the facet costs energy, growth of a new layer can occur only when the energy barrier for forming a critical terrace has been overcome. So, kinetics is extremely slow, and these slow processes can be described in terms of step motion (the steps being either terrace edges or preexisting ones, due to the emergence of dislocations)[10].

(ii) Saying that the interface is rough means that, at the melting temperature, it gains more in entropy than it loses energy by creating a lot of defects (steps, terraces, vacancies). These are sites where it is easy for a new unit of the solid to attach, the kinetics is much faster. This is the case, for example, of metals.

(a) Think first of a pure material : the "resistance" to instantaneous attachment, not being due to the topology of the solid surface, is only due to the fact that the attaching elementary unit must "push its liquid neighbors around" so as to worm into the right position and orientation. This can be modelled[11], in analogy with diffusion, as creeping above an activation barrier the top of which corresponds to the "biggest effort". We thus get the picture sketched on Figure 1.

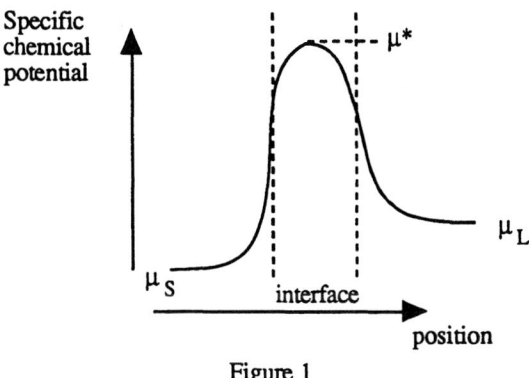

It is immediately clear that, if $\mu_L = \mu_S$, the mass currents in both directions, $J_\rightarrow$ and $J_\leftarrow$, compensate.

$$J_{tot}^{mass} = \rho V_{fr} = 0$$

It takes a finite positive $\delta\mu = \mu_L - \mu_S$ for the solid to grow. The __kinetic law__ has, in this model, the form :

$$\rho V_{fr} = J_{tot}^{mass} = Cst \cdot e^{-(\mu^* - \mu_L)/kT} \left[1 - e^{-\delta\mu/kT} \right]$$

so that, at small enough $\delta\mu$,

$$V_{fr} = \alpha\delta\mu$$

(b) Consider now a binary alloy of the two species A, B. If A-A, A-B and B-B interactions were the same (or negligible), we could repeat the same scheme for each species, getting for example:

$$J_{L\rightarrow S}^{(A)} \propto C_A^{(L)} \exp\left[- \frac{\mu^* - \mu_{LA}(C_A)}{kT} \right]$$

However, A and B interactions are often quite different, e.g. A and B sizes are different, or they are molecules with different shapes... This means that, in the above picture, the activation energy is different for each species, but may also depend on the local concentration of the other component.

It may also happen that A and B, which are independent in the melt, need to associate into specific molecular combinations in the solid.

At this stage, it is clear that specific models must be built, and I will sketch two of them, which we think may have some degree of generality :

1 Catalytic kinetics[12]

We assume that μ^* for the majority species A depends on the local value of C_B, while attachment of B is instantaneous (i.e. B is a consumed catalyst). This model could be relevant to electrodeposition in the presence of what are called in electrochemistry growth inhibitor or activators.

One can then perform a linear analysis of the stability of a planar front growing at constant velocity V against space homogeneous fluctuations. It is found that : if the B "solute" is selectively rejected (resp. incorporated) from the solid, and if B is an activator (resp. inhibitor) of A attachment, beyond a threshold V_c the front undergoes an oscillatory instability at a finite frequency Ω_c. V_c and Ω_c depend on the B concentration in the melt and on its segregation coefficient. Such an oscillation gets frozen into concentration modulations of the solid parallel to the front.

The instability mechanism is easily understood : assume that at some moment a fluctuation increases the front velocity. More B is rejected, since diffusion is not instantaneous, C_B increases on the (L) side of the front, attachment of A becomes more easy, more A is incorporated, which again increases C_B, etc.. That is, such a catalytic kinetics has a destabilizing effect counteracted by the homogeneizing effect of diffusion.

2   Competing molecular kinetics[13]

Such a model was first suggested by Haase et al[9] for the growth of plagioclases, although these authors did not look for the possibility of sustained oscillatory growth.

Plagioclases are feldspaths with a very complicated chemical composition. So we have reduced the initial model to what we think are its essential ingredients, i.e. :
- the liquid is a mixture of (atomic or ionic) species one of which, A, diffuses more slowly than the others (A- limited diffusion).
- these species condense, at the interface, into two molecular species (here complex alumino-silicates) X and Y :

X contains 2 A
Y contains 1 A

- the surface is facetted, so the kinetics is very slow, i. e. $\mu_{LA} \gg \mu_{SA}$, we can neglect the inverse chemical reactions, so the rates of formation of X and Y are :

$$G_X = N_X \left[A_{fr} \right]^2$$

$$G_Y = N_Y \, A_{fr}$$

Conservation of A mass at the interface then reads :

$$\left[D \frac{\partial A}{\partial n} + VA \right]_{fr} = 2 \, G_X + G_Y$$

$$\frac{\text{mass of A which}}{\text{disappears from the liquid}} = \frac{\text{rate of consumption by}}{\text{solidification}}$$

We considerer here free growth from the supercooled liquid.

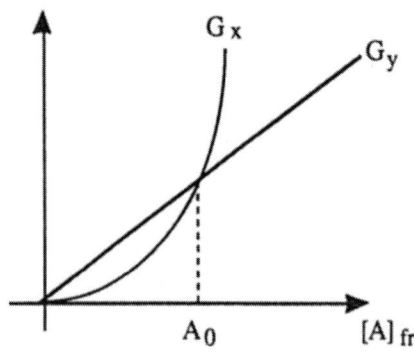

Figure 2

Here again, a Hopf bifurcation from steady to oscillatory growth is found when N_Y/N_X exceeds a threshold value.The instability mechanism is clear : if at some t_0 $A > A_0$, the formation of X is favored. If it is fast enough compared with diffusion, A decreases to below A_0 (see Figure 2), then Y formation takes over, consumption of A is slow, diffusion thus increases A etc...

What we think should be concluded from this sketchy discussion is the following :

- While diffusion gives a wealth of growth patterns which are, at leastunder directional solidification conditions, roughly stationary in their gross features, competing kinetics in multicomponent systems is a source of oscillating growth instabilities. At larger velocities, these modes of growth can evolve into temporal chaos, as is now being studied on the electrodeposition of Zn by Argoul et al[14].

- Understanding more about these phenomena means, in the long run, improving our qualitative understanding of molecular processes at interfaces. An example of such open questions is whether (or when) attachment kinetics can be related systematically to a cluster description of dynamic correlations in the liquid.

We believe that such work should, for the moment, concentrate more on qualitative questions and simple illustrative models than on detailed physico-chemical description of particular cases, so as to encourage the development of more numerous systematic experimental studies of these phenomena.

References

(1) W.W. Mullins, R.F. Sekerka : J. Appl. Phys. 33 : 323 (1963).
(2) J.S. Langer : Rev. Mod. Phys. 52 : 1 (1980) .
(3) B. Caroli, C. Caroli, B. Roulet : "Instabilities of planar solidification fronts" in Solids Far From Equilibrium, C. Godrèche ed., Cambridge University Press (1991).
(4) H. Müller-Krumbhaar, W. Kurz : "Solidification" in Phase Transformations in Materials, P. Haasen ed., VCH Verlag, Weinheim (1991).
(5) J.M. Flesselles, A.J. Simon, A.J. Libchaber : Adv. Phys. 40 : 1 (1991) .
(6) M. Carrard, M. Gremaud, M. Zimmermann, W. Kurz : Acta Metall. Mater. (USA) 39 : 1431 (1991) .
(7) W.J. Boettinger, D. Shechtman, R.J. Schaefer, F.S. Biancaniello : Met. Trans A15 : 55 (1984).
(8) See for example : F.W. Schlitter, G. Eichkorn, J. Fischer : Electrochim. Acta 13 : 2063 (1968)
(9) C.S. Haase, J. Chadam, D. Feinn, P. Ortoleva : Science 209 : 272 (1980) and references therein.
(10) W.K. Burton, N. Cabrera, F.C. Frank : Phil. Trans. Roy. Soc. 243 : 299 (1951).
(11) M.J. Aziz : J. Appl. Phys : 53 : 1158 (1982).
(12) B. Caroli, C. Caroli, B. Roulet : Acta Metall. 34 : 1867 (1986).
(13) B. Caroli, C. Caroli, B. Roulet : J. Physique 44 : 945 (1983).
(14) F. Argoul, J. Huth, P. Merzeau, A. Arneodo, H. Swinney : to be published.

LOCALLY INTERACTING CELL SYSTEMS

AS MODELS FOR CARCINOGENESIS

Petre Tautu

German Cancer Research Centre
Department of Mathematical Models
Im Neuenheimer Feld 280
6900 Heidelberg, FRG

INTRODUCTION

There are two main purposes of this paper. The first is to give some results concerning the growth patterns of a stochastic spatial model for carcinogenesis (Sections 2 and 3); the second is to introduce an urn model 'which should specify the local dynamics assumed in the model (Section 4). The attempt to compare the qualitative behaviour of this model with the behaviour of some related stochastic processes, as for instance the d-dimensional simple random walk, pervades the last three sections with the intention to get suggestions and to gain insight into the complexity of the process of carcinogenesis. A malignant tumour develops itself into a large system of normal cells; one has to derive the macroscopic parameters of this observable tumour from a probability distribution that describes the complicated interactions among the constituent heterogeneous cells. This seems to be one of the decisive ways to fully characterize a growth model. The rest of this introduction is devoted to a brief description of the considered model.

A spatial stochastic model for carcinogenesis has been described (Schürger and Tautu, 1976a) as a multitype infinite cell system with interactions; formally, it is a Hunt process with pregenerator $\mathcal{G}$,

$$\mathcal{G}f(\xi) = \sum_{j=1}^{m-1} a_j \sum_y [f(\xi_y^{(j)}) - f(\xi)] + \frac{1}{2d} \sum_{i=1}^{k} \sum_{i'} d_i d_{ii'} \sum_y \sum_{z \in \mathcal{N}(y)} [f(\xi_{yz}^{(ii')}) - f(\xi)], \qquad (0.1)$$

where

[1] $a_j > 0$ is the rate a cell runs through its cycle phases $j \in (1, \ldots, m-1)$.

[2] $d_i > 0$ is the rate a cell of type $i \in (1, \ldots, k)$ divides into two daughter cells, that is, it moves from phase $(m-1)$ to phase m, the mitosis.

[3] $d_{ii'}$ is the probability that the daughter cells both inherit their 'mother' phenotype i or change it with a certain probability in phenotypes ('colours') $(i+1)$ or $(i-1)$, that is, $i' \in (i-1, i, i+1)$ and $d_{i,i+1} + d_{ii} + d_{i,i-1} = 1$. The probabilities $d_{ii'}$ represent the specific

cellular responses (transformation of cell phenotype to the action of carcinogens and other factors which are concerned in carcinogenesis).

[4] ξ_y is the configuration (state) at site y of the d-dimensional integer lattice Z^d, $d \geq 2$.

[5] $\mathcal{N}(y)$ is the set of nearest neighbours of a cell at site $y \in Z^d$, $|\mathcal{N}(y)| = 2d$, with which this cell interacts.

The notations and definitions above correspond with the assumptions on which the model is based, namely (1) spatial structure with unique occupancy and fixed nearest neighbours (short range interactions), (2) cycling cells which may reversibly change their phenotype chosen from an ordered subset of colours, and (3) permanent multiplication with cell loss. Thus, carcinogenesis is represented as a multivariate stochastic growth process expanding on a regular geometrical structure through local interactions. In this model it is assumed that the interaction has two aspects, i.e. the 'connectivity' one (e.g., intercellular junctional communication) and the 'invasivity' one, because a daughter cell from the neighbourhood $\mathcal{N}(z)$ can with probability $[2d]^{-1}$ occupy site $z \in Z^d$ and the old occupant (if it exists) vanishes (interpretation: it is pushed on a temporary vacant neighbour site or it is 'killed'). Two consequences must particularly be pointed out: (a) the growth process at each lattice site is then a non-Markovian birth-and-death process, and (b) the genealogy of a cell at a fixed site may be broken by invasive local interaction, so that when the population density increases, the sites might be occupied only by the most recent families, the extinction of the initial ones being slowly completed.

It is clear that this stochastic approach fundamentally differs from a non-spatial growth model; it shows, for example, that there exists a critical spatial behaviour for $d = 3$, a theoretical statement which has been confirmed experimentally (Folkman and Greenspan, 1975). Also, the above probability assumptions yields the apparition of some malignant cells at different distant sites, in contradistinction to those models which start with one single tumour cell, like a contagious process.

The finite size version of this model (Schürger and Tautu, 1976b) as well as a discrete time model (Tautu, 1975) were built up in the same breath. By using a coupling technique (see Liggett, 1985, Chapter II), W.J. Braun and R.J. Kulperger (1991) introduced a new model where a linear birth process is coupled with a k-type interaction process.

1. PRELIMINARIES

In order to describe formally the infinite interacting cell system as a continuous time Markov process, some definitions and notations are required. The reader is referred to T.M. Liggett (1985) and to S.N. Ethier and T.G. Kurtz (1986) for all unexplained notations and terminology. Let us consider $M = \{1, \ldots, m\}$, the set of cell cycle phases and $K = \{1, \ldots, k\}$, the set of cell types or 'colours'. Further consider the product $W = \{0\} \cup K \times M$ topologized by the discrete topology and $\Xi = W^{Z^d}$ by the product topology. The topology of W is assumed to have a metric in which W is separable and complete. Elements of Ξ will be denoted by $\xi(= \xi(z), z \in Z^d, d \geq 2)$ and should be thought of as the configuration (or 'microstate') of a (continuous time) infinite system of cells located on the integer lattice.

In terms of probability theory, our system is modelled microscopically by a collection of random variables $\{X(z), z \in Z^d\}$ which are defined on a probability space $(\Xi, \mathcal{F}, P)$, taking values in a space W. In other words, the random variables X are indexed by the sites of the integer lattice; one defines Ξ as the product space W^{Z^d} and the collection $\{X(z), z \in Z^d\}$ to be the coordinate representation process. That is, given a point $\xi = \{\xi(z), z \in Z\}$ in Ξ, one defines $X(z, \xi) = \xi(z)$, the z-th coordinate of ξ. Throughout this paragraph, Ξ will denote a locally compact Hausdorff space with countable base; it is a separable, metrizable space with measurable structure.

For the three-steps construction of an interacting particle system the reader is referred to T.M. Liggett (1985, Chapter I). The Hunt process $\{\xi_t\}_{t \geq 0}$ possesses useful properties such as right continuity of trajectories, quasi-left continuity and strong Markov property (see, e.g., Sharpe, 1988). The existence of the multivariate carcinogenetic system (originally called 'Markov configuration process') as a Hunt process, by using (1.2) below and (0.1), has been proven by K. Schürger and P. Tautu (1976a, Th. 2.2).

Let $f \in C_0(\Xi)$, C_0 being the space of all functions depending on finitely many coordinates, and write $\Delta_z f(\xi) = f(\xi^{(z)}) - f(\xi)$, where $\xi^{(z)}$ is the configuration which is equal to ξ at all sites $z \in Z^d$ except for $z = y$, and $f \in C(\Xi)$ with supremum norm $\| f \| = \sup_\xi |f(\xi)|, \xi \in \Xi$. Consider now the set D of cylinder (local) functions on Ξ, i.e. $f : \Xi \to R$ and $f(\xi)$ depends only on finitely many coordinates of $\xi \in \Xi$. The infinitesimal generator of a semigroup $(T_t)_{t \geq 0}$ when restricted to D is given by

$$\mathcal{A}f(\xi) = \sum_z c(z, \xi) \Delta_z f(\xi), \xi \in \Xi, z \in Z^d, \tag{1.1}$$

where the jump rates $c(z, \xi)$ are positive functions which satisfy the detailed balance equation (which means that when started in equilibrium the process is time reversible). It is assumed that $c(z, \xi)$ satisfies the following conditions:

(i) for every $z \in Z^d$, $c(z, \cdot)$ is a non-negative continuous function on Ξ;

(ii) $\sup_z \sup_\xi c(z, \xi) < \infty$, $z \in Z^d, \xi \in \Xi$;

(iii) $\sup_z \sum_y \sup_\xi |c(z, \xi^{(y)}) - c(z, \xi)| < \infty, y \in Z^d$.

Let $\mathfrak{S}$ be the set of all jump functions which satisfy the above conditions. It is known that for every $c \in \mathfrak{S}$, there exists a unique Markov process whose generator is of the form (1.1) with

$$\mathcal{A}f(\xi) = \sum_z \mathcal{G}_z f(\xi), f \in C_0(\Xi). \tag{1.2}$$

$\mathcal{A}$ is a closed densely-defined operator (Hille-Yosida theorem). If the Hille-Yosida theorem is not applicable, the Markov semigroups can be constructed directly for the class of attractive jump functions (Helms, 1984). The assumption of attractiveness means that the semigroup preserves monotone functions (see Ligett, 1985, Th. 2.2, p. 134).

Let us introduce an order $\geq$ in the state space Ξ which is defined by the componentwise inequality

(*) Two configurations ξ and η satisfy the relation $\xi \geq \eta$ if for every $z \in Z^d$, the inequality $\xi(z) \geq \eta(z)$ holds.

If $\xi \geq \eta$, one defines a relation $\triangleright$ in $\mathfrak{S}$ as follows: Let $c_1, c_2 \in \mathfrak{S}$. If

$$\{\xi(z) + \eta(z)\}[c_1(z, \xi) - c_2(z, \eta)] \leq 0,$$

then $c_1 \triangleright c_2$. It follows that a jump rate is *attractive* if $c \triangleright c$ holds, i.e. for every $z \in Z^d, \eta, \xi \in \Xi$,

$$\{\xi(z) + \eta(z)\}[c(z, \xi) - c(z, \eta)] \leq 0 \tag{1.3}$$

whenever $\eta \geq \xi$ (Higuchi, 1991).

In the sequel it will be assumed that the jump rate is attractive. Actually, the spatial stochastic growth systems have attractive jump rates (Durrett and Griffeath, 1982). According to R. Holley (1985, Th. 0.1), finite range, translation invariant and attractive d-dimensional interacting processes converge to equilibrium either exponentially fast or at a rate at most t^{-d}.

Let us consider the following jump function

$$c(x,\xi) = \begin{cases} \sum_y p(x,y)\xi(y), & \text{if } \xi(x) = 0 \\ \sum_y p(x,y)[1-\xi(y)], & \text{if } \xi(x) = 1 \end{cases}$$

where $p(x,y) \geq 0$ for $x,y \in Z^d$ and $\sum_y p(x,y) = 1$ is the transition probability function of a Markov chain, say, of a simple (irreducible) random walk on Z^d. This is the simplest jump (speed) function of the so-called *basic voter model*. The interpretation of this process is that at each site $x \in Z^d$ an individual is located and that he has to choose one of the two possible colours { white, black }. At exponential times, the individual reassesses his state with rate $c(x)$ by choosing a neighbour site y with probability $p(x,y)$ and then changing his colour to that of the individual at site y. This means that an individual at x samples the colours of his $2d$ neighbours and $\xi(x)$ is the colour at site x. In invasion terms (Clifford and Sudbury, 1973), a site x which belongs to a white population will be invaded by a black one at a rate which is the sum of $p(x,y)$ over those sites y which belong to the black population.

The carcinogenetic cell system has a similar jump function, namely

$$c(x,\xi) = (2d)^{-1} \sum_A \sum_{M \backslash m} \sum_K \sum_{i'} p(x,y; A_{j,ii'})\chi_A(\xi)a_j d_i d_{ii'} , \qquad (1.4)$$

where $A \in Z$ (Z is the collection of finite subsets of Z^d), χ is an indicator function, $y \in \mathcal{N}(x), \xi \in \Xi$. Because of self-transformations and of invasive local interactions, the variation of colours at a site has under certain conditions a quasi-permanent character in the $3D$ space. One might say that the site promotes a phenotypic 'instability'. This dynamic population heterogeneity reveals the illusion of "killing tumour cells" therapy.

The descendants of a cell located at a site x will perform a simple random walk on the lattice, how long no invasive local interaction should stop it at a certain site: the 'collision rule' compels that whenever two random walks attempt to occupy the same site at the same time they merge into one, i.e. they coalesce. If coalescence is thought of as the absorption of one walking cell, this phenomen is analogue to that of reaching a random trapping site. Let β be the probability that a given site is a trapping point; the probability that trapping is not occured up to the n-th step is

$$B_n = [1-\beta]^{R_n - 1} ,$$

if one assumes that the origin is not a trap (Zumofen and Blumen, 1982). R_n is the range of the random walk, that is, the number of distinct sites visited (occupied) by a simple walker, satisfying the inequality

$$R_n \leq \exp\{2(\log n)^{1/2} \log\log\log n\} a.s.$$

for all but finitely many n (Révész, 1989, Th. 1). If R_n has a normal distribution, the averaged probability that trapping occurs after the n-th step is asymptotically

$$\overline{B}_n \approx \int_0^\infty f(R_n)[1-\beta]^{R_n}\,dR_n,$$

where f is the probability density for R_n (Weiss, 1980). It is clear that the knowledge of R_n is sufficient for the exact evaluation of the survival probability of a random walker in the presence of traps. Additionally, one must take notice that quantities like R_n depend on the neighbourhood, i.e. the coordination number $|\mathcal{N}(x)|$, and also on the range of interactions: an increase of this range or of the neighbourhood increases the number of distinct sites visited, and the probability of stepping to an already visited site decreases (see also den Hollander, 1984).

The averaged survival probability of an n-step walk is asymptotically proportional to $\exp\{-\alpha n^{\alpha/(\alpha+2)}\}$, where α is a constant that depends on trap concentration.

Clearly, one deals with a simple d-dimensional (nearest neighbour) random walk $\{S^d, d \geq 1\}$ with transition probability $p(0,x) = (2d)^{-1}$ when $|x| = 1$; the sequence $\{S_n^d, n \geq 0\}$ represents the position at the n-th step of this walk. (The superscript d will be eliminated in the sequel.) Let $p = P\{S_1 \neq 0, S_2 \neq 0, \ldots\}$: the random walk is called transient if $p > 0$ or, equivalently, if $\sum_{n=1}^\infty P\{S_n = 0\}$ converges. Otherwise it is called recurrent. There exists a duality equation (see, e.g., Cox and Griffeath, 1986, Eq. 3.1) which connects the voter model with a system of coalescing random walks: recurrence of the individual walks conditions *clustering* of the corresponding voter model ($d \leq 2$), while transience corresponds to a *stable* voter model which approaches a nontrivial equilibrium if $d \geq 3$ (see, e.g., Cox and Griffeath, 1986, Th. 1). Because of the analytical difficulties in investigating the behaviour of associated (dual) processes (i.e. coalescing random walks and coalescing branching random walks), results about simple random walks or branching random walks (e.g., Durrett, 1979) are taken in as reasonable first approximations. It must, however, be pointed out that between the original process and its dual may exist significant differences: for example, the limiting distribution for the basic voter model is sensitive to the distribution of the initial configuration, whereas for the coalescing random walk it is not (Bramson and Griffeath, 1980a). Suggestions about the behaviour of the multivariate carcinogenesis model are deduced by applying some properties of simple random walks (Tautu, 1978; Röthinger and Tautu, 1990). For instance, $E[N_t] = E[R_{2t}]$, where N_t is the number of occupied sites at time t by a voter (invasion) model and R_t is the range of a continuous time (rate 1) of a simple random walk on Z^d, $d \geq 1$, up to time t.

2. RATE OF CONSOLIDATION IN THE MULTIVARIATE CARCINOGENETIC SYSTEM

This section is devoted to the process of consolidation of a malignant (black) cell subpopulation growing in a two-dimensional space. Consolidation means persistent colour uniformity (i.e. consensus in voter's terms) showing a clustering pattern which appears to be a common pattern realised by different processes developing in homogeneous 2D random media. In order to make more clear the process of consolidation, a new model will be discussed, namely the 'stepping stone' model, which describes the evolution of $k < \infty$ allelic types ('colours') in a geographically structured population. This population is located on the integer lattice Z^2, and at each of its sites a colony of N individuals is appointed. During each generation the population (1) undergoes random mating within each colony, with the new generation of individuals replacing the old, (2) its individuals migrate independently among sites, following an isotropic random walk, and (3) they are subject to possible mutations into alleles which are wholly new to the population; they have no effect on their reproductive or migratory behaviour, but can be distinguished otherwise (Sawyer, 1976). An appropriate 'measure of consolidation' would be the probability $q_n(x,y)$ that two different individuals chosen randomly from colonies x and y at epoch n are of the same colour (Sawyer, 1977). This corresponds to the intuitive interpretation that individuals living nearby tend to be more alike than those living for apart, an idea which is analogous to voters consensus. If the mutation rate $\alpha > 0$ then $q(x, y) < 1$. So, the phenomenon of local differentiation may be interpreted in terms of change in correlation with distance.

If $d \leq 2$, the considered population tends to consolidate into larger and larger 'patches' of individuals of identical types – in the sense that any bounded set of colonies becomes homogeneously the same type for any preassigned number of generations with probability converging to one (Sawyer, 1977). Accordingly, one can measure the size of such a patch by $N(n)$, the total number of individuals in the n-th generation being of the same type as an individual chosen at random in some fixed colony. It has been noticed (Sawyer, 1979) that $E[N(n)]$ shows the increase in size of clusters but not their eventual decay and extinction. Indeed, a theorem given by S. Sawyer (1976, Th. 3.5) states that the descendants of any given gene in an infinite system of colonies eventually become extinct if the migration random walk has the step distribution $Q(0) < 1$. This eventual extinction can show up as a limiting exponential distribution: if $d \geq 2$ and $\alpha > 0 (\approx 10^{-6})$,

$$\lim_{\alpha \downarrow 0} P\{ \frac{N(\infty, \alpha)}{E[N(\infty, \alpha)]} \leq z \} = \int_0^z e^{-u} du, \forall z \geq 0$$

(Sawyer, 1979, Th. 1.2). The reader is referred to M. Bramson and D. Griffeath (1980b) and to M. Bramson, J.T. Cox and D. Griffeath (1986) for corresponding results in the case of a basic voter model.

It appears that in the multivariate carcinogenesis model only the malignant cells manifest a tendency to consolidation. If $d = 2$, the clustering pattern can be shown by numerical experiments (see the colour plates in Röthinger and Tautu, 1990). Small black clusters are formed at different places of Z^2 and have the tendency to confluence.

In the case of the basic voter model with $|K| = 2$, regions of white sites coexist with regions of black sites, both of size $t^{\alpha/2}$ for all $0 < \alpha < 1$. In the case of the k-coloured voter model (stepping stone) the colour densities on a region (block) of side $t^{\alpha/2}$ converge as a process in $\alpha \in [0, 1]$ to a time change of a Wright-Fisher diffusion (Cox and Griffeath, 1987, Th. 3). A similar qualitative behaviour in the case of the carcinogenesis model is purely conjectural. If $d = 3$, it is plausible that small clusters of different colour can coexist (interpenetrating, intertwining). The single available result is a proposition by J.T. Cox and D. Griffeath (1987) which states that if a box of side n centered at x has $S^i(x)$ sites of colour $i \in K$, then this field of block sums normalized by $n^{(\alpha+2)/2}$ converges to a limiting Gaussian field. As it is known (Georgii, 1988, Chapter 13), a Gauss random field can be viewed as a Gibbs measure for a suitable local specification.

3. THE GROWTH PATTERN

It is already known that the family of those interacting cell systems to which the multivariate carcinogenesis model belongs produces nonfractal growth clusters with a rough surface. The nonfractal growth is supported by the well-known Richardson's theorem (1973) which states that for a given class of spatial growth processes on the lattice the set of inhabited sites tends to be round and compact. This is proven for the Eden model, the Williams-Bjerknes model and for the Richardson model, the simplest growth model.

For the infinite cell system of carcinogenesis a conjecture is made in a similar sense (Schürger and Tautu, 1976a, Conj. 3.2): the cell system as a whole grows like a solid blob. Let $\tau_1(x)$, $x \in R^d$, denote the first epoch when x is occupied, and consider the random variable

$\tau_h(z) = h\tau_1(z/h)$, $h > 0$. For all $\xi \in \Xi$ and $z \in R^d$, the limit

$$\lim_{h\downarrow 0} E_\xi[\tau_h(z)] = N(z), \xi \in \Xi \tag{3.1}$$

exists for all $z \in R^d$ and defines a norm equivalent to the Euclidean norm $\|z\|$ of $z \in R^d$.

Furthermore for all $\epsilon > 0$,

$$\lim_{t\to\infty} P_\xi[\{z|N(z) \leq (1-\epsilon)t\} \subset \{z|\tau_1(z) \leq t\} \subset \{z|N(z) \leq (1+\epsilon)t\}] = 1, \xi \in \Xi.$$

Yet, as J.M. Hammersley (1977) pointed out, the set is not circular "in the ordinary sense of the word": the norm depends upon the shape of the distribution of the travel times between adjacent sites such that the circle becomes a diamond when this distribution has little dispersion, but becomes round as the dispersion increases. Indeed, the conditions given by H. Kesten (1986) about the inclusion in a diamond meet the value of $\lambda(F)$, where F is the distribution of the passage times of edge $e, (t_1^d, \ldots, t_{2d}^d)$, and $\lambda(F) = \inf\{z : F(z) > 0\}$: if $\lambda(F) > 0$ and F is not concentrated on $\{\lambda\}$, the nonrandom convex set $B_0 \subseteq R^d$ which represents the asymptotic shape is strictly contained in the "diamond" $\{z \in R^d : |z| = 1/\lambda\}$ (Prop. 6.15). Moreover, for 'sufficiently large' d, the shape is not a Euclidean ball (Coroll. 8.4), a statement which confirms and generalizes the result of D. Dhar (1986) about the shape of Eden clusters.

The role of *local* structures in the realization of compact clusters has been suggested by P. Meakin (1986): in all the models in which the growth probability depends on the local structure in the vicinity of a growth site, compact structures are obtained.

The surface roughness of the carcinogenesis model has been expressed by calculating the 'crinkliness' $C(\xi)$ (Mollison 1972, 1974); the dynamic scaling approach of surface growth (see Family, 1990) was not studied. On the basis of numerical experiments, it has been conjectured (Wolf and Kertész, 1987) that the roughness exponent α might be equal to $1/d$.

Let assign to each $z \in Z^d$ a cube ('box') $B(z)$,

$$B(z) = \{y \in R^d | z^i - 1/2 < y^i \leq z^i + 1/2, 1 \leq i \leq d\}.$$

The site $z \in R^d$ is called occupied if the center z of the (unique) box $B(z)$ to which z belongs is occupied. Let $b(\xi), \xi \in \Xi$, denote the boundary (in the usual topological sense) of the union of all $z \in R^d$ that are occupied in ξ; the length of this boundary (if $d = 2$) will be denoted by $l[b(\xi)]$. The idea behind the use of crinkliness as a measure of irregularity of finite configurations is the comparison of $l[b(\xi)]$ with the length of the boundary of a square array having the same area as ξ, so that

$$C(\xi) = \frac{l[b(\xi)]}{4\sqrt{N(\xi)}}, \tag{3.2}$$

where $N(\xi)$ is the number of sites in the support of ξ, $\mathrm{supp}(\xi) = \{z : z \in Z^d, \xi(z) \neq 0\}$. This implies that $C(\xi) = 1$ iff the set of all $z \in R^d$ that are occupied in ξ is a square, and $C(\xi) \leq 1$ for all $\xi \in \Xi_0$, the set of all configurations ξ having a finite nonvoid support. It has been conjectured (Schürger and Tautu, 1976a, Conj. 3.1) that

$$\lim_{t\to\infty} C(\xi_t) = c_1 \quad \text{almost everywhere } (P_\xi), \tag{3.3}$$

where c_1 is a positive constant depending on a_j, d_i and $d_{ii'}$ but possibly not on ξ.

D.Y. Downham and D.H. Green (1976) showed that the crinkliness of the simulated Williams-Bjerknes model displays great variability for small values of n (number of malignant cells); for n increasing, the crinkliness tends to a limit which depends upon the 'carcinogenetic advantage' κ, a specific parameter of the model. As D. Mollison noticed (1974), the possibility of multiple origin of a tumour would artificially increase C; in this case the introduction of a 'local' crinkliness would be required.

4. THE POLYCHROMATIC SCENERY

This section gives over to investigate the connection between a particular urn scheme and the local interaction process. The lattice sites are considered to be urns and the cells are referred to as balls. Initially one starts with a finite number N of balls which are distributed uniformly among urns in such a way that the set of occupied urns is connected. In each connected urn there is a white ball. A draw is performed as follows: At a given instant, a ball is taken from urn z and its colour $i \in K$ is inspected. There are two possibilities for a drawing with replacement: (a) the ball is returned to the urn, or (b) it is changed with a ball of colour $i' = (i - 1, i + 1)$ and then returned. The choice is made with probability $d_{ii'}$. The next step is the selection of one ball of the colour drawn and its introduction into the nearest neighbour urn chosen with probability $1/4$. Let $Y_n = (Y_{n1}, \ldots, Y_{nk})$ denote the composition of the urn after n consecutive draws, where Y_{ni} is the number of balls of colour i. Yet, the urn process is not Markov; the composition of the urn z at draw $n + 1$ does not depend only on the colour of the last ball of draw n, but also on the colours of the last balls in the four neighbouring urns as well as on the random mechanism which might change the colour of one of the five balls in question.

Clearly, this urn scheme has a mechanism of generating randomness different than the classical models: a peculiar randomized scheme is introduced in which the possible outcomes may change from drawing to drawing. The corresponding algebra becomes complicated (Johnson and Kotz, 1977, p. 204), and only a limiting case independent of the spatial structure would be closer to the specialized but familiar schemes (e.g., an urn model for stepping stone). However, the observation of the sequence $\{Y_n\}_{n \geq 0}$ of coloured balls in one urn will give the occupation time by colour $i \in K$, say, as well as the number of colours not obtained or the fluctuation of colours at a site. One can imagine such an urn having a tubular form so that the replacement drawing order can be strictly preserved. Urn schemes for special compartment models have also been suggested (e.g., Bernard, 1977).

A 'polychromatic stochastic landscape' is thus created (see Kasteleyn, 1985, for $k = 2$); the 'local scenery' σ (den Hollander, 1988) will be defined by the finite set Λ_σ of lattice sites together with the k-colouring random process. The colouring probabilities will be described by a joint probability distribution, namely by a probability measure on the measure space $(C, \mathcal{F}_C)$: C is the set of all possible colourings and $\mathcal{F}_C$ is the σ-field generated by the cylinder set of colourings, i.e. by the local colourings. As P.W. Kasteleyn (1985) mentioned, painting the sites of a lattice destroys its homogeneity, but is is assumed that the system is instead statistically homogeneous in the sense that the colour distribution μ is invariant under lattice translations. The most convenient probability measure would be in this case the translation invariant Gibbs measure.

REFERENCES

[1] Bernard, S.R. (1977). *An urn model study of variability within a compartment*. Bull. Math. Biol. **39**, 463–470.
[2] Bramson, M., Cox, J.T., Griffeath, D. (1986). *Consolidation rates for two interacting systems in the plane*. Probab. Theory Rel. Fields **73**, 613–625.

[3] Bramson, M., Griffeath, D. (1980a). *Clustering and dispersion rates for some interacting particle systems on Z^1.* Ann. Probab. **8**, 183–213.

[4] Bramson, M., Griffeath, D. (1980b). *Asymptotics for interacting particle systems on Z^d.* Z. Wahrscheinlichkeitstheorie verw. Gebiete **53**, 183–196.

[5] Braun, W.J., Kulperger, R.J. (1991). *Coupling results for a family of multi-type interacting particle systems.* Techn. Rep. 91–10, Univ. of Western Ontario.

[6] Clifford, P., Sudbury, A. (1973). *A model of spatial conflict.* Biometrika **60**, 581–588.

[7] Cox, J.T. (1989). *Coalescing random walks and voter model consensus times on the torus in Z^d.* Ann. Probab. **17**, 1333–1366.

[8] Cox, J.T., Griffeath, D. (1986). *Diffusive clustering in the two dimensional voter model.* Ann. Probab. **14**, 347–370.

[9] Cox, J.T., Griffeath, D. (1987). *Recent results for the stepping stone model.* In: Percolation Theory and Ergodic Theory of Infinite Particle Systems (H. Kesten ed.), pp. 73–83. Springer-Verlag: New York.

[10] Den Hollander(1984). *Random walks on lattices with randomly distributed traps. I. The average number of steps until trapping.* J. Statist. Phys. **37**, 331–367.

[11] Den Hollander, W. Th. F. (1988). *Mixing properties for random walk in random scenery.* Ann. Probab. **16**, 1788–1802.

[12] Dhar, D. (1986). *Asymptotic shape of Eden clusters.* In: On Growth and Form. Fractal and Non-Fractal Patterns in Physics (H.E. Stanley, N. Ostrowsky eds), pp 288–292. Dordrecht: M. Nijhoff.

[13] Downham, D.Y., Green D.H. (1976). *Inference for a two-dimensional stochastic growth model.* Biometrika **63**, 551–554.

[14] Durrett, R. (1979). *An infinite particle system with additive interactions.* Adv. Appl. Probab. **11**, 355–383.

[15] Durrett, R., Griffeath, D. (1982). *Contact processes in several dimensions.* Z. Wahrscheinlichkeitstheorie verw. Gebiete **59**, 535–552.

[16] Ethier, S.N., Kurtz, T.G. (1986). *Markov Processes. Characterization and Convergence.* New York: Wiley.

[17] Family, F. (1990). *Dynamic scaling and phase transitions in interface growth.* Physica A **168**, 561–580.

[18] Folkman, J., Greenspan, H.P. (1975). *Influence of geometry on control of cell growth.* Biochim. Biophys. Acta **417**, 211–236.

[19] Georgii, H.-O. (1988). *Gibbs Measures and Phase Transitions.* Berlin: de Gruyter.

[20] Hammersley, J.M. (1977). *Discussion to D. Mollison: Spatial contact models for ecological and epidemic spread.* J. Roy. Statist. Soc. Ser. B. **39**, 319.

[21] Helms, L.L. (1984). *Order properties of attractive spin systems.* Acta Appl. Math. **2**, 379–390.

[22] Higuchi, Y. (1991). *Level set representation for the Gibbs states of the ferromagnetic Ising model.* Probab. Theory Rel. Fields **90**, 203–221.

[23] Holley, R. (1985). *Possible rates of convergence in finite range, attractive spin systems.* In: Particle Systems, Random Media and Large Deviations (R. Durrett ed.) [Contemporary Mathematics 41, pp. 215–234. Providence: Amer. Math. Soc.

[24] Johnson, N.L., Kotz, S. (1977). *Urn Models and Their Application.* New York: Wiley.

[25] Kasteleyn, P.W. (1985). *Random walks through a stochastic landscape.* Bull. Intern. Statist. Inst. 45, 27-I, 1–13.

[26] Kesten, H. (1986). *Aspects of first passage percolation.* Lecture Notes in Math. Vol. 1180 [École d'Été de Probabilités de Saint-Flour XIV-1984], pp. 125–264. Berlin: Springer-Verlag.

[27] Liggett, T.M. (1985). *Interacting Particle Systems.* New York: Springer-Verlag.

[28] Meakin, P. (1986). *Computer simulation of growth and aggregation processes.* In:

On Growth and Form. Fractal and Non-Fractal Patterns in Physics (H.E. Stanley, N. Ostrowsky eds), pp 111–135. Dordrecht: M. Nijhoff.

[29] Mollison, D. (1974). *Percolation processes and tumour growth (Abstract)*. Adv. Appl. Probab. **6**, 233–235.

[30] Mollison, D. (1977). *Spatial contact models for ecological and epidemic spread*. J. Roy. Statist. Soc. Ser. B **39**, 283–313.

[31] Révész, P. (1989). *Simple symmetric random walk in Z^d*. In: Almost Everywhere Convergence (G.A. Edgar, L. Sucheston eds), pp. 369–392. Boston: Academic Press.

[32] Richardson, D. (1973). *Random growth in a tessellation*. Proc. Cambridge Philos. Soc. **74**, 515–528.

[33] Röthinger, B., Tautu, P. (1990). *On the genealogy of large cell populations*. In: Stochastic Modelling in Biology (P. Tautu ed.), pp. 166–235. Singapore: World Scientific.

[34] Sawyer, S. (1976). *Results for the stepping stone model for migration in population genetics*. Ann. Probab. **4**, 699–728.

[35] Sawyer, S. (1977). *Rates of consolidation in a selectively neutral migration model*. Ann. Probab. **5**, 486–493.

[36] Sawyer S. (1979). *A limit theorem for patch sizes in a selectively-neutral migration model*. J. Appl. Probab. **16**, 482–495.

[37] Schürger, K., Tautu, P. (1976a). *A Markov configuration model for carcinogenesis*. In: Mathematical Models in Medicine (J. Berger, W. Bühler, R. Repges, P. Tautu eds) [Lecture Notes in Biomathematics 11], pp. 92–108. Berlin: Springer-Verlag.

[38] Schürger, K., Tautu, P. (1976b) *Markov configuration processes on a lattice*. Rev. Roumaine Math. Pures et Appl. **21**, 233–244.

[39] Sharpe, M. (1988). *General Theory of Markov Processes*. Boston: Academic Press.

[40] Tautu, P. (1975). *A stochastic automaton model for interacting systems*. In: Perspectives in Probability and Statistics (J. Gani ed.), pp. 403–415. Sheffield: Applied Probability Trust.

[41] Tautu, P. (1978). *Blackening a d-dimensional lattice*. Rev. Roumaine Math. Pures et Appl. **23**, 141–152.

[42] Tautu, P. (1988). *On the qualitative behaviour of interacting biological cell systems*. In: Stochastic Processes in Physics and Engineering (S. Albeverio et al eds), pp. 381–402. Dordrecht: Reidel.

[43] Weiss, G.H. (1980). *Asymptotic form for random walk survival probabilities on three-dimensional lattices with traps*. Proc. Natl. Acad. Sci. USA **77**, 4391–4392.

[44] Wolf, D.E., Kertész, J. (1987). *Surface width exponents for three- and four-dimensional Eden growth*. Europhys. Lett. **4**, 651–656.

[45] Zumofen, G., Blumen, A. (1982). *Energy transfer as a random walk on regular lattices. II. Two-dimensional regular lattices*. J. Chem. Phys. **76**. 3713–3731.

IONIC CONCENTRATION AND ELECTRIC FIELD IN FRACTAL

ELECTRODEPOSITION

M. Rosso, V. Fleury, J.-N. Chazalviel, and B. Sapoval

Laboratoire de Physique de la Matière Condensée
Ecole Polytechnique, 91128 Palaiseau, France

INTRODUCTION

The development of fractal geometry, initiated by B. Mandelbrot in the 70's,[1] has substantially improved the description and study of many systems with a complex structure, such as filamentary, porous or rough materials. Among these systems, dendritic or ramified electrodeposited aggregates have received much attention in the past decade.[2-7] The early works in the field were mainly motivated by the close similarity between the aggregates obtained in a variety of experimental conditions and the clusters grown by the Diffusion Limited Aggregation (DLA) model of Witten and Sander.[8]

In relation to the wide interest on DLA, much attention has been paid to *morphological* aspects of filamentary growth in electrodeposition.[2-5] Recent studies have revealed the more subtle and complex electrochemical effects involved in these electrodeposition experiments,[6,7] showing the interplay of diffusion and migration of the *two kinds of ions*, anions and cations, present in the solution. The discovery of additional phenomena, such as oscillatory response of the cell[9] or electroconvective motion[10] in the solution, still makes the whole picture more complicated, and a global understanding of the fractal electrodeposition process is still to be found.

MORPHOLOGICAL ASPECTS OF RAMIFIED ELECTRODEPOSITION

Pioneer works have revealed that electrodeposition may generate aggregates with a structure similar to DLA clusters, both in two and three dimensions.[2,3] Brady and Ball[2] have deposited 3d aggregates on a point-like cathode from a very viscous solution in presence of a supporting electrolyte (copper sulfate and sodium sulfate in polyethylene oxide aqueous solution). In these conditions, the deposition reaction was in effect limited by the diffusion of Cu^{2+} ions, and the obtained aggregate had a fractal dimension $D_{exp} = 2.43$, very close to the value $D_{th} \approx 2.5$, which is expected for 3d DLA.[11]

On the other hand, Matsushita et al.[3] have obtained quasi-2d deposits, grown at the interface between a zinc sulfate solution and n-butyl acetate. The observed aggregates had again a structure close to that of 2d DLA clusters, with a dimension 1.66, although, in this case, the aggregation was probably not driven by diffusion, but rather by the electric field.

Since then, most electrodeposition experiments have been done in quasi-2d radial cells. A major result of these studies was the observation of "morphology transitions",[4,5,12] very similar to those obtained in different physical or biological systems displaying DLA-like structures.[13] In particular, Grier et al.[4] and Sawada et al.[5], have reported that, depending on the experimental conditions (solution concentration, applied voltage), several morphologies could be observed : a DLA-like structure at low voltage, low concentration, dendrites in the high voltage high concentration region, and dense morphology in intermediate regimes. The transitions between these regions are not very sharp, and all authors do not agree on their exact location in the concentration-voltage diagram.

In the absence of a supporting electrolyte, electrodeposition is not expected to be limited by diffusion, hence cannot be described by the usual DLA model. However, in the absence of electric charges in the solution, the electric potential V obeys Laplace equation $\Delta V=0$, with boundary conditions similar to DLA. A DLA-like structure is then expected.[4,5] Halsey and Leibig[14] have shown that the DLA model still remains valid within more refined (and more realistic) conditions, taking into account ionic diffusion and moderate departure from electroneutrality in the double layer near the cathode. Besides, the dense radial morphology was ascribed to a stabilizing effect due to the finite resistance of the deposit.[15]

A puzzling effect was discovered by Nancy Hecker[16] : a morphology transition was observed to occur during the growth of dense deposits. This transition appeared as a sudden change in the characteristic features of the deposit, which could be its color, roughness, number of branches, shape, velocity, etc… Changes in several of these parameters could appear simultaneously. The most striking property of these changes was that the place where it occurred scaled with the size of the cell. Moreover, the line formed by the transition points was shown to be almost a homothetical picture of the anode.[17]

It is now known that impurity cations and/or protons are responsible for the Hecker effect.[18,19] Charged cations (mostly protons) are released at the anode during the experiment : when they hit the growing deposit, they change its morphology. Fleury et al. have shown that the astonishing shape observed for this transition is due to a purely geometrical effect, related to the relative velocities of the different ions involved in the cell.[19]

Apart from these purely morphological aspects, several studies have revealed interesting features of ramified electrodeposition. Impedance analysis experiments have related the Constant Phase Angle (CPA) exponent with the fractal dimension of the deposit.[20] Argoul and Arneodo have observed a chaotic evolution of the voltage across the cell during the growth of dendritic deposits.[9] Melrose and Hibbert have studied electrodeposition in filter paper support.[6] In their experiments, the electrical resistance of the filaments was shown to depend on applied voltage : this was related to changes in microscopic structure revealed by electron microscopy.

ELECTROCHEMICAL ASPECTS

It is only recently that the basic electrochemistry of these systems has been, at least partly, understood. Both experimental[7,21] and theoretical[22] studies have pointed out the essential role of the coupling between two charge flows (anions and cations) in the electric and diffusion fields around the deposit. The use of well controlled electrochemical conditions has permitted to clearly demonstrate the interplay of these different phenomena, and how they govern the ramified electro-deposition process.

In particular, using cells with parallel geometry and galvanostatic conditions, Fleury et al.[7] have shown that the governing parameter of deposition is the current density, as it is usual in an electrochemical process. In particular, it is an important parameter for the morphology of the deposit in a given cell : for copper deposition from copper sulfate solution, the so-called fractal-dense transition was found to occur at current densities around 20-40 mA/cm^2.

In the dense regime, where the growing filaments exhibit a well defined, rectilinear front, this front was shown to advance at a constant speed, equal to the mean electrical migration velocity of the anions in the copper-salt solution. This has been evidenced both for copper sulfate and copper acetate, where the anions have different mobilities.[7]

Using very thin copper potential probes, spaced about 0.5 mm apart between the anode and the cathode (see inset of Fig.1), one can measure intermediate potentials in the cell. Figure 1 shows the typical variation of the potential between these probes and the cathode as a function of time, for a given applied current.[21,23]

One observes four distinct behaviors :
- When the current is switched on, the potential between the intermediate electrodes and the cathode reaches an initial value which varies linearly with the position of the intermediate electrode. Then, the potential first rapidly increases of about 1 V to 3 V.
- While the deposit grows (at a constant speed) between the cathode and the intermediate electrode, the potential decreases linearly with time.
- An abrupt decay, of the order of 1 V to 3 V, occurs when the filaments reach the probe (a similar behavior has been reported by Melrose and Hibbert[6]).
- finally one observes a very small, almost constant potential.

This suggests the following picture (see Fig.2) : in zone A near the cathode, the growing filaments have very low resistivity. In zone B, the concentration remains almost constant, equal to its initial values (except in the neighborhood of the anode, region C, as will be discussed

155

later). The voltage between the cathode and an intermediate electrode ahead of the deposit is thus roughly proportional to the distance between the end of the growing filaments and this electrode. As mentioned above, this distance decreases linearly with time, at a speed equal to the migration velocity of the anions in the applied electric field. The voltage drop observed when the deposit reaches the probe is ascribed to an almost point-like space charge region near the tips of the filaments.

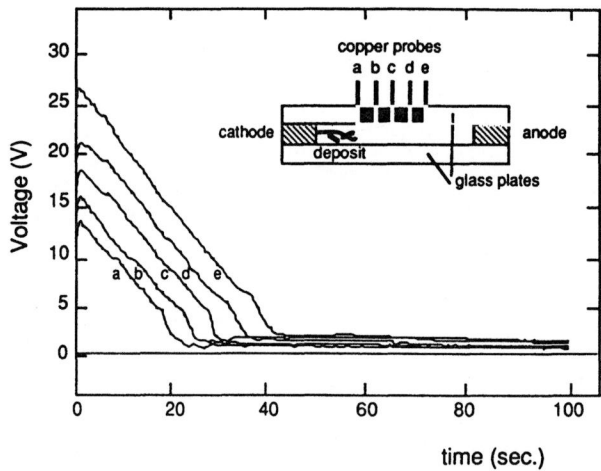

Fig.1 Variation of the voltage between the probes (shown in the inset) and the cathode, during the growth.

Chazalviel has proposed a model to explain this behavior, which may be summarized as follows.[22] Homogeneous deposition is first postulated and calculated in a 1d approximation : it is shown to induce the formation of a space charge and a very high electric field in a narrow region close to the cathode. The system very probably tends to escape this situation. Convective motion in the solution and/or ramified growth are two possible ways for that. Then Chazalviel has computed a simple approximation for ramified electrodeposition in the absence of convection : the growing deposit is modeled as a set of rectilinear, equally spaced, parallel filaments. Due to aggregation, the zone behind the growing front (zone A in Fig.1) must be depleted of cations;[24] correspondingly, the anions must be expelled from this region, otherwise they would create a large space charge region. In the absence of a supporting electrolyte, the front must then progress at the same velocity as the anions, which leave this region.

The growth can thus be regarded as the advance of the front of an array of copper branches at a speed equal to the drift velocity of the anions. As the total amount of anions must remain constant, the anions accumulate near the anode, where their charge is balanced by anodic generation of an equal amount of cations. This results in building a high concentration, low resistivity region (zone C in Fig.1). The width of this region is limited by the diffusion of the ions. A width of 1mm is found for the typical duration of an experiment.

Optical absorption has been measured during electro-deposition, in thin rectangular cells.[25] A video CCD camera mounted with an infrared filter permits to visualize the absorption due to Cu^{2+} ions. A copper sulfate solution with relatively high concentration (0.1 M/l) was used for better contrast. Observation at the very early stages of the deposition shows that the system first evolves, trying to grow an homogeneous, dense copper deposit on the cathode.[25,26] In

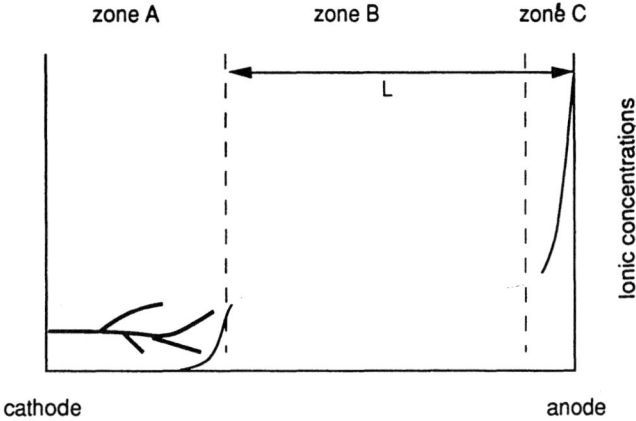

Fig. 2 Schematic picture of the growth. The resistive part of the impedance is essentially due to the region ahead of the deposit, where the concentrations of the ions in the solution remain constant, equal to their initial value. The resistance of the cell is thus roughly proportional to the distance L between the tips of the filaments and the anode.

agreement with the 1d calculation described above, one observes the building up of a narrow region with high electric field ahead of the deposit. When the potential drop in this region reaches a value of a few volts, a ramified deposit starts growing, advancing at the velocity of the anions. A steady state is attained very rapidly, in which the electric field at the tip of the deposit is constant, with a value much lower than the steady-state value calculated in the 1d approximation.[25]

At later times, when the ramified deposit has grown, one obtains a map of the light absorption in the cell (see Fig.3), giving a picture of the concentration distribution around the deposit. Our results confirm that the space between the filaments is completely depleted from the ions which tend to accumulate near the anode.[25] One clearly sees in Fig.3 that, close to the deposit, the isoconcentration lines form arches. Such a shape of the concentration distribution is very different from distributions calculated from Chazalviel's model in the absence of convection.[22] We want to emphasize that concentration profiles similar to Fig.3 have been observed for a colored anion solution.[21] On the other hand, concentration profiles obtained from shadowgraph or Schlieren methods are slightly different :[26,27] concentration depletion is also observed inside the deposit, but the concentration contours follow the global shape of the deposit, and do not form arches between nearby branches. The reason for this difference is not clear at present.

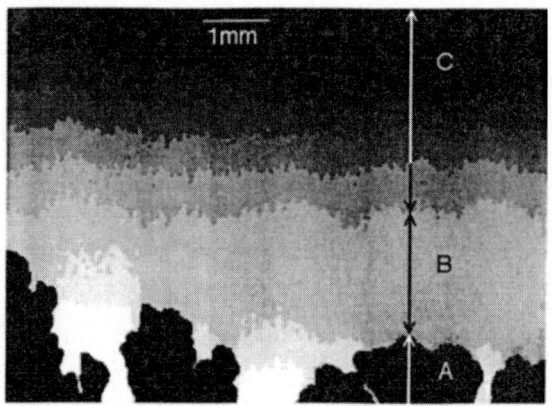

Fig.3 Optical absorption from a 6 x 8 mm^2 region in the electrochemical cell between the deposit and the anode. The number of grey levels is reduced to show some contours of constant optical absorption. The growing deposit is seen at the bottom of the picture. Near the deposit, the decreased optical absorption evidences a depletion of the cations (zone A in Fig.1), whereas, near the anode (top), the increased absorption corresponds to an increase of the copper ion concentration (zone C in Fig.1). In between, a plateau is observed, corresponding to the region where the concentration remains constant (zone B in Fig.1). Here the deposit has invaded a large part of the cell and zone C is very narrow.

Convective motion has been evidenced experimentally (see Fig.4), and electroconvection calculated in the framework of a simple model.[10] Considering again the deposit as a set of rectilinear, equally spaced, parallel filaments, Fleury et al.[10] have calculated the convective flow due to the action of an almost point-like force at the tip of the branches. This force is actually related to the migration of ions which, in this region, leaves a net positive charge together with a high electric field. The fluid flow pattern obtained from this model is in qualitative agreement with the observation (Fig.4) : in particular, one observes vortices between nearby branches of the deposit, which should be empty of ions, whereas in the region ahead of these vortices the concentration should keep a constant value, close to its initial value. One then expects an arch shaped boundary between the two regions, very similar to that observed for the iso-concentration lines of Fig.3.

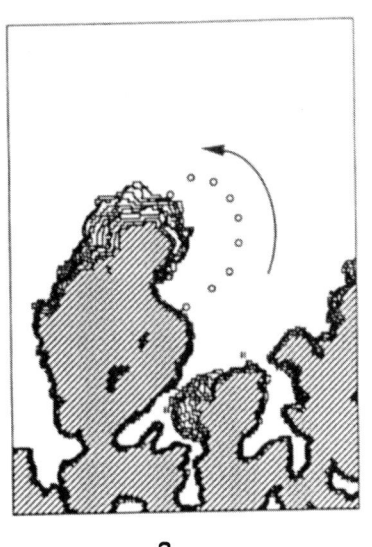 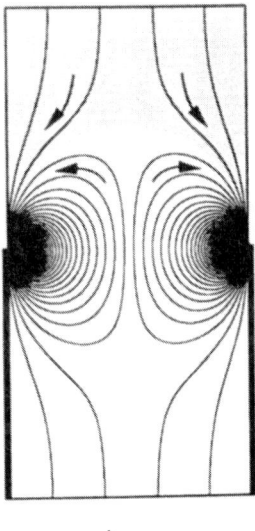

a   b

Fig.4 a) A simple evidence of electroconvective motion around the branches of the deposit : the position of an impurity at nine successive times is shown, together with the growth of the aggregate. b) fluid flow pattern calculated by Fleury et al.[10] Two nearby parallel branches of the deposit are shown as thick vertical lines on both sides of the picture. The direction of the flow is shown by the arrows. The region inside the vortices and between the branches (in white) is empty of ions, whereas, in the region ahead of the deposit (in grey), the concentration remains constant, almost equal to its initial value (in both pictures, the aggregate grows upward).

CONCLUSION

Several authors have shown that electrodeposition in dilute binary electrolytes produces ramified, fractal deposits. A better understanding of this behavior is obtained by solving the equations governing the ionic motion in the solution : electrodeposition is shown to result in ion depletion and formation of a strong space charge in the region close to the deposit. Ramified deposition and electroconvective motion inside the solution then appear as a natural consequence of this effect. The front of the deposit is predicted to advance at the velocity of the anions in the electric field ahead of the space-charge region. This prediction, the existence of a potential drop associated with the space-charge, and ion depletion near the deposit, are confirmed by optical and electrical studies of electrodeposition of copper from copper sulfate solution. Experimental observations of convection are qualitatively explained from a simple model. However, the existence of a DLA-like structure in a system highly perturbed by space-charge and electro-convection remains a puzzling question.

ACKNOWLEDGEMENT

This work was supported by the "Centre National d'Etudes Spatiales" under contract 89/1229.

REFERENCES

1. B.B. Mandelbrot, "Les Objets Fractals : Forme, Hasard et Dimension", Flammarion, Paris (1975, 1989); "The Fractal Geometry of Nature", Freeman, San Francisco (1982).
2. R.M. Brady and R.C. Ball, Nature, 309:225 (1984).
3. M. Matsushita, M. Sano, Y. Hayakawa, H. Honjo and Y. Sawada, Phys. Rev. Lett. 53:286 (1984).
4. D. Grier, E. Ben-Jacob, R. Clarke, and L.M. Sander, Phys. Rev. Lett. 56:1264 (1986).
5. Y. Sawada, A. Dougherty, and J.P. Gollub, Phys. Lett. 56:1260 (1986).
6. J.R. Melrose and D.B. Hibbert, Phys. Rev. A40:1727 (1989).
7. V. Fleury, J.-N. Chazalviel, M. Rosso and B. Sapoval, J. Electroanal. Chem., 290:249 (1990).
8. T.A. Witten and L.M. Sander, Phys. Rev. Lett. 47:1400 (1981).
9. G.L.M.K.S. Kahanda and M. Tomkiewicz, J. Electrochem. Soc., 136:1497 (1989); F. Argoul and A. Arneodo, J. Phys. France, 51:2477 (1990); F. Argoul, J. Huth, P. Merzeau, A. Arneodo, to be published.
10. V. Fleury, J.-N. Chazalviel and M. Rosso, to be published.
11. P. Meakin, Phys. Rev. A27:604 (1983).
12. P.P. Trigueros, J. Claret, F. Mas and F. Sagués, J. Electroanal. Chem., to be published.
13. See, for example : E. Ben-Jacob and P. Garik, Nature, 343:523 (1990) and references therein; H. Fujikawa and M. Matsushita, J. Phys. Soc. Japan, 60:88 (1991).
14. T.C. Halsey and M. Leibig, J. Chem. Phys. 92:3756 (1990).
15. D.G. Grier, D.A. Kessler, and L.M. Sander, Phys. Rev. Lett. 59:2315 (1987).
16. N. Hecker, D.G. Grier and L.M. Sander, in "Fractal aspects of materials", R.B. Laibovitz, B.B. Mandelbrot and D.E. Passoja, ed., Material Research Society extended abstracts (1985).
17. P. Garik, D. Barkey, E. Ben-Jacob, E. Bochner, N. Broxholm, B. Miller, B. Orr and R. Zamir, Phys. Rev. Lett. 62:2703 (1989).

18. J.R. Melrose, D.B. Hibbert and R.C. Ball, Phys. Rev. Lett. 65:3009 (1990).

19. V. Fleury, M. Rosso and J.-N. Chazalviel, Phys. Rev. A43:6908 (1991).

20. G.L.M.K.S. Kahanda and M. Tomkiewicz, J. Electrochem. Soc. 137:3423 (1990).

21. V. Fleury, M. Rosso, J.-N. Chazalviel and B. Sapoval, Phys. Rev. A44:6693 (1991).

22. J.-N. Chazalviel, Phys. Rev. A42:7355 (1990).

23. M. Rosso, V. Fleury, J.-N. Chazalviel, B. Sapoval and E. Chassaing, in "Scaling in disordered materials : fractal structure and dynamics", J.P. Stokes, M.O. Robbins and T.A. Witten, ed., Material Research Society extended abstracts (1990).

24. Cation depletion around the deposit has also been observed in growth experiments performed in filter paper, in the presence of a supporting electrolyte : D.B. Hibbert and J.R. Melrose, Phys. Rev. A38:1036 (1988).

25. M. Rosso, V. Fleury, J.-N. Chazalviel and B. Sapoval, to be published.

26. R.H. Cork, D.C. Pritchard and W.Y. Tam, Phys. Rev. A44:6940 (1991).

27. D. Barkey, J. Electrochem. Soc. 138:2912 (1991).

PROPERTIES OF THE MORPHOLOGIES ENVELOPE IN A DIFFUSION LIMITED GROWTH

O. Shochet, R. Kupferman and E. Ben-Jacob

School of Physics and Astronomy
Raymond & Beverly Faculty of Exact Sciences
Tel-Aviv University, Tel-Aviv 69978, Israel

INTRODUCTION

Despite the major progress with the discovery of the microscopic solvability criterion [1, 2, 3, 4, 5, 6, 7, 8, 9], the problem of interfacial pattern formation is not yet solved [8, 10]. The new existence principle for dendritic growth predicts that, whenever anisotropy is present, a specific dendrite corresponding to the fastest-growing needle crystal exists and is linearly stable. However, both in time evolution studies of various models [8, 11, 12, 13, 14, 15] and in experimental observations [16, 17, 18, 19, 20, 21] even with anisotropy present, dendrites are not always observed: the dense branching morphology (tip-splitting) occurs instead.

The observation of the DBM under growth conditions suitable for dendrites means that the two morphologies can coexist. Hence, the *"Microscopic solvability" can clearly be merely a part of the picture, and a more general principle is needed to select one of the morphologies.* Ben-Jacob et al.[8, 11, 12, 13] proposed that the selected morphology is (in many cases) the fastest-growing one. They have argued that not only does dendritic growth have constant velocity, but also the DBM has a well defined envelope that propagates at constant velocity. Similar arguments have been proposed by Goldenfeld [22] for spherulitic growth. A morphology selection principle leads to the concept of a morphology diagram, which has been confirmed experimentally in various systems [16, 17, 18, 21, 23]. Our main goal in this paper is to use the diffusion-transition scheme [10] to test the range of validity of the "fastest growing morphology" hypothesis and the nature of the morphology transition.

For systems in equilibrium, the phase that minimizes the free energy, for a given set of state variables, is the selected one, irrespective of the prior history of the system (although it can take long time to reach this state). The concepts of a selection principle and a phase diagram go hand in hand. By contrast, non-equilibrium growth processes are time dependent, so it is not clear *a priori* that a morphology diagram can exist. The question is whether appropriate control parameters describing the growth conditions, analogous to the state variables of thermodynamics, can be identified, that the late stage morphology will depend only on these parameters and neither on initial conditions nor on the size. The existence of a morphology diagram has been confirmed experimentally in various systems [16, 17, 23, 18, 21], suggesting that a morphology selection principle must exist.

When a system is driven out of equilibrium by the imposition of a gradient of one of the thermodynamic variables (e.g. the temperature or the concentration), the response of the system is described by the conjugate flux (the heat flux and particle flux, respectively). These fluxes may, in general, be viewed as the rate of entropy production, or the rate of approach towards global equilibrium. In growth processes specifically, the driving force (e.g. the undercooling in solidification) is the equivalent of the thermodynamic gradient. The average velocity measures the rate of approach towards equilibrium, and serves, naturally, as a response function. (by the term "average velocity" one should refer to the velocity weighted according to the geometry of the interface, and thus take into account the global shape of the object as will be discussed in section 6). This was the motivation for the "fastest growing morphology" hypothesis proposed by Ben-Jacob et al. [8, 11, 12, 13]. We expect this "average velocity" to be an important variable, but by no means the only one. It should have two counterparts representing the branching dynamics and scaling, and on the microscopic level, the equilibrium properties of the interface and the selected growing stable phase.

The analogy with equilibrium systems may be carried even further: Two types of morphology transitions, dependent on the growth conditions, have been proposed in analogy to phase transitions in equilibrium [8, 13]. The first kind shows a discontinuous jump in the velocity at the transition point (hence classified as a first order morphology transition). In the other type (characterized as second order), the velocity itself is continuous as the morphology changes, but shows discontinuity in its derivative. Here we will focus on the second kind.

In this paper, we show that both the DBM and dendritic growth have shape preserving envelope that propagates at constant velocity. However, the velocity scales differently for the two morphologies. Moreover, the DBM envelope is convex while the dendritic envelope is concave. Hence, the envelope provides a useful characterization of the growing morphology and the transition between morphologies.

NUMERICAL SIMULATIONS OF THE DIFFUSION TRANSITION SCHEME

Recently, we have developed a new diffusion-transition approach to study pattern formation in systems described by a conserved order-parameter (keeping in mind a specific example of solidification from supersaturated solution). We assume an ideal solution, so that the chemical potential of the liquid phase is given by

$$\mu_l = k_B T \ln c. \tag{1}$$

Hence, the time evolution of the concentration field obeys the linear diffusion equation:

$$\frac{\partial c}{\partial t} = D \nabla^2 c \tag{2}$$

where D is the diffusion constant.

The second part of the dynamics in the model is that of the phase transition at the interface. We divide the phase transition process into local processes of solidification and melting of single cells. Only liquid cells adjacent to the solid can solidify and only solid perimeter cells can melt.

The processes of phase transition and diffusion are executed sequentially on a square lattice. We first solve the diffusion equation over a time interval Δt, which is small relative to the diffusion characteristic time ((lattice size)$^2/D$) and the phase transition time (discussed below). Zero derivative boundary conditions are used in order to guarantee conservation of order parameter (material) during this stage. Next we execute melting and solidification along the interface according to the probabilities which are discussed below. The concentration at the perimeter sites along the interface changes during the phase transition. Then a new cycle of diffusion and phase transition starts, and so on.

The rate of phase transition of a cell is calculated by assuming local equilibrium. In this case we know that the entropy of a microscopic state of the system is

$$S(s^1) = -k_B \ln p_{s_1}, \tag{3}$$

where p_{s_1} is the probability that it is in the microscopic state s_1. We are interested in transitions between states and present their rates as:

$$\omega_m = melting\ rate = \omega_0 p(\Delta S_{melting}), \tag{4}$$

$$\omega_s = solidification\ rate = \omega_0 p(\Delta S_{solidification}), \tag{5}$$

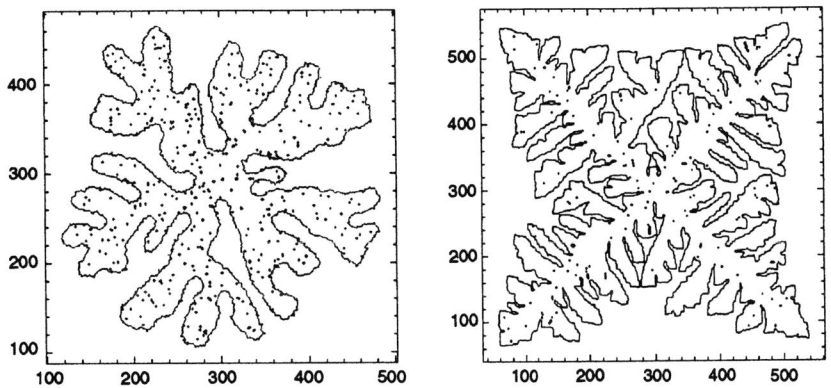

Figure 1. (a) DBM growth ($\omega = 80000, \mu_s/T = -1.25, E_B/T = 1.25, D = 10^4, c_\infty = 0.7$, the system size is 600×600). (b) Dendritic growth ($\omega = 8000, \mu_s/T = -8.33, E_B/T = 5.33, D = 10^4, c_\infty = 0.7$ the system size is 600×600).

where

$$\frac{p(\Delta S_{melting})}{p(\Delta S_{solidification})} = e^{\frac{-(\Delta S_{melting} - \Delta S_{solidification})}{k_B}}, \tag{6}$$

and ω_0 is the characteristic rate of the phase transition.

The entropy change during the solidification and melting processes is calculated by:

$$\Delta S = \frac{\Delta E}{T} - \frac{\mu_s - \mu_l}{T}\Delta N \tag{7}$$

where μ_s is the chemical potential of a single solid cell of size a^2. During solidification, the new solid cell has $c = 1$, and $\Delta N = 1$. During melting, the concentration in the new liquid cell should remain 1. The only contribution to the difference in energy is from the interfacial energy. This interfacial energy is calculated by assuming that each border between a liquid cell and a solid cell contributes an energy E_B, the bond energy, to the total interfacial energy.

To complete the model we have to specify the functional form of the probability function (equation 6). We emphasize that this form can not be derived at present from first principles. One natural form is:

$$p = (1 + e^{\frac{\Delta S}{k_B}})^{-1}, \tag{8}$$

which is the form used in [10].

One of the advantages of the model is that is can readily relate the model parameters to the parameters of the continuum picture (such as surface tension, the capillary length and the surface kinetics).

In figure 1 we show one realization of a late stage growth of the DBM (after it has reached a steady state growth), and of the dendritic growth. At this stage, the larger scale density is constant in time and is given by

$$\rho_g = \frac{N_s}{N_s + N_l} = \Delta \equiv \frac{c_\infty - c_{eq}}{1 - c_{eq}} \tag{9}$$

where N_s and N_l is the number of solid and liquid sites respectively, ρ_g is the global mass density and Δ is the dimensionless supersaturation.

In figure 2 we show the time evolution of the averaged envelope. The averaging is over 30 different realizations for the same parameters each for a different seed of the random number generator. First it is obvious that the envelope is shape preserving and propagates at constant velocity. It also has

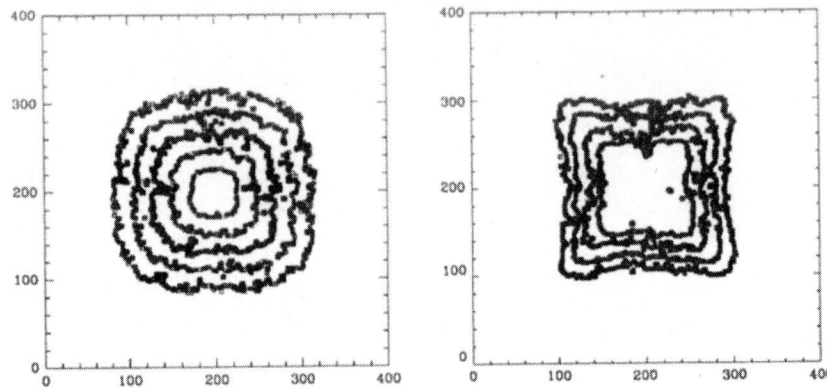

Figure 2. Average envelope at different time steps. The envelope is shape preserving and advances with constant velocity. $\omega = 8000, \mu_s/T = -4.75, E_B/T = 2.66, D = 10^4, c_\infty = 0.7$ the system size is 400×400.

a pronounced 4-fold symmetry, concave for DBM and convex for dendrites. Hence, the propagation of the interface can be viewed as the propagation of a solid at unit undercooling with effective surface tension and surface kinetics. Both can be derived from an effective free energy, using the phase field approach [24].

In figure 3 we show the scaling of the velocity of the DBM and the dendrites with $\Delta\mu$. Here $\Delta\mu$ is the difference between the solid chemical potential and the liquid chemical potential at infinity. Each point indicates the averaged velocity over about 30 realizations. The error bars indicate the standard deviation of the velocities measured for different realizations (for given $\Delta\mu$). Although the range of $\Delta\mu$ is too small for accurate measurements of the exponent, it appears that the velocity scales as

$$v \propto \begin{cases} (\Delta\mu)^3 & \text{DBM} \\ (\Delta\mu)^{1.5} & \text{dendrites} \end{cases} \tag{10}$$

Measurements of the dependence of the velocity on the supersaturation Δ indicates that for DBM: $v \propto \Delta^3$. Uwaha and Saito [25] also found that the averaged velocity (for finite concentration DLA) scales approximately as Δ^3. To explain their observation they have related the velocity to the small scale fractal dimension D_f. Brener et al. [26] also obtained (using scaling arguments) that $v \propto \Delta^{\frac{1}{2-b_f}}$ in the limit where the kinetic term is dominant. Both results are based on the small length structure being fractal, which is not the case in our simulations. At this point it is not clear either the similarity is merely a coincidence or there is a more general principle that determines the scaling of the velocity.

Our preliminary measurements of dendritic growth show oscillations in the velocity and in the concentration field in front of the four main dendrites. These oscillations are associated with the emission of side branches. The results are consistent with experimental observation of Raz et al. [27].

We have studied the transitions between the DBM and the dendritic growth as function of the bond energy E_B, the temperature T, the supersaturation Δ and the chemical potential difference $\Delta\mu$. In all cases we observed sharp transitions in agreement with the assumption that morphology selection principle exists as proposed by Ben-Jacob et al. [8, 13, 28]. Here we present in details the results for the case when $\Delta\mu$ is varied (other cases will be presented elsewhere [29]). As was mentioned in section 2, the crucial step is to define an appropriate response function to characterize the transition. We found that the growth velocity is indeed constant, so in principle is can be used to characterize the growth. The finite sizes of the crystals introduces relatively large fluctuations in the

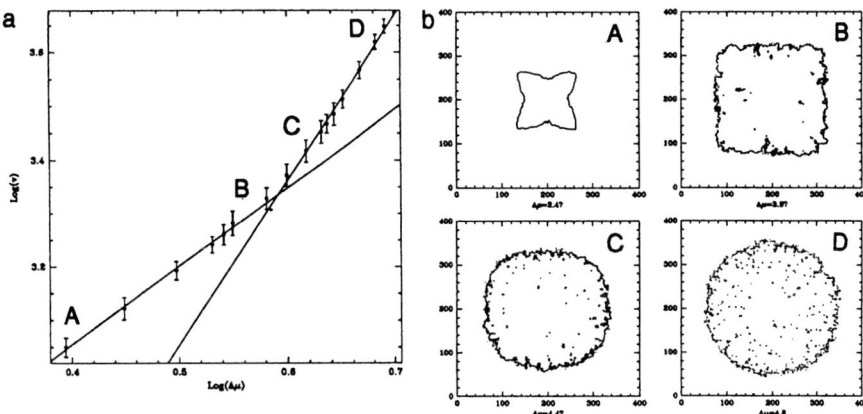

Figure 3. (a) The transition on a Log-Log plot of the velocity v versus the difference in chemical potential $\Delta\mu$. The slope of least squares fit changes from 1.5 for dendrites to 3.0 for DBM. (b) The ensemble average envelope for different chemical potentials. The distance from the transition is shown in (a). At the transition, the envelope changes from convex (for DBM) to concave (for dendritc morphology). $\omega = 8000, -\mu_s/T = 2.83 - 5.26, E_B/T = 2.66, D = 10^4, c_\infty = 0.5 - 0.7$ the system size is 400×400.

measurements of the growth velocity. To overcome this problem we averaged over the growth velocity for 30 different realizations for the same parameters each for a different seed of the random number generator. In figure 3 we show a transition between the DBM and the dendritic growth as we vary the chemical potential. In forthcoming publication we will show that an orientation correlation length can be defined. This correlation length diverges as we approach the transition. We also studied the transition as we start with one morphology and than change the growth conditions to be in the range of another morphology. We observed a fast transition to the new morphology, indicating that the original morphology is not meta-stable. Both observations support the identification of the transition as a second order. In figure 3 we show that $v(\Delta\mu)$ (over the range of $\Delta\mu$ we have studied) can be fitted both with a log-log and log-linear scales (it can not be fitted with a linear-linear plot). A theoretical work is needed to determine the appropriate functional form which at present is not known.

As both dendrites and DBM have well defined envelope, its shape can be used to distinguish between the two morphologies. In figure 3 we show the averaged envelope as function of the chemical potential. At the transition point the envelope turns from concave to convex shape. Note that we present here the ensemble average of many realizations. Since we know that the envelope is shape preserving and advance at constant velocity we can also construct the envelope for a single growth realization by averaging over the scaled (by velocity) envelope at different time steps.

ENSEMBLE AVERAGING – THE ENVELOPE

The existence of the envelope motivates us to describe the system as two phase system where one phase is penetrating into the other, and the boundary between them is the envelope. One phase consists of a mixture of solid-liquid which are in local equilibrium and the other phase is the solution. The dynamics of the solution is of course diffusion of matter. We assume that there is no dynamics inside the envelope since the only effects are local (due to surface tension and not due to difference in chemical potential). In this case, a natural model for the dynamics of the envelope will be the free-boundary model in which we will include the diffusion properties of the solution while the growth will be describe by the boundary conditions which are at the envelope.

Keeping in mind the above motivation, we will consider the order-parameter, u, satisfying the following free boundary model

$$\frac{\partial u}{\partial t} = D\nabla^2 u \tag{11}$$

$$v_n = -D\nabla_n u \qquad u(\infty) = 0 \qquad u_i = \Delta - d_0\kappa - \beta v_n \tag{12}$$

D is the diffusion coefficient, v_n, the normal velocity of the envelope, u_i, the value of the field at the interface, Δ, a measure of the global distance from equilibrium, d_0, the effective surface tension length, κ, the local curvature of the envelope and β, the effective kinetic coefficient. Note that the boundary conditions on the velocity corresponds to a non-conserved order parameter.

How come that the envelope has an effective surface tension ? What are the stablizing effects which create the extra length-scale d_0 in the problem ? The answer is that these are kinetic effects which cause an effective surface tension of the averaged envelope. Assume a flat growing envelope where we introduce a perturbation to the envelope. There is no mechanism that will stop the growth of the bumps (on the contrary, it will grow faster due to the Mullins-Sekerka [30] instability), or will cause a faster growth for the holes. The tension effect is due to the fact that when a hole is created, a growth of the bumps over the holes will occur causing a movement of the average envelope and therefore results in an effective surface tension.

We are interested in shape preserving solutions, propagating with constant velocity. The simplest way to obtain the scaling of the velocity, and the stability properties of the envelope, is by studying the case of a flat envelope. This problem was extensively studied for a real solid-liquid interface. For that case, simplifying approximations were done, using the fact that the surface tension capillary length d_0, is much shorter than the diffusion length D/v. As we consider effective parameters, a-priori unknown, we perform this calculation for the general case.

A steady state solution of eqs. (11)-(12) exists only if $\Delta \geq 1$, given by

$$u(z) = (\Delta - \beta v)e^{-(v/D)z} \qquad z = x - vt, \tag{13}$$

with the velocity given by

$$v = \frac{1}{\beta}(\Delta - 1). \tag{14}$$

To study the stability of the envelope, we perturb the location of the envelope by a periodic function

$$\xi(y,t) = \delta_k e^{iky+\omega_k t}, \tag{15}$$

where the y-axis is perpendicular to the direction of propagation. The dispersion relation $\omega_k(k)$, is given implicitly by

$$\bar{\omega} = -\bar{k}^2/\epsilon + \bar{q}^2 - \bar{q} \tag{16}$$

$$\bar{\omega}[1 + (\Delta - 1)\bar{q}] = -\bar{k}^2\bar{q} + \bar{q} - 1 \tag{17}$$

where $\bar{k} = \sqrt{(D/v)d_0}\,k$, $\bar{\omega} = D/v^2\,\omega$ and $\epsilon = (v/D)d_0$. Useful information is obtained by calculating the so-called Mullins-Sekerka mode k_{MS}, i.e. the mode for which $\omega = 0$ (except $k = 0$). Combining eqs.(16)-(17), we find that

$$k_{MS} = \sqrt{\frac{D}{v}d_0}\sqrt{1 - \sqrt{\epsilon}}. \tag{18}$$

The envelope is unstable against perturbation with $k < k_{MS}$ (long wavelength), and stable against perturbation with $k > k_{MS}$ (short wavelength). However, if $\epsilon \geq 1$, the marginally stable mode disappears, and the envelope becomes stable against perturbations of all wavelengths. This point was overlooked in the study of solid-liquid interface, where $\epsilon \ll 1$. The turning point from a non-stable growth to stable growth is when the diffusion length is equal to the effective capillary length.

DISCUSSION

We have demonstrated the existence of the dense branching morphology and dendritic growth in a detailed model inspired by solidification from supersaturated solution. That is, for a conserved order parameter. The model includes explicit surface tension anisotropy, explicit surface kinetic and implicit anisotropy in the surface kinetics. Our observation of DBM in the presense of anisotropy (so that needle crystal solution do exist) strongly support the assumption about the existence of morphology selection principle. We have shown that the DBM also grow at a constant velocity, and

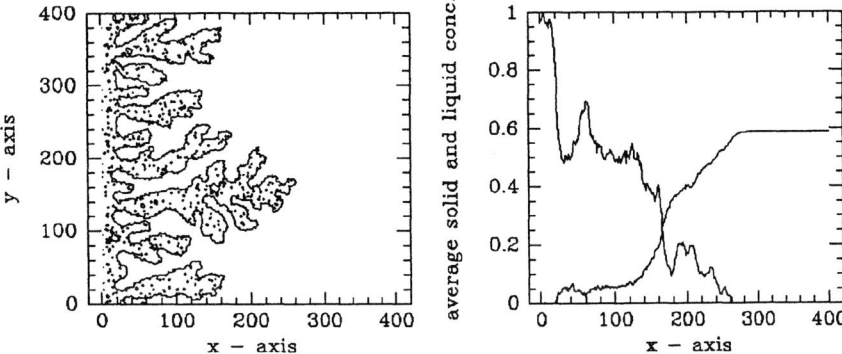

Figure 4. Simulations results of growth in a channel. In the top, we observe the shape of the growing fingers. On the bottom graph, the average over the y direction concentration of solid (starting from 1 at the left size) and of liquid (0.59 at the right size). The similarity between this picture to the phase-field will be discused in a consequence publication. $\omega = 8000, \mu_s/T = -2.22, E_B/T = 1.33, D = 10^4, c_\infty = 0.59$ the system size is 400×400.

that the later cane be used as a possible response function to describe the growth. Moreover, our results support the "fastest growing morphology" selection hypothesis and the characterization of the transition as a second order morphology transition.

In many experimental systems as well as in the boundary layer model the common scenario is that DBM is observed at low undercooling while dendritic growth is characterized of large undercooling [31, 13, 32]. On the other hand in the present model the scenario is reversed. There is no contradiction; It has to do with the fact that our model describes conserved order parameter and that here the dominant anisotropy is in the surface tension. In a forthcoming publication we will present studies of surface kinetic anisotropy and competition between surface tension and surface kinetic anisotropies using our "diffusion transition scheme". There are other experimental systems like liquid crystals [18] and Ammonium chloride [19] and succinonitrle [20] in which there is also a conserved order parameter and the scenario is the same as in the model.

To complete our understanding, the most important next step is to find a method to calculate the growth velocity of the morphologies. At present we only know how to calculate the velocity of a free dendrite assuming constant velocity. It is not clear that the dendritic morphology has the same velocity. In particular our preliminary results indicate oscillations in the growth of the tip (similar to those observed in experiments[27]). To find such steady but oscillatory needle crystal one has to formulate a time dependent solvability conditions. In addition there are no calculations of the DBM velocity (which will also require time dependent solvability criterion). Recently, scaling arguments (in the limits of low undercooling) to calculate the velocity have been proposed. However, these arguments relate the velocity to the geometrical structure of the morphology (the small scale fractal dimension) which in turn has to be obtained from the numerical simulations. In addition, these arguments are valid in the range of parameters where two length scales can be identified. At large supersaturation we found no regime of fractal like structure. Instead, we have observed that the distribution of the branches width scales with D/v.

Our simulations suggest that the envelope of the morphology plays a crucial role in the growth process. We believe that the existence of a well defined shape preserving envelope is the clue for the determination of the growth velocity. One can write a phase-field model to describe the dynamics of the envelope (e.g. the phase being the average over the structure and the field the averaged concentration shown in figure 4). Such a model has an effective surface tension and surface kinetic that describe the dynamics of branching. The velocity is determined self-consistently when these effective surface parameters corresponds to a marginally stable envelope [33].

ACKNOWLEDGEMENTS

Part of the results which are presented here were done in collaboration with K. Kassner, S. G. Lipson and H. Müller-Krumbhaar [10]. We thank E. Brener and D. E. Temkin for useful discussions. This research was supported in part by a grant from the G.I.F., the German-Israeli Foundation for Scientific Research and Development, and by the Program for Alternative Thinking at Tel-Aviv university.

REFERENCES

[1] D. A. Kessler, J. Koplik and H Levine, Adv. Phys. 37 (1988) 255.

[2] E. Ben-Jacob, N. D. Goldenfeld, B. G. Kotliar and J. S. Langer, Phys. Rev. Lett. 53 (1984) 2110.

[3] D. A. Kessler, J. Koplik and H Levine, Phys. Rev. A 33 (1986) 3352.

[4] M. Ben-Amar and B. Moussallam, Physica D 25 (1987) 155.

[5] R. Combescot, T. Dombre, V. Hakim, Y. Pomeau and A. Pumir, Phys. Rev Lett. 56 (1986) 2036 ; Phys. Rev. A 37 (1984) 1270.

[6] M. Kruskal and H. Segur, Aeronautical Research Associated of Princeton, Technical Memo, 85-25 (1985) unpublished.

[7] J. S. Langer, Science 243 (1989) 1150.

[8] E. Ben-Jacob and P. Garik, Nature 343 (1990) 523.

[9] E. Brener and V.I. Melnikov, Adv. Phys. 40 (1991) 53.

[10] O. Shochet, K. Kassner, E. Ben-Jacob, S.G. Lipson and H.M. Müller-Krumbhaar, to be published in Physica A.

[11] E. Ben-Jacob, P Garik and D. Grier, Superlattices and Microstructures 3 (1987) 599.

[12] E. Ben-Jacob, P Garik, T. Muller and D. Grier, Phys. Rev A 38 (1988) 1370.

[13] E. Ben-Jacob and P. Garik, Physica D 38 (1989) 16.

[14] T. Vicsek , Phys. Rev A 32 (1985) 3084.

[15] J. Nittmann and H.E. Stanley, Nature 321 (1986) 663.

[16] D. G. Grier, E. Ben-Jacob, R. Clarke and L. M. Sander, Phys. Rev. Lett. 56 (1986) 1264.

[17] E. Ben–Jacob, R. Godbey, N. D. Goldenfeld, J. Koplik, H. Levin, T. Mueller and L. M. Sander, Phys. Rev. Lett. 55 (1985) 1315.

[18] P. Oswald, J. Bechhoefer and F. Melo, MRS Bul. (Jan. 1991) 38.

[19] H. Honjo, S. Ohta and M. Matsushita J. hys. Soc. Jpn. 55 (1986) 2487.

[20] H. Honjo, S. Ohta and M. Matsuhsita Phys. Rev. A 36 (1987) 4555.

[21] V. Horvath, T. Vicsek and J. Kertesz, Phys. Rev A 35 (1987) 2353.

[22] N.D. Goldenfeld, J. Cryst. Growth. 84 (1987) 601.

[23] S. K. Chan, H. H. Reimer and M. Kahlweit, J. Crystal Growth 32 (1976) 303.

[24] R. Kupferman and E. Ben-Jacob, in preparation.

[25] M. Uwaha and Y. Saito, Phys. Rev. A 40 (1989) 4716.

[26] E. Brener H.M. Müller-Krumbhaar and D.E.Temkin, submited to Eeurophys. Lett.

[27] E. Raz, S. G. Lipson and E. Polturak, Phys. Rev. A 40 (1989) 1088.

[28] E. Ben-Jacob, G. Deutscher, P. Garik, N. D. Goldenfeld and Y. Lareah, Phys. Rev. Lett. 57 (1986) 1903.

[29] O. Shochet, K. Kassner, E. Ben-Jacob, S.G. Lipson and H.M. Müller-Krumbhaar, in preparation.

[30] W. W. Mullins and R. F. Sekerka, J. Appl. Phys. 35, 1964 444..

[31] E. Ben–Jacob, N. Goldenfeld, J. S. Langer and G. Schon, Phys. Rev. A. 29 (1984) 330.

[32] S. H. Tirmizi and W. N. Gill, J. Crystal Growth 96 (1989) 277.

[33] R. Kupferman, O. Shochet and E. Ben-Jacob, in preparation.

GROWTH PATTERNS IN ZINC ELECTRODEPOSITION

Francesc Sagués, Francesc Mas, Josep Claret, Pedro P. Trigueros
and Laura López-Tomàs

Departament de Química Física, Universitat de Barcelona
Facultat de Química, c/ Martí i Franquès 1
E-08028-Barcelona, Spain

I. INTRODUCTION

Non-equilibrium growth processes are conspicuous in nature. Electrodeposition constitutes a particularly fascinating example belonging to such general class of phenomena[1]. By slightly changing the experimental conditions under which electrodeposits are formed, they may become shaped either in a anisotropic dendritic form or without any apparent regularity as typical fractal aggregates[2-4]. A full understanding of the actual mechanism involved in electrodeposition is still lacking, and it is well recognized that this goal can only be achieved by working complementarly from the theory, simulation and experimental practice of electrodeposition.

In this respect we report here on some of our most recent progress in elucidating some theoretical, simulation and experimental aspects of the problem. Obviously one of the most important theoretical questions concerning electrodeposits is to know whether or not their growth is Laplacian controlled. As a first step in this direction, an experimental pattern is tested against a Laplacian growth rule by directly computing the scaling properties of the growth probability distribution (GPD)[5]. It is clear from the results reported here that, at least on respect to the distribution support, fractal-like electrodeposits behave in front of the harmonic measure in a way compatible with DLA-Laplacian predictions. Further work is presently done to use direct experimental procedures to compute the GPD[6].

On the other hand experimental results will be here presented reporting on our work to classify pattern morphologies for quasi two-dimensional Zn electrodeposits grown in a parallel cell. In the course of this research, new and interesting features of cell dimensions effects have been also detected[7]. Finally, the effect of a different transport regime on the deposit morphology is here illustrated by referring to electrodeposition experiments

Growth Patterns in Physical Sciences and Biology, Edited
by J. M. Garcia-Ruiz *et al.*, Plenum Press, New York, 1993

conducted with a radial cell on an aqueous / organic interface[8]. The loss of the deposit radial symmetry and a certain degree of branch bending are the main observed effects. These last conditions have been also simulated in terms of a multiparticle DLA with superimposed drift, and results of the simulations are briefly sketched[9].

II. SCALING PROPERTIES OF THE MASS AND GROWTH PROBABILITY DISTRIBUTION IN ELECTROCHEMICAL DEPOSITION

Taken for granted from the theory of dynamical systems, an extended analysis of the fractal characteristics of an aggregation pattern is presently formulated in terms of what are called the generalized fractal dimensions D_q [10]. Several numerical routines have been developed to compute the spectrum of dimensions D_q [11]. The one that will be used below, previously employed for electrodeposition aggregates in refs. 4, 12, and 13, is based on a box-counting algorithm especially appropriate to the evaluation of D_q with $q \geq 0$. It consists in covering the pattern with a mobile grid composed of square boxes of variable size $0 < \varepsilon \equiv$ $(l/L) < 1$, where l and L are respectively the box size and a typical length of the aggregate. Following Grassberger, Hentschel and Procaccia[10] one then introduces the hierarchy of exponents τ_q according to

$$Z_q(\varepsilon) \equiv \sum_{i=1}^{N(\varepsilon)} p_i^q \quad \underset{\varepsilon \to 0}{\alpha} \quad \varepsilon^{\tau_q} \tag{1}$$

The set of generalized fractal dimensions D_q are easily obtained from τ_q:

$$D_q \equiv \frac{\tau_q}{(q-1)} \tag{2}$$

where p_i is the relative portion of the cluster cointained in the i-th cell, with $\sum_{i=1}^{N(\varepsilon)} p_i = 1$ and $N(\varepsilon)$ is the total number of boxes necessary to cover the aggregate. The generalized dimensions are then obtained from the relation:

$$D_q = \lim_{\varepsilon \to 0} (q-1)^{-1} \frac{\log \left\{ \sum_{i=1}^{N(\varepsilon)} [p_i(\varepsilon)]^q \right\}}{\log(\varepsilon)} \quad ; \quad D_1 = \lim_{\varepsilon \to 0} \frac{\left(\sum_{i=1}^{N(\varepsilon)} p_i(\varepsilon) \log p_i(\varepsilon) \right)}{\log(\varepsilon)} \tag{3}$$

When applying these relations to an electrodeposited pattern of Zn[13] obtained in a strip geometry, the set of generalized fractal dimension D_q show a notorious q-independent behaviour ($0 \leq q \leq 6$), except maybe for D_0 in the zinc electrodeposit. This seems to indicate that this aggregate shows an appreciable degree of self-similarity, at least over a characteristic and intermediate range of length scales. In fact, the whole set of obtained D_q values may be all approximated in the range: $2^{-5.5} < \varepsilon < 2^{-3.0}$ to $D_q = 1.60 \pm 0.05$. Actually, these errors bars are of the same order of magnitude than those reported in ref. 12, and might be understood on the basis of the limitations of the experimental and digitization procedures here utilized.

This value is in good agreement with previous reported results on experimental electrodeposition patterns D_q ($q \geq 0$) = 1.66 ± 0.04 and D_q (all q) = 1.66 ± 0.08 [12]. Finally, let us compare these results on real electrodeposits with simulated DLA clusters ($M \approx 5.10^4$ particles) generated in a strip geometry using an on-square lattice algorithm with periodic boundary conditions. The value quoted is D_q ($q \geq 0$)= 1.60 ± 0.02, independent of q for a quite large range of length scales: $2^{-7} \leq \varepsilon \leq 2^{-3}$ [12], and for an off-lattice cluster D_q ($0 \leq q \leq 20$) = 1.69 ± 0.03 [14].

As it is nowadays well recognized the proper description of the fractal nature of a growth pattern exceeds the simple characterization of its mass distribution. A deeper insigth into the growth dynamics itself requires the analysis of the growth probability distribution (GPD) measured on the perimeter of an experimental aggregate (Fig. 1). As stated in the introduction section the GPD on every perimeter site of the metallic aggregate will be here computed according to the simplest proportionality assumption i.e., in terms of the normal gradient of a Laplacian field ($P_g(\vec{r}_s) \propto \left| \vec{\nabla}_n \Phi(\vec{r}_s) \right|$), which solves Laplace equation ($\Delta \Phi = 0$) in the region limited by two equipontentials, one set at the surface of the aggregate and the outer one separated from the cathode more than three times the largest height of the deposit. Periodic boundary conditions are imposed on the lateral directions.

The scaling properties of the GPD are analyzed in relation with its moments Z_q, here calculated in terms of the growth probabilities accumulated within square boxes of lattice constant ε, composing a grid which covers the cluster boundary. Plotts of D_q are respectively shown in Fig. 2.

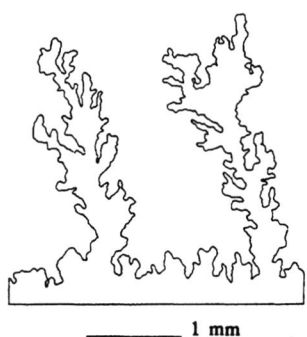

_____ 1 mm

Fig. 1. Perimeter of a digitized image corresponding to an electrodeposit of Zn [13].

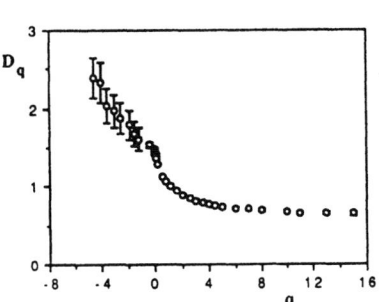

Fig. 2. The generalized dimensions D_q vs. q.

The different growth properties corresponding to the more visited ($q \to \infty$) and more ramified ($q \to -\infty$) regions of the electrodeposit are clearly exhibited. The expected non-trivial scaling of the GPD, as evidenced in previous related studies[15-17], is here apparent. Correspondingly, a entire spectrum of generalized dimensions D_q is clearly necessary to analyze the growth probability measure in fractal terms.

The next step in this standard strategy consists in resolving the exponents τ_q into a density of singularities f_q with singularity strengths α_q. In doing so we avoid the common use of a Legendre transformation, and its numerical complications, to resort to a more direct procedure[18] based on the following equations:

$$\alpha_q \equiv \frac{d}{dq}\tau_q = \lim_{\varepsilon \to 0}\left(\frac{\sum_{i=1}^{N(\varepsilon)} \hat{p}_i(q,\varepsilon)\log p_i(\varepsilon)}{\log \varepsilon}\right) \qquad (4)$$

$$f_q \equiv q\,\alpha_q - \tau_q = \lim_{\varepsilon \to 0}\left(\frac{\sum_{i=1}^{N(\varepsilon)} \hat{p}_i(q,\varepsilon)\log \hat{p}_i(q,\varepsilon)}{\log \varepsilon}\right) \qquad (5)$$

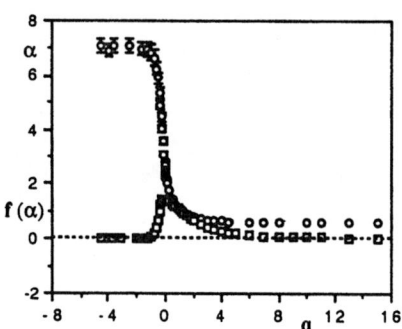

Fig. 3. Plots of α (circles) and $f(\alpha)$ (squares) vs. q.

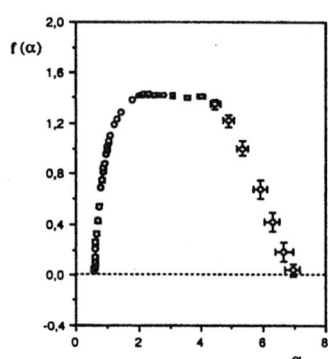

Fig. 4. Plots of the function $f(\alpha)$ vs. q.

which follows from eqs. (2) and (3) and the definition: $\hat{p}_i(q,\varepsilon) \equiv \dfrac{p_i^q(\varepsilon)}{\sum_{j=1}^{N(\varepsilon)} p_j^q(\varepsilon)}$. Results for α_q and f_q are plotted in Fig. 3. Finally the well-known representation in terms of the convex shaped $f(\alpha)$ function is shown in fig. 4.

In the practical computation of the above mentioned measures one is compelled to use a limited range of about two decades on a pixel based scale in order to circumvent both lower and upper cutt offs directly originated in the image acquisition techniques.

The most representative values of the above multifractal analysis applied to the electrodeposition growth process, together with related ones existing in the literature are summarized in the following table:

Table 1

singularity exponents	ECD	CG	VF	DLA
$D_{-\infty} = \alpha_{-\infty}$	7.1±0.3	9.4±0.2	~5.3	~9.0
$f(\alpha_0) = D_0$	1.43±0.00	1.63±0.01	~1.1	1.64±0.01
α_0	2.32±0.03	2.97±0.04	~1.6	~4.0
$f(\alpha_1) = \alpha_1 = D_1 = 1$	1.02±0.01	1.13±0.02	-	1.04±0.01
$D_\infty = \alpha_\infty$	0.59±0.01	0.60±0.04	~1.0	0.64-0.70

where the columns have the following respective meanings: ECD gives the results for electrochemical deposition[5], CG refers to the cristal growth of NH_4Cl[15], VF summarizes the results of a viscous fingering experiment[16] and DLA stands for the results of the numerical solution of Laplace equation, on a DLA aggregate[17].

From their comparison it apears clearly that the scaling behavior of the GPD in fractal electrodeposition is, also from a quantitative point of view, similar to those reported earlier for both experimental[15,16] and DLA[17] fractal like patterns. This is particularly true when referring to the singularities of positive indices q. Actually, peculiar DLA features in the scaling behavior of the GPD corresponding to the more active zones of the electrodeposit are directly evidenced through the independent results: $1 + \alpha_\infty = 1.59±0.01 \approx d_{f(DLA)} \approx 5/3$ (D_q (DLA) $= D_0 = 1.60±0.02$, in ref. 12), and $\tau_3 = 2 D_3 = 1.63±0.02 \approx d_{f(DLA)}$, as respectively proposed by Turkevich and Sher in ref. 19 and by Halsey in ref. 20. Contrarily for $q \leq 0$, as corresponds to the scaling structure of the GPD for the more screened regions, some discrepancies are found and even the value of $D_0 = 1.43$ falls sensibly lower than expected ($D_0 \approx D_q = 1.60±0.05$ [13]). To our understanding, however, these deviations are somewhat spurious and can be interpreted in two different ways. First of all, the linear regression procedures used in relation with eqns. (3-5) become rapidly inaccurate in going to progressively lower values of q (see Fig. 2). On the other hand, and in order to facilitate the numerical resolution of Laplace equation a smoothing procedure aimed to connect the very irregular boundary of our experimental electrodeposit was necessary. This would also explain the appearance of a flat cusp in the $f(\alpha)$ representation (Fig. 4).

In summary, the scaling structure of the growth probability distribution for fractal-like electrodeposited patterns has been investigated. The general behaviour of the different scaling indices in the GPD is well-understood and qualitatively agrees with the general multifractal predictions. From a quantitative point of view the representative results for less screened regions are compatible with the well-known DLA results. Whether or not electrodeposition is indeed Laplacian governed appears from these results as a valid pausible conjecture on which further experimental work is in progress[5].

III. ELECTRODEPOSITION EXPERIMENTS IN A PARALLEL GEOMETRY

Quasi two-dimensional deposits were grown from a thin electrolyte ($ZnSO_4$) solution layer sandwiched between two glass plates.Cell thickness is determined by the diameter of two copper wires, one acting as a cathode and the other one as a spacer in the anodic side. An external anode is used consisting of a zinc bar immersed in the anodic compartment dug out in the lower plate parallel to the cathode. Two main topics have been studied with regard to the morphology of zinc electrodeposits: first, its dependence on concentration and applied potential, and afterwards, its dependence on cell dimensions.

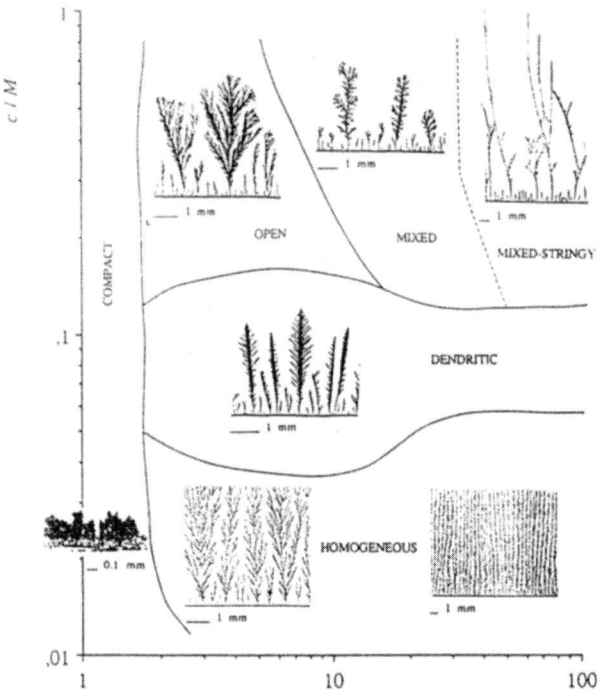

Fig. 5. Diagram of morphologies for zinc electrodeposition.

The different morphologies of zinc electrodeposits for various electrolyte concentration (0.01-0.75 M) and constant applied potentials (1-70 V), at constant cell dimensions (electrode length l 3.5 cm, electrode separation d 3 cm and cell thickness s 70 μm), are summarized in the morphological diagram of Fig. 5. We can observe that deposit morphology strongly depends on potential and concentration values, as reported previously for radial geometry[2,3]. However, besides of the well-known anisotropic dendrites and homogeneous patterns (characterized by a growing front parallel to the cathode), some new features concerning the patterns obtained at high concentration values (c > 0.15 M) should be stressed[4]. First, a growth regime has been distinctively identified at low potential values, characterized by a continuous tip-splitting process, giving rise to the formation of self-similar fractal deposits, as can be clearly deduced from the very similar values of the calculated generalized fractal dimensions. This result allows us to assign to this region, known as open fractal, an average fractal dimension equal to 1.61±0.02, in good agreement with other results[12,13] and with DLA simulations[12,14]. At high potential values, a new mixed pattern is obtained being formed by anisotropic dendritic backbones with open fractal ramification. In this case, the trees grow as a typical dendrite keeping their tips sharp, until they split and stop growing after a given development degree is attained in such a way that only a few trees can reach the anode region. Finally, at higher enough potentials, a dynamic morphological transition from mixed to non ramified stringy structures is observed[7], although a transition front is not seen as reported for other experimental conditions (homogeneous to dendritic transitions in zinc electrodeposits)[4] and other systems (Hecker transitions in copper electrodeposits)[21].

The reported morphological diagrams for various cell dimensions show some differences[2-4] at high concentration values which can not be attributed to changes in cell geometry[4]. For this reason, the effect of cell thickness and electrode separation on mixed and homogeneous patterns under constant potential growth conditions is here reported[7]. At high concentrations, a transition from mixed to thicker and non-ramified stringy patterns is mainly observed as cell thickness (s) is increased[4]. This behaviour which has also been observed under galvanostatic conditions[7] (Fig. 6), could explain the above quoted differences between the diagrams of morphologies reported in references 2-4.

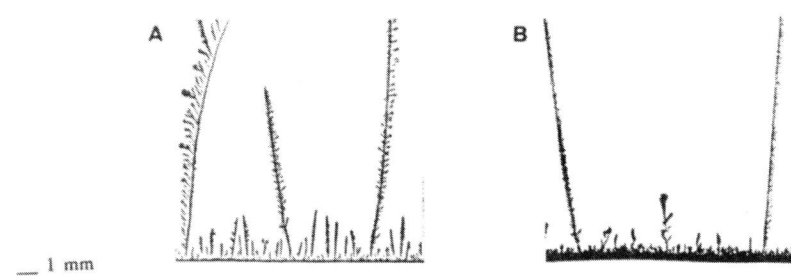

Fig. 6. Effect of cell thickness on galvanostatic patterns at high concentration values. $[Zn^{2+}] = 0.4$ M, I = 30 mA, l = 9 cm, d = 3 cm. A) s = 70 μm, B) s = 240 μm.

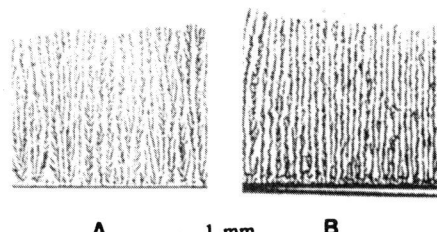

Fig. 7. Effect of cell thickness on homogeneous patterns. $[Zn^{2+}] = 0.02$ M, $\Delta V = 20$ V, l = 9 cm, d = 2 cm. A) s = 70 μm, B) s = 240 μm.

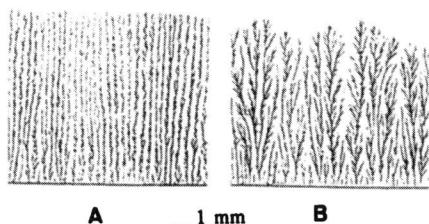

Fig. 8. Effect of electrode separation on homogene ous patterns: $[Zn^{2+}] = 0.02$ M, $\Delta V = 40$ V, l = 9 cm, s = 70 μm, A) d = 3 cm, B) d = 7 cm.

179

On the other hand, the branching degree of an homogeneous deposit is also affected by cell dimensions. In these conditions, a tendency to non-ramified chanel-like deposits is clearly observed as cell thickness (s) is increased and electrode separation (d) is decreased (Fig. 7-8). The electrode separation effect can be understood as a consequence of the change of the electric field at the interface since the deposit is formed at constant potential conditions[22-24].

IV. ELECTRODEPOSITION UNDER FORCED CONVECTION: EXPERIMENTS AND SIMULATIONS.

Transport conditions different from the common pure diffusive ones are worth examining both from the point of view of the simulations and the electrodeposition experiments.

The effect of a convection (advection) flow upon diffusion-limited deposition is investigated using a biased-random walk multiparticle simulation routine, where a superimposed drift represents the convective flow. Radial and parallel geometries have been considered and in both cases very distinct morphologies are found with varying the strength and direction of the flow[9].

Neglecting surface kinetics effects, the growth process is governed by a convection-diffusion law which under a quasistationary approximation and in dimensionless form reads

$$u_x\frac{\partial c}{\partial x} + u_y\frac{\partial c}{\partial y} = \frac{\partial^2 c}{\partial x^2} + \frac{\partial^2 c}{\partial y^2} \tag{6}$$

where $u_{x,y} = U_{x,y}(a/D)$, $c = C/C_0$, a and C_0 being respectively the length and concentration units, and D the diffusion coefficient. Boundary conditions are given by $c = 0$ at the deposit surface and $c = 1$ on the external boundary. In the language of a biased random walker, the central quantity is the probability $P(i,j)$ to visit the site (i,j), this quantity satisfying

$$P(i,j) = \left[\frac{1-P_x-P_y}{4}\right]P(i+1,j) + \left[P_x + \frac{1-P_x-P_y}{4}\right]P(i-1,j) + \left[\frac{1-P_x-P_y}{4}\right]P(i,j+1) + \left[P_y + \frac{1-P_x-P_y}{4}\right]P(i,j-1) \tag{7}$$

where

$$P_{x,y} = \frac{u_{x,y}}{1 + u_x + u_y} \tag{8}$$

In the strip geometry deposition occurs on the bottom line of a square lattice of dimensions (256 x 256). Boundary conditions in the horizontal direction are periodic. We start the simulation filling at random the square lattice until a fraction of lattice sites equivalent to C_0 is occupied. At the end of each iteration we remove the particles that remain at the upper row, the bulk region, and we fill again this row at random to complet the density C_0. In the radial case one proceeds analogously by using a central seed and a lattice of dimensions (512 x 512) where the bulk is represented by a circle of radius 256 lattice units. In this situation the original flow conditions are better represented in terms of tangential an radial drifts which are conveniently transformed to the pairs $P_{x,y}$ in terms of appropriate trigonometric relations.

Fig. 9 illustrates, for the strip geometry, the morphological changes with varying both the density of particles and the strength of the horizontal flow. With increasing P_x the typical multitree structure converts into a single needle or a columnar structure depending on the value of C_0. In any case the aggregates clearly grow against the horizontal flow in a way strongly reminiscent of the simulations of balistic deposition onto inclined surfaces. Analogous conclusions apply for the deposits grown radially as show in Fig 10 .

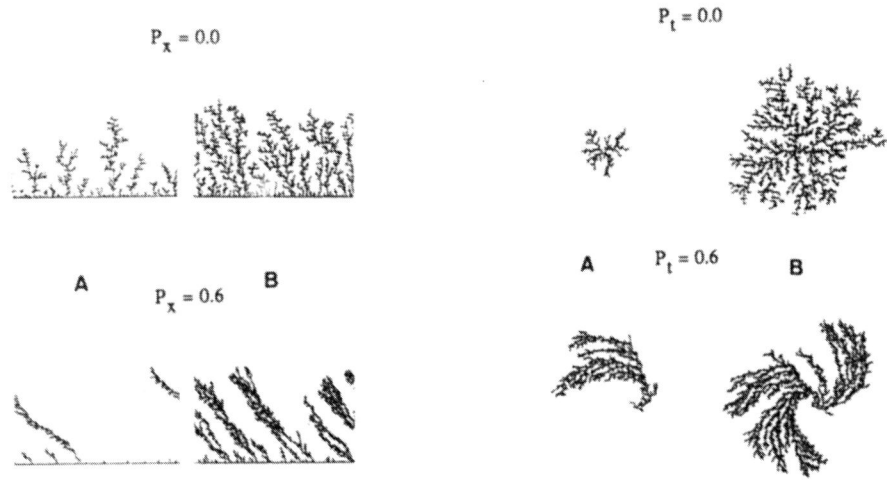

Fig. 9. Effect of an horinzontal drift on a multiparticle DLA pattern. A) 0.03 and B) 0.2 particles/site.

Fig. 10. Effect of a tangential drift on a multiparticle DLA pattern. A) 0.05 and B) 0.2 particles/site.

In order to investigate the effect of an external forced convection into the deposit morphology, an interface radial electrochemical cell has been used. In this case, the electrodeposit grows in the interface between the electrolite solution and an organics (n-butyl acetate)[25]. The cathode is a graphite wire (pencil core of 0.5 mm Ø) inserted perpendicular to the interface and central to the zinc anodic ring of 8 cm of diameter. Convection is superimposed to the cell as a rotation of the interface in relation to the fixed cathode, with an angular velocity ranging from 0 to 12 r.p.m.

Our very preliminar results show that similarly to the simulation results reported above, a distintive loss of the deposits radial simmetry and a certain degree of branch bending are clearly exhibited (Figs. 11). Further interesting aspects of this experiment are presently being investigated, especially those related to the most favorable conditions of concentration and applied potential values for which those effects are enhanced and the role of the forced convection strength on the number, morphology and dynamics of the branching development. These results will be published elsewhere[8].

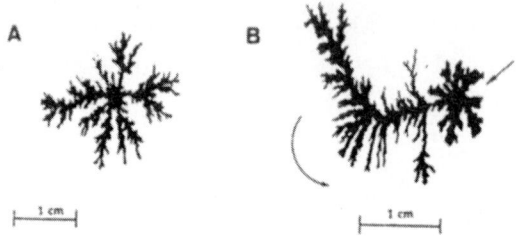

Fig. 11. Effect of forced convection on zinc electrodeposits. Bend arrow shows flow direction. Straight arrow shows the cathode position. $[Zn^{2+}] = 1$ M, $\Delta V = 4$ V; A) 0 r.p.m. and B) 6 r.p.m.

V. ACKNOWLEDGEMENT

Financial support from DGICYT under Project PB90-455 is acknowledged.

VI. REFERENCES

1.- T. Vicsek, "Fractal Growth Phenomena", World Scientific, Singapore, 1989.

2.- Y. Sawada, A. Dougherty and J.P. Gollub, *Phys. Rev. Lett.* 56 (1986) 1260.

3.- D. Grier, E. Ben-Jacob, R. Clarke and L.M. Sander, *Phys. Rev. Lett.* 56 (1986) 1264.

4.- P.P. Trigueros, J. Claret, F. Mas and F. Sagués, *J. Electroanal. Chem.* 312 (1991) 219.

5.- F. Mas and F. Sagués, *Europhys. Lett.* in press.

6.- J. Mach, F. Mas and F. Sagués, to be published.

7.- P.P. Trigueros, J. Claret, F. Mas and F. Sagués, *J. Electroanal. Chem.* in press.

8.- L. López-Tomàs, J. Claret and F. Sagués, to be published.

9.- L. López-Tomàs *et al.*, to be published.

10.-P. Grassberger, *Phys. Lett.* A 97 (1983) 227 ; H.G.E. Hentschel and I. Procaccia, *Physica* D 8 (1983) 435 ; P. Grassberger and I. Procaccia, *Physica* D 13 (1984) 34 .

11.-G. Grasseau, Thèse de Doctorat, Bordeaux (1989); T. Tél, A. Fülöp and T. Vicsek, *Physica* A 159 (1989) 155.

12.-F. Argoul, A. Arneodo, G. Grasseau and H.L. Swinney, *Phys. Rev. Lett.* 61 (1988) 2558. F. Argoul, A. Arneodo, G. Grasseau and H.L. Swinney, *Phys. Rev. Lett.* 63 (1989) 1323

13.-F. Sagués, F. Mas, M. Vilarrasa and J.M. Costa, *J. Electroanal. Chem.* 278 (1990) 351 .

14.-G. Li, L.M. Sander and P. Meakin, *Phys. Rev. Lett.* 63 (1989) 1322..

15.-S. Otha and H. Honjo, *Phys. Rev. Lett.* 60 (1988) 611.

16.-J. Nittmann, H.E. Stanley, E. Touboul and G. Daccord, *Phys. Rev. Lett.* 58 (1987) 619.

17.-Y. Hayakawa, S. Sato and M. Matsushita, *Phys. Rev.* A 36 (1987) 1963.

18.-A. Chhabra and R.V. Jensen, *Phys. Rev. Lett.* 62 (1989) 1327

19.-L.A. Turkevich and H. Scher, *Phys. Rev. Lett.* 55 (1985) 1026.

20.-T.C. Halsey, *Phys. Rev. Lett.*, 59 (1987) 2067.

21.-P. Garik, D. Barkey, E. Ben-Jacob, E. Bochner, N. Broxholm, B. Miller, B. Orr and R. Zamir, *Phys.Rev. Lett.* 62 (1989) 2703.

22.-V. Fleury, J.N. Chazalviel, M. Rosso and B.J. Sapoval, *J. Electroanal. Chem.* 290 (1990) 149.

23.-J.N. Chazalviel, *Phys. Rev.* A 42 (1990) 7355.

24.-J.R. Melrose, D.B. Hibbert and R.C. Ball, *Phys. Rev. Lett.* 65 (1990) 3009.

25.-R. Tamamushi and H. Kaneko, *Electrochim. Acta* 25 (1980) 391.

NATURAL VISCOUS FINGERING

Juan Manuel García-Ruiz

Instituto Andaluz de Geología Mediterránea
CSIC-Universidad de Granada
Av. Fuentenueva s/n   Granada 18002   Spain

INTRODUCTION

The fingering phenomenon that occurs when a low-viscosity fluid displaces another with a higher viscosity[1] has been extensively studied in the last few years from a fractal point of view. Excellent reviews of this phenomenon have been recently published[2-4]. Oil recovery is by far the most classical case of naturally occurring viscous fingering. I will discuss several features observed on an unplanned large scale "experiment" of viscous fingering and I will try to apply these new features to two natural cases for which a morphogenetical explanation linked to viscous fingering has been proposed.

THE PATTERNS OF "LOS LEBREROS"

An anti-crack window pane is usually made up of three parallel glass plates separated from each other by a very thin film of polyvinyl. Once properly sealed with silicone, the system is very stable and does not present any modification over time. However, if there are any small holes in the paraffin sealing the window edges, air is able to penetrate and gain access to the polyvinyl film between the glass plates, creating a potential Hele-Shaw cell[5]. If such a system is heated, thermal constraint separates the glass plates and if the temperature goes above the viscous transition of the polyvinyl film, air invades the gaps created and pushes the viscous thin film. I have observed the results of such a phenomenon at "Los Lebreros" Hotel in Seville. In the coffee shop of this hotel one can observe three beautiful fractal fans (one of them reaching seventy centimeters in diameter) which formed over the last eleven years (Figure 1). The larger fan has a semicircular canopy enveloping the fractal pattern. The mean velocity of the tip propagation (assuming the process to be continuous) is estimated to be 0.1 mm/day. Therefore, the viscous film between the glass plates should be immiscible with air because the characteristic length of the fluid displacement is similar to the characteristic diffusion length, but no diffusion patterns are observed.

This example of "dynamic fractal self-decoration" is

Figure 1.
A pattern observed at the Hotel Los Lebreros. The fractal dimension of this pattern, measured by the box-counting method, is about 70 cm.

interesting beyond simple curiosity and beauty because it presents several features that extend the field of application of viscous fingering to natural processes. Note that the time and space scales are unusually large. This means that when considering applications to natural phenomena, one cannot restrict the viscous fingering phenomenon to the classical short time scale used in laboratory experiments. Moreover, because of the slow rate of development of the pattern, it is possible to observe different stages of development, which have different values for their fractal dimensions when measured by the box-counting method. When deciphering the growth history of a natural process, this last observation, trivial as it may seem, is of sufficient interest to merit further study. For instance, identification of the growth mechanism of a highly complicated pattern emerging from a natural process by fractal geometry can obviously be an excellent tool in the morphogenetical study of both biological and geological systems. The protocol of these studies consists of 1) looking for a self-similar natural structure, 2) measuring its fractal dimension d_f, 3) searching for the existence of physical or chemical instabilities eventually leading to the formation of growth patterns with known d_f values 4) comparing d_f values for the natural structures and the model suspected of generating it. One difficulty arises from the fact that biological and geological structures are the result of a growth history, in many cases without any known reference frame. These structures are the last stage of a morphogenetical sequence generally governed by a decreasing driving force. When dealing with natural patterns, there is no reason to assume that the observed structures should have the d_f value characteristic of their triggering physical mechanism. On the contrary, they are expected to be one of the pre-fractal or simply precursor structures with a lower level of complexity. Let me illustrate the problem using the case of ammonite sutures.

THE CASE OF AMMONITE SUTURES

Ammonites are fossil cephalopods with planispiral shells divided into several camerae by septa[6]. The animal lived in the last-formed chamber and displaced itself forward, segregating at discrete intervals of time a mineralized replica of the rear mantle of its organic body, these replica being called septa. The intersection of the septa and the inner shell forms the so-called suture lines, which in many cases are intrincate fractal patterns. See for instance reference [6], or any other textbook on Paleontilogy, for a detailed description of these suture lines. Based on the fractal characterization and on the physics of the last chamber, which is supposed to work as a vessel pressure, it has recently been proposed[7] that the suture lines are the result of a morphogenetical process conducted by a Saffman-Taylor instability. This theory explains several important characteristics of the suture lines. The large scale patterns observed at "Los Lebreros" is experimental evidence that viscous fingering may be a long morphogenetical process, such as the one which occurred during the life of the ammonite. Moreover, the different suture lines generated by the ammonites when forming septa discretally became more and more convoluted, increasing their complexity as the animal aged, sometimes adopting complicated patterns. This development of complexity is the one recorded by the different patterns observed at "Los Lebreros".

A more surprising characteristic observed in the window panes is that the formation of the pattern should have worked discontinuously. This pulsating mechanism may be the one responsible for the banding structure shown in Figure 2. Temperature is the only parameter that one could imagine to be involved in this system. Whatever the case, the main corollary is that the morphological information generated in a fingering process can be stored and recovered periodically, an important property that can be used in biological development. For instance, in the case of ammonite, the existence of fine lines between two consecutive suture lines has been recorded. According to Hewitt et al.[8], they represent cyclically secreted membranes called pseudosepta. Such intermediate stages could now be explained as the effect of a pulsating Saffman-Taylor instability.

Figure 2. Banding structure in viscous fingering.

The ammonites belonging to the same genus have not identical but very similar fractal suture lines. In particular, the main lobes and saddles always have the same location. This ability to reproduce similar patterns is, in my opinion, the main problem of a biophysical approach based on a fluid instability with high sensibility to initial conditions. Thus, it seems necessary to constrain the pattern formation in order to obtain the reproducibility obtained by ammonoidea in their suture lines. It is known that the behaviour of fluid instability is clearly modified when anisotropy is imposed (for example by engraving a linear groove in one of the glass plates of the Hele Shaw cell)[9]. A first glance at the picture[10] of the patterns formed in this way suggests that the suture line patterns in ammonoidea might be explained by the interplay between fluid instability and the inner shell anysotropy. Another possibility, is the anysotropic behaviour of the organic body of the animal.

One of the mineral structures that can be observed in many field studies are the fractal dendrites of manganese or iron oxides. One can also find these dendrites in any mineral shop, all of these clearly reminiscent of diffusion limited aggregation (DLA) patterns[11] and in many cases with a value of the fractal dimension close to the 1.70 value characteristic of DLA patterns. However, when collecting these manganese or iron oxides in the field, one can observe that the DLA-like patterns are not the general case. On the contrary, one finds many other dendritic patterns with different values for fractal dimensions, many non-dendritic patterns and even non-fractal patterns. Thus, the selection made by the mineral shops would indicate that people like the DLA pattern[12]. One can imagine that just as $(5^{1/2}+1)/2$ is considered to be, as an aesthetic canon, the "*golden number*" of Euclidean geometry, $d_f = 1.70$ could be considered the "*golden number*" of Fractal geometry. The second conclusion, more relevant to our case, is that any explanation of these manganese patterns should explain such a diversity of patterns.

In spite of this common occurrence, the origin of these dendrites is still an open problem. Recently, a modified diffusion limited aggregation model with two counterdiffusing reactants has been proposed by Chopard et al[13]. In our laboratory we have arrived at an alternative explanation for these dendrites as the mineral record of fluid structures. In our view, these dendrites are random fractals where the structures created in the growth environment dominate crystal anisotropy. In order to have a mineral record of such fluid structures, it is necessary that the kinetics of the phase transition be faster than the kinetics of the pattern formation and consequently, the precipitation process must be a highly irreversible one occurring at conditions far from equilibrium. It follows from this consideration that the precipitate formed in this way must be colloidal or in any case, have very low crystallinity and that the whole growth pattern must lack any geometrical relationship derived from the crystal structure. We have observed and evaluated the lack of even short range order in manganese dendrites and also the absence of manganese in the part of the rock surrounding the coating dendrites. I will not go into the details of the process that will be published elsewhere[14] but only briefly discuss the pattern diversity.

When a rising Mn^{2+} solution invades a natural Hele Shaw cell with rough surfaces filled with a colloidal suspension, both percolation and dendritic DLA-like patterns can be expected[15]. The fractal dimension of the patterns depends on the thickness of the cell, the viscosity of the pushed and pushing fluids and the injection rates[3-4]. If, as previously mentioned, we also consider the existence of patterns frozen at early stages of development, it is clear that a wide range of fractal patterns can be expected, as occurs in manganese dendrites[14]. Such a range of fractal behaviour can be extended when geometrical constraints are considered. Returning to the "Los Lebreros" patterns, note that in spite of the semicircular canopy enveloping the "Los Lebreros" pattern, typical of experiments with two-dimensional geometry , they are best described as one plus one. The semicircular shape is due to the large dimensions on the window panes which impede the wall effect currently found in laboratory experiments on a smaller scale. Nevertheless, the larger pattern is close to one of the lateral sides of the pane. The clear fractal pattern that developed over the years becomes a dense branching morphology (DBM)[16] with a non-fractal character when it approached the window edge. Such DBM patterns are also displayed by manganese dendrites (Figure 3).

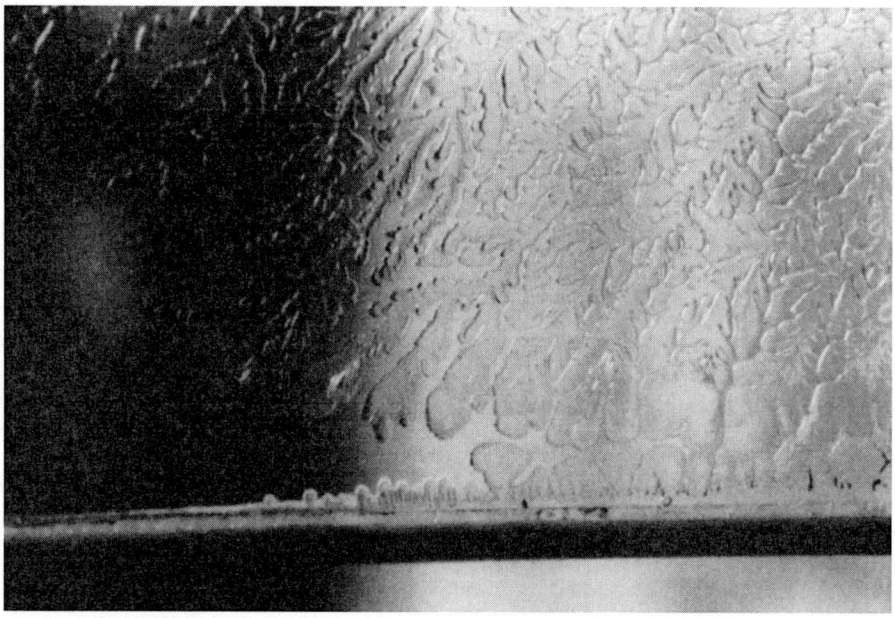

Figure 3. The transition from fractal to non-fractal pattern.

GRAVITY EFFECTS ON VISCOUS FINGERING

Finally, I would like to mention another observation in the "Los Lebreros" patterns. When the injection point is on the lateral sides of the panes, it can be observed that gravity influences the shape of the fractal pattern, sometimes so drastically that it becomes the main morphological control. Consequently, it should be considered that when the fingering process is long enough to be altered by the gravity field, the effect of the latter on the morphology may be important. Several organisms, for instance the slime mold Physarum polycephalum, grows and takes the shape of fractal patterns by a pulsating mechanism consisting of streaming endoplasm into the outer medium. Although a mechanism of viscous fluid has not yet been proposed for these organisms, it merits further consideration, especially as these molds are currently the subject of experiments conducted under microgravity conditions[17].

ACKNOWLEDGMENT

I would like to thank Fermín Otálora for the measurement of the fractal dimensions of the "Los Lebreros" patterns. The work was financed by CICYT Project PA89-0119 and the Junta de Andalucía.

REFERENCES

1. P.G. Saffman and G. Taylor, Proc. R. Soc. London, A245:312 (1958).
2. J. Feder, "Fractals," Plenum Press. New York (1988).
3. H. Van Damme, in "The Fractal Approach to Heterogeneous Chemistry," D. Avnir, ed., John Wiley & Sons, Chichester. (1989). pp 199-226.
4. T. Vicsek, "Fractal Growth Phenomena," World Scientific, Singapore (1989).
5. H.J.S. Hele Shaw, Nature 58:34 (1898).
6. D.M. Raup and S.M. Stanley, "Principles of Paleontology," W.H. Freeman and Co., San Francisco (1971).
7. J.M. García-Ruiz, A. Checa and P. Rivas. Paleobiology 16:349 (1990).
8. R.S. Hewitt, A. Checa, G.G.G. Westermann and P.M. Zaborski. Lethaia 24:271 (1991).
9. K. Horváth, J. Kertész and T. Vicsek, to be published.
10. E. Guyon and H.E. Stanley (ed.), "Fractal Forms," Elsevier-Palais de la Decouverte (1991).
11. T.A. Witten and L.M. Sander, Phys. Rev. B 27:5686 (1983).
12. I enjoyed sharing this idea with Prof. Vicsek during the Workshop.
13. B. Chopard, H.J. Herrmann and T. Vicsek, Nature 353:409 (1991).
14. J.M. García-Ruiz, F. Otálora, A. Sanchez-Navas and F.J. Higes Rolando, Submitted for publication.
15. R. Lenormand. in: "Fractals In Natural Sciences," Princeton University Press, Princeton (1989). pp 159-168.
16. R. Lenormand and G. Daccord, in: "Random Fluctuations and Pattern Growth," H.E. Stanley and N. Ostrowsky, eds., Kluwer Academic Publishers, Dordrecht (1988). pp 69-74.
17. I. Block and W. Briegleb, Adv. Space Res. 9:1175 (1989) and personal communication.

FIBONACCI SEQUENCES IN DIFFUSION-LIMITED AGGREGATION

A. Arneodo[1], F. Argoul[1], E. Bacry[2], J.F. Muzy[1] and M. Tabbard[1]

[1]Centre de Recherche Paul Pascal, Av. Schweitzer, 33600 Pessac, France
[2]Ecole Normale Supérieure, 45 rue d'Ulm, 75230 Paris Cedex 05, France

INTRODUCTION

Pattern formation in systems far from equilibrium is a subject of considerable current interest[1-6]. Recently, much effort has been directed towards the study of fractal growth phenomena in physical, chemical and biological systems[7,8]. Unfortunately, the understanding of phenomena like viscous fingering in Hele-Shaw cells[9] and electrochemical deposition[10] is hampered by the mathematical complexity of the problem. Highly ramified structures are generally produced in the zero surface tension limit. In this limit, both processes are equivalent to a Stefan problem[11]: a diffusion problem for the pressure or the electrochemical potential, with boundary values specified on the moving interface, whose local velocity is in turn determined by the normal gradient of the Laplace field. This highly nonlinear problem is not readily amenable even to modern numerical simulations. When solving the Stefan problem by direct means, the interface develops unphysical cusps in a finite time[9]. One is thus led to introduce some short-distance cutoff which in some sense mimics surface tension[12,13]. Thus far no computer simulations of the equations of motion achieve the necessary size to make definite conclusions about the deterministic character of the fractal patterns observed in the experiments[1-6].

An alternative to solving the Stefan problem consists of simulating the diffusion limited aggregation (DLA) model introduced by Witten and Sander[14] in the early eighties. In this prototype model, an aggregate is grown by the successive accretion of random walkers to the perimeter sites of the cluster. On lattice and off-lattice computer investigations[14,15] have produced complex, apparently randomly branched fractals that bear a striking resemblance to the tenuous tree-like structures observed in viscous fingering, electrodeposition, bacterial growth and neuronal outgrowth[1-8]. The structure of these aggregates has been analyzed by computational[14-23] and analytical methods[24-28]. But despite its appealing simplicity, the DLA model has resisted all attempts at a full physical understanding and there is still no rigorous theory for diffusion-limited growth processes. One of the main obstacles to theoretical progress is the lack of structural characterization of the growing clusters. Most of the numerical analysis have focused on the determination of the so-called generalized fractal dimensions which all coincide to the fractal dimension[29-31]: $D_q = D_F^{DLA} = 1.60 \pm 0.02, \forall q$. Even though these results definitely establish the statistical self-similarity of the DLA clusters, they are insufficient because they bring only limited information about the puzzling DLA architecture. In particular, it is still an open question whether or not some structural order is hidden in the apparently disordered DLA morphology. In this communication, we use the wavelet transform microscope[32,33] to explore the intricate fractal geometry of large-mass off-lattice DLA clusters. This analysis reveals the existence of Fibonacci sequences in the internal "extinct" region of these clusters[34,35]. It also indicates that this underlying hierarchy is likely to be intimately related to a predominant structural five-fold symmetry.

Growth Patterns in Physical Sciences and Biology, Edited
by J. M. Garcia-Ruiz *et al.*, Plenum Press, New York, 1993

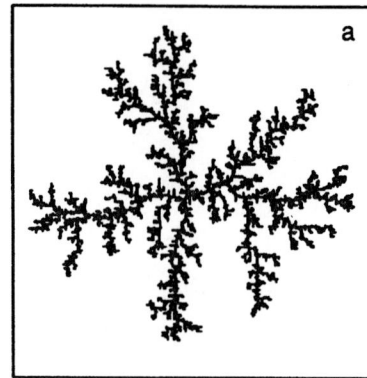
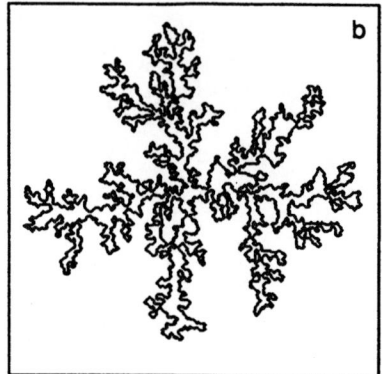

Fig. 1 (a) A 5000 particles on-lattice DLA cluster computed with the random walker model of Witten and Sander[14]. (b) The active zone as defined by the perimeter sites that are accessible to the random walkers.

We suggest an interpretation of the DLA geometry as a "quasifractal" counterpart of the well-ordered snowflake fractal morphology.

FIVE-FOLD SYMMETRY IN THE FRACTAL MORPHOLOGY OF DLA CLUSTERS

Most activity in the context of irreversible growth processes has been focused on the geometrical properties of growing aggregates[1-6,14-32]. In the early numerical studies, on-lattice DLA clusters were found to have different scaling properties in the radial and the azimuthal directions, raising the question of self-affinity (rather than self-similarity) for these fractal aggregates[20,23,36-38]. Further large scale simulations of off-lattice clusters have shown that the existence of two different scaling exponents is only a cross-over (finite-size) effect that vanishes in the asymptotic limit of large mass[31,34,39]. The statistical "mono-fractality" of DLA clusters is now well admitted and rather accurately established[30-34] by the measurement of the generalized fractal dimensions D_q using box-counting and fixed-mass algorithms. But these dimensions are statistical quantities (thermodynamical functions) that do not provide deep insight into the geometrical complexity of the aggregate[33]. To achieve a more refined structural analysis we thus need a tool which is well adapted to the large hierarchy of scales involved in fractal patterns[7]. A known method which comes close to satisfying this requirement is the wavelet transform[40-42].

Wavelet analysis is a mathematical technique introduced recently for analyzing seismic data and accoustic signals[40,41]. Since then, the wavelet transform has been the subject of considerable theoretical developments and practical applications in a wide variety of fields[43-45]. The wavelet transform of a one-dimensional signal consists in decomposing the signal into elementary contributions, the so-called wavelets, which are constructed from one single function g by means of dilations and translations[40-42]. The generalization to higher dimensions involves rotations as well[46]. Here, we will concentrate on the analysis of fractal aggregates embedded in a two-dimensional space[32]. Hence, let $d\mu(\vec{x}) = \rho(\vec{x})d\vec{x}$ be the measure with density $\rho(\vec{x})$ (in practice, a mass density); for the sake of simplicity, let g be a regular radially symmetric function that is localized around the origin. The wavelet transform of μ with respect to the analyzing wavelet g is defined as[32]:

$$T_g[\mu](a, \vec{b}) = \int \overline{g}\left(\frac{\vec{x} - \vec{b}}{a}\right) d\mu(\vec{x}) . \tag{1}$$

This transformation may be inverted provided $< g >= 0$. The wavelet transform can thus be regarded as a mathematical microscope, for which position and magnification correspond to $\vec{b}$ and a^{-1}, respectively, and the performance of the optics is determined

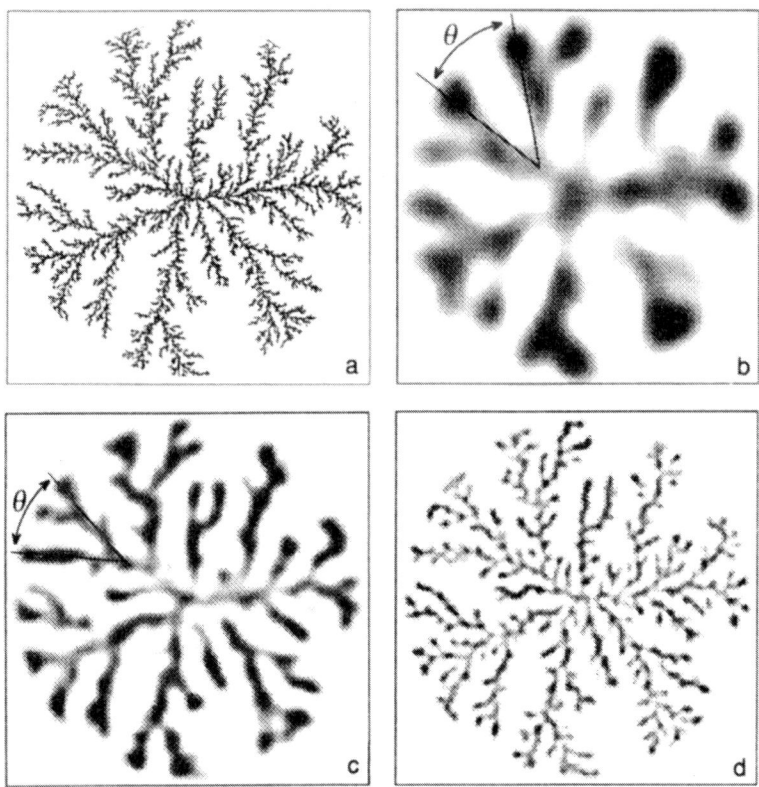

Fig. 2 Wavelet analysis of an off-lattice DLA cluster. (a) The inner frozen region of an off-lattice DLA cluster of mass $M = 10^6$; about $8 \, 10^4$ particles are contained in a disk of radius $R = 480$ particle sizes. In (b), (c) and (d), this frozen region is explored with the Mexican hat microscope $(g(\vec{x}) = (2 - |\vec{x}|^2)e^{-|\vec{x}|^2/2})$, for magnification a^{-1}, $(2.2) \, a^{-1}$ and $(2.2)^2 \, a^{-1}$ respectively. The intensity of the wavelet transform is coded using 32 shades from white $(T_g \leq 0)$ to black $(\max T_g)$.

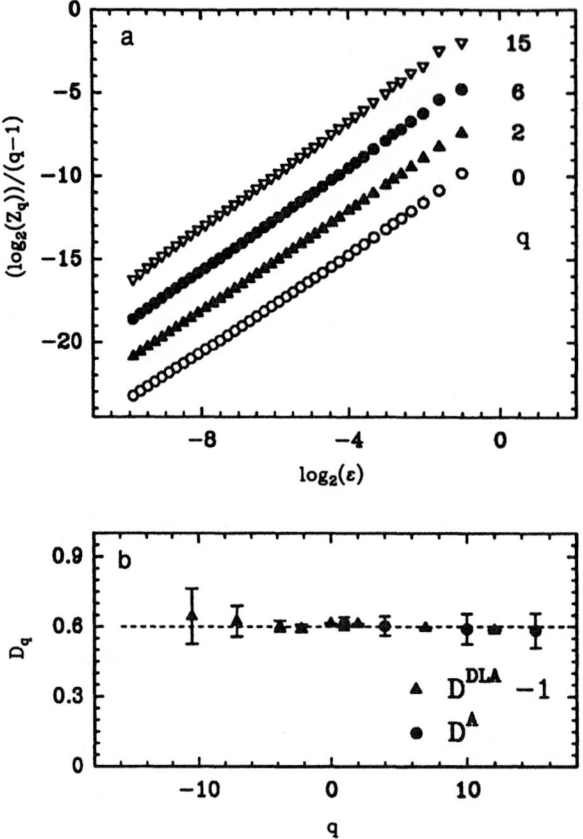

Fig. 3 (a) These graphs illustrate the determination of the generalized fractal dimensions D_q for the inner frozen region of large mass off-lattice DLA clusters; the D_q are estimated from linear regression fits of $log_2 Z_q(\epsilon)/(q-1)$ vs $log_2(\epsilon)$, where $Z_q = \sum_i \mu_i^q(\epsilon)$. (b) Box-counting ($q > 0$) and fixed mass ($q < 0$) computation of the D_q's of the frozen region of the DLA cluster ($\bullet$) and of the azimuthal Cantor set ($\blacktriangle$) (Fig. 5b).

by the choice of the analyzing wavelet g. Of course, this microscope is isotropic only if g is radially symmetric. The wavelet transform microscope has proven to be well suited for studying the scaling properties of fractal objects[33,47-51]. Here, however, we will mainly exploit its amazing ability to reveal the structural hierarchy of fractal aggregates, as recently tested on snowflake patterns that display a well organized "crystalline" fractal architecture[32].

As illustrated in Fig. 1, a characteristic feature of diffusion-limited aggregation is the fact that most of the growth takes place in an "active" zone, near the outer radius of the cluster, which collects practically all the new particles. This active zone moves outward, leaving behind an "extinct" region[52-54] that can be considered as asymptotic in the sense that it will no longer be modified by further growth. This screening of the inner region by the tips is the basic physical reason for the fractal branching in DLA growth. The investigation of the fractal DLA edifice requires this inner frozen region to be large enough to contain several generations of branching. In Fig. 2, we show the inner central region (corresponding to inaccessible perimeter sites for the random walkers) of a 10^6 particles off-lattice cluster generated using an efficient algorithm[34,35] which combines the simplicity of the off-lattice algorithm designed in Ref. 55, to the rapidity of on-lattice hierarchical algorithms[19]. As illustrated in Figs 2b, 2c and 2d

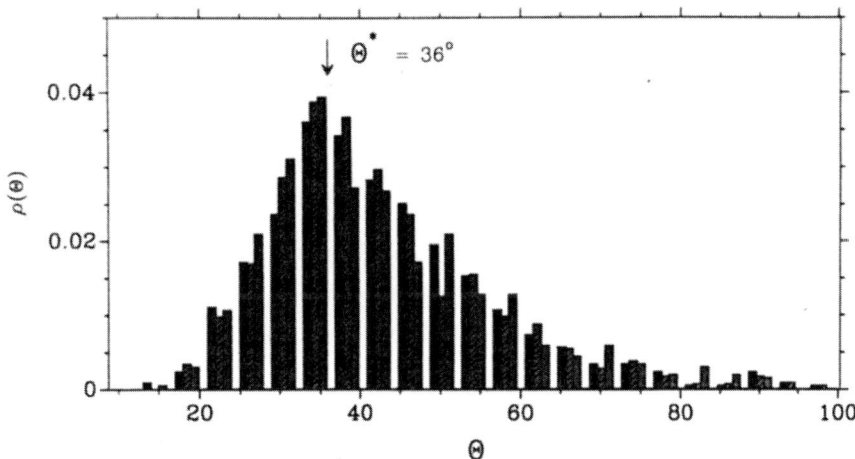

Fig. 4 Histogram of screening angle values at the branching bifurcations in the wavelet
transform representation (Figs 2b and 2c) of 4 off-lattice DLA clusters; three
magnifications a^{-1} (black), $(2.2)\,a^{-1}$ (dashed) and $(2.2)^2\,a^{-1}$ (grey) are shown,
corresponding respectively to three successive generations of branching. A single
maximum is observed for $\theta^* \sim 36°$.

respectively, when increasing the magnification of the wavelet transform microscope,
one reveals progessively the successive generations of branching. A first indication of
the statistical self-similarity of the DLA clusters is that these branchings occur rather
uniformly in space without any preferential location and this at all scales. The results
of box-counting and fixed-mass dimension measurement in Fig. 3 provide a quantitative
confirmation of this qualitative finding: the generalized fractal dimensions are found
to be equal to the fractal dimension[34]: $D_q = D_F^{DLA} = 1.60 \pm 0.02, \forall q$. Note that this
numerical value is the same, up to the numerical uncertainty, as for the entire aggregate.
This is consistent with the recent numerical demonstration that the subset of inaccessible
sites is a fat fractal that involves $\sim 37\%$ of the total perimeter sites[53,54] (Fig. 1b).

Moreover, at each generation of branching, the wavelet transform provides an efficient way to measure the screening angle between bifurcating branches (Figs 2b and
2c). In Fig. 4, we present the results of a systematic screening angle investigation of
4 off-lattice DLA clusters, similar to the one shown in Fig. 2a. Three shades are used
to differentiate the distributions obtained for three values of the magnification corresponding to successive branching generations. The three histograms are actually almost
indistinguishable, which clearly suggests that the statistical self-similarity of DLA clusters is intimately related to the existence of a screening angle distribution that is scale
invariant[35]. This distribution diplays a (unique) maximum at the value $\theta^* \sim 36° = \pi/5$.
The presence of a pentagonal symmetry at a macroscopic level, in diffusion-limited aggregation, has already been suggested in previous works[25,26,56]. The existence of this
symmetry at all scales, however, is likely to be a clue to a structural hierarchical fractal
ordering.

QUASIFRACTALITY IN DLA CLUSTERS

The intimate relationship between regular pentagons and Fibonacci numbers and
the golden mean $\phi = 2\cos(\pi/5) = 1.618\ldots$ has been well known for a long time[57]. The
proportions of a pentagon approximate the proportions between adjacent Fibonacci
numbers; the higher the numbers are, the more exact the approximation to the golden
mean becomes. The angle defined by the sides of the star and the regular pentagons
is $\theta^* = 36°$, while the ratio of their length is a Fibonacci ratio (F_{n+1}/F_n). The recent
discovery of "quasicrystals"[58] in solid state physics, is a spectacular manifestation of

this relationship. This new organization of atoms in solids, intermediate between perfect order and disorder, generalizes to the crystalline "forbidden" symmetries, the properties of incommensurate structures. Similarly, there is room for "quasifractals" between the well-ordered fractal hierarchy of snowflakes and the disordered structure of chaotic or random aggregates[5-8]. This section is devoted to the demonstration that DLA clusters are possible "quasifractal" candidates[34,35].

Fibonacci sequences are naturally generated by the recursive process[57]:

$$A \to AB \quad , \quad B \to A . \tag{2}$$

If one starts with the species B at the generation $n = 0$, one gets A at the generation $n = 1$, and successively AB, ABA, $ABAAB$, $ABAABABA$, ... In other words, the population F_n at the generation n can be deduced from the two preceeding populations, F_{n-1} and F_{n-2}, according to the iterative law:

$$F_n = F_{n-1} + F_{n-2} \quad , \quad F_0 = F_1 = 1 . \tag{3}$$

Note that F_{n-1} and F_{n-2} are also the respective populations of A and B at step n. A remarkable property of the Fibonacci series $\{F_n\} = \{1, 1, 2, 3, 5, 8, 13, 21, 34 \ldots\}$ is that the ratio of two consecutive Fibonacci numbers converges to the magic irrational number, the golden mean:

$$\lim_{n \to +\infty} F_{n+1}/F_n = \phi = (1 + \sqrt{5})/2 = 1.618\ldots \tag{4}$$

Evidence of Fibonacci sequences has been reported in various contexts as diverse as mathematics, art, architecture, sciences and technology[57]. In particular, the golden mean arithmetic has been shown to play a fundamental role in the growth of phyllotactic patterns in the botanical world[59,60]. But there are more intriguing geometries in nature that involve Fibonacci numbers. Some trees, root systems, algae, blood vessels and the bronchial architecture do appear to exhibit morphologies that are strikingly similar to DLA fractal patterns[1-8]. It is thus tempting to speculate how far the search for Fibonacci ordering can be pushed in the context of fractal growth phenomena. As a first step of our demonstration, we will focus our wavelet analysis on the azimuthal Cantor set[34] defined by the intersection of the DLA cluster with the circle of radius R that delimits the extinct region (Fig. 2a). As recently addressed in various theoretical studies[61-64], the wavelet analysis of singular measures actually does not require the analyzing wavelet g to be of zero mean. In this section, we will use the one-dimensional version of the wavelet transform defined in Eq. (1), with a Gaussian function $g(x) = e^{-x^2/2}$.

For a pedagogical purpose, we first show in Fig. 5a, the wavelet transform of the uniform triadic Cantor set[33,48-51]. Indeed, only the skeleton defined by the positions of the local maxima of $|T_g(a,x)|$ are represented in this figure[61,62]. Although we have reduced considerably the amount of data for the representation, the so-obtained tree-like structure reveals the construction rule of the self-similar triadic Cantor set. (We refer the reader to Refs 63, 65, 66 for rigourous results concerning the modulus maxima wavelet transform representation). At the scale $a = a_0 3^{-n}$, each one of the $k_0 2^n$ modulus maxima simultaneously bifurcates into 2 new maxima, giving rise to a cascade of symmetric pitchfork branchings in the limit $a \to 0$ (a_0 and k_0 are constants that depend on the specific shape of g). The fractal dimension $D_F = \ln 2/\ln 3$ of the uniform triadic Cantor set can be directly obtained from the branching ratio $r_B = 2$ and the scale factor (length ratio) $r_L = 3$, according to the general formula[61-63]:

$$D_F = \ln r_B / \ln r_L . \tag{5}$$

The wavelet transform modulus maxima representation of the azimuthal Cantor set[34] of a 10^6 particles off-lattice DLA cluster is shown in Fig. 5b. At first sight, one does not see any conspicuous recursive structure in the wavelet transform skeleton.

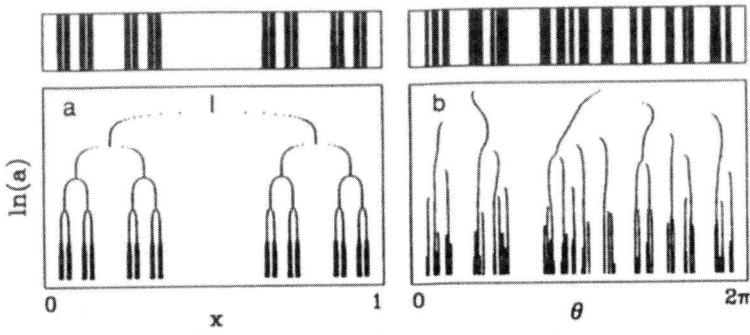

Fig. 5 Wavelet transform skeleton defined from the local maxima of $|T_g(a,x)|$ considered as a function of x. (a) The uniform triadic Cantor set. (b) The azimuthal DLA Cantor set. The analyzing wavelet $g(x)$ is the Gaussian function.

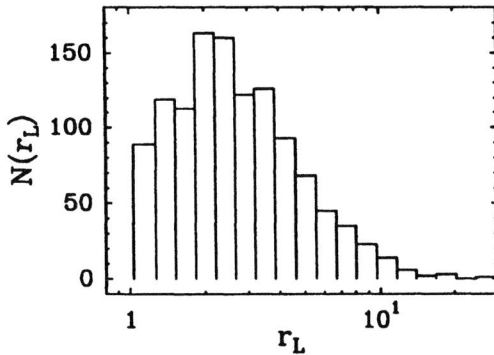

Fig. 6 Histogram of values of the scale factor r_L separating two successive bifurcations (($\bullet$) in Fig. 5b) in the wavelet transform modulus maxima skeleton of 23 DLA azimuthal Cantor sets. A single maximum is observed for $r_L^* = 2.2 \pm 0.2$.

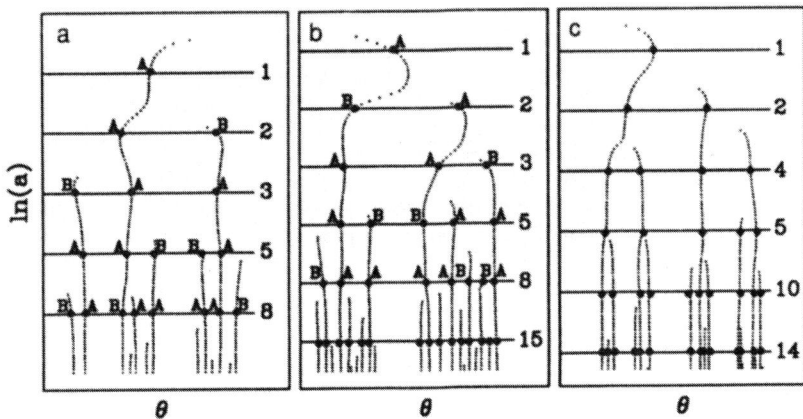

Fig. 7 Enlargements of the wavelet transform skeleton of the azimuthal Cantor set in Fig. 5b, corresponding to three distinct main branches of a large mass off-lattice DLA cluster. The horizontal lines mark the scales $a_n = a_0 r_L^{*-n}$ with $r_L^* = 2.2$. In (a) and (b), the number of wavelet transform modulus maxima at each generation follows the Fibonacci series (3) (see text); moreover, a symbol A or B can be assigned to each of these maxima according to the Fibonacci resursive process (2). (c) illustrates some local departure from the Fibonacci structural ordering.

One can, however, proceed to a systematic investigation of the value of the scale factor r_L between two successive bifurcations (black dots in Fig. 5b). The results of a statistical analysis of 23 azimuthal Cantor sets, similar to the one in Fig. 5b, are reported in Fig. 6. The distribution of scale factors displays a (unique) maximum at the value $r_L^* = 2.2 \pm 0.2$. Now, with the additional numerical information obtained by box-counting and fixed mass computation of the generalized fractal dimensions of the azimuthal Cantor set in Fig. 3b, one can insert the numerical values $D_F^A = 0.60 \pm 0.02$ (Note that the observation that $D_F^A = D_F^{DLA} - 1$ is in good agreement with Mandelbrot's argumentation[67] concerning one-dimensional cuts of homogeneous fractals embedded in a two-dimensional space.) and $r_L^* = 2.2 \pm 0.2$ into Eq. (5); doing so, one gets a branching ratio $r_B^* \sim (2.2)^{0.6} \sim 1.61$. This numerical value is significantly different from 2, which unambiguously discards the possibility of an exact binary branching process. The most striking feature is that the value r_B^* is remarkably close to the golden mean ϕ.

To further establish the relevance of the golden mean arithmetic to the statistical self-similarity of DLA clusters, we have magnified in Fig. 7 three regions of the wavelet transform skeleton (Fig. 5b), corresponding to three well-separated regions of the azimuthal Cantor set issued from three distinct main branches of the considered off-lattice DLA cluster. The horizontal lines in the (a, θ) half-plane are drawn as guide marks for the successive "generations" of wavelet transform modulus maxima. From the histogram in Fig. 6, these generations are (in a statistical sense) expected to occur preferentially at scales $a_n = a_0 r_L^{*-n} = a_0 (2.2)^{-n}$, where a_0 is a constant that depends on the size of the DLA branch under study. As seen in Fig. 7a, the number of wavelet transform modulus maxima at each generation follows closely the Fibonacci series defined in Eq. (3). As indicated in Fig. 7b, some deviations from the Fibonacci ordering are observed at small scales, but this is not surprising since at scales a of the order of a few particle sizes, the azimuthal Cantor set is very sensitive to small changes in the radius R of the circle which delimits the frozen region of the DLA clusters. The overall Fibonacci ordering is, however, rather robust with respect to the arbitrariness of the choice of this circle. As illustrated in Figs 7a and 7b, by assigning a symbol A or B to each maxima line issued from a bifurcation, one obtains a coding of the wavelet transform skeleton that complies with the recursive law (2). However, a systematic investigation of this coding for our statistical sample reveals some randomness in the relative position of symbols A and B at the bifurcations $A \to AB$. Apparently B is equally likely to be found on the right or on the left of A. There exists also some arbitrariness in the spatial location of A when B proceeds to $B \to A$; as discussed in the next section, this arbitrariness is likely to result from local fluctuations in the value of the screening angle (about $\theta^* = 36°$) in the DLA branching morphology. These fluctuations can produce some local departures from the Fibonacci ordering as shown in Fig. 7c. A close examination of the wavelet transform skeleton in Fig. 5b reveals the presence of many of these defects. But the Fibonacci sequences are statistically predominant in the wavelet transform modulus maxima representation of the 23 off-lattice DLA azimuthal Cantor sets investigated in this study. In that respect, our wavelet analysis[34] provides the first numerical evidence for the existence of a "Fibonaccian" quasifractal structural ordering in DLA clusters.

FIBONACCI SEQUENCES IN THE FRACTAL BRANCHING OF DLA CLUSTERS

A fundamental step in our demonstration is now to return to the DLA cluster itself and to point out Fibonacci sequences in its disordered branched morphology[35]. In Fig. 8, we use the two-dimensional wavelet transform microscope (Eq. (1)) to explore the internal structure of two main branches of a 10^6 particles off-lattice DLA cluster. The analyzing wavelet is the so-called Mexican hat and the magnification is such that three successive significant branchings are resolved. These branchings proceed according to the Fibonacci recursion law (2), as identified by assigning a symbol A or B to the branches of successive generations. In Fig. 8a, the original branch A gives two branches A and B; both these branches bifurcate into two new branches, but one of these branches, issued from B, dies before reaching the reference circle that delimits the frozen region of the DLA cluster. This peculiar electrostatic screening actually governs the growth process and originates in a statistical Fibonacci structural hierarchy with a branching

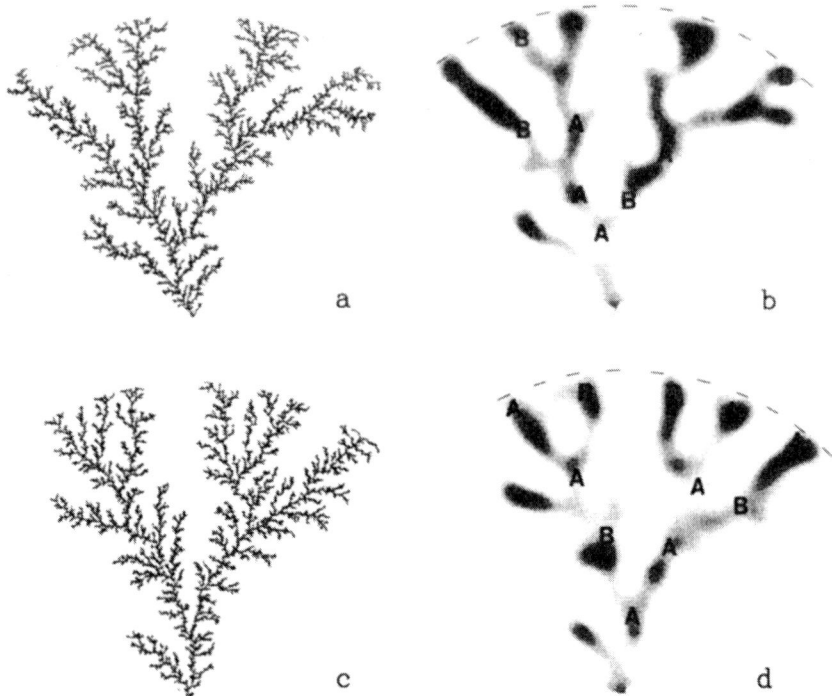

Fig. 8 Two main extinct branches of a 10^6 particles off-lattice DLA cluster ((a) and (c)) as seen through the Mexican hat microscope ((b) and (d) respectively). In (b) and (d), T_g is coded using 32 shades from white ($T_g \leq 0$) to black (max T_g). The magnification a^{-1} is such that three successive generations of branching are identified. At each branching, a symbol A or B can be assigned to the new branches according to the Fibonacci recursive process (2).

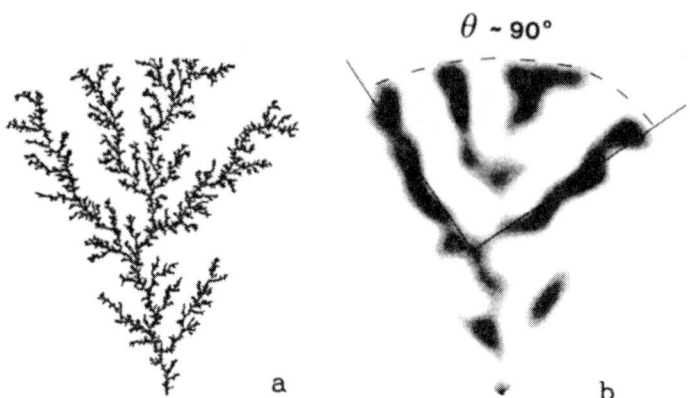

$\theta \sim 90°$

Fig. 9 One main extinct branch of a 10^6 particles off-lattice DLA cluster as seen through the Mexican hat microscope. The wavelet transform reveals a structural defect to the Fibonacci ordering induced by some local fluctuation of the screening angle ($\theta \sim 90°$) away from $\theta^* = 36°$.

ratio that converges asymptotically to the golden mean[34,35]. As previously noticed in the wavelet transform skeleton of the azimuthal Cantor set, this Fibonacci architecture contains, however, some randomness in the relative position of the branches A and B at the bifurcations $A \to AB$. Note that the observation of a perfect Fibonacci ordering in Figs 8a and 8b, coincides with a succession of screening angles that do not significantly deviate from $\theta^* = 36°$.

But the histogram in Fig. 4 is rather widely spread around $\theta^* = 36°$, which clearly indicates the existence of important fluctuations in the screening angle value. As illustrated in Fig. 9, these fluctuations can produce some local departure from the Fibonacci structural ordering. The screening angle $\theta^* \sim 90°$ at the primary branching is so large that there is enough space between the two outcoming branches for a new branch to grow and split almost immediatly into two new branches with screening angle $\theta \sim 36°$. Despite the presence of many of these local structural defects, the DLA branching organization does exhibit a fascinating prevalent tendancy to be Fibonaccian. A systematic investigation of the actual relationship between the Fibonacci branching ordering and the structural five-fold symmetry is currently in progress.

CONCLUSION

To summarize, we have reported the discovery of Fibonacci sequences in the frozen morphology of large-mass off-lattice DLA clusters. This observation is consistent with a preferential branching ratio that converges asymptotically to the golden mean. As seen through the wavelet transform microscope, this "quasifractal" hierarchy is likely to be related to a structural five-fold symmetry. These results provide an important clue to the theoretical understanding of diffusion-limited aggregation and should guide future work addressing the crucial issue of the selection mechanism of the DLA morphology.

ACKNOWLEDGEMENTS

We are very grateful to Y. Couder and V. Hakim for interesting discussions. This work was supported by the Direction des Recherches Etudes et Techniques under contrat (DRET N°89/196) and the Centre National des Etudes Spatiales under contrat (N°91/CNES/0323).

REFERENCES

1. H. E. Stanley and N. Ostrowsky, eds., "On Growth and Form: Fractal and Non-Fractal Patterns in Physics", Martinus Nijhof, Dordrecht (1986).
2. L. Pietronero and E. Tosati, eds., "Fractals in Physics", North-Holland, Amsterdam (1986).
3. W. Guttinger and D. Dangelmayr, eds., "The Physics of Structure Formation", Springer-Verlag, Berlin (1987).
4. H. E. Stanley and N. Ostrowsky, eds., "Random Fluctuations and Patterns Growth", Kluwer, Dordrecht (1988).
5. J. Feder, "Fractals", Pergamon, New York (1988).
6. T. Vicsek, "Fractal Growth Phenomena", World Scientific, Singapore (1989).
7. B. B. Mandelbrot, "The Fractal Geometry of Nature", Freeman, San Francisco (1982).
8. A. Aharony and J. Feder, "Fractals in Physics", Essays in honour of B.B. Mandelbrot, Physica D 38 (1989).
9. D. Bensimon, L.P. Kadanoff, S. Liang, B.I. Shraiman and C. Tang, Rev. Mod. Phys. 58:977 (1986).
10. L. M. Sander, in Ref. 3, p.257.
11. L. Rubinstein, "The Stefan Problem", A.M.S., Providence (1971).
12. L. M. Sander, P. Ramanlal and E. Ben Jacob, Phys. Rev. A 32:3160 (1985).
13. P. Ramanlal and L.M. Sander, J. Phys. A 21:L995 (1988).

14. T. Witten and L.M. Sander, Phys. Rev. Lett. 47:1400 (1981); Phys. Rev. B 27:5686 (1983).
15. P. Meakin, in "Phase Transitions and Critical Phenomena", Vol. 12, C. Domb and J.L. Lebowitz, eds., Academic Press, Orlando (1988).
16. P. Meakin, Phys. Rev. A 27:1495 (1983).
17. P. Meakin and Z.R. Wasserman, Chem. Phys. 91:391 (1984).
18. P. Meakin and L.M. Sander, Phys. Rev. Lett. 54:2053 (1985).
19. R. C. Ball and R.M. Brady, J. Phys. A 18:L809 (1985).
20. P. Meakin, Phys. Rev. A 33:3371 (1986).
21. J. Nittmann and H.E. Stanley, Nature 321:663 (1986).
22. J. Nittmann and H.E. Stanley, J. Phys. A 20:L1185 (1987).
23. P. Meakin, R.C. Ball, P. Ramanlal and L.M. Sander, Phys. Rev. A 35:5233 (1987).
24. L. A. Turkevich and H. Scher, Phys. Rev. Lett. 55:1026 (1985).
25. R. C. Ball, Physica A 140:62 (1986).
26. T. C. Halsey, P. Meakin and I. Procaccia, Phys. Rev. Lett. 56:854 (1986).
27. M. Matsushita, K. Konda, H. Toyoki, Y. Hayakawa and H. Kondo, J. Phys. Soc. Jpn. 55:2618 (1986).
28. L. Pietronero, A. Erzan and C. Evertz, Phys. Rev. Lett. 61:861 (1988); Physica A 151:207 (1988).
29. P. Meakin and S. Havlin, Phys. Rev. A 36:4428 (1987).
30. F. Argoul, A. Arneodo, G. Grasseau and H.L. Swinney, Phys. Rev. Lett. 61:2558 (1988); 63:1323 (1989).
31. G. Li, L.M. Sander and P. Meakin, Phys. Rev. Lett. 63:1322 (1989).
32. F. Argoul, A. Arneodo, J. Elezgaray, G. Grasseau and R. Murenzi, Phys. Lett. A 135:327 (1989); Phys. Rev. A 41:5537 (1990).
33. A. Arneodo, F. Argoul, J. Elezgaray and G. Grasseau, in "Nonlinear Dynamics", G. Turchetti, ed., World Scientific, Singapore (1989) p. 130.
34. A. Arneodo, F. Argoul, E. Bacry, J.F. Muzy and M. Tabard, "Golden mean arithmetic in the fractal branching of diffusion-limited aggregates", preprint (December 1991) submitted to Phys. Rev. Letters.
35. A. Arneodo, F. Argoul, J.F. Muzy and M. Tabard, "Uncovering Fibonacci sequences in the fractal morphology of diffusion-limited aggregates", preprint (January 1992) submitted to Nature.
36. R. C. Ball, R.M. Brady, G. Rossi and B.R. Thompson, Phys. Rev. Lett. 55:1406 (1985).
37. M. Kolb, J. Phys. Lett. 46:L631 (1985).
38. P. Meakin and T. Viscek, Phys. Rev. A 32:685 (1985).
39. P. Ossadnik, "Branch order and ramification analysis of large DLA clusters", preprint (September 1991).
40. A. Grossmann and J. Morlet, S.I.A.M. J. Math. Anal. 15:723 (1984).
41. A. Grossmann and J. Morlet, in "Mathematics and Physics, Lectures on Recent Results", L. Streit, ed., World Scientific, Singapore (1987).
42. I. Daubechies, A. Grossmann and Y. Meyer, J. Math. Phys. 27:1271 (1986).
43. J. M. Combes, A. Grossmann and P. Tchamitchian, eds., "Wavelets", Springer-Verlag, Berlin (1988).
44. Y. Meyer, "Ondelettes", Herman, Paris (1990).
45. P. G. Lemarié, "Les ondelettes en 1989", Springer-Verlag, Berlin (1990).
46. R. Murenzi, Thesis, University Catholique de Louvain (1990).
47. M. Holschneider, J. Stat. Phys. 50:963 (1988); Thesis, University of Aix-Marseille II (1988).
48. A. Arneodo, G. Grasseau and M. Holschneider, Phys. Rev. Lett. 61:2281 (1988).
49. A. Arneodo, G. Grasseau and M. Holschneider, in ref. 43, p. 182.
50. A. Arneodo, F. Argoul and G. Grasseau, in ref. 45, p. 125.
51. A. Arneodo, F. Argoul, E. Bacry, J. Elezgaray, E. Freysz, G. Grasseau, J.F. Muzy and B. Pouligny, "Wavelet transform of fractals: I. From the transition to chaos to fully developed turbulence. II. Optical wavelet transform of fractal growth phenomena", to appear in "Wavelets and some of their Applications", Springer-Verlag, Berlin (1991).
52. M. Plischke and Z. Rácz, Phys. Rev. Lett. 53:415 (1984).

53. F. Argoul, A. Arneodo, J. Elezgaray and G. Grasseau, in "Measure of Complexity and Chaos", N.B. Abraham, A.M. Albano, A. Passamante and P.E. Rapp, eds., Plenum, New York (1989).

54. C. Amitrano, P. Meakin and H.E. Stanley, Phys. Rev. A 40:1713 (1989).

55. B. Derrida, V. Hakim and J. Vannimenus, Phys. Rev. A 43:888 (1991).

56. I. Procaccia and R. Zeitak, Phys. Rev. Lett. 60:2511 (1988).

57. T. H. Garland, "Fascinating Fibonacci: Mystery and Magic in Numbers", Dayle Seymour, Palo Alto (1987).

58. L. Michel and D. Gratias, eds., Proceedings of the International Workshop on "Aperiodic Crystals" (Les Houches, 1986), J. Phys. (Paris) 47:C3 (1986).

59. R. V. Jean, "Mathematical Approach to Patterns and Forms in Plant Growth", Wiley, New York (1984).

60. T. A. Steeves and I.M. Sussex, "Patterns in Plan Development", Cambridge Univ. Press, Cambridge (1984).

61. J. F. Muzy, E. Bacry and A. Arneodo, Phys. Rev. Lett. 67:3515 (1991).

62. A. Arneodo, E. Bacry and J.F. Muzy, "Wavelet analysis of fractal signals: direct determination of the singularity spectrum of fully developed turbulence data", Springer-Verlag, Berlin (1991) to appear.

63. E. Bacry, J.F. Muzy and A. Arneodo, "Singularity spectrum of fractal signals from wavelet analysis: exact results", preprint (January 1992) submitted to J. Stat. Phys.

64. J. M. Ghez and S. Vaienti, "Integrated wavelets and fractal sets. The generalized fractal dimensions", preprint (1991).

65. S. Mallat and S. Zhong, "Complete signal representation with multi-scale edges", Technical Report 483 (Comput. Sci. Dept., NYU, 1989) submitted to IEEE Transactions on Pattern Analysis and Machine Intelligence.

66. S. Mallat and W.L. Hwang, "Singularity detection and processing with wavelets", Technical Report 549 (Comput. Sci. Dept., NYU, 1990).

67. B. B. Mandelbrot, J. Stat. Phys. 34:895 (1984); in "Fractals: Physical Origin and Properties", L. Pietronero, ed., Plenum, New York (1989).

PATTERN FORMATION IN SCREENED ELECTROSTATIC FIELDS: GROWTH IN A CHANNEL AND IN TWO DIMENSIONS

O. Plá[†,§], J. Castellá[‡], F. Guinea[†], E. Louis[‡],
and L. M. Sander[§]

[†] Instituto de Ciencia de Materiales (CSIC)
Facultad de Ciencias. C-XII. Universidad Autónoma
28049 Madrid. Spain
[‡] Departamento de Física Aplicada
Universidad de Alicante
Apartado 99. 03080 Alicante. Spain
[§] Department of Physics
University of Michigan
Ann Arbor MI 48109-1120

INTRODUCTION

The Diffusion-Limited-Aggregation (DLA)[1,2] and the Dielectric Breakdown (DB)[3] models have been very successful in illustrating the possibility of fractal growth[4] in Laplacian fields. Nature does offer, however, a much richer scenario in which both fractal and non-fractal patterns may grow. The dependence on the boundary conditions in DB discussed in ref. 5 illustrates this point: a change in the shape of electrodes induces drastic changes in the growing patterns which evolve into a rather dense multi-branched structure with fractal dimension $D \sim 2$. Although several reasons have been suggested[5] to explain this dependence on the boundary conditions, among which we mention the existence of a threshold field and the internal resistance of the breakdown pattern (plasma channels in the case of a discharge in a gas), in very few instances their effects have been analysed in any depth. Only the possibility of a different growth law in which the growth rate is assumed to be proportional to a power η of the local field, different in general from unity, has been examined in detail in the DB context[3], and by utilizing an equivalent approach in DLA[6], and used to explain the more diluted than DLA patterns that may occur in Nature. Note, however, that there are microscopic reasons for expecting $\eta = 1$ in DB[7]. More recently, the possibility of a crossover from a DLA pattern to a more diluted one has also been investigated by using more complicated growth laws, both in DB[8] and in the somewhat similar phenomena of mechanical

Growth Patterns in Physical Sciences and Biology, Edited
by J. M. Garcia-Ruiz *et al.*, Plenum Press, New York, 1993

breakdown[9-11]. The variety of structures further increases for the growth of metallic aggregates through Electrochemical Deposition (ECD). These may have a fractal character like in DLA, be dendritic crystals, or give rise to dense radial structures[12-16]; the stability of the latter has been ascribed to the finite resistivity of the aggregate[12], or to the anion migration between the electrodes[13,16]. Besides, a transition from a dense pattern to a more diluted branched structure has been observed[17,18] and referred to as the Hecker transition[17].

In this paper we investigate the effects of screening[19] on structures growing in electrostatic fields. The origin of screening might lie on the presence of free charges such as in the case of ECD and DB. From elementary considerations of thermal equilibrium, the Debye-Hückel theory deduces the existence of a screening length which depends on the total density of charges and the temperature[20]. The same situation may arise in DLA, in this case screening might be due to the presence of sinks (screening) or an ambient of particles (antiscreening). We have carried out numerical simulations and an analytical study along the lines proposed by Mullins and Sekerka[21]. The results show that screening leads to a much richer variety of patterns. It introduces a new length scale and a non-trivial dependence on the boundary conditions which, as discussed below, is responsible for a transition that resembles the Hecker transition. Patterns may have a fractal character at shorter scales than the screening length, be Eden-like or grow dense. The mentioned transition (from dense to multibranched growth) is shown to occur at a point that depends on the potentials at the two boundaries, the distance between them and the screening length.

MODEL AND NUMERICAL METHODS

We concentrate on the DB model[3]. In that model, a breakdown pattern is allowed to grow in a dielectric medium placed between two electrodes at different potentials. The aggregate is assumed to be a perfect conductor, and, thus, at constant potential, whereas fields in the dielectric follow the Laplace equation. To account for screening we replace the Laplace equation by

$$\nabla^2 \phi = \lambda^2 \phi \qquad (1)$$

where λ^{-1} is the screening length. Antiscreening would correspond to a minus sign on the right hand side of Eq. (1); its effects will be briefly discussed at the end of this paper. We shall consider a planar geometry (growth in a channel) and growth in two dimensions.

To originate an aggregate the standard growth procedure was followed assuming a growth rate proportional to the absolute value of the local field at the surface of the aggregate. The potential was fixed at the outer (ϕ^o) and the inner (ϕ^i) electrodes, the latter also gave the potential at the aggregate, assumed to be a perfect metal at constant potential. In the following we shall consider λ and the potentials ϕ^o and ϕ^i as tunable positive parameters. In an actual experiment, ϕ^o and ϕ^i will be determined by the external circuit, which fixes the potential drop, and by the requirement of charge conservation. As a consequence Eq. (1) is not gauge invariant and it only has physical meaning for the gauge implied by the choice of potentials just mentioned. We also note that care should be taken that the linearization implicit in the Debye-Hückel theory can be applied.

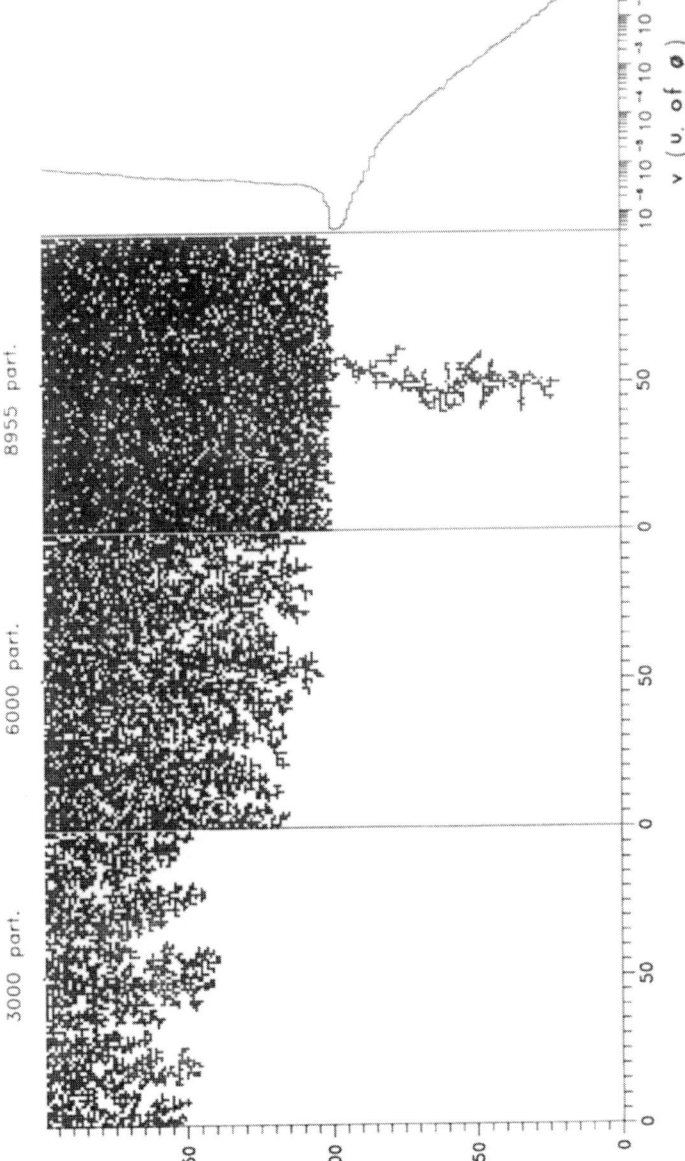

Figure 1. Growth in a channel: Patterns at three stages of growing and growth velocity (average of the absolute value of the electric field at the surface of the aggregate) for $\lambda^{-1} = 10$, $\phi^{o} = 1$ and $\phi^{i} = 10^{-4}$.

Numerical simulations were carried out on samples of the square lattice of sizes 100×200 for the case of growth in a channel, and on samples of the square and the triangular lattices with radii up to 100, for two-dimensional growth. In the former case the boundaries along the longest direction were taken as the electrodes, whereas periodic boundary conditions were used in the shorter direction. On the other hand simulations in $2D$ were carried out for a variety of shapes of the outer electrode, in particular circular, triangular and square electrodes were considered. As usual Eq. (1) was solved iteratively. It has to be remarked that screening can strongly decrease the potential at the surface, changing quickly (see Fig. 1) as the pattern evolves, thus the error used to stop the iteration process should be decreased or increased properly. Our criterion was that the maximum error at each node was less than one per cent the average value of the electric field at the boundary of the pattern. This gave around 50 iterations to relax the electrostatic field.

ANALYSIS OF THE INSTABILITY

To get a qualitative idea of the effects of screening on the growth process, we first analyze the stability of a slightly deformed smooth surface by following the treatment first discussed in ref. 17. We discuss both the stability of a flat surface growing in a channel, and that of a circle growing in a circular cell.

Growth in a channel

Let us consider a flat boundary, say $y = l$, growing in the y-direction between two electrodes at potentials ϕ^o and ϕ^i respectively and itself at a constant potential ϕ^i. We then deform the surface as $y^n = l + \delta \cos(mx)$, δ being very small. In the screened case the potential takes the form (setting $\phi = \phi^i$ at $y = y^n$)

$$\phi(x, y) = \phi_o(y) + E(l)\delta \exp(-\sqrt{\lambda^2 + m^2}\,(y - l))\cos(mx) \qquad (2)$$

where

$$\phi_o(y) = \frac{\phi^o \sinh[\lambda(y - l)] + \phi^i \sinh[\lambda(L - y)]}{\sinh[\lambda(L - l)]}, \qquad (3)$$

L is the length of the cell in the growing direction (y), and $E(l)$ is the electric field at the flat surface $(y = l)$

$$E(l) = \frac{\phi^i \cosh[\lambda(L - l)] - \phi^o}{\sinh[\lambda(L - l)]}\,\lambda\,. \qquad (4)$$

Then, assuming that the growth rate v is proportional to the absolute value of the field at the surface of the aggregate and writing $v = \dot{l} + \dot{\delta}\cos(mx)$, we find for the ratio between the instantaneous rates of growth of the perturbation (δ) and that of the flat surface (l) the following expression

$$\alpha_m = \frac{\dot{\delta}/\delta}{\dot{l}/l} = \left| \sqrt{\lambda^2 + m^2} - \frac{\lambda^2 \phi^i}{E(l)} \right| l \qquad (5)$$

In the case of no screening Eq. (5) reduces to the known result $\alpha_m = ml$. When

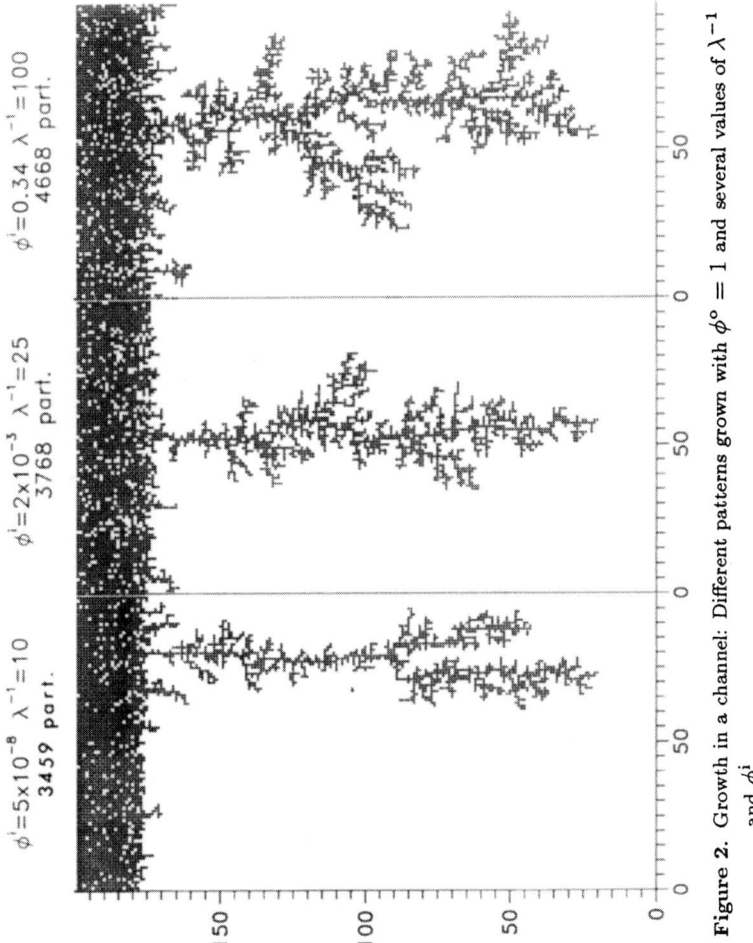

Figure 2. Growth in a channel: Different patterns grown with $\phi^o = 1$ and several values of λ^{-1} and ϕ^i.

screening is present the instantaneous growth rate depends on the potentials at the electrodes, the gap between them $(L - l)$ and the screening length (λ^{-1}). Two cases should be differentiated. For $\phi^o < \phi^i$, the field at the surface of the aggregate has the same polarization of the electrodes for all values of l $(E(l) > 0)$. Thus, the 2nd term in the r.h.s. of Eq. (5) is always negative, decreasing in absolute value as the pattern evolves. Consequently, the effect of screening will be to create dense structures which become more dilute while growing, but still resembling the Eden model.

More interesting changes are found in the case of $\phi^o > \phi^i$. In this case the most appealing feature is the possibility that $E(l)$ vanishes. For $L \gg l$ and provided that $\phi^o/\phi^i < \cosh(\lambda L)$, $E(l)$ is opposite to the polarization of the electrodes $(E(l) > 0)$. It goes to zero at a given value of the length of the aggregate (l_c) and, beyond this point, it is always negative. As a consequence, the second term in α_m is initially negative, as in the previous case, but now it increases in absolute value. As the length of the aggregate increases, and the zero of the denominator is approached, the screening term slows down the perturbations of every wavelength, and the growth rate is reduced. The α_m will vanish at different values of l, favouring dense growth. Then, at a distance which depends on the parameters of the problem, (ϕ^o, ϕ^i and the screening length λ^{-1}) $E(l)$ vanishes and the growth rates for all m go to infinity. The sharpness of the transition from dense to ramified growth will depend on the width of the range of l over which the α_m go from zero to infinity. For $\lambda \ll 1$, $\alpha_m = 0$ ($m \geq 1$) at a length of the aggregate (l) similar to that for which $E(l)$ vanishes and the transition will be very abrupt. Beyond l_c point all wavelengths become unstable. Once a sharp tip develops, it will be amplified. This behavior is a consequence of the potential and its associated charge distribution. Before the transition, there is a small screening layer near the growing electrode. In the intermediate region the potential decreases to a value close to zero, to rise again near the external electrodes. Beyond the transition these two layers merge and the potential increases monotonously between the aggregate and the outer electrode.

Growth in two dimensions

We now consider a perfectly metallic aggregate of circular shape, i.e., a disc of radius r^i at constant potential ϕ^i, growing in a dielectric medium, and an outer electrode also circular $(r = r^o > r^i)$ at potential ϕ^o. Here we shall only discuss the case $\phi^o > \phi^i$. The field at a point $z = \lambda r$ in the dielectric, takes the form

$$E(z) = \frac{\phi^o[K_0(z^i)I_1(z) + I_0(z^i)K_1(z)] - \phi^i[K_1(z)I_0(z^o) + I_1(z)K_0(z^o)]}{I_0(z^i)K_0(z^o) - I_0(z^o)K_0(z^i)}\lambda \qquad (6)$$

where $z^{i,o} = \lambda r^{i,o}$ and I_n and K_n are the modified Bessel functions. As in the 1D case the field vanishes at a point z^c. For $z^i < z^c$, $E(z)$ is first ($z \sim z^i$) opposite to the polarization of the electrodes, and goes through zero at a point between the two electrodes. Instead, for $z^i > z^c$ the field has for all z the same polarization of the electrodes. At $z^i = z^c$ the field vanishes at the surface of the aggregate. This behavior determines the stability of the circularly shaped aggregate, as found in the one-dimensional case. To carry out the stability analysis we deform the circular surface of the aggregate as $z_n = z^i + \lambda \delta \cos(n\theta)$ where δ is small. The result for the ratio between the instantaneous rates of growth of the perturbation (δ) and that of the disc (z^{i}) is

$$\alpha_n = \frac{\dot{\delta}/\delta}{\dot{z}^{\mathrm{i}}/z^{\mathrm{i}}} = \left| n - 1 + z^{\mathrm{i}} \left\{ \frac{K_{n-1}(z^{\mathrm{i}})}{K_n(z^{\mathrm{i}})} - \frac{\lambda}{E(z^{\mathrm{i}})} \frac{\phi^{\mathrm{i}}}{} \right\} \right| \tag{7}$$

The behavior of α_n is similar to that obtained in the previous case. Again the α_n vanish at different values of z^{i}, favoring dense growth. At the point where $E(z)$ vanish, α_n for all n goes to infinity, and the transition from dense to ramified growth takes place. The discussion that follows Eq. (5) is also valid here.

NUMERICAL SIMULATIONS

The results of the numerical simulations carried out in this work, illustrated in Figs. 1-4, fully coincide with the predictions of the analysis of the instability discussed in the preceeding section. We first comment on the results for the one-dimensional case. Fig. 1 shows the aggregate at different stages during the growing process. As predicted, growth sharply changes from dense to ramified . We have calculated, from Eq. (5), the length at which the transition should occur for the parameters of the figure, resulting $l \sim 101$, in excellent agreement with the numerical results. We also note that $\alpha_m = 0$ for $m \geq 1$ at $l = 100 - 101$, explaining the sharpness of the transition displayed in Fig. 1. It should be pointed out that this is a remarkable demonstration of the validity of the analysis first suggested by Mullins and Sekerka[21]. The occurrence of the transition in a wide range of the parameters is further illustrated in Fig. 2, again it takes place at the points predicted by the analytical study. In Fig. 1 we have also plotted the average growth speed, that is the average of the absolute value of the field at the aggregate surface. We note that, as discussed above, the velocity nearly vanishes at the transition. Finally we refer to the width of the branches that grow beyond the transition point.

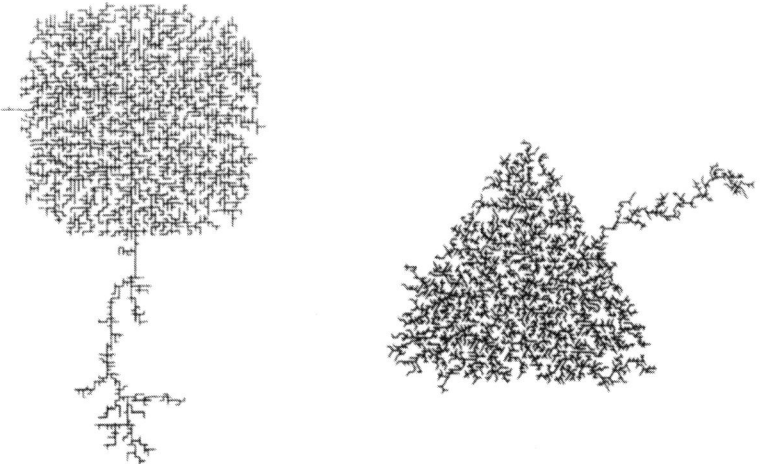

Figure 3. Growth in two-dimensions for $\lambda^{-1} = 10$ and $\phi^{\circ} = 1$. Patterns grown with a square electrode ($\phi^{\mathrm{i}} = 0.002$) and a triangular electrode ($\phi^{\mathrm{i}} = 0.062$) are shown.

As shown in Fig. 2 its width increases as λ decreases, being DLA-like at scales shorter than the screening length λ^{-1}. It is remarkable that screening produces dilute patterns without the need of using a growth rate proportional to a power η of the electric field different from unity.

We have also considered under which conditions several branches may develop. Screening reduces the range of the interaction, and, therefore, should pose no problems to the growth of parallel branches. In the simulations outlined above, however, particles are added one at a time. This effect induces a sharp threshold in the velocity of growth, so that points where the field exceed this threshold will grow and not others. As the fields in this screened situation have an exponential dependence on the separation between electrodes, this artificial cut off prevents most of the front from growing. To overcome this difficulty a front of particles can be attached stochastically at different sites without rearranging the potential in the dielectric and several branches can grow simultaneously[19].

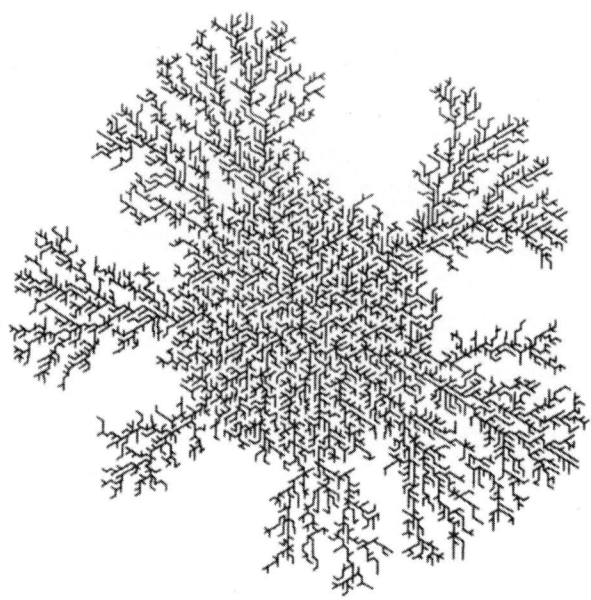

Figure 4. Growth in two dimensions: Circular electrode, $\lambda^{-1} = 10$, $\phi^o = 1$ and $\phi^i = 0.006$. 75% of the pixels contained in the average perimeter were attached at each step.

Figs. 3 and 4 show some patterns obtained in the two dimensional case. The transition is also well defined, due to the choice of parameters. The most outstanding feature of the aggregates shown in Fig. 3 is their shape before the transition, that mimics the shape of the outer electrode. This is a consequence of screening, as in the case of no screening aggregates grow isotropic and do only reflect the shape of the outer electrode when they get very close to it. This effect could be easily understood by looking at the equipotential lines between two electrodes at different potentials, the inner a single point at the centre and the outer of an arbitrary shape. In the unscreened case the equipotential lines are almost circular up to very near the outer electrode. In the presence of screening and due to the much shorter range of the interaction, the equipotential lines reflect the shape of the outer electrode even far from it, as only the zones that are at the shortest distances contribute appreciably to the local potential. Although this explanation is rather obvious, we have checked that the results of Fig. 3 are not lattice effects by considering the case of a triangular electrode in the square lattice and viceversa. The results demonstrate that the shape of the aggregate only depends on the shape of the electrode and not on the choice of a particular lattice. Finally we note that the effect of allowing a front of particles to stick simultaneously to the aggregate surface is similar to what can be found for growth in a channel (Fig. 4). In this case we have kept constant the flux of particles (instead of the number as is done when dealing with the $1D$ case).

DISCUSSION AND CONCLUDING REMARKS

At first glance the transition mentioned above share many common features with the so-called Hecker transition[17] observed in ECD. In both cases dense and filamentary patterns develop at different times. The complexity of the real experiments greatly exceeds the simple model described here, and, presumably, other effects like those derived from the propagation of fronts of charged impurities towards the cathode, as discussed in refs. 13 and 16, shall also be taken into account in a complete theory. However, some aspects of the Hecker transition are similar to those of the present model. The transition described before is characterized by a change in the sign of the electrostatic field at the surface of the aggregate. If that takes place in the Hecker transition, a change in the charge of the chemical species being accumulated near the cathode should also take place. This conclusion seems to agree with experimental findings, where a change in color associated with a change in the material being deposited (metal oxides are replaced by metallic ions in going through the transition) has been reported. A similar change in color of the solution has been identified as a change in the pH, which occurs simultaneously with the transition. Moreover, assuming the screening length to be much smaller than the size of the cell, the occurrence of the transition in the centre of the cell, as observed experimentally, requires that $\phi^o \gg \phi^i$ (Fig. 1).This implies that, before the transition, the field near the aggregate will be opposite to the polarization of the electrodes (as outlined above), although very low. Hence this electrostatic barrier can easily be overcome by cations through diffusive process.

We turn now to comment briefly on the effects of antiscreening. As remarked above this would correspond to the presence of an ambient (or sources) of particles in DLA, and might be relevant in ECD as far as ions could be generated anywhere between the electrodes. A procedure similar to that described above gives the following expression for the instantaneous growth rate in the one dimensional case

$$\alpha_m = \left| \theta(m^2 - \lambda^2)\sqrt{m^2 - \lambda^2} + \frac{\lambda^2 \phi^i}{E(l)} \right| l \tag{8}$$

where θ is the step function. Now the field at the surface of the aggregate is an oscillating function of l and, as a consequence, the growth rate also oscillates. This feature originates a behavior which is even richer than that found in the screening case. For instance, a transition similar to that described above may also take place, although in this case it occurs for $\phi^i > \phi^o$.

In conclusion, we have presented an investigation of the effects of screening on growth phenomena in screened electrostatic fields. Screening strongly increases the diversity of patterns, giving rise, under certain conditions and in a very simple way to a transition from dense to multibranched growth. On the other hand the shape of the aggregate before this transition closely reflects that of the outer electrode. Some of these features are similar to those of the so-called Hecker transition observed in ECD.

Acknowledgments

Two of us (JC and OP) wish to acknowledge finantial support from the Ministerio de Educación y Ciencia, Spain. We are also grateful to A. Aldaz, J.A. Vallés-Abarca and J. Vazquéz for useful suggestions and comments. LMS is supported by the USA National Science Foundation grant DMR 88-15908, and FG and EL by the spanish CICYT grant MAT91-0905-C02-02.

REFERENCES

1. T. A. Witten and L. M. Sander, Phys. Rev. Lett. **47**, 1400 (1981).
2. T. A. Witten and L. M. Sander, Phys. Rev. B **27**, 2586 (1983).
3. L. Niemeyer, L. Pietronero and H. J. Wiesmann, Phys. Rev. Lett. **52**, 1033 (1984).
4. B. B. Mandelbrot, "Fractal Geometry of Nature," Freeman, New York (1982).
5. L. Niemeyer, L. Pietronero and H. J. Wiesmann, Phys. Rev. Lett. **57**, 650 (1986).
6. J. H. Kaufman, G. M. Dimino and P. Meakin, Physica A, **157**, 656 (1989).
7. I. Gallimberti, J. Phys. (Paris) Colloq. **40**, c7-1936 (1979).
8. E. Arian, P. Alstrom, A. Aharony and H. E. Stanley, Phys. Rev. Lett. **63**, 3670 (1990).
9. E. Louis and F. Guinea, Europhys. Lett. **3**, 871 (1987).
10. E. Louis and F. Guinea, Physica D **38**, 235 (1989).
11. O. Pla, F. Guinea, E. Louis, G. Li, L. M. Sander, H. Yan and P. Meakin, Phys. Rev. A **42**, 3670 (1990).
12. D. G. Grier, D. A. Kessler and L. M. Sander, Phys. Rev. Lett. **59**, 2315 (1987).
13. V. Fleury, M. Rosso and J.-N. Chazalviel, Phys. Rev. B **43**, 690, (1991).
14. P. Garik, D. Barkley, E. Ben-Jacob, E. Bochner, N. Broxholm, B. Miller, B. Orr and R. Zamir, Phys. Rev. Lett. **62**, 2703 (1989).
15. V. Fleury, J.-N. Chazalviel, M. Rosso and B. Sapoval, J. Electroanal. Chem. **290**, 249 (1990).
16. J. R. Melrose, D. B. Hibbert and R. C. Ball, Phys. Rev. Lett. **65**, 3009(1990).
17. N. Hecker, D. G. Grier and L. M. Sander, in: "Fractal Aspects Of Materials", R. B. Laibowitz, B. B. Mandelbrot and D. E. Passoja, ed., Materials Research Society, University Park, PA (1985).
18. L. M. Sander, in "The Physics of Structure Formation", W. Guttinger and G. Dangelmayr, ed., Springer-Verlag, Berlin (1987).
19. E. Louis, F. Guinea, O. Pla and L. M. Sander, to be published.
20. W. J. Moore, "Physical Chemistry", vol. 1, Prentice-Hall Inc., New Jersey (1962).
21. W. W. Mullins and R. F. Sekerka, J. Appl. Phys. **34**, 323 (1963).

SELF ORGANIZED CRITICALITY IN SIMPLE GROWTH MODELS

O. Pla[†]*, F. Guinea[†], and E. Louis[‡]

[†] Instituto de Ciencia de Materiales (CSIC)
Facultad de Ciencias, Universidad Autónoma
28049 Madrid, Spain
[‡] Departamento de Física Aplicada
Universidad de Alicante
Apartado 99, 03080 Alicante, Spain

A stimulating development in the analysis of complex systems has been the hypothesis of *self organized criticality*[1,2]. The main assumption is that such systems evolve towards a state where small perturbations can give rise to changes (catastrophes) of all sizes. This state describes the most unstable situation compatible with some kind of equilibrium. The distribution of catastrophes, because of its inherent scale invariance, is characterized by power laws. In previous work, we have checked that this hypothesis is well satisfied in systems which evolve into a steady state far from equilibrium[3]. The model we analyzed is Diffusion Limited Aggregation[4], for which a comprehensive amount of work is already available[4].

In the present work we study the mechanisms through which the self organized critical regime is established, as the complexity of the system is switched on. The SOC hypothesis is only applicable to situations which cannot be described by simple, deterministic laws. When the behavior of the system is very predictable, it will show only certain kinds of catastrophes. The basic assumption required for the existence of SOC, a scale invariant noise spectrum, must break down.

To analyze in detail this question, we study a simple generalization of the DLA model: the Dielectric Breakdown (DB) model with a scale invariant growth law[5]. As in DLA, the growth is determined by a scalar field, ϕ, which obeys the Laplace equation outside the aggregate. The velocity of growth, however, depends on the gradient of this field raised to some power, $v \propto |\nabla\phi|^{\eta}$. When $\eta = 1$, we recover DLA, and $\eta = 0$ describes the Eden model. For $\eta \gg 1$, growth takes place only in those points at the surface of the aggregate where the field, $\nabla\phi$, has a maximum. Thus, the system behaves in a simple and deterministic way. In this limit, only narrow needles can grow, with a

---

* present address Department of Physics, University of Michigan, Ann Arbor MI 48109-1120

Growth Patterns in Physical Sciences and Biology, Edited
by J. M. Garcia-Ruiz *et al.*, Plenum Press, New York, 1993

Figure 1. Shapes of different patterns for different values of η.

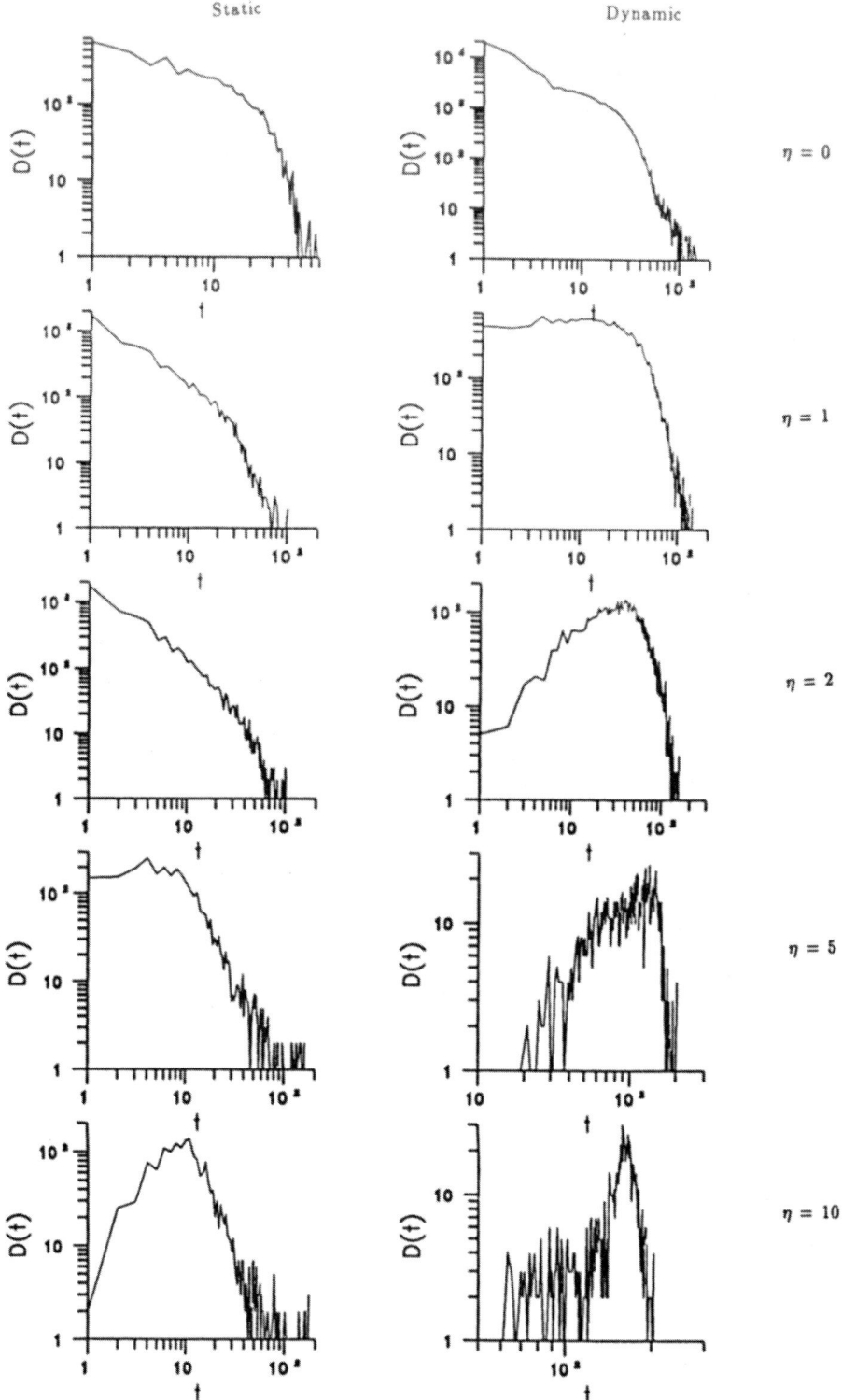

Figure 2. Static and dynamic distributions of times of duration of catastrophes.

fractal dimension close to one. It is clear the the change in the growth law does not introduce any new scale in the system. Extensive calculations show that the aggregates are, indeed, self similar. Figure (1) shows typical patterns generated for different values of η.

We now investigate the changes in the noise spectrum as function of η. The calculations reported below were obtained averaging results for eight aggregates, grown in a circular cell of radius 108.

We define catastrophes following our previous work on DLA. We first analyze aggregates of a given size. We calculate, once growth has take place at certain site, in how many lattice nodes outside the aggregate the diffusion field changes above a given threshold, ϵ. This procedure allows us to obtain a definition of the spatial extent of the rearrangement which follows a growth event. Alternatively, we also study how many iterations are required to make the diffussion field converge to its new equilibrium value, given an overall error tolerancy in the calculations. In this way, we can define the duration of the catastrophe. We have made detailed calculations which show that the SOC regime, as measured by its critical exponents, is independent of the thresholds used. Also, modified definitions of the size and duration of the catastrophes do not alter the main results, that is, the existence of a power law distribution of catastrophes, and only change slightly the exponents.

From the calculations reported above we can easily infer the noise spectrum, assuming that the aggregate grows by one particle par unit of time. There is well defined correlation between the size, or duration, of a catastrophe which occurs when the aggregate grows at a given site, and the value of the field at that site. The latter quantity determines the likelyhood of growth at that point. Hence, the probability that a given catastrophe will take place in the growth process is given by the distribution of catastrophes calculated for an aggregate of fixed size, times the field associated with that type of catastrophes. This procedure is consistent with the direct calculation of the number of catastrophes which occur for a particular evolution of the system. The advantage of calculations done for an aggregate of fixed size is that they allow for better numerical accuracy. Moreover, we can relate in this way dynamical properties, like the noise spectrum, to static features of the aggregate, which have been extensively studied.

Our results are summarized in figures (2) and (3). In figure (2), we show the distribution of catastrophes, according to their duration in time, as measured statically and dynamically. As discussed above, both calculations are related through the correlation between catastrophe size and probability of growth. As the latter depends on the value of the field, we present in figure (3) the distribution of fields, also calculated by both methods. As in figure (2), large fields, or large catastrophes, have a bigger share in the time evolution of the system. In a given aggregate, there are many possibilities to induce small catastrophes, or, correspondingly, there are many points in the perimeter with associated small fields. As the system grows, however, these sites are less likely to be active than those which correspond to large catastrophes or fields.

To complete our analysis, we present in figure (4) a study of the noise spectrum, as measured by the energy dissipated after each growth event, in real time. The Fourier transform of this quantity shows a deceptively good power law behavior. The accuracy of this calculation, however, is not as reliable as in the previous ones.

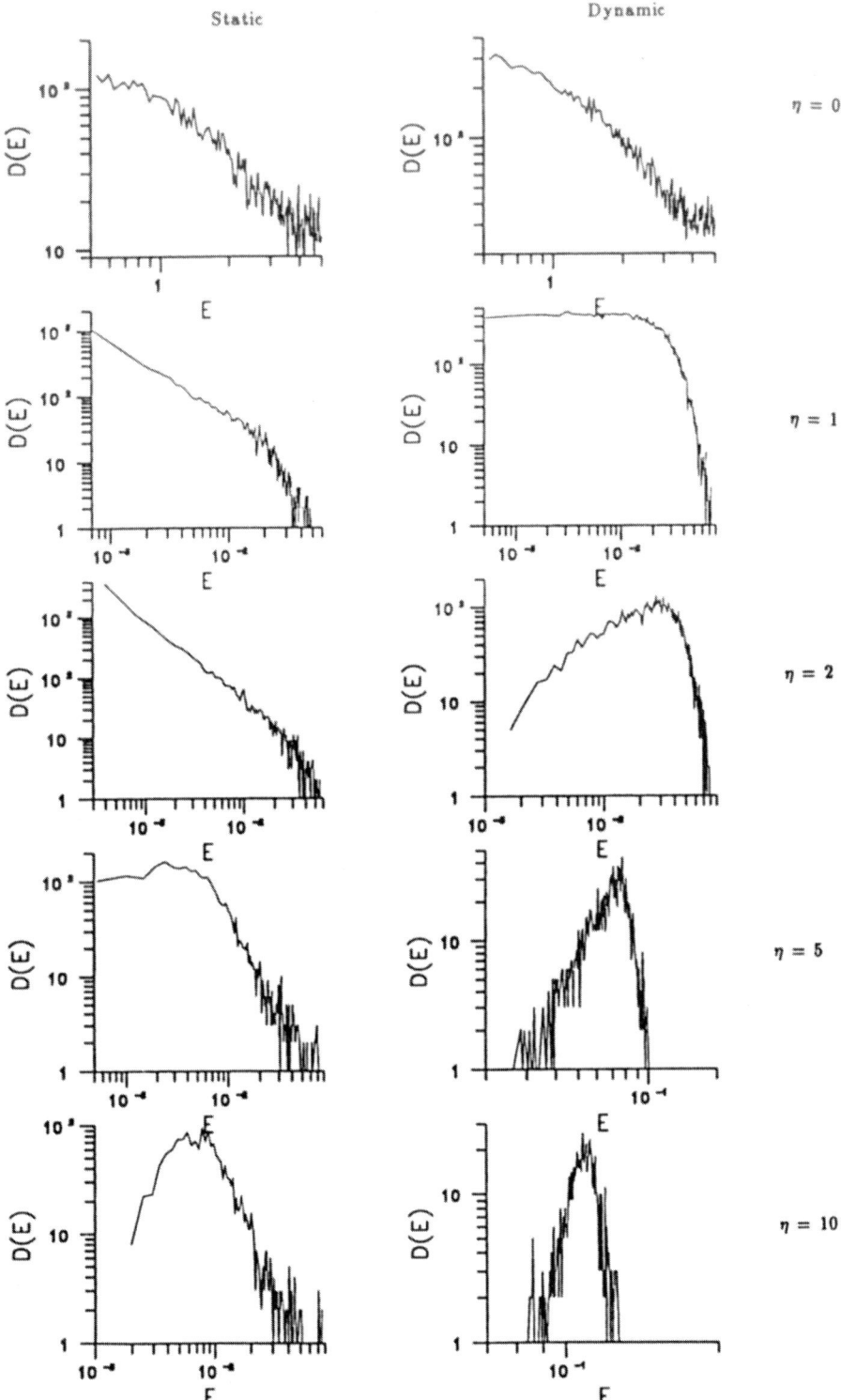

Figure 3. Same distributions as in Fig. 2 but for the electric field E.

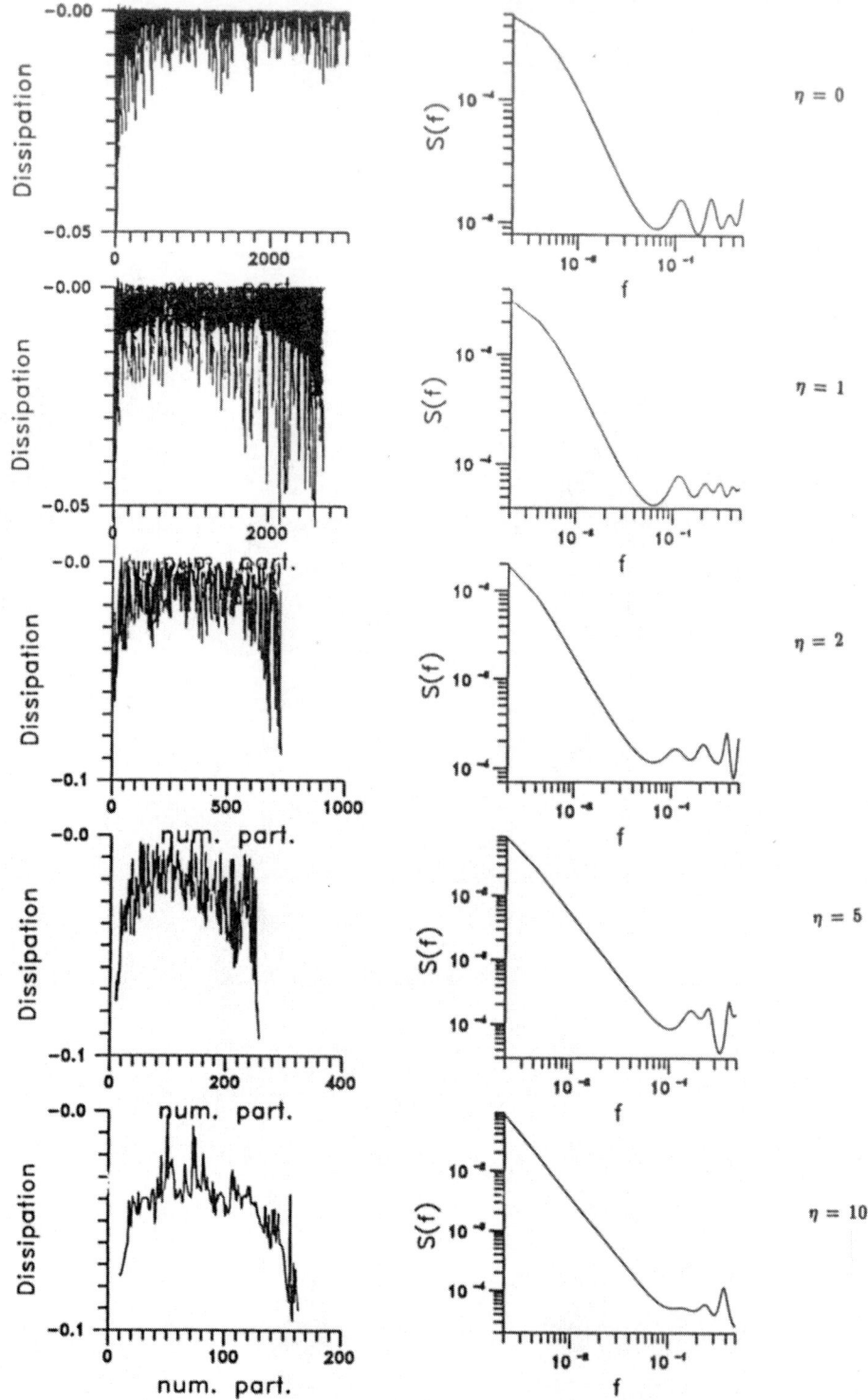

Figure 4. Dissipation in a particular sample for each η as the pattern grows, and its power spectrum as function of frecuency.

In figures (2) and (3), the deterministic situations correspond to the bottom graphs. The difference with the upper ones, which clearly show a SOC regime, is remarkable. The main new feature is a peak in the distribution, which indicates that large catastrophes become more likely. This peak is at the upper side of the curves, that is, the largest possible catastrophes, those of size similar to the aggregate itself. The difference between the two types of behavior, small η and large η, suggests the possibility of an abrupt transition between the two.

We adscribe this difference in the type of distribution of catastrophes to the change in the growth law, which becomes quasi deterministic when $\eta \gg 1$. As mentioned in the introduction, we expect that, in this regime, only certain catastrophes will occur. Our results suggest that these processes take place at the largest scale available, given by the size of the growing aggregate. Thus, the scale invariance implicit in SOC and fractal growth is broken for $\eta \gg 1$. On the other hand, the new scale which appears is not intermediate between the upper and the lower cutoff, as in conventional phase transitions. The transition which we observe from complex (SOC) behavior to deterministic behavior, is associated with the fact that the largest scale becomes relevant and determines the evolution of the system.

These features are very different from those calculated for other systems which exhibit power law noise[6,7]. In these systems, SOC has been associated to the existence of a conserved order parameter. It implies that any perturbation gives rise to long range effects (basically, any excess of material has to be removed at the edges). DLA, and related models, do not exhibit this behavior. We do not adscribe SOC to the stochasticity in the equations which govern the system, but to the complexity of the equations themselves. Note that, for all values of η, the choice of growth site is made by means of a probability distribution.

REFERENCES

1. P. Bak, C. Tang and K. Wisenfeld, Phys. Rev. Lett. **59**, 381 (1987), and Phys. Rev.
   A **38**, 364 (1988).
2. P. Bak, K. Chen and M. Creutz, Nature **342**, 780 (1989).
3. O. Pla, F. Guinea and E. Louis, Phys. Rev. A **42**, 6270 (1990).
4. T. A. Witten and L. M. Sander, Phys. Rev. Lett. **47** 1400 (1981); Phys. Rev. **B27** 5686 (1983).
5. L. Niemeyer, L. Pietronero and H. J. Wiesmann, Phys. Rev. Lett. **52** 1033 (1984).
6. T. Hwa and M. Kardar, Phys. Rev. Lett. **62**, 1813 (1989), G. Grinstein, D.-H. Lee and S. Sachdev, *ibid* **64**, 1927 (1990).
7. P. L. Garrido, J. L. Lebowitz, C. Maes and H. Spohn, Phys. Rev. A **42**, 1954 (1990).

SCALING PROPERTIES OF AVERAGE DIFFUSION

LIMITED AGGREGATION CLUSTERS

M. Kolb

Laboratoire Chimie Théorique
Ecole Normale Supérieure
69364 Lyon, France

INTRODUCTION

Diffusion limited aggregation (DLA) intrigues by its complex structure which results from an extremely simple irreversible growth algorithm[1]. Despite a considerable effort there is no complete theory that can explain the many features observed in experimental realizations and calculated in numerical simulations[2].

A method that has been used to simplify the study of the disordered growth patterns is to consider the global envelopes of such structures. This approach has been the starting point for the cone-angle treatment by Turkevich and Scher[3] which advanced the understanding of DLA and produced a relation between the fractal dimension D and the angle at the tips of a growing cluster. Global features also prove useful to show the influence of the underlying lattice on the pattern; this idea was first applied by Ball et al.[4,5] and more recently extended by Meakin[2]. Considering average profiles is particularly useful when correlating the envelope of a growing cluster to the boundary conditions. This relation was pointed out in the film on aggregation processes[6]; it has been explored quantitatively in a series of careful viscous fingering experiments[7-9] (and compared with two-dimensional DLA simulations). In view of the long range nature of the Laplacian field it is not surprising to find a connection between viscous fingering and DLA growth envelopes - it is nevertheless remarkable to find a quantitative agreement between the two systems. In particular, the average density of fingers is surprisingly well approximated by the $\lambda=0.5$ Saffman-Taylor (S-T) solution for zero surface tension.

Here we wish to present further results on average DLA clusters and compare them with analogous data from viscous fingering experiments.

Standard off-lattice DLA simulations were performed in a channel with reflecting or periodic boundaries and either with a

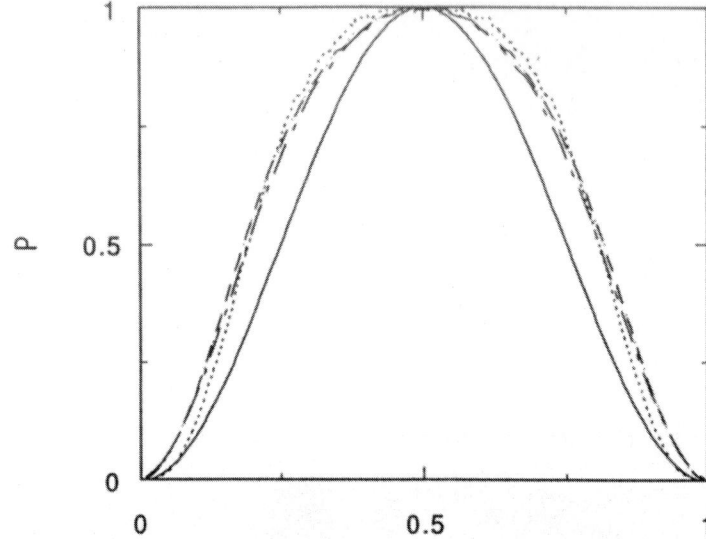

Fig.1 Normalized transverse profile of average density of off-lattice DLA clusters grown in a channel. The density P is normalized by its maximum value, the position across the channel by its width. Data is shown for different channel widths L (L=32, dashed-dotted; L=64, dashed; L=256, dotted). The solid line is the analytical result $P=\sin^2(\pi x/L)$.

line of seeds or a single seed on an absorbing base line.

RESULTS

First the steady state transverse profile was calculated from the intermediate portion of long fingers. In Fig. 1 the normalized profile is shown for different channel widths L and is compared with the analytical form which was used to fit the experimental data and the square lattice DLA results[8]. One notices that the off-lattice fingers are broader than the analytical form. Furthermore the finger in the L=256 channel has a wider tip than the one in the narrower L=32,64 channels.

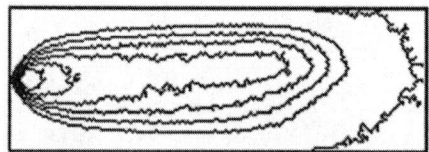

Fig.2 Asymmetry of a single finger in the steady state regime (growth from left to right). There is a close resemblance with one of the noisy experimental modes[12].

Apparently the limiting profile has not yet been reached.

When comparing the S-T finger with DLA density profiles there is the ambiguity of how to choose the iso-density line. The present results suggest P≅0.7 for the λ=0.5 S-T solution. The experiments[8] however are better modeled by P≅0.5. For the lattice simulations, one expects a crossover towards an anisotropic behavior for large channel widths.

The present results can also be compared with mean-field result[10]. Dependent on the cutoff parameter the mean field result is closer to the narrow analytical form (γ=2) or closer the wide DLA result (γ=4). A clear difference between mean field result and simulations is that for the former the density goes to zero linearly near the channel walls, for the latter it approaches the walls horizontally.

DLA is an inherently noisy phenomenon. Averaging DLA fingers in a channel suppresses the fluctuations, by construction. By breaking the symmetry of the channel, modes other than the S-T solution may become accessible. In order to check this we have averaged a large number of DLA fingers, but imposing that the tip always lies in the left half of the

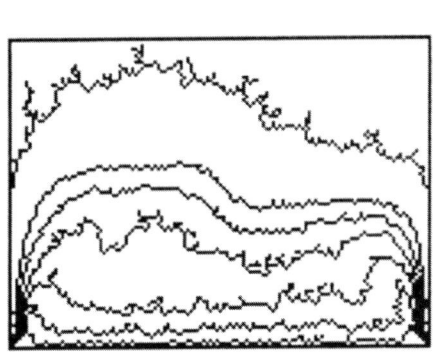
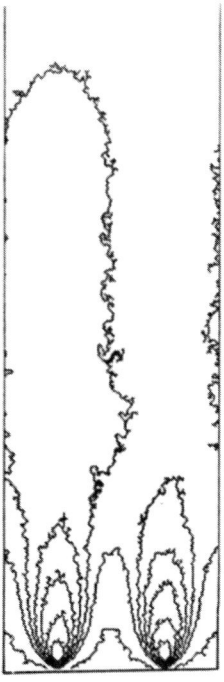

Fig.3 Finger instability of average DLA growth (growth from bottom to top). On the left the iso-density lines for growth from a base line at an intermediate stage. On the right growth from two tips on an absorbing base line with periodic boundaries, at a much later stage.

channel. The shape of the finger, for intermediate iso-density lines, closely resembles the hump mode predicted in ref. 11 and also observed experimentally[12].

The destabilization and the competition between two tips can also been computed by this method and compared with experiments. The result are shown in Fig. 3, for two boundary conditions and at two stages of the growth. On the left one sees the early stage of the (average) tip competition. While at the lowest densities (corresponding to the outermost tips) the competition is not apparent, at intermediate densities one clearly sees the effect (the left tip wins the competition, by construction). On the right, a similar phenomenon is visible, at a much later stage. In this case the boundary condition has been changed to enhance the visibility of the competition. Two seeds are placed on an otherwise absorbing baseline, and periodic boundary conditions are imposed. The different iso-density lines roughly correspond to different stages of the growth (the internal density of the fingers decreases with time as they are fractal). The results resemble the finger structure in corresponding Hele-Shaw experiments[13,14]. In particular, one notices a a bulging of the dominant finger in both cases.

DISCUSSION

The comparison between average properties of DLA clusters and viscous fingering can be developed beyond the average shape of the S-T finger. Features such as asymmetry and finger competition are qualitatively similar in both cases. The key difficulty for a rigorous quantitative comparison is the lack of a clear criterion for choosing the best iso-density line.

REFERENCES

1   T.A.Witten and L.M.Sander, Phys. Rev. Lett. 47,1400 (1981)
2   P.Meakin,'Fractal Aggregates and their Fractal Measures', in 'Phase Transitions and Critical Phenomena', C.Domb and J.L.Lebowitz (eds.), Vol. 12, Academic 1988
3   L.A.Turkevich and H.Scher, Phys. Rev. Lett. 55, 1026 (1985), see also ref. 4
4   R.C.Ball and R.M.Brady, CECAM workshop on 'Kinetic models for cluster formation', Orsay 1984
5   R.C.Ball, R.M.Brady, G.Rossi and B.R.Thompson, Phys. Rev. Lett. 55, 1406 (1985)
6   'Aggregation',film on growth processes, author: M.Kolb, production: ZEAM, Berlin
7   Y.Couder, O.Cardoso, D.Dupuy, P.Tavernier and W.Thom, Europhys. Lett. 2,437 (1986)
8   A.Arnéodo, Y.Couder, G.Grasseau, V.Hakim and M.Rabaud, in 'Nonlinear evolution of spatio-temporal Structures in Dissipative continuous systems', F.H.Busse and L.Kramer (eds.), Proceedings NATO ARW, Streitberg, 1989
9   Y.Couder, F.Argoul, A.Arnéodo, J.Maurer and M.Rabaud, Phys. Rev. A42, 3499 (1990)
10   E.Brener,H.Levine and Y.Tu, Phys. Rev. Lett. 66, 1978 (1991)
11   D.Bensimon, Phys. Rev. A33, 1302 (1986)11
12   P.Tabeling and A.Libchaber, reproduced in ref. 11
13   S.A.Curtis and J.V.Maher, Phys. Rev. Lett. 63,2729 (1989)
14   H.Zhao and J.V.Maher, Phys. Rev. A42, 5894 (1990)

TOPOLOGICAL CONSIDERATIONS ON FINGER DYNAMICS

IN THE SAFFMAN-TAYLOR PROBLEM

J. Casademunt and David Jasnow

Department of Physics and Astronomy
University of Pittsburgh
Pittsburgh, PA 15260

INTRODUCTION

The study of viscous fingering in Hele-Shaw cells[1], has been playing an important role in the understanding of interfacial pattern formation. As a prototype system for the study of diffusion-controlled growth, it is relevant for a variety of non-equilibrium phenomena, such as dendritic growth, directional solidification, chemical electrodeposition and flame propagation[2].

In this paper we are going to focus on the case of a Hele-Shaw cell with channel (or rectangular) geometry. This case is sometimes referred to as the Saffman-Taylor problem[1], and it involves the emergence of a steady, single-finger propagating solution out of complicated multi-finger patterns generated by the morphological instability of the planar interface.

Although the initial stages of the linear instability [1], and the selection and stability of the single-finger steady state[3] are well understood, the understanding of the dynamical mechanisms by which a multifinger structure progresses toward the single finger solution is still relatively poor and at a very qualitative level. Moreover, a deeper insight into this dynamical process could lead to a better understanding of the dynamics of a wide class of related systems.

Our main purposes here are to show how a topological point of view may be useful in the study of finger competition in general, and to illustrate such an approach. We address some interesting questions raised by the sensitivity of fingering dynamics to the viscosity contrast parameter in the Saffman-Taylor problem, particularly questions concerning the domain of attraction of the single-finger stationary solution.

We consider a Hele-Shaw cell of width W in the x direction and of infinite extent in the y direction, with a gap b between the plates and tilted at an angle α

to the vertical. The two-dimensional velocity field obeys Darcy's law, $\mathbf{v} = \nabla\phi$, where the velocity potential in each phase $i = 1, 2$ is related to the pressure p as $\phi_i = -(p + \rho_i gy \cos\alpha)b^2/12\mu_i$, where μ_i are the viscosities, and ρ_i the densities. Assuming incompressibility, ϕ obeys Laplace's equation in the bulk. The problem then becomes completely specified by two boundary conditions at the interface: the pressure drop, taken to be proportional to the curvature κ, $p_1 - p_2 = \sigma\kappa$, where σ is surface tension, and the continuity condition $v_n = \mathbf{n}.\nabla\phi_1 = \mathbf{n}.\nabla\phi_2$. The two dimensionless parameters of the problem[4] are the viscosity contrast $c = (\mu_1 - \mu_2)/(\mu_1 + \mu_2)$, and a dimensionless surface tension $d_0 = 12b^2\sigma/(UW^2(\mu_1 + \mu_2))$, where $U \equiv V_G + cV_\infty$ is assumed positive, with $V_G = 12b^2 g \cos\alpha(\rho_1 - \rho_2)/(\mu_1 + \mu_2)$, and V_∞ is the velocity at $y \to \pm\infty$. For convenience and without loss of generality we will formally take $V_\infty = 0$. This means that if the problem is not purely gravity-driven ($V_\infty \neq 0$), we reformulate the problem in a reference frame moving with velocity V_∞; that is, all velocities are taken relative to V_∞.

In the last decade, the study of the Saffman-Taylor problem has concentrated, to some degree, on the role of surface tension as a mechanism of selection of a unique steady state out of a continuum of possible single-finger solutions[3]. Although the current microscopic solvability scenario[3] for selection has not been explicitly applied for arbitrary viscosity contrast, numerical evidence indicates that the sensitivity of the steady state to c is very weak[4]. However, both simulations [4], and experiments[5], [6] have shown that in the transient non-linear regime, far from the two well-understood limits of the problem (that is, the linear instability of the planar interface, and the linear relaxation in a neighborhood of the single-finger solution), the viscosity contrast can play a major role.

The competition of parallel arrays of fingers in the high contrast limit $c = 1$, has been described in terms of a global instability, which refers to the linear instability of the envelope of the finger front[7]. Consistently with experiments[6], the current qualitative understanding of the competition regime is that this global instability leads to coarsening, as longer fingers grow at the expense of smaller ones, giving rise to a cascade into large length scales ending up with a single finger. Here we will show that this scenario needs to be modified depending, in general, on the viscosity contrast, and that the global instability of the finger front, which is reminiscent of the underlying Saffman-Taylor instability, does not necessarily lead to the single-finger steady state.

The existence of different types of behavior in the dynamics of finger competition has been clearly illustrated in simulations of two-finger configurations [8],[9]. The qualitative observations can be summarized as follows. For the high-contrast limit $c = 1$, the global instability proceeds very rapidly until the velocity of the smaller finger reverses its sign (see below). For $c = 0$, however, even very large differences between the heights of two competing fingers are not sufficient to reverse the velocity of the small finger. Furthermore, although the velocity at the tip of the long finger is greater, the smaller finger is somehow growing at the expense of the larger one, as the former widens whereas the latter tends to develop a narrow neck. This behavior is radically different from the case $c = 1$ and clearly exhibits a lack of progress toward the single-finger steady state. In order to go beyond these qualitative observations, in the next section we introduce a theoretical framework which provides a more precise characterization of the underlying mechanisms of finger competition, and which will allow us to address the question of the domain of attraction of the Saffman-Taylor finger.

TOPOLOGICAL APPROACH TO FINGERING DYNAMICS

The convenience of a global point of view focussing on topological propeties of the flow field in the bulk, becomes apparent when the problem is formulated for the velocity field relative to V_∞. By doing so, the sink and source respectively at $\pm\infty$ have been removed, and the remaining flow field can be seen as describing a redistribution of area. The streamlines become closed, oriented lines which always cross the interface, where all the vorticity is confined. The flow appears then naturally divided into different regions according to the sense of flow circulation. Accordingly, the competition can be pictured by realizing that neighboring fingers generate flows with opposite sense of circulation. Periodic arrays of fingers define stripes with alternate circulation (see next section), with the particular case of the single-finger steady state having only two such stripes. This clearly suggests that the existence of defects in the flow structure, occurring as vertices of streamlines separating domains of circulation, must play a central role in the evolution toward the steady state. The separatrices of the flow defined by the streamlines connected to defects, provide a useful diagrammatic characterization of the flow structure (see examples in next section).

The only topological defects which are (topologically) stable for our problem once formulated in the frame with zero flux at infinity, are point-defects of two kinds, corresponding to saddle points and local extrema of the stream function ψ. This function is defined as the harmonic conjugate of the velocity potential ϕ so that the complex potential $\Phi(z) = \phi(x,y) + i\psi(x,y)$, is analytic in $z = x + iy$ in the bulk. The stream function plays a central role in the topological considerations that follow due to the fact that, unlike the velocity potential ϕ, the stream function ψ is continuous at the interface, though non-differentiable. The topological stability of those defects implies that they must be locally preserved by the dynamics, and that they can only be created or annihilated in pairs satisfying the appropriate conservation rules (see below). Furthermore, the fact that ψ is harmonic in the bulk implies that its extrema can only occur at the interface. The saddle-points, however, can occur in the bulk of both phases as zeroes of the complex velocity field $\Omega(z) = d\Phi/dz = v_x - iv_y$. The saddle-points are precisely the point-defects anticipated above as vertices of streamlines. The order of the zeroes of Ω defines point defects of corresponding order. The local structure of Φ around a saddle-point defect of order n at $z_i(t)$ in the bulk is thus given by

$$\Phi_i(z,t) = A_i(t) + B_i(t)(z - z_i(t))^{n+1} + ... \qquad (1)$$

defining a generalized saddle point flow. These point defects are singular only in a topological sense, describing phase singularities of the complex velocity field Ω, since the phase of Ω changes by an amount $\pm 2n\pi$ along any path enclosing the defect. Physically, defects in the bulk correspond to (non-stationary) stagnation points of the flow, i.e. $\mathbf{v}(x_i(t), y_i(t)) = \mathbf{0}$ [10].

It is possible to attribute an integer "charge" (the winding number[10]) $Q = +1$ to each extremum (either maximum or minimum) and $Q = -n$ to each saddle-point defect of order n. The charge corresponding to the stagnation point at infinity can be obtained from the number of streamlines reaching infinity. For intance, $2(n + 1)$ streamlines connected with infinity define a defect with charge $-n$ at infinity.

The conservation of this topological charge makes possible a global characterization in terms of the motion of these representative points. The localization of the extrema at the interface implies that saddle-point defects can only be annihilated when hitting the interface and coalescing with extrema. New saddle-point defects can also be created only at the interface as extremum/saddle-point pairs.

This reduced description of the dynamics in terms of a few degrees of freedom is a way of exploiting the highly non-local nature of the problem, which somehow complements the formally similar characterization of the problem in terms of the singularities of the conformal mapping of the interface into the unit circle of the complex plane[1]. A more detailed discussion of the differences between the two approaches will be presented elsewhere[8]. Apart from the obvious ones related to each specific description (i.e. zeros of the complex velocity field vs. singularities of the conformal map) we would like to emphasize here the *topological* rather than *analytical* nature of the approach presented, which makes it very general and potentially applicable to a great variety of situations. We will show in the next sections that, for Saffman-Taylor fingering, it becomes particularly useful for the comparison of the dynamical behavior associated with the variation of the viscosity contrast.

Finally, there is another crucial implication of the topological analysis, which is the existence of a global invariant of the problem, which constrains the total topological charge to be a characteristic constant of the problem, independent of initial conditions and parameters. This can be seen by recognizing that the stream function is continuous under compactification of the infinite strip with periodic boundary conditions into a sphere. That is, it is possible to map the infinite channel into a sphere where the north and south poles are the images of the points $y \to \pm\infty$, and the equator, for instance, is the image of the interface. (The same considerations can be applied for the case with rigid walls, since this case is reducible to the periodic case in an enlarged system with the appropriate symmetries). The fact that the stream function becomes a continuous function in a compact space automatically leads to the existence of a global invariant of the problem. This implies, for instance, that if ψ has M local maxima and m local minima, they satisfy $M + m - S = \chi = 2$ where S is the total weight of saddle points of ψ. (For example, a saddle point like Eq.(1) contributes as $S_i = n$)[11]. This means that not only is the charge Q defined above locally preserved by the dynamics but also that the *total* charge for an arbitrary configuration must equal the Euler characteristic $\chi = 2$ of the sphere[11]. The interest of this observation is that it provides a natural classification of all possible flow configurations in terms of S, which can now be redefined as

$$S = \frac{1}{2}\Sigma_i |Q_i| - 1 \tag{2}$$

In particular, the steady state configuration corresponds to the class $S = 0$, that is, containing a maximum, a minimum and no saddle-point defects of the stream function. Therefore, the quantity S, which essentially counts the number of saddle-point defects, is not only a global measure of the complexity of the flow structure but also defines a "distance" from the steady state configuration. As it will become clear in the next sections, in general, the positive integer S is not trivially related to the number of fingers in a given interface configuration.

FINGER COMPETITION IN THE HIGH-CONTRAST LIMIT

In this section we apply the ideas of the previous section to characterize the competition for two-finger configurations for $c = 1$. This is the most commonly studied case, for which the global instability has been formulated, and for which the competition is known to be very efficient. By computing the streamfunction $\psi(s)$ as a function of arclength s along the interface, which can be obtained using standard boundary integral methods[2], [12] one can infer the qualitative structure of the flow field in the bulk. In Fig.1 we schematically show the resulting diagrams. The topology of these diagrams depends solely on global features of $\psi(s)$, typically the number of zeroes of $\psi(s)$ and $\psi'(s) = v_n(s)$ [8]. The corresponding interface shape has also been depicted

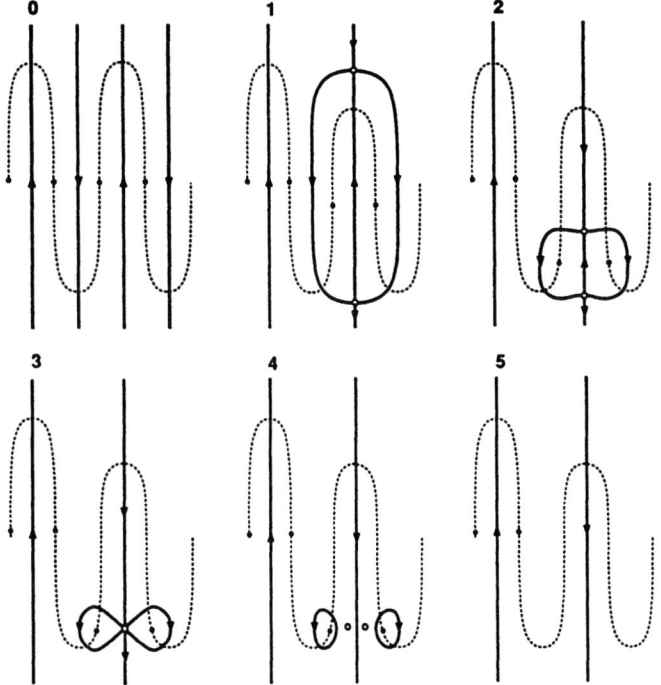

Fig.1. Qualitative flow diagrams for two-finger competition for $c = 1$. Dotted line, interface; solid lines, streamlines with $\psi = 0$; black dots, extrema of ψ; open circles, saddle-point defects.

only qualitatively for a better visualization of the process. Actual fingers differ by approximately 10% in height, and do not change substantially during the process described in Fig.1.

The explicit mechanism for defect annihilation that we obtain from these results can thus be summarized as follows. The dynamical elimination of a finger is characterized by a diagram of two connected defects, one in each phase, and located on the central separatrix of the finger to be eliminated (Fig.1.1). These defects have to be annihilated at the maximum and minimum of the streamfunction occurring at both sides of the finger (Figs.1.4-5). To do so, the defects must first disconnect from each other, which implies that one of them has to cross the tip of the finger (Fig.1.2) and coalesce with the other one (Fig.1.3). The actual crossing of the defect through the fingertip may require a momentary creation and annihilation of defects in the tip region. This fact does not change the total S before and after the crossing, and therefore it is not essential for the present general discussion [8]. Notice, however that for Fig.2.5 $S = 0$ already, although the small finger still exists from a morphological point of view.

Finally, the global instability [7] of $n + 1$ equal fingers can be reinterpreted as the unstable nature of an n-defect at infinity. A perturbation in the periodic array breaks the defect at infinity bringing lower-order defects to finite distances (Figs.1.0-1).

FINGER COMPETITION IN THE LOW-CONTRAST LIMIT

We now return to the case $c = 0$. The time evolution of a two-finger configuration analogous to that of Fig.1.2 can be obtained numerically[8],[9] and shows that the flow diagram is also that of Fig.1.2, but now the two connected defects are *moving apart* from one another. This clearly indicates that the mechanism for defect annihilation described above (Fig.1) is not working. Furthermore, it can be shown that the increase in the number of zeroes of $\psi'(s) = v_n(s)$ [8] produced by the pinching-like morphology observed in the long finger implies the creation of extremum/saddle-point pairs. This means that the quantity S can actually *increase* as a result of competition, thereby moving away from the single-finger topology.

There is, however, an important exception to the creation of new defects for $c = 0$, which concerns initial configurations having only one finger. Numerical simulation shows that, in this case, the system relaxes to a steady single finger solution[4],[8]. Therefore, from the observed behavior of $S(t)$ we can conclude that, except for a very restricted class of configurations which are close enough to the steady state solution, the system in general evolves away from the single-finger steady state. This suggests that the domain of attraction of the Saffman-Taylor finger is drastically reduced for the case $c = 0$. Notice, however, that such domain of attraction is still finite, and therefore a solvability analysis for selection of the steady state could still be applicable, although it would not be sensitive to the non-linear dynamical effects discussed here. On the other hand, an interesting implication of the evidence reported here is the possible existence of other attractors in the Saffman-Taylor problem.

Finally, the behavior of the system in the range of intermediate viscosity contrast $0 < c < 1$, is still unclear. Although symmetry arguments could suggest that $c = 0$ is a special case, and that some kind of crossover to $c = 1$ behavior could be expected, the numerical[4],[8] and experimental[5] evidence available seems not to support this scenario. There is however some evidence for the predominance of the pinching-like morphology for a wide range of c [4], which, as we have seen, is the mechanism reponsible for the creation of new defects. For intermediate contrasts, a certain sensitivity to initial conditions of the fate of secondary fingers has also been reported [4]. Both facts seem to reinforce an alternative scenario consisting of a gradual dependence of the domain of attraction of the single-finger solution on the parameter c.

CONCLUSIONS

We have shown that topological features of the flow structure are useful points of focus in the study of finger competition. The great generality of these ideas makes the approach potentially applicable to a variety of related problems. The particular application to the characterization of finger competition in the Saffman-Taylor problem provides strong evidence of different behavior depending on the viscosity contrast, which may result in a drastic reduction of the domain of attraction of the Saffman-Taylor finger for a wide range of this parameter.

ACKNOWLEDGMENTS

J.C. acknowledges the MEC (Spain) for a Fulbright grant, and the DGICYT (Spain) Proj. No PB90-0030. D.J. is grateful to the NSF through the Division of Materials Research for support under grant DMR89-14621. Support from NATO Collaborative Research Grant No. 900328 is also acknowledged.

REFERENCES

1. P.G. Saffman and G.I. Taylor, Proc. Royal Soc. London A 245, 312 (1958); D. Bensimon, L. Kadanoff, S. Liang, B.I. Shraiman and C. Tang, Rev. Mod. Phys. 58, 977 (1986)

2. J.S. Langer, Rev. Mod. Phys. 52, 1 (1980); D. A. Kessler, J. Koplik and H. Levine, Adv. Phys. 35, 255 (1988); P. Pelcé, *Dynamics of Curved Fronts,* Persp. in Physics. Academic Press (1988)

3. J.W. McLean and P.G. Saffman, J. Fluid Mech. 192, 455 (1981); J.M. VandenBroeck, Phys. Fluids 26, 2033 (1983); D.A. Kessler and H. Levine, Phys. Rev. Lett. 57, 3069 (1983); B.I.Shraiman, Phys. Rev. Lett. 56, 2028 (1986); D.C. Hong and J.S. Langer, Phys. Rev. Lett. 56, 2032 (1986); R. Combescot, T. Dombre, V. Hakim, Y. Pomeau and A. Pumir, Phys. Rev. Lett. 56, 2036 (1986). S. Sarkar and D. Jasnow, Phys. Rev. A 35, 4900 (1987)

4. G. Triggvason and H. Aref, J. Fluid Mech. 136, 1 (1983); 154, 287 (1985)

5. J.V. Maher, Phys. Rev. Lett. 54, 1498 (1985); M.W. DiFrancesco and J.V. Maher, Phys. Rev. A 39, 4709 (1989), and 40, 295 (1989); H. Zhao and J.V. Maher, Phys. Rev. A 42, 5895 (1990)

6. S.A. Curtis and J.V. Maher, Phys. Rev. Lett. 63, 2729 (1989)

7. D.A. Kessler and H. Levine, Phys. Rev. A 33, 3625 (1986); A. Karma and P. Pelcé, Phys. Rev. A 41, 4507 (1990)

8. J. Casademunt and D. Jasnow (unpublished)

9. J. Casademunt, A. Hernández-Machado and D. Jasnow, Int. J. of Mod. Phys. B (in press)

10. For reviews on topological defects see N.D. Mermin, Rev. Mod. Phys. 51, 591 (1979) and *Physics of Defects*, edited by R. Balian, M. Kléman and J.P. Poirier, North-Holland (1980)

11. See Hopf-Poincaré index theorem for instance in J. W. Milnor, *Topology from the differentiable viewpoint,* The University Press of Virginia, Charlottesville (1965)

12. D. Jasnow and J. Viñals, Phys. Rev. A 40, 3864 (1989); 41, 6910 (1990)

ADAPTIVE CLUSTER GROWTH MODELS

Paul Meakin

Central Research and Development
The Du Pont Company
Wilmington, DE 19880-0356

INTRODUCTION

The introduction of the diffusion-limited aggregation (DLA) model by Witten and Sander[1], about a decade ago, stimulated a broad and still growing interest in simple models for non-equilibrium growth phenomena. This interest has been sustained by a wide range of applications[2-5] and the challenge of developing a general theoretical understanding of the irreversible growth of disorderly patterns. Much of this work has been concerned with models such as DLA (including its many variants such as the dielectric breakdown model[6], and fracture models[7,8]) and the screened growth models[9-11] which generate randomly branched fractal[12] structures. Models such as the Eden model[13] and ballistic deposition model[14] that generate dense structures that have rough (self-affine[15]) surfaces are also of considerable current interest.[16-18]

While models such as DLA can be used to describe quite successfully structures generated by well controlled laboratory experiments[4,19,20] they are not always completely realistic and the application to problems such as the growth of river networks[21] and vascular systems[22] has not been justified in detail. In some of these more complex cases growth is not completely irreversible and does not generate structures that continue to grow indefinitely. However, the configurations accessible to the evolving system may be severely constrained by earlier stages in the growth process so that equilibrium models such as lattice animals[23-25] may not be realistic either. Another important characteristic of most biological structures and systems such as river networks is that they "adapt" to changing conditions. Here some recent attempts[26-28] to include these features in simple growth models such as the DLA[1] and Eden[13] models are described.

Growth Patterns in Physical Sciences and Biology, Edited
by J. M. Garcia-Ruiz *et al.*, Plenum Press, New York, 1993

Computer Models

In the DLA model particles are added, one at a time, to a growing cluster or aggregate of similar particles via random walk trajectories that originate outside of the region occupied by the growing cluster (at "infinity"). When a particle first contacts the growing cluster it is immobilized and added permanently to the cluster to represent the irreversible growth process. A similar procedure is used in the corresponding adaptive growth model. After a particle has been added to the cluster, we can imagine drawing a vector from the center of the added particle to the center of the contacted particle in the cluster (in this work off-lattice models were used so there is only one such contact). A path of connected vectors of length ℓ (disc diameters) then leads from the added particle to the initial "seed" or "growth site". A score (σ) is

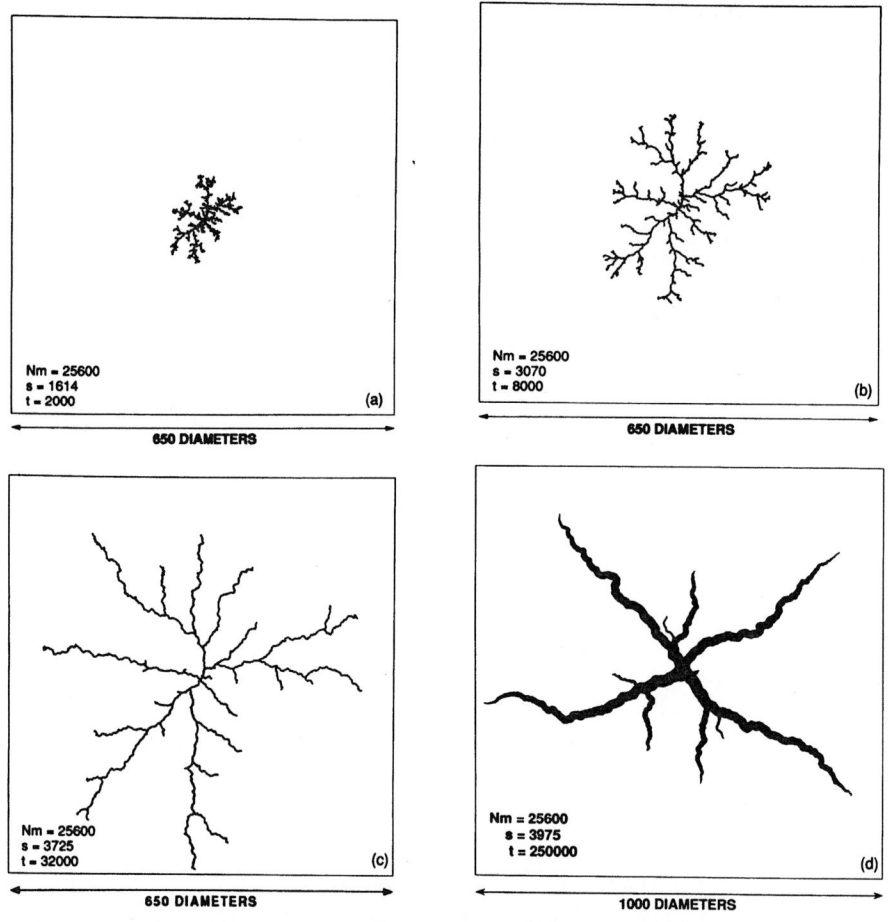

Fig. 1. Clusters generated using the modified DLA adaptive growth model with a value of 25,600 for the parameter N_m ($\delta_2 \simeq 3.9 \times 10^{-5}$). In Figs. 1a-1c the particles are represented by discs of equal diameter. In Fig. 1d the disc diameters are proportional to $\sigma^{1/2}$ where σ is the score associated with each of the particles.

associated with each of the particles in the cluster that provides a measure of the "contribution" of that particle to the growth process. After a particle has been added to the cluster the score of each of the $\ell+1$ particles associated with the path from the added particle to the origin of the cluster is increased by an amount δ_1 given by

$$\delta_1 = 1/(\ell+1)^\eta \tag{1}$$

and the score associated with all of the particles in the cluster is decreased by an amount δ_2 given by

$$\delta_2 = 1/N_m \tag{2}$$

where N_m is a parameter in the model. If the score associated with any particle in the cluster falls below zero, then it is removed from the cluster. In this way particles that do not connect growing regions of the cluster to the cluster origin are removed.

Figure 1 shows some clusters generated by simulations carried out using values of 25,600 for the parameter N_m in equation (2) and 1 for the exponent η in equation (1). In this figure t is the "time" (the total number of particles added to the cluster) and S is the number of particles remaining. In the early

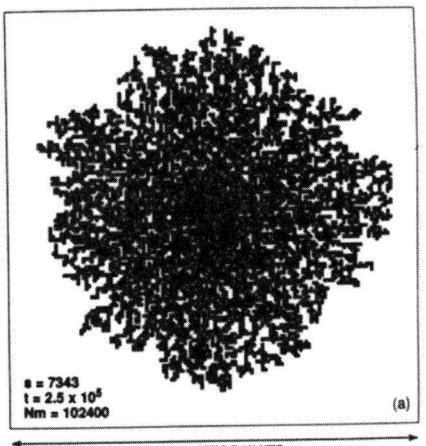
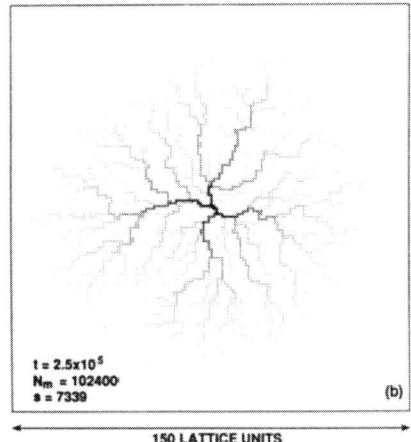

Fig. 2. Two clusters generated using the adaptive Eden growth model with the parameter N_m set to a value of 1.024×10^5. In Fig. 2a all of the sites in the cluster are shown and in Fig. 2b each site is occupied by a circle of radius $\sigma_i^{1/2}$ where σ_i is the score associated with the ith site. These clusters are shown at late "times" ($t = 2.5 \times 10^5$ added sites) well into the steady state regime.

stages (Fig. 1a) $S \simeq t$ and the structure is like that of an ordinary DLA cluster. At later stages S approaches and fluctuates about an asymptotic value that depends on N_m. Figure 1d shows a cluster at a relatively late stage $t = 2.5$ $\times 10^5$; in this figure the particles are shown as overlapping discs with areas that are proportional to their net scores.

Similar adaptive growth models can be developed based on other growth algorithms. For example, an adaptive growth model based on the original Eden growth[13] model has been developed.[28] In this model unoccupied sites on the perimeter of a growing cluster are selected at random with probabilities that are proportional to the number of occupied nearest neighbors. The selected site is then filled and new unoccupied perimeter sites are identified. After an unoccupied perimeter site has been filled, one of its occupied nearest neighbors is selected at random and we can imagine drawing a vector from the center of the newly occupied site to the randomly selected nearest neighbor. In this way a path of length ℓ lattice units consisting of these unit vectors leads from each site in the cluster to the original "seed" or growth site. Equations (1) and (2) can then be used to update the running score (σ_i) associated with each of the occupied sites and sites with negative scores can be removed.

Figure 2 shows clusters generated using this model with values of 102,400 for the parameter N_m and 1 for the exponent η in equation (1) after $t = 2.5 \times 10^5$ sites have been added (and most of them subsequently removed). Under these conditions the system appears to be well into the steady state regime and the cluster size (the number of sites, S, exhibits small fluctuations about a steady state value of about 7350. Figure 2a shows a typical cluster of 7343 sites and Figure 2b shows a similar cluster represented by filled circles with a radius proportional to $(\sigma_i)^{1/2}$ where σ_i is the score associated with the ith site.

Adaption to Changing Boundary Conditions and Model Parameters

The adaptive nature of these models can be illustrated in several ways. In Figure 3a the time (number of added particles) dependence of the net cluster size (S) is shown for two simulations carried out using the ($\eta = 1$) adaptive DLA model. In simulation A the cluster was grown up to $t = 10^5$ with a value of 12,800 for the N_m and the process was then continued for an additional 9×10^5 trajectories with a value of 800 for N_m. In the other simulation (B) the first 10^5 trajectories were carried out using a value of 800 for N_m and this parameter was then switched to a value of 12,800. Figure 3b shows similar results obtained using the $\eta = 1$ adaptive Eden growth model. The results shown in Figure 3 suggest that the steady state cluster size depends on the current value of N_m but not on the value(s) of N_m at much earlier times. The relaxation towards the new steady state appears to be essentially exponential, but the

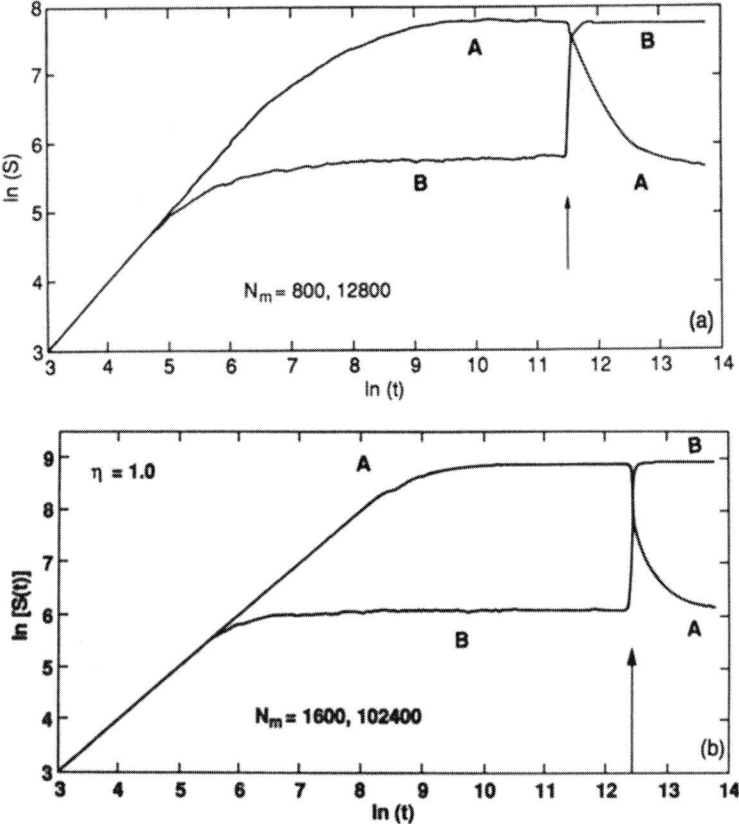

Fig. 3. Fig. 3a shows the evolution of the cluster size (S) during simulations carried out using the (η = 1) adaptive DLA model in which the parameter N_m is changed abruptly at t = 10^5 trajectories. For curve A N_m changes from 12,800 to 800 while for curve B N_m changes from 800 to 12,800.

Fig. 3b shows similar results obtained using the η = 1 adaptive Eden growth model. In this case N_m was changed abruptly at t = 2.5×10^5.

characteristic relaxation time may be quite long, particularly when the parameter N_m is decreased abruptly.

Figure 4 shows a cluster grown with a value of 4000 for N_m. Each particle is launched from the position (0, 550) with respect to the original seed or growth site. The clusters are shown after 64,000 trajectories have contacted the cluster (t = 64,000). At this stage the cluster has a size of S = 871 particles and one arm extends towards the source of particles. At this stage the position of the launch site for the random was changed and all of the subsequent particles were launched from the position (0, -550). Figure 4b shows the cluster after an additional 64,000 trajectories. Figure 4c shows the cluster after a total of 1,024,000 trajectories. The response of the cluster shape to the position of the particle source is quite evident in these figures.

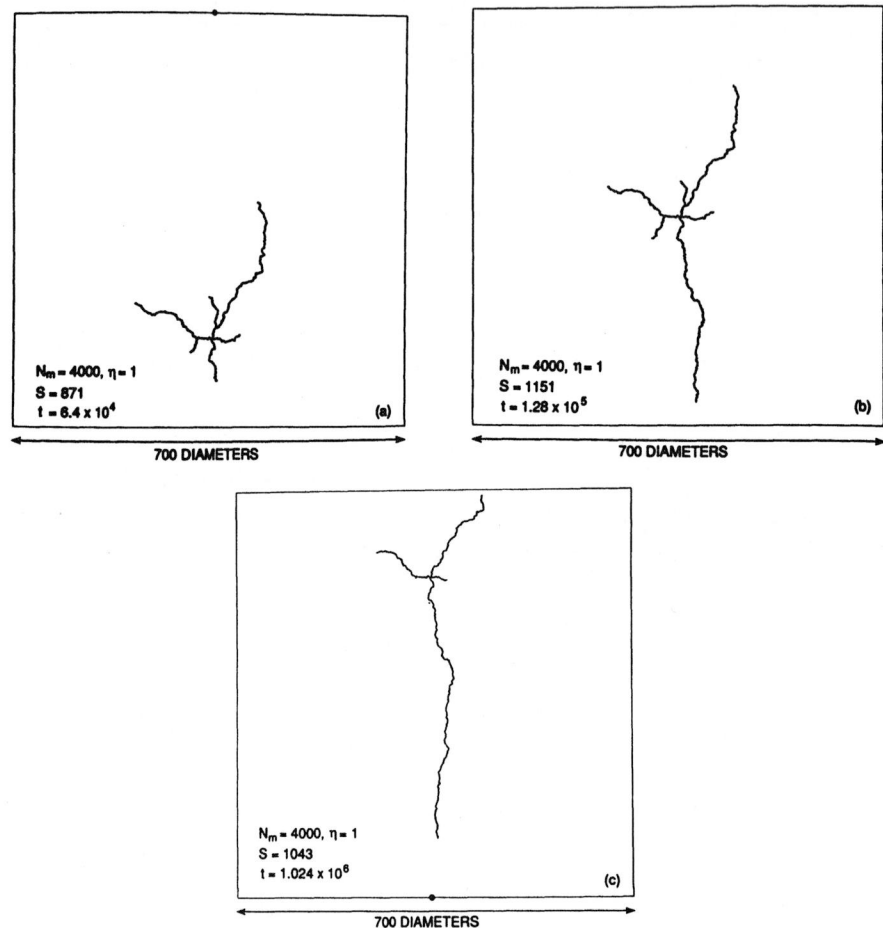

Fig. 4. Clusters generated during a simulation in which particles are released from a position with coordinates (0, 550) for the first 64,000 trajectories and from (0, -550) for the remaining trajectories. Fig. 4a shows the cluster after 64,000 trajectories (t = 64,000). Fig. 4b shows the cluster after an additional 64,000 trajectories (t = 1.28×10^5) and Fig. 4c shows the cluster after t = 1.024×10^6 trajectories. A value of 4000 was used for the parameter N_m. In Figs. 4a and 4c the "launch points" at (0, 550) and (0, -550) respectively are indicated by large dots at the edge of the box of size 700 d_0 x700 d_0 which encloses the clusters.

The Steady State Regime and Approach to the Steady State

For both the Eden and DLA adaptive growth models the net cluster size (S) grows linearly with "time" for short times and approaches a constant value (with small fluctuations) at long times. The constant, steady state, value (S_∞) depends on the model parameters N_m and η. Figure 5 shows the dependence of S_∞ on N_m for the $\eta = 1$ adaptive DLA model. In general S_∞ and N_m are related

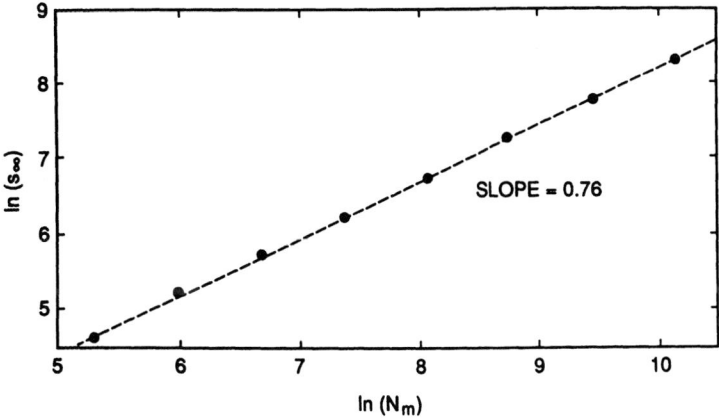

Fig. 5. Dependence of the asymptotic (steady state) cluster size on the parameter N_m for the $\eta = 1$ adaptive DLA model.

by

$$S_\infty \sim N_m{}^\upsilon \tag{3}$$

and for the case illustrated in Figure 5 the exponent υ has a value of about 0.75.

In Figure 6a the dependence of $\ln[\overline{S}(t)/N_m{}^\upsilon]$ ($\upsilon = 0.75$) on $\ln(t)$ is shown for the 8 different values of N_m where $\overline{S}(t)$ is the mean cluster size at time t obtained by averaging over 10 simulations for each value of N_m. The results shown in Figures 5 and 6 indicate that it might be possible to represent $\overline{S}(t, N_m, \eta = 1)$ by the scaling form

$$\overline{S}(t) = N_m{}^\upsilon f(t/N_m{}^\upsilon) \tag{4}$$

where the scaling function f(x) has the form f(x) = x for x<<1 (so that $\overline{S}(t) \sim t$) and f(x) = const. for x>>1 (so that $\overline{S}(\infty) \sim N_m{}^\upsilon$). Figure 6b shows the results of an attempt to scale the $\overline{S}(t)$ curves shown in Figure 6a using the scaling form given in equation (4) with $\upsilon = 0.75$. A good data collapse is obtained for x<<1 and x>>1 where $x = t/N_m{}^\upsilon$ is the argument of the scaling function f(x). The data collapse is not so good for x ~ 1 but the results are consistent with the idea that the scaling form given in equation (4) describes the dependence of the mean cluster size, $\overline{S}$, on t and N_m with $\upsilon \sim 0.75$ (for $\eta = 1$). For other values of the exponent η in equation (1) the convergence onto the asymptotic scaling function (as N_m is increased) is usually slower. However, the simulation results are consistent with equations (3) and (4) with values for υ of about 0.5, 0.6, 0.75, 1.0 and 2.0 for $\eta = 2$, 1.5, 1.0, 0.5 and 0 respectively.

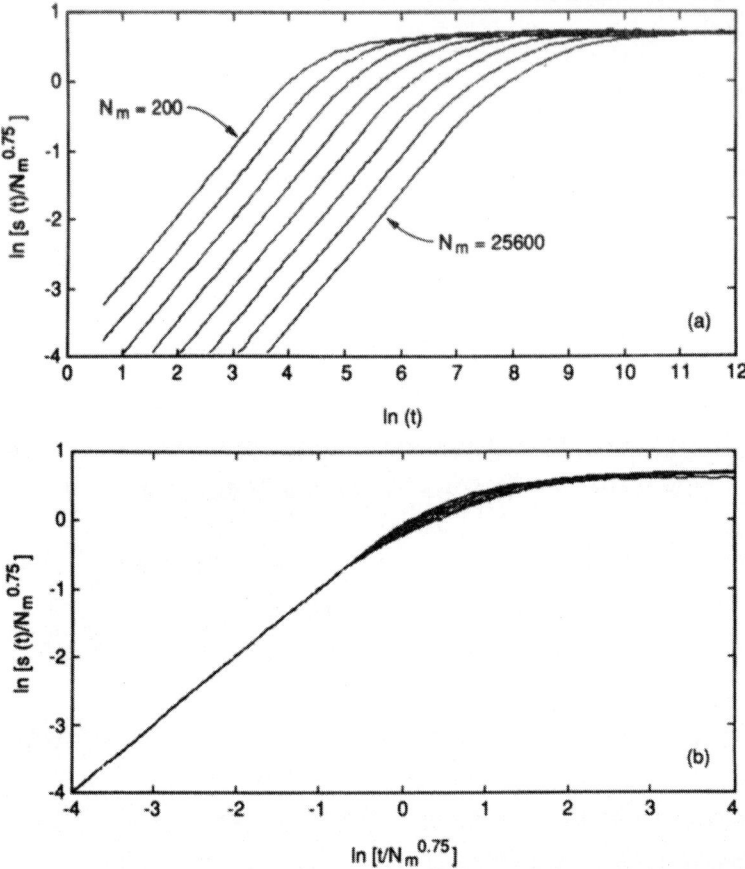

Fig. 6. Time dependence of the mean cluster size for 8 different values of N_m (200, 400, 800, 1600, 3200, 6400, 12,800 and 25,600). Fig. 6a shows the dependence of $\ln[\overline{S}(t)/N_m^{0.75}]$ on $\ln(t)$. Fig. 6b shows the results of attempts to scale the curves shown in Fig. 6a using the scaling form given in equation (4) with $\upsilon = 0.75$.

Very similar results were obtained using the adaptive Eden growth model. Figure 7a shows the growth of $\overline{S}(t)$ to its steady state values obtained from simulations carried out using a value of 0.0 for the exponent η with 9 values of the parameter N_m ($N_m = 10 \times 2^n$, $n = 0$-8). The behavior is qualitatively similar to that obtained from the models with other values of the exponent η. Figure 7b shows the result of an attempt to scale the curves shown in Figure 7a using the scaling form of equation (5). Except for quite small values of the parameter N_m a good data collapse can be obtained with a value of about 1.0 for the exponent υ. Values of about 1.0, 0.85, 0.7, 0.6 and 0.5 respectively were obtained for the exponent υ in equations (3) and (4) from simulations carried out using values of 0.0, 0.5, 1.0, 1.5 and 2.0 for η. These results suggest that the scaling exponent υ

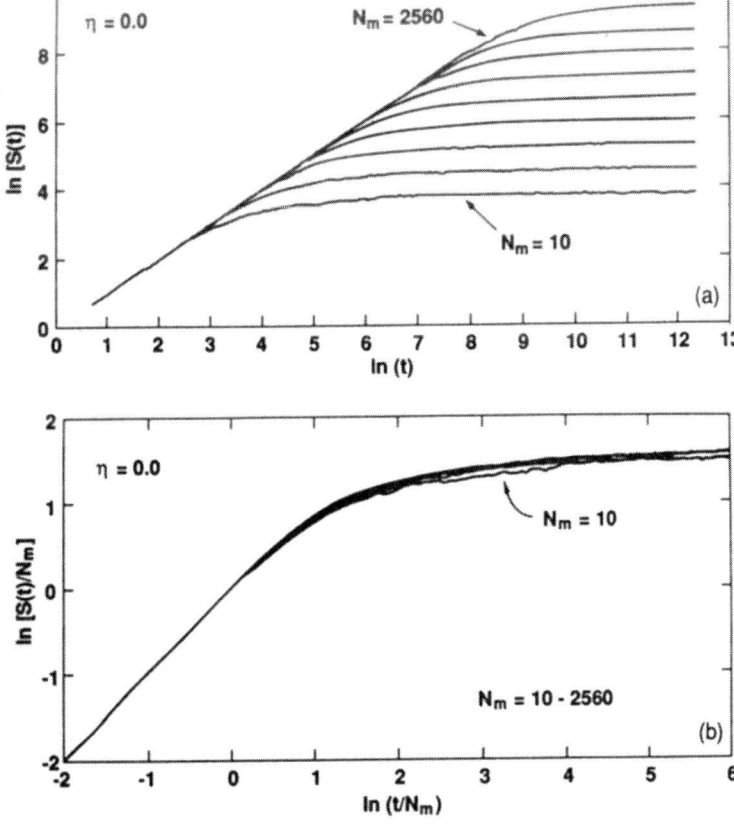

Fig. 7. Fig. 7a shows the growth of the net cluster size obtained from simulations carried out with $\eta = 0$ adaptive Eden growth models and 9 different values for the parameter N_m. Fig. 7b shows the results of an attempt to scale these curves using the scaling form given in equation (4). It is apparent that a reasonably good data collapse is obtained with a value of about 1.0 for υ.

may be given by

$$\upsilon = 2/(2+\eta). \tag{5}$$

Cluster Geometry

Figure 1 indicates that in the steady state ($t \rightarrow \infty$) clusters generated by the adaptive DLA model have a much more open, lower density structure than DLA clusters of comparable size. The structure of the clusters was characterized more quantitatively by measuring the number of particles $N(r)$ within a distance r from the cluster origin. For a self-similar fractal $N(r)$ will have the form

$$N(r) \sim r^{\gamma} \tag{6}$$

where the exponent γ is equal to the fractal dimensionality (D_γ). For the case $\eta = 1$ an effective value of about 1.25 was found for the exponent γ in the steady state regime. Figure 8 shows the dependence of $\ln[N(r)/r]$ found for clusters generated with large values for N_m after $t = 2.5 \times 10^5$ trajectories. For $\eta = 1/2$, 3/2 and 2 a slope of about 0.25 was found (Fig. 8) corresponding to a fractal dimensionality of about 1.25. For $\eta = 0$ ($N_m = 40$) the dependence of $N(r)$ on r could not be well represented by equation (6). For $\eta = 1/2$, 3/2 and 2 simulations carried out using values for N_m smaller than those illustrated in Figure 8 show an essentially linear dependence of $\ln[N(r)/r]$ on $\ln(r)$ with a slope of about 0.25. However, for small values of N_m the linear behavior does not extend to such large values of r because of the smaller overall cluster size.

For the adaptive Eden growth models the asymptotic structure is uniform on all but quite small length scales. However, for the larger value of η the clusters are noticeably more compact near to the center than near to the outer regions. Clusters appear to consist of a more or less permanent skeleton of large score sites decorated by fluctuating low score sites. In this respect these clusters resemble a variety of biological structures. Since the total score continues to grow, the "older" parts of the cluster become less and less "adaptive" with increasing time.

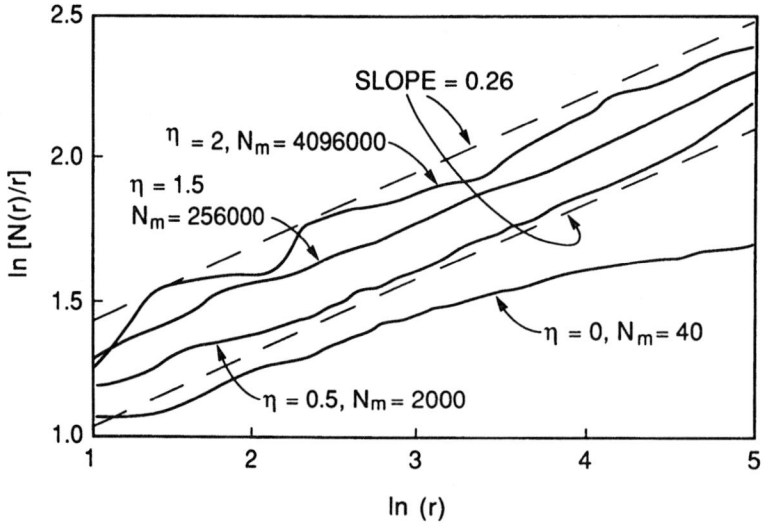

Fig. 8. Dependence of $\ln[N(r)/r]$ on $\ln(r)$ obtained from clusters generated with values of 0.0, 1/2, 3/2 and 2.0 for the exponent η with large values of N_m. Results are shown for clusters after $t = 2.5 \times 10^5$ successful random walk trajectories.

SUMMARY

The simple "adaptive growth" models studied here lead to the formation of structures that are quite different from those of the parent DLA or Eden growth models and also different from the equilibrium lattice animals. The approach to the steady state cluster size can be described quite well by equation (4). In the case of the adaptive Eden growth model simple theoretical arguments based on the ideas that the clusters are "porous" but uniform on all but quite short length scales and that the steady state corresponds to equal growth and decay rates at the cluster periphery lead to equation (5). Consequently, these simple theoretical ideas and the computer simulation results are in good agreement.

A similar approach can be used to estimate the scaling exponent (υ) for the adaptive DLA model using the approach developed for DLA by Turkevich and Sher[29] and Ball et al.[30] in which the shapes of the ends of the cluster arms are approximated by a wedge with a characteristic internal angle θ. The growth probabilities at the arm tips can then be obtained from the analytical solution of the Laplace equation with absorbing boundary conditions on the surface of the wedge. Assuming that $\theta \simeq 0$ the result

$$\upsilon = D/(\eta+1/2) \tag{7}$$

is obtained where D is the fractal dimensionality of steady state clusters. Using a value of 1.25 for D values of 0.5, 0.625, 0.8333---, 1.25 and 2.5 are predicted for the exponent υ in the cases η = 2.0, 1.5, 1.0, 0.5 and 0 respectively. These should be compared with the measured values of 0.5, 0.60, 0.75, 1.0 and 2.0. For the larger values of η equation (7) with a value of 1.25 for D gives results that are consistent with the measured values of υ. For the case η = 0 Figure 8 indicates that D $\simeq$ 1.0. Using this value for D in equation (7) a value of 2.0 is predicted for υ in good agreement with the simulation results.

ACKNOWLEDGMENT

Much of the work described here was carried out at the University of Oslo in collaboration with Jens Feder and Torstein Jøssang.

REFERENCES

1. T. A. Witten and L. M. Sander, Phys. Rev. Lett. 47:1400 (1981).
2. H. E. Stanley and N. Ostrowsky, eds., "On Growth and Form: Fractal and Non-Fractal Patterns in Physics", NATO ASI Series E100, Martinus Nijhoff, Dordrecht (1986).
3. J. Feder, Fractals, Plenum, New York (1988).

4. P. Meakin in "Phase Transition and Critical Phenomena", C. Domb and J. L. Lebowitz, eds., Vol. 12, p. 335. Academic, New York (1988).
5. "The Fractal Approach to Heterogeneous Chemistry: Surfaces, Colloids, Polymers", D. Avnir, ed., Wiley, Chichester (1989).
6. L. Niemeyer, L. Pietronero and H. J. Wiesmann, Phys. Rev. Lett. 52:1033 (1984).
7. E. Louis and F. Guinea, Europhys. Lett. 3:871 (1987).
8. E. Louis. F. Guinea and F. Flores in "Fractals in Physics" (Proceedings of the Sixth International Symposium on Fractals in Physics ICTP) L. Pietronero and E. Tosatti, eds., Elsevier-North Holland, Amsterdam p. 177 (1986).
9. P. A. Rikvold, Phys. Rev. A26:647 (1982).
10. P. Meakin, Phys. Rev. B28:6718 (1983).
11. P. Meakin, F. Leyvraz and H. E Stanley, Phys. Rev. A32:1195 (1985).
12. B. B. Mandelbrot "The Fractal Geometry of Nature", W. H. Freeman and Company, New York (1982).
13. M. Eden "Proceedings of 4th Berkeley Symposium on Mathematics, Statistics and Probability" Vol. IV, F. Neyman, ed., University of California Press, Berkeley p. 233 (1960).
14. M. J. Vold, J. Colloid Sci. 14:168 (1959); J. Phys. Chem. 63:1608 (1959).
15. B. B. Mandelbrot, Physica Scripta, 32:257 (1985).
16. J. Krug and H. Spohn in "Solids Far From Equilibrium: Growth Morphology and Defects", C. Godreche, ed.
17. F. Family and T. Vicsek, "Dynamics of Fractal Surfaces, World Scientific, Singapore (1991).
18. P. Meakin, Progr. Solid State Chem. 20:135 (1990).
19. R. C. Ball in ref. 2, p. (1986).
20. M. Matsushita in ref. 5, p. 161 (1989)
21. P. Meakin, Rev. Geophys. (1991).
22. F. Family, B. R. Masters and D. E. Platt, Physica D38:98 (1989).
23. D. Stauffer, Phys. Rep. 54:1 (1979)
24. T. C. Lubensky and J. Isaacson, Phys. Rev. A20:2130 (1979).
25. F. Family, J. Phys. A13:L325 (1980).
26. P. Meakin, J. Feder and T. Jossang, Phys. Rev. A44:5104 (1991).
27. P. Meakin, J. Feder and T. Jossang, preprint.
28. P. Meakin, Physica A179:167 (1991).
29. L. Turkevich and H. Sher, Phys. Rev. Lett. 55:1026 (1985).
30. R. C. Ball, R. M. Brady, G. Rossi and B. R. Thompson, Phys. Rev. Lett. 55:1408 (1985).

THE DOUBLE LAYER IMPEDANCE AT SELF-SIMILAR SURFACES

Thomas C. Halsey

The James Franck Institute and the Department of Physics
The University of Chicago
Chicago, Illinois 60637

INTRODUCTION

Unlike the other contributions in this volume, this chapter will not address the formation of complex patterns in physical or biological systems. Instead, I shall be addressing a complementary problem: given such patterns, what are their physical properties? From what sorts of physical measurements on a system can one infer information about its geometrical properties?

In both physics and biology, electrical measurements are a standard approach to structure determination. Here one is often in a regime dominated by surface effects–if media have an appreciable conductivity, then double layer relaxation at charged or conducting interfaces will strongly influence the macroscopic conductivity, impedance, or dielectric constant of a system. Such interfaces might be charged cell membranes in a biological system, or metallic surfaces in contact with a semiconductor or an electrolyte in a physical system.

Of course, this is a large field. In this contribution, we will restrict ourselves to discussing the double layer impedance between an electrolyte and a perfectly conducting surface, in the limit where diffusive effects are negligible.

It has long been recognized that the double layer impedance of solid electrodes can show puzzling low-frequency behavior[1]. If an electrode is perfectly polarizable, (if no current can pass from the electrolyte to the electrode), then the impedance of an electrolyte in contact with the electrode will be the impedance of a resistance R (corresponding to the solution) in series with a capacitance C (corresponding to a double layer of size λ_D, the Debye-Hückel length of the electrolyte). As a function of frequency ω, this impedance will be[2]

$$Z(\omega) = R - \frac{1}{\imath \omega C}.$$

(1)

Frequently, solid electrodes show different behavior at low frequencies[3,4]. The most notable example of anomalous low frequency behavior is "constant phase angle" (CPA) behavior, for which the low frequency impedance behaves as

$$(Z(\omega) - R)^{-1} \propto (\imath\omega)^p, \qquad (2)$$

where $0 < p < 1$. A number of studies have shown that the exponent p varies with the roughness of the electrode surface, smaller values of p being characteristic of rougher surfaces[5]. Of course, at extremely low frequencies purely capacitive behavior ($p = 1$) should be recovered.

A number of discussions of the exponent p have appeared in the literature. Liu and collaborators and Sapoval and collaborators have introduced transmission line models for the behavior of the surface impedance[6,7]. Several authors, commencing with Le Mehaute and Crepy, have introduced scaling arguments, variously claiming that the exponent p is related to the Hausdorff dimension D_s of the surface, or alternatively to the "self-affine" dimension of the surface[8-10]. No theoretical consensus appears to have been reached; however, and recent experiments have cast doubt on the idea that simple statistical measures of surface roughness[11] can be related to p.

We can loosely classify surfaces into three types[12]. Surfaces that have features much deeper than their width are essentially porous in character; it is these surfaces that are modelled by the transmission line approach. We call these "deep" surfaces. The opposite case, in which the surface features are much shallower than their width, is amenable to a perturbative treatment[13]. This distinction between shallow and deep surfaces is somewhat oversimplified; statistically "self-affine" surfaces can be shallow on large scales but deep on small scales.

Here I shall concentrate on the case of surfaces intermediate between deep and shallow surfaces, in which the depth and width of features are comparable on all length scales. This is the case of self-similar surfaces, with a non-trivial Hausdorff dimension. Since we are in neither a deep nor a shallow limit, we can apply neither transmission line nor perturbative approximations to treat this case. Below we will review a method involving the statistics of random walks, which can be used to numerically compute the double layer impedance for any surface; with the assistance of scaling arguments it can also be used to analytically calculate the double layer impedance for self-similar surfaces.

THE IMPEDANCE AND RANDOM WALKS

We consider a perfectly polarizable (blocking) electrode, the BE, connected in a circuit across an electrolyte with a counter-electrode, the CE (see Figure 1). The counter electrode will be assumed to be perfectly non-polarizable; its effective resistance to charge transfer with the electrolyte is zero. Its potential will be $V(t)$, which will thus also be the potential of the electrolyte in its vicinity (no screening layer forms at the CE). The BE is at potential zero. The current arriving in the double

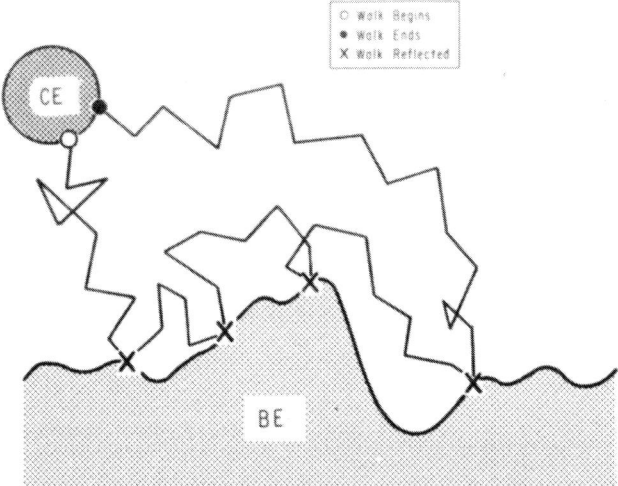

Figure 1. The random walk problem discussed. Random walkers are emitted from the counter-electrode, and are destroyed when they return. They are reflected at each contact with the blocking electrode. The probability that a walker, which has struck the blocking electrode at least once, returns to the counter-electrode after n contacts with the blocking electrode defines the coefficient b_n.

layer at the surface point s of the BE will be determined by the potential $V(t)$ and · by the charge in the double layer. We write[14]

$$\frac{dQ(s,t)}{dt} = \sigma_e \{P(s)V(t) + \int_{BE} ds' H(s,s') Q(s',t)\}. \tag{3}$$

Here $Q(s,t)$ is the charge density in the double layer at position s and time t. The electrolyte conductivity is σ_e. $P(s)V(t)$ is the normal electric field at s given that the BE is at potential zero and the CE is at potential $V(t)$; $H(s,s')$ is the normal electric field at s given that both electrodes are at the same potential and there is a unit charge at a distance λ_D from the BE at s'. The integral is over the surface of the BE. The form of Eq. (3) is determined by the linearity of electrostatics[14]. Since Eq. (3) is linear, we can directly solve for the impedance,

$$Z^{-1}(\omega) = \sigma_e \int_{BE} \int_{BE} dsds' \left[1 + \frac{\sigma_e H}{\iota\omega}\right]^{-1}(s,s') P(s'), \tag{4}$$

where the quantity in brackets is understood as an operator inverse.

It is well known that $P(s)$ can be represented in terms of random walks[15]. If a random walker is launched from the surface of the CE, then the probability that it will strike at the point s before either returning to the CE or striking elsewhere on the BE is proportional to $P(s)$. Similarly, $H(s, s')$ may be represented in terms of random walks, although some caution is necessary in order to obtain the correct result[16].

Consider the function $\Pi(s, s')$, the probability that a random walker commencing a distance λ_D from the surface at s' will first strike the BE at s, without having previously struck the CE. It can be shown that this gives the electric field at the BE, given that there is a unit charge a distance λ_D from the surface point s', and both electrodes are at the same potential. But we are interested not in the field at the surface of the BE, but rather in the field at the surface of the double layer, a distance of λ_D from the BE. This will differ from $\Pi(s, s')$ only within a microscopic distance λ_D of s', thus we write this correction in the form of a δ-function,

$$H(s, s') = \frac{1}{\epsilon}(\Pi(s, s') - \delta(s - s')), \tag{5}$$

where the coefficient of the δ-function is fixed by Gauss' law applied to a small region enclosing the charge, and ϵ is the dielectric constant of the electrolyte.

Since the inverse operator in Eq. (4) consists of an identity, or δ-function, added to $\Pi(s, s')$, it is natural to expand the operator in series. Some simple manipulations lead to[17]

$$\frac{Z^{-1}(\omega)}{Z_0^{-1}} = (1 - \lambda) \sum_{n=0}^{\infty} B_n \lambda^n, \tag{6}$$

where $\lambda = [1 - (\imath\omega/\omega_0)]^{-1}$, with $\omega_0 = \sigma_e/\epsilon$, and $Z_0^{-1} = \sigma_e \int_{BE} ds P(s)$. The coefficients B_n are defined by

$$B_n = \frac{1}{\int ds P(s)} \int ds_0 \int ds_1 \Pi(s_0, s_1) \cdots \int ds_n \Pi(s_{n-1}, s_n) P(s_n). \tag{7}$$

Thus B_n is the total probability that a walk commencing at the CE, which strikes the BE at least once before returning to the CE, will strike the BE more than n times before returning to B. The impedance is a *generating function* for the probability fluxes $\{B_n\}$. Since $|\lambda| < 1$ for all physical ω, and $B_n > B_{n+1}$ for all n, the series in Eq. (6) is convergent for physical ω.

It is important to remember that the random walks do not represent the physical diffusion of the ions in the solution. They instead provide a mathematical method of modelling the Laplace behavior of the electric field, which leads to ionic transport even in the absence of the concentration gradients that would drive diffusion. Diffusion effects can be important at lower frequencies than those that concern us here, especially for insulating surfaces[18].

The B_n can be directly calculated numerically. We use random walk algorithms that have proven useful in the study of diffusion-limited aggregation (DLA), and of related problems[17]. We release particles from a boundary representing the CE, which surrounds the electrode of interest. We then follow the random walk. If the particle strikes the surface of the BE at s', its walk is restarted a distance of the order of the particle diameter from s'. When the particle strikes the CE, it is destroyed (see Figure 1). By recording the number of times the walk strikes the BE before its destruction, we obtain numerical estimates for $b_n = B_{n-1} - B_n$, the total probability for the walk to return to the CE after n bounces at the BE.

THEORETICAL RESULTS FOR FRACTAL ELECTRODES

We now introduce the quantities $\nu_n = B_n/B_{n+1} - 1$. ν_n is the probability that a walker returns after exactly n bounces divided by the probability that the walker bounces at least n times.

In two limits the $\{\nu_n\}$ can be related to *multifractal* properties of the surface of the blocking electrode. ν_1 is the probability that a random walker that has travelled from the CE to the BE without returning will return after exactly one bounce. From Eq. (7) above,

$$\nu_1 = 1 - B_1 = \frac{\int ds\, ds' [\delta(s, s') - \Pi(s, s')] P(s')}{\int ds\, P(s)}. \tag{10}$$

A simple identity connects $P(s)$ and $H(s, s')$, $P(s) + (1/\mathcal{C}) \int ds' H(s, s') = 0$, where $\mathcal{C}$ is the capacitance per unit area of the BE surface. This can be derived from Eq. (3). Using this, we can show that[17] $\nu_1 = \epsilon \int ds\, P^2(s)/\mathcal{C} \int ds\, P(s)$.

If the BE is self-similar, then the scaling of the moments of $P(s)$ is determined by the multifractal exponents $\tau(q)$ of the probability measure defined by $P(s)$ (sometimes called the "harmonic measure"), and by the length scales a, for the smallest length scale of the BE, r, for the size of this electrode, and R, for the distance to the CE[19]. Now, and for the remainder of this chapter, I will discuss only two dimensional results. In two dimensions, $\nu_1 \approx 2\pi\epsilon(a/r)^{\tau(2)}/\log(R/r)a\mathcal{C}$.

In the opposite limit, of $n \to \infty$, we expect a different simplification. If a particle diffuses for an extremely long time in the neighborhood of the BE, striking it many times, then we expect that its probability to next strike the BE will become independent of the surface position s (at least to lowest order in $P(s)$, which is a small quantity for large aspect ratio R/r). We then obtain, using the same identity connecting H and P that we used following Eq. (10),

$$\nu_\infty \equiv \lim_{n \to \infty} \nu_n \approx \frac{1}{S} \int\int ds\, ds' [\Pi(s, s') - \delta(s - s')] = \frac{\epsilon \int ds\, P(s)}{\mathcal{C}S}. \tag{11}$$

where S is the surface area of the BE. For a self-similar surface, S is determined by the Hausdorff dimension D_s of the surface, and we have $\nu_\infty \approx 2\pi\epsilon(a/r)^{D_s}/\log(R/r)a\mathcal{C}$. For multifractal surfaces, we have in general $\tau(2) < 1 < D_s$, which implies $\nu_\infty \ll \nu_1$.

We can use similar ideas, plus a "blob" argument[19], to estimate the behavior of the $\{\nu_n\}$ for intermediate values of n. We assume that, typically, a particle that has struck the surface n times has wandered a distance x_n from its initial point of contact on the surface. We further assume that the particle is equally likely to be found anywhere within a distance x_n of its initial collision with the surface. Its probability of then returning to the CE is given by two factors: by the probability of a particle returning from the blob of size x_n, multiplied by the local probability density of the particle in this region. The first will scale with $\tau(2)$, with the ultraviolet scale given by x_n; the second will scale with D_s, with the infrared scale given by x_n. Thus we obtain

$$\nu_n \approx \frac{2\pi\epsilon}{\log(R/r)a\mathcal{C}}(\frac{x_n}{r})^{\tau(2)}(\frac{a}{x_n})^{D_s}. \tag{12}$$

Clearly this will match onto the previously calculated values for ν_1 and ν_∞ as the blob size decreases to a or increases to r respectively. Also, this result already implies the scaling of ν_n with r for intermediate n, a scaling in good agreement with numerical results.

Of course, this still leaves undetermined the scaling of x_n with n. We assume that this scaling gives a power law, with $x_n \propto n^\gamma$. The "spreading" exponent γ can be directly calculated for a flat surface; the result is $\gamma = 1$. For fractal surfaces, we use a simple mean-field argument. Consider a random walk of length w starting from the surface. If the walk does not interact with the surface, then it will extend a distance $x \propto \sqrt{w}$ from its initial starting point, and will intersect the surface $n = x^{D_s}$ times. This gives the scaling $x \propto n^{1/D_s}$. We will thus take $\gamma = 1/D_s$ as the spreading exponent. More sophisticated arguments give the same result[12]. Taking $x_n \approx an^{1/D_s}$, we thus have

$$\nu_n \approx \frac{2\pi\epsilon(a/r)^{\tau(2)}n^{\beta-1}}{\log(R/r)a\mathcal{C}}, \tag{13}$$

with $\beta = \tau(2)/D_s$.

It is now a simple matter to estimate the $\{B_n\}$. Noting that ν_n is essentially a logarithmic derivative of B_n with respect to n, we obtain

$$B_n \approx \exp[-\frac{2\pi\epsilon(a/r)^{\tau(2)}n^\beta}{\log(R/r)\beta a\mathcal{C}}]. \tag{14}$$

Thus the $\{B_n\}$ will display stretched exponential relaxation. From Eq. (4), we see that at low frequencies $\omega \ll \omega_0$, we can approximate

$$(1-\lambda)\sum_n B_n\lambda^n \approx (-\imath\omega/\omega_0)\int_0^\infty dnB_ne^{(\imath\omega n/\omega_0)}. \tag{15}$$

It is instructive to consider the response in the time domain. Suppose that a voltage step occurs at $t = 0$. Then the current $I(t) = \int d\omega e^{-\omega t}(Z^{-1}(\omega)/\omega)$. This current will be

$$I(t) \propto \exp[-\frac{2\pi\epsilon(a/r)^{\tau(2)}(\omega_0 t)^\beta}{\log(R/r)\beta a\mathcal{C}}], \qquad (16)$$

so that we expect to see stretched-exponential relaxation. Alternatively, we can easily show that at frequencies satisfying $\omega_0(a/r)^D \ll \omega \ll \omega_0$, the impedance $Z(\omega)$ is well approximated by

$$\frac{Z(\omega)}{Z_0} \approx 1 - \frac{2\pi\epsilon(a/r)^{\tau(2)}}{\log(R/r)\beta a\mathcal{C}}(\imath\omega)^{-\beta}, \qquad (17)$$

where $Z_0 = 1/Z_0^{-1}$, yielding a high-frequency CPA exponent $\beta = \tau(2)/D_s$. Note that Eq. (17) will be subject to stretched-exponential corrections, so the actual power law observed experimentally might be somewhat different. However, we believe that Eq. (16) for the time-domain behavior should be quite accurate.

NUMERICAL RESULTS FOR FRACTAL ELECTRODES

We have studied the behavior of the coefficients B_n for a variety of systems. Here we shall review these results for two self-similar structures–a regular structure (the Koch curve) and a stochastic fractal (the DLA cluster)[12,20].

The Hausdorff dimension for a Koch curve is[20,21] $D_s = \log(4)/\log(3)$. The CE was a circle entirely enclosing the curve. We define an aspect ratio AR as the ratio between the radius of the CE and that of the Koch curve. We typically chose an aspect ratio $AR \approx 3.0$, (For both the Koch curve and the DLA clusters, we used the radius of gyration of the surface to define its scale.)

Diffusion-limited aggregates possess a Hausdorff dimension $D_s \approx 1.71$ in two dimensions, and thus have extremely rough surfaces[22]. To create a DLA, a particle of radius a begins a random walk at a large distance from the growing cluster. When the particle strikes the cluster, it sticks to it at the point of contact. The procedure may be repeated for an arbitrary number of particles. The CE was a circle centered on the first particle in the DLA, which totally enclosed the cluster.

The first parameter we determined was $\tau(2)$. We obtained the values

$$\tau(2) = \begin{cases} 0.95 \pm 0.01 & \text{for the Koch curve;} \\ 0.89 \pm 0.02 & \text{for the DLA cluster.} \end{cases} \qquad (18)$$

The inset in Figure 2 shows our results for ν_n as a function of $1/n$ for various DLA clusters. Also shown in this figure is the collapse of the data, showing a power law regime for large n. I use the radius of gyration for the scale of the cluster r.

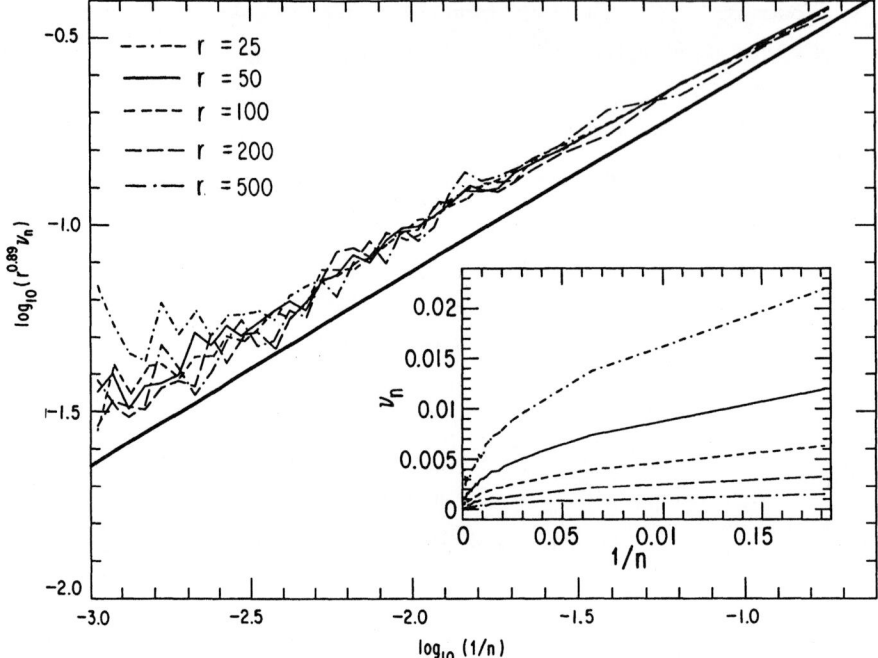

Figure 2. $\log_{10}(r^{0.89}\nu_n)$ vs. $\log_{10}(1/n)$ for DLA in $d = 2$. Inset in the figure is ν_n vs. $1/n$. The values of r are 25 (dot-short dash), 50 (solid), 100 (short dash), 200 (long dash), and 500 (dot-long dash). The heavy solid line has a slope of 0.48, which is the slope predicted by the scaling theory, and is shown for comparison.

The plots of $\log_{10}(\nu_n r^{\tau(2)})$ vs. $\log_{10}(1/n)$ show a distinct linear regime, the slope of which determines the value of $\beta - 1$. We obtained comparable results for the Koch curve.

The observed values for β were

$$\beta = \begin{cases} 0.72 \pm 0.01 & \text{for the Koch curve;} \\ 0.48 \pm 0.01 & \text{for the DLA cluster.} \end{cases} \tag{19}$$

These values can be compared with the theoretical prediction $\beta = \tau(2)/D_s$. Using the values quoted above for D_s, and $\tau(2)$ from Eq. (18), we predict

$$\beta = \begin{cases} 0.75 & \text{for the Koch curve;} \\ 0.52 & \text{for the DLA cluster.} \end{cases} \tag{20}$$

in adequate agreement with the numerical results from Eq. (19). The tendency for the numerical values to be slightly smaller than the theoretical results is systematic.

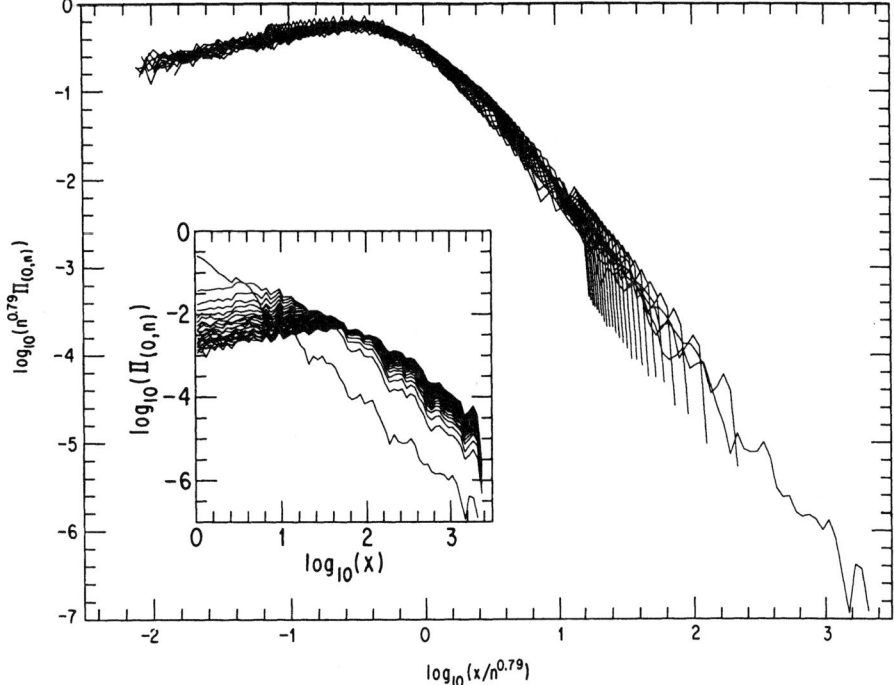

Figure 3. $\log_{10}(n^{0.79}\Pi_{(0,n)})$ vs. $\log_{10}(x/n^{0.79})$ for $n = 1, 21, 41, \ldots, 501$ for the Koch curve. Inset in the figure is the unscaled data.

We do not understand the origin of this discrepancy, although it may lie in the rather simple form of our scaling hypothesis.

We can also study the "spreading" of a walker's probability on the surface, thereby determining the exponent γ. Consider the probability that a walker has diffused a distance x after $n + 1$ contacts with the surface. This defines the function $\Pi_{(0,n)}(x)$, which can easily be extracted using our numerical method. The inset of Figure 3 shows $\log_{10}(\Pi_{(0,n)}(x))$ as a function of $\log_{10}(x)$ for $n = 1, 21, 41, \ldots, 501$ for the Koch curve. These curves can be collapsed into one function by plotting $\log_{10}(n^{\gamma}\Pi_n(x))$ vs. $\log_{10}(x/n^{\gamma})$ for each surface. Again, we obtained similar results for the DLA cluster. The range of acceptable values (as determined by the data collapse) for these scaling exponents is $0.74 \leq \gamma \leq 0.84$ for the Koch curve, or $0.57 \leq \gamma \leq 0.64$ for the DLA cluster. This compares with the prediction

$$\gamma = \frac{1}{D_s} = \begin{cases} 0.79 & \text{for the Koch curve;} \\ 0.58 & \text{for the DLA cluster.} \end{cases} \tag{21}$$

Again, the agreement is satisfactory.

253

DISCUSSION

In conclusion, we have demonstrated a new method of analyzing the properties of the double layer impedance at a rough surface, which uses the fact that this impedance is a generating function for a random walk problem. This approach allows relatively easy numerical computation of the double layer impedance at an arbitrary surface. For fractal aggregates or fractal surfaces, this results in the stretched-exponential relaxation commonly associated with glasses[23]. Although the last two sections of this chapter were confined to two dimensions, all of the results can be generalized to higher dimensionalities[12].

Acknowledgements

The work reported here was conducted in collaboration with Michael Leibig. We are grateful to B. Duplantier, P. Meakin, V. Privman, B. Sapoval, R.M. Townsend, and T. Witten for stimulating and useful conversations. This work was supported by the Materials Research Laboratory of the University of Chicago. Acknowledgement is made to the donors of the Petroleum Research Fund for the partial support of this research. T.C.H. is also grateful for the support of a Presidential Young Investigator grant from the National Science Foundation, DMR-9057156.

REFERENCES

1. T. Borisova and B. Ershler, *Zh. Fiz. Khim.* **24**:337 (1950).

2. G. Kortüm, "Treatise on Electrochemistry" (Elsevier, Amsterdam, 1965), p. 389ff; J. O'M. Bockris and A.K.N. Reddy, "Modern Electrochemistry, Vol.2" (Plenum, New York, 1970).

3. W. Scheider, *J. Phys. Chem.* **79**:127 (1975).

4. P.H. Bottelberghs and G.H.J. Broers, *J. Electroanal. Chem.* **67**:155 (1976); P.H. Bottelberghs, in "Solid Electrolytes," edited by P. Hagenmuller and W. Van Gool (Academic, New York, 1978) Chap. 10.

5. R.D. Armstrong and R.A. Burnham, *J. Electroanal. Chem.* **72**:257 (1976).

6. S.H. Liu, *Phys. Rev. Letts.* **55**:529 (1985); T. Kaplan, and L.J. Gray, *Phys. Rev. B* **32**:7360 (1985); T. Kaplan, L.J. Gray, and S.H. Liu, *Phys. Rev. B* **35**:5379 (1987).

7. B. Sapoval, *Solid State Ionics* **23**:253 (1987); B. Sapoval, J.W. Chazalviel, and J. Peyrière, *Solid State Ionics* **28-30**:1441 (1988). B. Sapoval and E. Chassaing, *Physica A* **157**, 610 (1989); E. Chaissang, B. Sapoval, G. Daccord, and R. Lenormand, *J. Electroanal. Chem.* **279**:67 (1990).

8. A. Le Mehaute and G. Crepy, *Solid State Ionics* **9**:17 (1983).

9. L. Nyikos and T. Pajkossy, *Electrochim. Acta* **30**:1533 (1985).

10. R. Ball and M. Blunt, *J. Phys. A* **21**:197 (1988); M. Blunt, *J. Phys. A* **22**:1179 (1989).

11. J.B. Bates, Y.T. Chu, and W.T. Stribling, *Phys. Rev. Letts.* **60**:627 (1988); M. Keddam and H. Takenouti, *Electrochim. Acta* **33**:445 (1988).

12. T.C. Halsey and M. Leibig, unpublished; M. Leibig and T.C. Halsey, unpublished.

13. T.C. Halsey, *Phys. Rev. A* **36**:5877 (1987).

14. T.C. Halsey, *Phys. Rev. A* **35**:3512 (1987).

15. S. Chandrasekhar, *Rev. Mod. Phys.* **15**:1 (1943); L. Pietronero and H.J. Wiesmann, *J. Stat. Phys.* **36**:909 (1984).

16. T.C. Halsey and M. Leibig, *J. Chem. Phys.* **92**:3756 (1990).

17. T.C. Halsey and M. Leibig, *Europhys. Letts.* **14**:815 (1991); *Electrochim. Acta* **36**:1699 (1991); a related method may be found in P. Meakin and B. Sapoval, *Phys. Rev. A* **43**:2993 (1991).

18. W.C. Chew and P.N. Sen, *J. Chem. Phys.* **77**:4683 (1982).

19. T.C. Halsey, M.H. Jensen, L.P. Kadanoff, I. Procaccia, and B. Shraiman, *Phys. Rev. A* **33**:1141 (1986); T.C. Halsey, *Phys. Rev. A* **38**:4789 (1989).

20. T.C. Halsey and M. Leibig, *Phys. Rev. A* **43**:7087 (1991).

21. B. Mandelbrot, "The Fractal Geometry of Nature," (W.H. Freeman, San Francisco, 1986).

22. T.A. Witten, Jr. and L.M. Sander, *Phys. Rev. Letts.* **47**:1400 (1981); P. Meakin, *Phys. Rev. A* **27**:1495 (1983).

23. For a discussion of stretched-exponential relaxation in glassy systems, see P.K. Dixon, L. Wu, S.R. Nagel, B.D. Williams, and J.P. Carini, *Phys. Rev. Letts.* **65**:1108 (1990); and references therein. For a theoretical model, see R.G. Palmer, D.L. Stein, E. Abrahams, and P.W. Anderson, *Phys. Rev. Letts.* **53**:958 (1984).

MULTIFRACTALS

Amnon Aharony

School of Physics and Astronomy, Raymond and Beverly Sackler Faculty of Exact Sciences, Tel Aviv University, Ramat Aviv, Tel Aviv 69978, Israel

and

A. Brooks Harris

Department of Physics, University of Pennsylvania, PA 19104, USA

INTRODUCTION

Multifractal measures have become a common tool in characterizing growing patterns. After reviewing the general formalism, we use the moments of the growth probabilities for obtaining new results for the *finite size corrections* to the asymptotic multifractal function $f(\alpha)$, which basically describes the distribution of these growth probabilities. These corrections, of relative order $1/\ell nL$ (where L is the linear size of the pattern, may be quite large, particularly in regimes which are dominated by small growth probabilities.

In the second part of the paper we consider various scenarios for the shape of the function $f(\alpha)$, with and without "phase transitions." Finally, we present evidence that if there is a phase transition, it must happen only below a strictly negative (and not below the zeroth) moment.

FORMALISM

Growth patterns look *self-similar*, and for large sizes their "mass" M seems to behave as a power of their linear size L, $M \sim L^D$, with a non-integer *fractal dimension* D. In practice, one rarely reaches the asymptotically large sizes needed for such a pure power-law behavior. The deviations are then described by *corrections to scaling*, of the form

$$M(L) = AL^D + A_1 L^{D_1} + A_2 L^{D_2} + ...,\tag{1}$$

where $D > D_1 > D_2 ...$ is a hierarchy of fractal dimensions.[1] Realistically it is difficult to measure more than a few of these dimensions, and these are not sufficient for characterizing the pattern in a unique way.

An alternative characterization, which uses an infinite continuum of exponents, is based on superimposing *measures* on top of the growing fractal set.[2-7] In the context of growth, a natural choice for this measure is the probability p_i that the pattern will grow at the site i on its external perimeter. At present there are many models, involving different growth rules.[6,7] Perhaps the most well known of these is diffusion limited aggregation (DLA),[8] which is also referred to as Laplacian growth (arising also e.g. in the dielectric breakdown model.[9]) In these models, p_i is proportional to the gradient of the Lapalcian field at the growth site. Much of our discussion does not depend on specific details of the growth rules.

If the distribution of these growth probabilities, for an aggregate of linear size L, is $n(p, L)$, then one defines

$$f_L(\alpha) = \frac{\ell n\, n(p, L)}{\ell nL}, \quad \alpha = -\frac{\ell np}{\ell nL} \tag{2}$$

and one plots f_L versus α for different sizes L. The asymptotic $f(\alpha) = \lim_{L \to \infty} f_L(\alpha)$ curve is identified as the multifractal spectrum of the growth pattern.

When the asymptotic $f(\alpha)$ is finite, one may invert Eq. (2), and say that asymptotically for large L

$$n(p, L) \sim L^{f(\alpha)}, \quad p \sim L^{-\alpha}. \tag{3}$$

Such power-law relations will be considered here as the "standard" multifractal behavior,[10] where $f(\alpha)$ may be looked at as the fractal dimension of the subset of growth sites whose growth probability scales as $L^{-\alpha}$.

In the above, $n(p, L)$ was defined as the distribution of the p_i's for a specific *single* aggregate. A priori, one might expect a significant variation of $n(p, L)$ among different aggregates. In numerical simulations, sampling over the ensemble of aggregates of a given size is normally extemely far from complete. In such a case, one expects to realize only "typical" aggregates, and observed properties will tend to assume their most probable, rather than their average values.[11] For some theoretical considerations it is more convenient to average over all the possible aggregates of a given size. However, unless explicitly specified, our arguments below apply to both the "typical" and the average distributions.

Power-law relations as in Eq. (3) have indeed been shown to be valid for small α (large p). Specifically, since the largest growth probability p_{max} occurs on the "tips" of the branches, one can show[12] that indeed p_{max} obeys the power law $p_{max} \sim L^{-\alpha_{min}}$, with $\alpha_{min} = D-1$. Also, the actual growth occurs when $n(p, L)p \sim L^{f(\alpha)-\alpha}$ is maximal and of order unity. This happens on the subset with the *information dimension* $\alpha_I = f(\alpha_I)$, where $df/d\alpha|_{(\alpha = \alpha_I)} = 1$. For Laplacian growth, like in diffusion limited aggregation (DLA),[8] in two dimensions, one has[13] $\alpha_I = 1$. Indeed, these two special values of α have recently been shown to determine the *crossover* from "normal" DLA to "spiky" growth when one imposes a lower cutoff on p or on the local gradient of the Laplacian field.[14]

At present, it is not clear if the power laws (3) also hold for large α (small p). Some recent studies[15-17] indicate a "phase transition" at some critical value of α, such that for larger α (smaller p) $n(p, L)$ and /or $p(L)$ decrease with L faster than a power law. However, the existence of such a transition requires that these small values of p do not occur only on very rare realizations, which may never be observed on "typical" aggregates and whose contribution to the average $n(p, L)$ is negligible.[18] Recently, Mandelbrot[10] presented a generalized multifractal formalism, which allows non - power - law dependences. In the present paper we restrict ourselves to the "standard" approach, and discuss possible scenarios for different size dependences of $f(\alpha)$.

At present, our main sources for $f(\alpha)$ curves are numerical simulations of DLA,[7] or of the equivalent Laplacian dielectric breakdown model.[9] Sizes are limited, and therefore the resulting functions $f_L(\alpha)$ still exhibit a significant size-dependence. Usually,[7,17,19] $f_L(\alpha)$ has a cap-like shape (with a negative second derivative, as in ∩), and its right hand side shows a strong size dependence, moving upwards with increasing L. Similar behavior is observed for the distribution of currents on percolating random resistor networks.[20] One of the aims of the present paper is to discuss scenarios which may explain this behavior.

MOMENTS

For many purposes it is convenient to consider the unnormalized *moments* of the distribution $n(p,L)$, defined via

$$M_q(L) = \sum_i p_i^q = \int dp\, n(p,L) p^q . \tag{4}$$

(Again, one may consider either M_q on "typical" aggregates, or the average $[M_q]_{av}$ over all aggregates). These moments can be demonstrated to have power law dependences on L for all $q \geq 0$: At $q = 0$, $M_0(L)$ is simply the "mass" of the growth perimeter sites, which form a fractal. For DLA, this mass is of the same order as the mass of the aggregate itself, hence $M_0(L) \sim L^D$. For $q = 1$, $M_1(L) = \Sigma_i p_i = 1 \propto L^0$. For $q \rightarrow \infty$, $M_q(L)$ becomes dominated by the largest p_i, which scales as $L^{-\alpha_{min}}$, hence $M_q(L) \sim n(p_{max},L)p_{max}^q \sim L^{f(\alpha_{min})-q\alpha_{min}}$. Since $M_q(L)$ is monotonic in q, this proves that for all $q \geq 0$ $\ell n M_q / \ell n L$ approaches a finite value as $L \rightarrow \infty$, and one may write (asymptotically for large L)

$$M_q(L) \simeq A_q L^{\tau(q)} , \tag{5}$$

with $\tau(q)$ decreasing monotonically from D at $q = 0$, through 0 at $q = 1$ to a linear behavior with slope $-\alpha_{min} = -(D-1)$ at large q (see positive q parts of Fig. 1). Note that similar to Eq. (1), Eq. (5) may have power law corrections, with powers smaller than $\tau(q)$. These are ignored here, since they are much smaller than the logarithmic corrections discussed below. Thus, we assume that A_q is indpendent of L.

The situation for $q < 0$ depends on the L-dependence of the small p's. For large negative q, $M_q(L)$ is dominated by the smallest p, p_{min}. If $p_{min} \sim L^{-\alpha_{max}}$, as implied by the scenario of Ref. 18, and if also $n(p_{min},L) \sim L^{f(\alpha_{max})}$, with a finite $f(\alpha_{max})$, then for $q \rightarrow -\infty$ $\tau(q)$ approaches a straight line, with slope $-\alpha_{max}$ (Fig. 1a). Alternatively, p_{min} may decrease with L faster than a power-law. In this case, $\ell n\, M_q(L)/\ell n L$ will diverge as $L \rightarrow \infty$ below some negative critical threshold q_c, implying a "phase transition."[15-17] We return to this possibility below.

The structure of the sum in Eq. (4) is similar to that of a *partition function* in statistical physics,[21] $p_i^q = e^{-q\alpha\ell n L}$ being analogous to the Boltzmann factor $e^{-\beta E}$. Here, q represents the "inverse temperature" β, and $\alpha\ell n L$ represents the energy. The "thermodynamic" size of the system is thus represented by $\ell n L$ (and not by the usual volume). It is thus not surprising that $\tau(q)\ell n L$, which is similar to the thermodynamic free energy, may be obtained from $f_L(\alpha)$ via a Legendre transformation. This arises from the Laplace transform relation between $n(p,L)$ and $M_q(L)$,

$$M_q(L) = \int_{p_{min}}^{p_{max}} dp\, n(p,L)\, e^{q\ell n p}$$

$$= \ell n L \int_{\alpha_{min}}^{\alpha_{max}} d\alpha\, \exp\{[f_L(\alpha) - q\alpha]\, \ell n L\} . \tag{6}$$

259

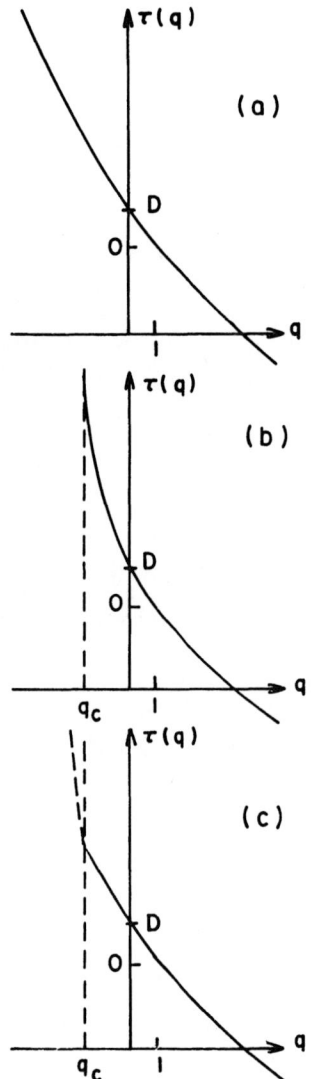

Fig. 1. Three scenarios for $\tau(q)$:
(a) Analytic. (b) Divergent as $q \to q_c^+$.
(ç) Finite with finite derivative as $q \to q_c^+$. Dashed lines dorrespond to finite L.

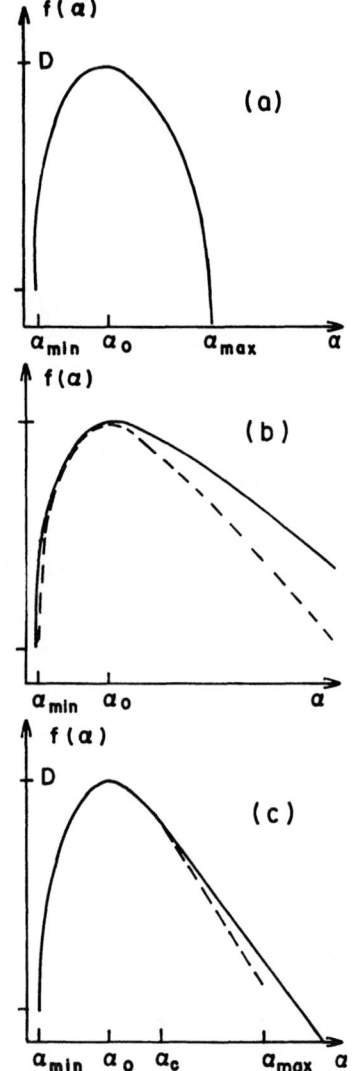

Fig. 2. Three scenarios for $f(\alpha)$, corresponding to those in Fig. 1. Dashed lines correspond to $f_L(\alpha)$ for finite L.

In Eq. (6), we have emphasized the size dependence of $f_L(\alpha)$, since we wish to include finite size corrections.

If the function $[f_L(\alpha)-q\alpha]$ has a maximum at $\alpha^*(L,q)$, where

$$\left.\frac{\partial f_L(\alpha)}{\partial \alpha}\right|_{\alpha^*} = q ,\tag{7}$$

then we may use the steepest descent method and find that

$$\frac{\ell n\, M_q(L)}{\ell nL} = f_L(\alpha^*) - q\alpha^* - \frac{\ell n\left[\left.\partial^2 f_L/\partial\alpha^2\right|_{\alpha^*}\right]}{2\ell nL} + O[(\ell nL)^{-2}],\tag{8}$$

where the last term arises from the Gaussian integral over the parabolic deviations from the maximum[20,22] and where we ignore q-independent contributions.

For $L\to\infty$ we thus identify

$$\tau(q) = f(\alpha) - q\alpha,\tag{9}$$

where $\alpha(q)$ is the solution of

$$\frac{\partial f}{\partial \alpha} = q .\tag{10}$$

In this limit, we see that the left hand side of the $\cap$-shaped $f(\alpha)$ corresponds to $q>0$, and the right hand side to $q<0$. The former is dominated by the large growth probabilities, or small α, and the latter by small p. From our earlier discussion it follows that if there is a phase transition, it should occur at $q<0$, or at $\alpha>\alpha_0 = \displaystyle\lim_{L\to\infty}\,\alpha^*(L, q=0)$.

For finite L, the left hand side of Eq. (8) is given, from Eq. (5), by $\tau(q) + \ell nA_q/\ell nL$. Even if Eq. (5) holds for all q, and $\tau(q)$ has no explicit dependence on L, the counterpart of Eq. (8) (arising from the inverse Laplace transform) yields

$$f_L(\alpha) = \tau(q^*) + q^*\alpha + \frac{\ell nA_{q^*}}{\ell nL} - \frac{\ell n\left[\left.\partial^2\tau/\partial q^2\right|_{q^*}\right]}{2\ell nL} + O[(\ell nL)^{-2}],\tag{11}$$

where $q^*(\alpha)$ is the solution of

$$\alpha = -\left.\left(\frac{\partial\tau}{\partial q} + \frac{\partial\ell nA_q}{\partial q}\frac{1}{\ell nL}\right)\right|_{q^*} .\tag{12}$$

To first order in $1/\ell nL$ in Eq. (11), it is sufficent to use q^* from the equation $\alpha = -\partial\tau/\partial q$.

The last terms in Eqs. (11) and (12) thus represent *finite size corrections*. These corrections turn out to be of relative order $1/\ell nL$, and thus vary slowly with L. The left hand side of f_L is usually a steep curve, since $\alpha_{min} = D-1$ is finite and small, so that the whole range $0 < q < \infty$ maps onto the finite range $\alpha_{min} < \alpha < \alpha_0$. This steepness prevents one from seeing the deviations of $f_L(\alpha)$ from the asymptotic $f(\alpha)$. Since no such restrictions apply for $q < 0$, we expect large finite size corrections there, as indeed observed in the simulations.[7,17] We return to this issue below.

SCENARIOS FOR f(α)

In the simplest situation, $\tau(q)$ (as defined in Eq. (5)) remains finite for all finite q. For large negative q, $M_q(L)$ is dominated by the smallest p, and $\tau(q) \sim - q\alpha_{max}$. Thus, $\tau(q)$ evolves from a straight line with slope $- \alpha_{max}$ at $q \to - \infty$ to a straight line with slope $- \alpha_{min}$ at $q \to \infty$ (see Fig. 1a). Since $\alpha_{min} < \alpha < \alpha_{max}$, the function $f(\alpha)$ is limited to this finite range of α. By Eq. (10), $f(\alpha)$ has infinite slopes at the two bounds (Fig. 2a). In this situation, $d^2\tau/dq^2$ becomes very small near both α_{min} ($q \to \infty$) and α_{max} ($q \to - \infty$), implying large finite size corrections there (Eq. (11)). However, since $f(\alpha)$ is very steep there, these corrections are not easy to see.

Since typical curves[7,17,19] of $f_L(\alpha)$ for DLA do not look like Fig. 2a, we are led to alternative scenarios, which yield different behaviors for $q < 0$, or $\alpha > \alpha_0$. In the next situation, we expect $\ell nM_q(L)/\ell nL$ to diverge to infinity as $L \to \infty$ for all $q < q_c$. This divergence could be approached from $q \to q_c^+$ in one of several ways. In the first of these, the asymptotic $\tau(q)$ diverges continuously as $q \to q_c^+$, as shown in Fig. 1b.[23] This represents a "second order phase transition" at q_c. As $q \to q_c^+$, we also have a divergence of $\alpha = - (d\tau/dq)$. Thus, the finite range $q_c < q < 0$ maps onto the infinite range $\alpha_0 < \alpha < \infty$. Since $\tau(q)$ is analytic for $q > q_c$, we can still perform the Legendre transform, and derive $f(\alpha) = \tau(q) + q\alpha$, shown in Fig. 2b. As $\alpha \to \infty$, the slope $df/d\alpha = q$ approaches the finite value q_c. Since the slope of $f(\alpha)$ for all $\alpha_0 < \alpha < \infty$ is bounded between 0 and q_c, it changes very slowly and $f(\alpha)$ seems linear. Note that a linear $f(\alpha) \simeq q_c \alpha$ implies that

$$n(p, L) \sim L^{f(\alpha)} \sim L^{q_c \alpha} \sim (L^{-\alpha})^{-q_c} \sim p^{-q_c} , \qquad (13)$$

i.e. a power law dependence of n on p. Such a power law may in fact arise even if both $n(p, L)$ and p decrease faster than power laws, but with a similar L-dependence.[20]

Most divergent forms of $\tau(q)$ would also imply a divergence of $(d^2\tau/dq^2)$ as $q \to q_c^+$. By Eq. (11), this would imply large correction terms for large α. An example of $f_L(\alpha)$, for a finite L, is also drawn schematically in Fig. 2b (dashed curve).

An alternative "phase transition" arises if $\tau(q)$ remains *finite* as $q \to q_c^+$, although it is infinite for $q < q_c$ (Fig. 1c). If the slope $\alpha = - (d\tau/dq)$ diverges as $q \to q_c^+$, the resulting $f(\alpha)$ will still have the form shown in Fig. 2b. If, on the other hand, α approaches a finite value, α_c, as $q \to q_c^+$, then the range $q_c < q < 0$ maps onto the finite range $\alpha_0 < \alpha < \alpha_c$. To understand the behavior for $\alpha > \alpha_c$, consider a finite L. For any given L, there exists a value $q_1(L)$ such that for $q < q_1(L)$ the moment $M_q(L)$ is dominated by $p_{min}{}^q$, i.e. $\tau(q) \approx \ell nM_q(L)/\ell nL \approx q\ell np_{min}/\ell nL = -q\alpha_{max}(L)$. For $q < q_1(L)$, $\tau(q)$ is essentially a straight line with the very large slope $-\alpha_{max}$ (Fig. 1c). As L increases, $q_1(L)$, approaches q_c^-. Upon Laplace transformation, the dashed curve in Fig. 1c will map onto the dashed curve in Fig. 2c. The range $q < q_1(L)$ maps onto the single point $\alpha_{max}(L)$, with $\alpha_{max}(L) \to \infty$ as $L \to \infty$. The very narrow range $q_1(L) < q < q_c$ maps onto the very broad range $\alpha_c < \alpha < \alpha_{max}$. In this latter range, the slope $df/d\alpha = q$, is bounded between q_c and $q_1(L)$, and is very close to q_c. Thus, the $f(\alpha)$-curve is very close to being linear, with its slope approaching q_c from below as L increases. Asymptotically for $L \to \infty$ we thus expect a "phase transition" in $f(\alpha)$ at α_c, to an exactly straight line with slope q_c for $\alpha > \alpha_c$.

At the present time, the available data are insufficient to distinguish between these scenarios, except for apparently excluding the simplest Fig. 2a and qualitatively looking like Fig. 2b or 2c. We stress that a quantititative inclusion of the finite size corrections, of order $1/\ell nL$, in the analysis of simulation data should improve our ability to identify the correct asymptotic behavior.

MOMENTS OF LOGARITHMS

If the function $f(\alpha)$ indeed exists, it implies that $\ell n \, n(p,L)/\ell nL$ asymptotically depends only on the scaled variable $\alpha = -\ell np/\ell nL$ (see Eqs. (2), (3)). This leads one to expect a simpler behavior when one uses ℓnp instead of p as the basic variable. Specifically, consider the normalized moments of ℓnp_i,

$$\mu_k(L) = \sum_i |\ell np_i|^k / \sum_i 1 \tag{14}$$

and the corresponding cumulants, e.g.

$$\mu_1^c = \mu_1 \, , \ \mu_2^c = \mu_2 - \mu_1^2 \, ,$$

$$\mu_3^c = \mu_3 - 3\mu_1^2 \, \mu_2 + 2\mu_1^3 \, , \dots \ . \tag{15}$$

From the definition, Eq. (4), it immediately follows that[20]

$$\mu_k(L) = (-1)^k \left. \frac{\partial^k [M_q(L)/M_0(L)]}{\partial q^k} \right|_{q=0} \, , \tag{16}$$

$$\mu_k^c(L) = (-1)^k \left. \frac{\partial^k \ell n M_q(L)}{\partial q^k} \right|_{q=0} \, . \tag{17}$$

Assuming that Eq. (5) holds and that $\tau(q)$ is analytic at least for $q \to 0^+$, one immediately has

$$\mu_k^c = (-1)^k \left. \frac{\partial^k \tau(q)}{\partial q^k} \right|_{q=0} \ell nL + (-1)^k \left. \frac{\partial^k \ell n A_q(L)}{\partial q^k} \right|_{q=0} \, . \tag{18}$$

Substitution into Eq. (15) then yields

$$\mu_k(L) = (\alpha_0 \, \ell nL)^k \left\{ 1 + \left[kC_1 + \frac{1}{2} k(k-1)D_1 \right] (\ell nL)^{-1} + O[(\ell nL)^{-2}] \right\}, \tag{19}$$

with $\alpha_0 = -(d\tau/\partial q)\big|_{q=0}$, $D_1 = (\partial^2\tau/\partial q^2)\big|_{q=0}/\alpha_0^2$ and $C_1 = -(d\ell nA_q/dq)\big|_{q=0}/\alpha_0$.

Thus, the moments of ℓnp are indeed *unifractal* as function of (ℓnL). If $\tau(q)$ is *universal* (as is the case for the resistor network[23]), then α_0 and D_1 are also universal, whereas C_1 is not.

The moments of ℓnp can also be derived directly from integration of the form

$$\mu_k(L) \sim (\ell nL)^k \int d\alpha \; \alpha^k \; e^{f_L(\alpha)\ell nL} \quad , \tag{20}$$

using our results for $f_L(\alpha)$. This derivation,[20] which requires analyticity of $f(\alpha)$ on *both* sides of α_0, yields the same result Eq. (19), including the same finite size corrections. We interpret this fact as indicating that the singularity in $f(\alpha)$, or in $\tau(q)$, does not occur at $\alpha = \alpha_0$, or at $q = 0$, but rather at a finite distance away from it. The value q_c where the "phase transition" occurs must thus be strictly negative, $q_c < 0$.

CONCLUSIONS

We have identified the leading finite size corrections to the multifractal spectrum $f_L(\alpha)$, and found that these may be quite large for the right hand side of this function. It would be very illuminating to reanalyze existing and new numerical simulations including these corrections.

At present, it seems unlikely that finite size corrections are the sole explanation of the different size dependence on the two sides of $f(\alpha)$. Rather, it seems reasonable that this difference arises due to a "phase transition" in $\tau(q)$. Whether the singularity is or is not reflected in a singularity in $f(\alpha)$ (Fig. 2c or Fig. 2b) remains to be seen.

Our analysis of moments of the logarithms of the growth probabilities leads us to believe that the threshold q_c is strictly negative. In the simple scenario of Fig. 2b, q_c appears as the asymptotic slope of $f(\alpha)$ for large α. However, other scenarios are possible, and the actual value of q_c remains to be identified.

The moments and cumulants of log p turn out to be powerful tools for checking the theory and for measuring the (possibly) universal derivatives $\partial^k \tau(q) / \partial q^k |_{(q=0)}$. We urge the experts on numerical simulations to check our predictions, particularly Eqs. (18) and (19).

As mentioned, our whole analysis was based on the "standard" multifractal formalism, where any deviation from a power law is reflected by a "phase transition." It would be interesting to see if generalized formalisms, which allow for more complicated size dependences,[10] can be directly identified as occuring in DLA.

ACKNOWLEDGEMENTS

We enjoyed collaboration with R. Blumenfeld and discussions with B. B. Mandlebrot. This project was supported by grants from the Israel Academy of Sciences and Humanities and from the Materials Reserach Laboratory (MRL) program of the U.S. National Science Foundation under grant No. DMR-88-15469.

REFERENCES

1. e.g. A. Aharony, in *Advances on Phase Transitions and Disorder Phenomena*, edited by G. Busiello, L. De Cesare, F. Mancini and M. Marinaro (World Scientific, Singapore, 1986), p. 185.
2. B. B. Mandelbrot, J. Fluid Mech. 62, 331 (1974).
3. For a review see e.g. G. Paladin and A. Vulpiani, Phys. Rep. 156, 147 (1987).
4. T. C. Halsey, M. H. Jensen, L. P. Kadanoff, I. Proccacia and B. Shraiman, Phys. Rev. A33, 1141 (1986).
5. A. Aharony, Physica A168, 479 (1990).
6. A. Aharony, in *Fractals and Disordered Systems*, edited by A. Bunde and S. Havlin (Springer-Verlag, Berlin, 1991), p. 151.
7. P. Meakin, in *Phase Transitions and Critical Phenomena*, edited by C. Domb and J. L. Lebowitz (Adademic Press, New York 1988), Vol. 12, p. 335.
8. T. A. Witten and L. Sander, Phys. Rev. Lett. 47, 1400 (1981).
9. L. Niemeyer, L. Pietronero and H. J. Wiesmann, Phys. Rev. Lett. 52, 1033 (1984).

10. A generalized formalism was discussed by B. B. Mandelbrot, Physica A168, 95 (1990); B. B. Mandelbrot, C. J. G. Evertsz and Y. Hayakawa, Phys. Rev. A42 4528 (1990).

11. e.g. A. Aharony, Physica D38, 1 (1989) and references therein; A Aharony and A. B. Harris, Physica A163, 38 (1990).

12. L. Turkevich and H. Scher, Phys. Rev. Lett. 55, 1026 (1985).

13. T. C. Halsey, P. Meakin and I. Proccacia, Phys. Rev. Lett. 56, 854 (1986).

14. E. Arian, P. Alstrom, A. Aharony and H. E. Stanley, Phys. Rev. Lett. 63, 2005 (1989).

15. J. Lee and H. E. Stanley, Phys. Rev. Lett. 61, 2945 (1988).

16. R. Blumenfeld and A. Aharony, Phys. Rev. Lett. 62, 2977 (1989).

17. S. Schwarzer, J. Lee, S. Havlin, H. E. Stanley and P. Meakin, Phys. Rev. A43, 1134 (1991).

18. A. B. Harris, Phys. Rev. B39, 7292 (1989); A. B. Harris and M. Cohen, Phys. Rev. A41, 971 (1990).

19. B. B. Mandelbrot and C. J. G. Evertsz, Physica A177, 386 (1991).

20. A. Aharony, R. Blumenfeld and A. B. Harris, to be published.

21. M. J. Feigenbaum, M. H. Jensen and I. Proccacia, Phys. Rev. Lett. 57, 1503 (1986).

22. As far as we know, such finite size corrections were first discussed in the statistical context by H. E. Daniels, Ann. Math. Stat. 25, 631 (1954). Related corrections also appear in an analytical model presented in Ref. 10.

23. This happens for the distribution of currents on percolating resistor networks, where the analog of $\tau(q)$ has been calculated analytically as an ϵ-expansion in $d = 6-\epsilon$ dimensions. For a discussion see Ref. 20.

ANGIOGENESIS AND VASCULAR NETWORKS: COMPLEX ANATOMIES FROM DETERMINISTIC

NON-LINEAR PHYSIOLOGIES

Marc E. Gottlieb

Division of Biomedical Engineering
Arizona State University
Tempe, AZ

INTRODUCTION and A MODEL OF ANGIOGENESIS

Biological structures can have morphologies which are complex yet orderly, which vary in the fine details yet have an overall sameness, and which are seemingly impossible to describe geometrically yet have an unmistakeable form and esthetic. How do complex anatomies arise, and why do they have these features? Can the genome, which only encodes proteins, be a blueprint for complex anatomies, or do non-genetic forces control morphogenesis? For many biostructures, the answer is that structure arises passively from genetically encoded biochemical processes which have no prior concept of anatomy. This idea will be examined using vascular networks and a model of angiogenesis.

Embryonic angiogenesis is reducible to simple rules which can be simulated numerically. Each of these rules is a statement of known physiology or a general biophysical principle[1,2,3]: (1) host tissues are composed of cells growing and subdividing at some rate; (2) once a structure is established, including a blood vessel, it is part of the tissue and grows with the tissue at the common rate (isauxesis); (3) as the tissue grows, cells may become too far away from any existing blood vessel, becoming "ischemic"; (4) ischemic cells generate angiogenic factors (AF), diffusible polypeptides which cause existing blood vessels to sprout new branches; (5) nearby blood vessels, feeling a high concentration of AF, sprout new branches which grow toward the AF source; (6) nearby vessels which are equally close to the AF source sprout new branches in tandem; far vessels do not sprout, because AF concentrations are too low at a distance or because they are suppressed as sprouts from near vessels relieve the ischemia; (7) multiple sprouts which meet at an AF source anastomose end-to-end into a single vessel; (8) this process is continuous.

In the numerical implementation of these rules[4,5,6], the network exists as endpoint coordinates for each vessel segment. The process is a four step iterative loop (figure 1). *Host tissue growth [step 1]* is the primary and independent event (in software, a map of cell center coordinates is returned). The growth model is defined in advance, independent of the angiogenic rules, to recreate desired growth conditions. Next, existing cells undergo *isauxetic growth matching [2]* to the newly grown cell space (an isomorphic mapping which returns adjusted vessel coordinates). Each cell is *tested for ischemia [3]* by measuring metric D, the distance from cell center c to the closest point on the network, p(c). *New sprouts [4]* are created for each ischemic cell,

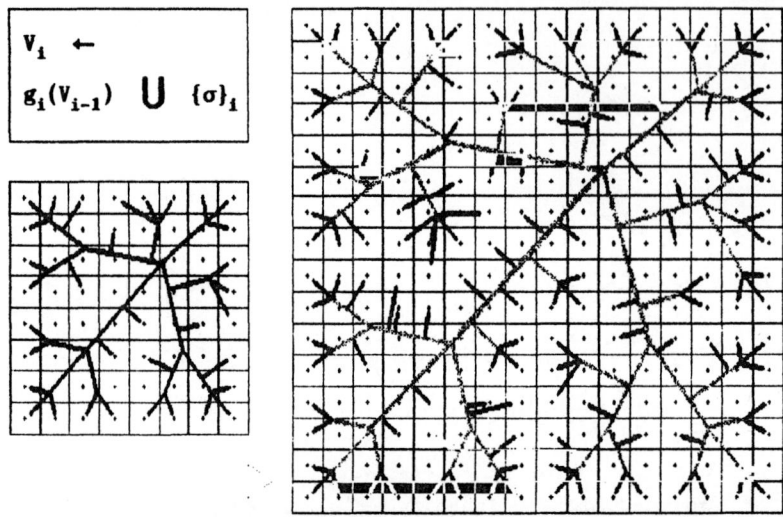

$V_i \leftarrow$

$g_i(V_{i-1}) \cup \{\sigma\}_i$

Fig. 1. The angiogenesis process. On each iteration, the cells grow (here, by subdivision into four daughters), then the existing network grows (stippled), then new vessels (solid) are added for each ischemic cell.

controlled by the three intrinsic angiogenic parameters τ, A, and R (figure 2). Whenever D exceeds the *ischemia threshold* τ, the cell is ischemic. τ is a distance, and a normalized distance scale is established by defining a resting cell as having width = 1 (the *unit cell*). A sprout arises from p(c), because this is where AF concentration, diffusing outward from c, is highest. It grows toward cell center c. The kinetics of this growth is represented by *reach R* ($0 \leq R \leq 1$, dimensionless) which determines how far sprouts grow from origin p(c) to target c. *Anastomosis limit A*, $A \geq 1$, is a coefficient of τ that fixes the outer bound beyond which AF is not felt. If a cell is

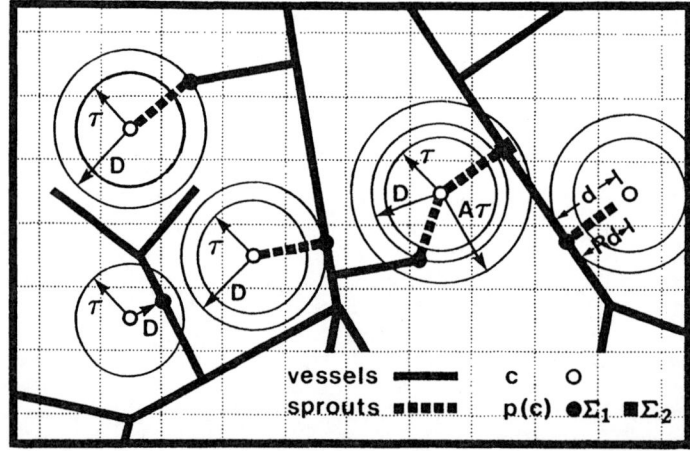

Fig. 2. The three angiogenic parameters: ischemia threshold τ is compared to cell-to-network distance D; anastomosis limit A sets an outer bound for spawning; reach R controls growth of new sprouts which arise from closest point p(c) (Σ: 1st and 2nd closest vessels). Note that right angle side branches and oblique terminal branches are the only constructs.

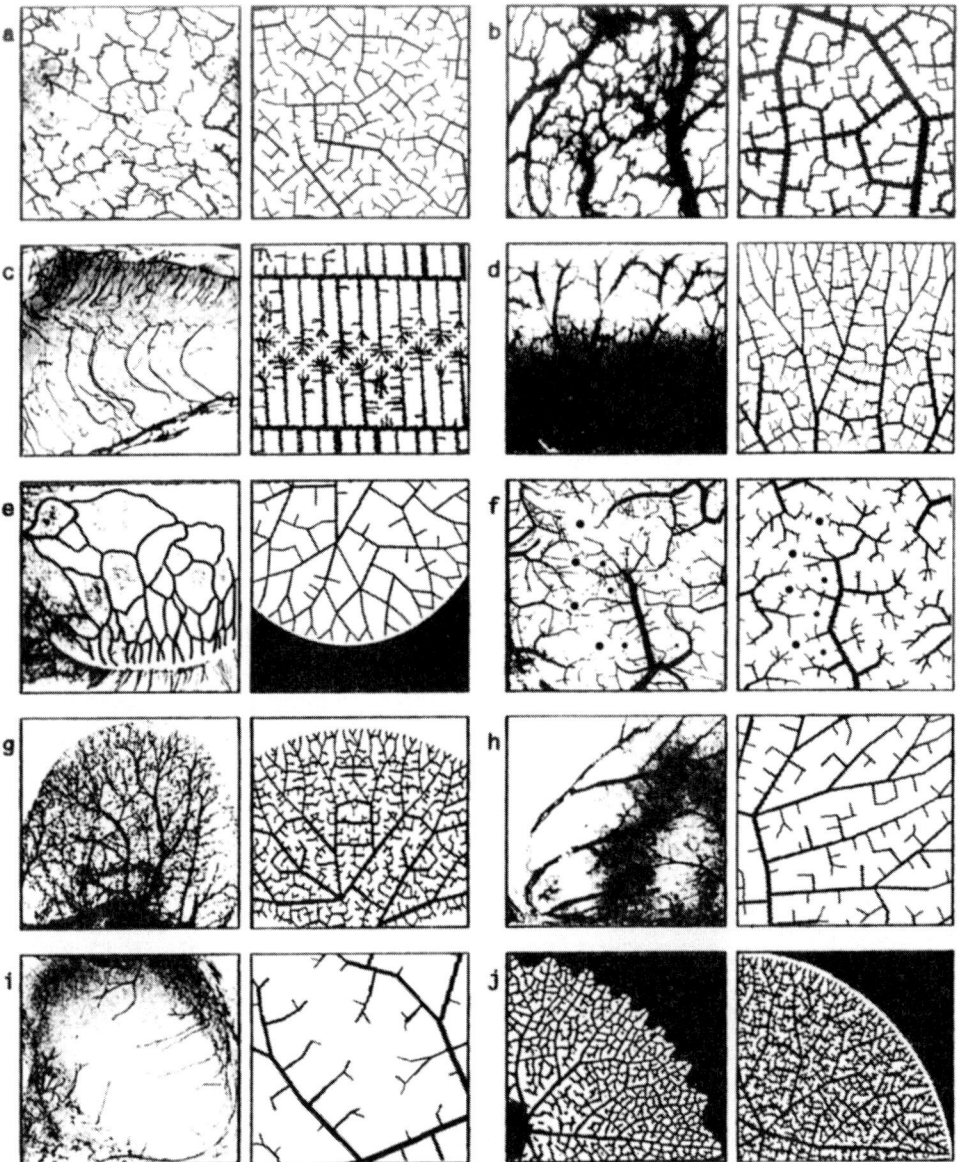

Fig. 3. A sampler of real and VT vessels. **(a)**, **(b)** Mesh patterns (human subcutaneous fascia). **(c)** Dense parallel arrays (human pons). **(d)** Ladder anastomoses (hog ileum). **(e)** Arcades (human eyelid). **(f)** Locality: vessels form locales in which new vessels grow centripetally (dots), creating smaller locales (human meninges). **(g)** Hierarchies or shells (mouse ear): the branch and loop patterns are repeated in layers with more and smaller vessels arising from larger central ones. **(h)** Axiality: adjacent vessels sire independent compact families of vessels (hog stomach). **(i)** A watershed between opposing vessels (human tarsal navicular; see also *c*). Axiality and watersheds together create uncollateralized vascular territories, the anatomical basis of stroke, heart attack, and other infarctions. **(j)** Leaf of *Alliaria officinalis*: this process is not limited to animal vessels. Note the diminishing arcade pattern with secondary sets piggyback to the first. This fractal property of pattern repetition at different scales also occurs in many of the other examples.

ischemic, $D > \tau$, then one new sprout must form. Another sprout arises from each existing vessel within radius $A\tau$ around c (each arising from the closest point on its parent vessel; the one-sprout-per-segment rule is based on other physiological principles). Multiple sprouts can arise whenever $A > 1$, but anastomoses form only when $R = 1$, because only then do sprouts meet at the cell centers. Outputs looking like real vessels arise within narrow parameter ranges, mostly within $1.0 \leq A \leq 1.8$, $0.6 \leq R \leq 1.0$, and $0.5 \leq \tau \leq 1.5$.

This system is purely deterministic. Once a tissue growth model, the three intrinsic parameters (τ, A, R), and arbitrary seed vessels are chosen, the output is uniquely determined. This system is non-linear in that the current state of the system becomes an input to the system in the next iteration:

$$V_i \leftarrow g_i(V_{i-1}) \bigcup \{\sigma\}_i \qquad (1)$$

Physical structure V on iteration i is the union of itself, rescaled from iteration i-1 by isauxetic growth matching function g, and the set of new vessels $\{\sigma\}$ forced by ischemic cells. Although numerical, this is a priori a biological model. It will be referred to as the VT (Vascular Tree) model.

BIOLOGICAL, DYNAMICAL, AND SYSTEMS FEATURES OF ANGIOGENESIS AND THE MODEL

Different VT outputs are obtained by varying the parameters and the growth model to reflect analogous differences in the growth of real tissues (figure 3). Geometric and topological features of real vessels, such as branch and anastomosis patterns, are readily created. Higher organizational features, such as axiality, watersheds, and territories, are also recreated. Pattern repetition at different scales, a fractal property, is easily identifiable. The model does not distinguish arteries and veins, and VT structures represent both, since real vessels are paired (the venae comitantes) and grow in tandem.

To understand why these anatomical features arise, some of the abstract, systems oriented aspects of these processes must be considered. This is a density invariant or density restoring system. As the host tissue grows, local areas become hypodense with vessels, creating ischemia which forces new vessel spawning (figure 4). This system is also an r-net (any point in the system space is no greater than some distance r from the embedded structure). Angiogenesis is a closed loop control system. Both the real and the VT systems are completely analogous, with blood supply or the r-net or density

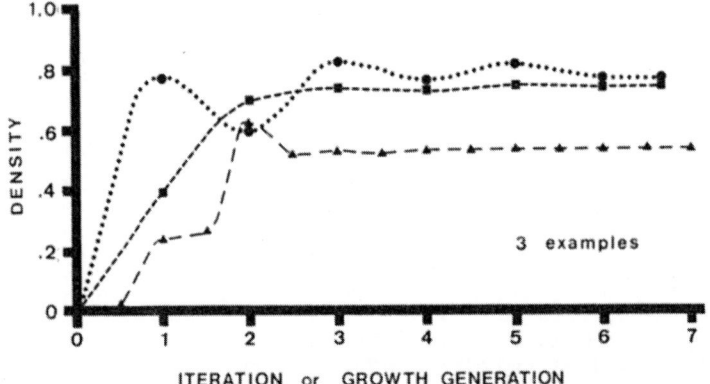

Fig. 4. Density (*total network length ÷ area of cell space*) illustrated for three typical VT structures. Stable densities are achieved within a few iterations.

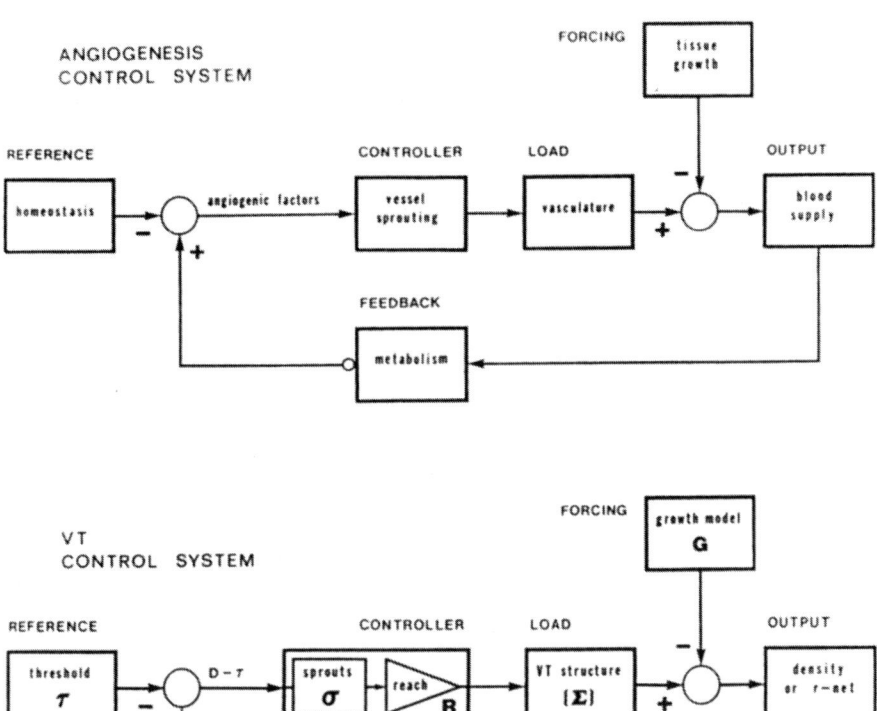

Fig. 5. Real and VT angiogenesis are analogous control systems. Both systems react only to tissue growth. Both control the vascular r-net, which translates to blood supply in the case of real vessels.

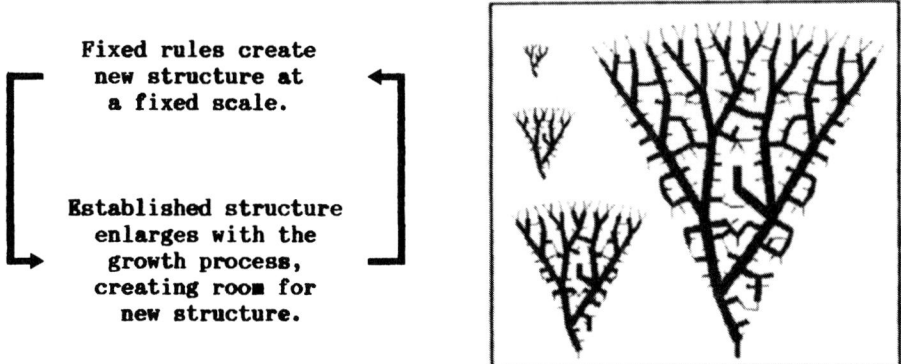

Fixed rules create new structure at a fixed scale.

Established structure enlarges with the growth process, creating room for new structure.

Fig. 6. This loop is equation 1 restated. Density restoration, the r-net, the control loop, space driving, non-linear growth, and fractal scaling are implicit. The output of the loop is shown over four iterations. Note fractons, such as a new tier of marginal vessels and "ladder" anastomoses across branches on the previous tier. Early structure has become the "gross anatomy" of later iterations.

being controlled and tissue growth forcing the system (figure 5). The vascular structure can likewise be viewed as a space driven system, growing reactively only in response to inputs from the host space (as opposed to a structure driven system growing actively and autonomously under its own rules). This entire system is summarized in the loop in figure 6.

Because of these system dynamics, angiogenesis is efficient and robust. Efficiency is twofold. First, minimum information is required to create a vascular network. In principle, as few as three system-specific genes, such as an AF gene, are sufficient. Second, the system is efficient in that, regardless of how intricate the structures look, there is little redundancy or excess. If an idealized minimum length distribution r-net ($r = \tau$) is drawn on a cell space, then the ratio of actual VT network length to ideal length approaches one (figure 7). This process is also fault tolerant. Angiogenesis is a life-and-death process, and maintenance of a valid distribution net is critical. However, random variability and environmental noise are inherent in biosystems. Figure 8 illustrates the system response to disruptive environmental conditions. The closed loop and reactive space driving of the system ensure that density is maintained and that the r-net is valid.

Vascular networks are fractal. Pattern repetition at different scales is seen in figure 3. Fractal measures are also valid, with extremely linear log(measure) vs log(scale) relationships (using five measures: ruler, box, and mass-per-box methods on VT vessel descriptors, and box and mass-per-box methods on bit-mapped graphic images of VTs and real vessels)[6]. It can be shown that formal fractal concepts, such as Borel sets, Cauchy sequences, and parameter spaces, are valid for VTs and real vascular networks[7].

Fractal structures arise from non-linear processes having system growth or rescaling. Blood vessels and angiogenesis are this type of structure/process pair. The network has two growth components, isauxetic or rescaling growth of existing structure, and new or fractal growth added at the baseline scale. This is implicit in the loop in figure 6: as established structure grows with the cell space, vessel-free space appears; the control loop reacts to this by creating new structure which reenters the cycle. If the system parameters do not change, then all new growth at the base scale is created with similar morphology or fractons (define *fracton* as an irreducible or tokenized pattern

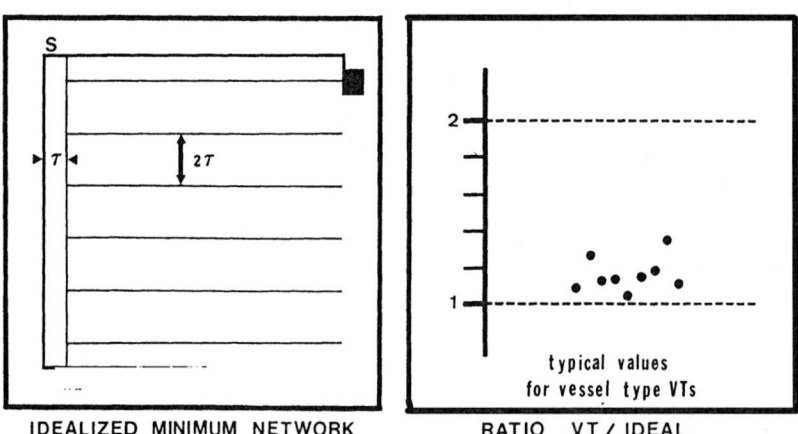

IDEALIZED MINIMUM NETWORK   RATIO   VT / IDEAL

Fig. 7. The efficiency of angiogenesis. For any VT structure, an idealized minimum distribution network can be drawn on the cell space S in which the structure grew (left). This is an r-net, with $r = \tau$. The ratio *actual VT length ÷ ideal length*, plotted for several typical structures, can approach 1 (right).

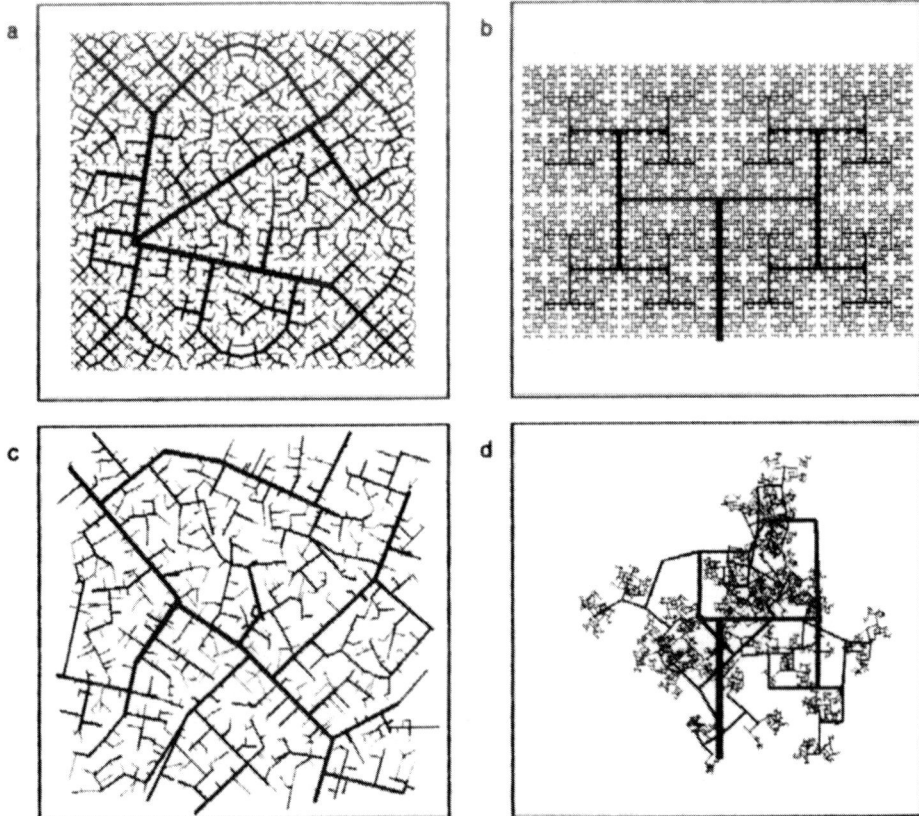

Fig. 8. (a) A typical VT. **(b)** A space filling tree (r-net equivalent under scale normalization) often used as a circulation analogue. Created by a branching algorithm whose parameters are left and right angles and scaling factors, this system is open loop and structure driven. **(c)** To simulate randomness and environmental hazards, fixed parameters have been replaced by broad frequency distributions, and spawning probability on D > τ is reduced from 1.0 to 0.25. Although disordered, the structure is still a valid distribution net on the cell space. Error tolerance is due to closed loop space driving. **(d)** Angles and scaling, one side only, replaced by narrow gaussian functions, and branching probability is 0.9: similar changes, but much much less severe. As an open loop system, this is a loss of calibration. The space is no longer served, a fatal flaw.

present at the smallest system scale). Viewed on a large scale, one sees older, larger copies of the fractons, whereas viewed on a small scale, newer, smaller copies are seen. This is fractal rescaling self-similarity. Linear log-log scaling relationships are also inherent in this system, and they are formally deducible from the premises of growth and density restoration.

VASCULAR MORPHOGENESIS and THE GENERALITY OF THE ANGIOGENIC PROCESS

Although vascular networks are intricate anatomies with many forms, their morphology and embryology are not arbitrary, and they are explainable. The overall structure is a fractal r-net. Fractalness is the geometry of a non-linear growth process, and the r-net (or the density restoration function) is the output of a control loop and reference. Any individual structure can be

Table 1. Process / Structure Features of Vascular Networks

ANGIOGENESIS *process, rules, dynamics* *physiology, embryology*	BLOOD VESSELS *structure, output, form* *anatomy*
biological rules	natural morphology
space driven	ordered complexity
process determined	efficiency
closed loop control	noise tolerance
system reference	r-net
non-linearity	fractalness

characterized by its parameter-dependent fractons, assembled and integrated by the dynamics of rescaling pattern repetition. The mathematical determinism of this system has a physical interpretation which explains overall vascular morphogenesis. Strict structural determinism, in which anatomy is genetically blueprinted, is not tenable. Reasons for this include inefficient use of the genome and poor error tolerance in an open loop system. Randomness without determination or control is equally inappropriate. Between these is process determinism, in which an anatomy arises passively from predefined process rules which operate without regard for or prior knowledge of the object's structure or morphology. Process predetermination models explain other complex anatomies[8,9,10]. Together, process determinism, space driving, and a closed loop and reference are responsible for the accurate, robust, and efficient self-organization of these vital distribution networks. The various dynamic or process features of angiogenic physiology and their physical or anatomical consequences are summarized in table 1.

Vascular networks are r-nets on their cell spaces. For all practical purposes, ischemia distance τ is the value of r for VTs. Maximum cell-to-vessel distances are the r values for real vessels. Because VT scale is normalized by the unit cell, τ represents multiples of a cell width. This allows these two sets of r values to be compared (figure 9)[6]. They are congruent, and this is not a coincidence. Regardless of their immediate physical natures, these are systems in which an embedded structure is

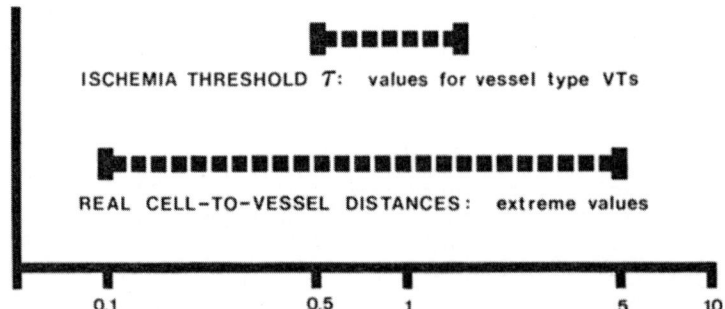

Fig. 9. Maximum cell-to-vessel distances for mammalian tissues compared to τ values for realistic VTs. Stated as cell widths (the horizontal scale), these are the r values for their respective systems. The congruence of values is the result of a common underlying dynamic organization which is independent of the physical nature of the system.

controlled by points of attraction distributed evenly throughout the host space, based upon distance from network to attracting points. The ratio *critical forcing distance / distance between attracting points* normalizes the system, and it largely determines morphology. Real vessels, having the r values that they do, look as they do. VT vessels can look like real vessels only when those values are recreated.

The concept that a structure embedded in a growing space is controlled by points of attraction distributed uniformly throughout the space implies all of the dynamics and geometry already discussed. It also implies a commonality of form for structures with similar *forcing threshold / intercellular distance* ratios, the normalized r value. If an object looks like blood or VT vessels, then its embryology can be inferred. For example, the dendrite morphology of certain neuroglial cells is vessel-like. It is a reasonable hypothesis that glial branching is controlled by an enlarging neuron space which makes growth

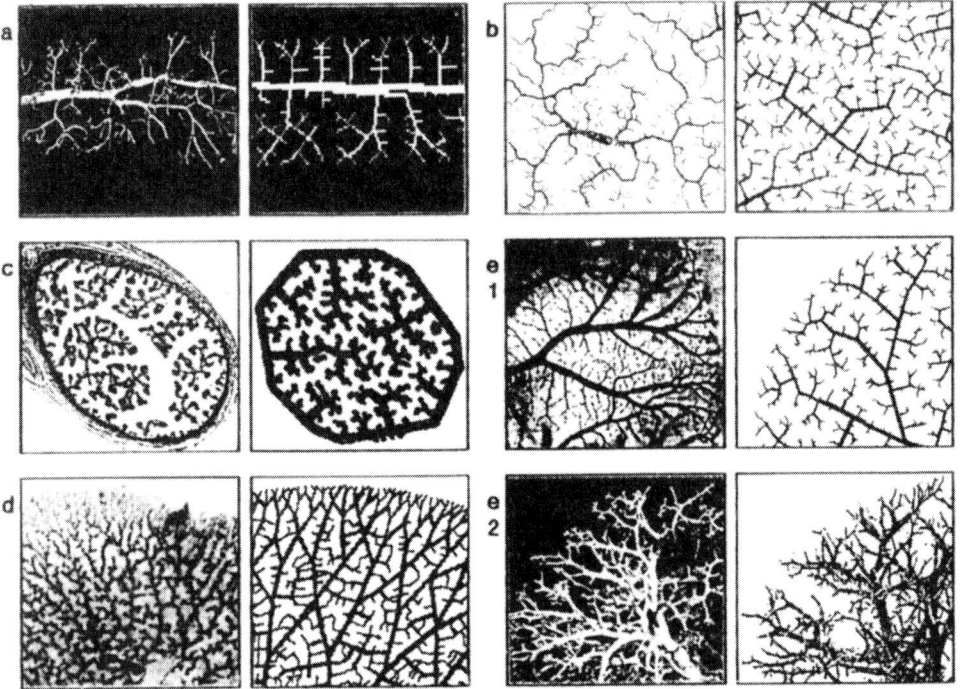

Fig. 10. Biostructures [and host space] with morphologies similar to blood vessels or VTs. **(a)** Pancreatic ducts [pancreas]. **(b)** Microglia [neuron space]. **(c)** Fallopian villi [fallopian duct, monkey]. **(d)** Digestive cavity [mesenchyme, polyclad flatworm]. **(e1)** Visual neuron [eye, house fly]. **(e2)** Bronchi [thorax, left]; hepatic ducts and paired vessels [liver, right]: the model is valid in 3 dimensions also, but for clarity is shown only in two. Figures *10e1, e2* (terminating, R < 1) and *10d & 3e, j* (anastomosing, R ≈ 1) are the common morphologies at curved margins. Closed loop control and distributed field forcing can be postulated for each example, based on its known physiology. (An alternative hypothesis may explain bronchi, bile ducts, and other ducts: to grow or differentiate, embryonic organ models must be vascularized; ducts form when ecto- or entoderm grows into mesoderm to form epithelia; ducts are accompanied by a virtually isomorphic vascular tree. This implies that ductal trees form along the paths of existing vessels, their morphology a result of the primary and independent angiogenic DFF system.)

factors if glial support is inadequate. There are many such biostructures. The systems which produce them may be physically very different, but their dynamics and integration are the same, representable by space driven control diagrams as in figure 5. In these examples, the host or control space is a *distributed forcing field* (DFF). Figure 10 illustrates probable DFF systems.

CONCLUSION

Vascular networks are morphologically complex anatomies with a vital function. A deterministic numerical model which recreates the physiology of angiogenesis explains how these intricate fractal r-nets arise necessarily from a predetermined simple process. Vascular structure itself is not determined, yet vascular geometry, topology, morphology, and functional organization arise predictably, accurately, and efficiently from the process. Angiogenesis is one case of a generalized process, distributed field forcing, in which growth of a stucture is controlled by attracting points distributed evenly throughout a host space. If a structure has a vessel-like morphology, then its system level dynamical organization can be inferred. Non-linearity, process determinism, space driving, density restoration, closed loop control, and isauxetic and fractal growth are the essential features of these systems.

REFERENCES AND ACKNOWLEDGEMENTS

1. B. M. Glaser, P. A. D'Amore, Adult tissues contain chemoattractants for vascular endothelial cells, Nature, 288:483 (1980).
2. C. M. Taylor, Effect of oxygen tension on the quantities of procollagenase-activating angiogenic factor present in the developing kitten, Br J Ophthalmol, 70:162 (1986).
3. D. A. Kessler, R. S. Langer, N. A. Pless, J. Folkman, Mast cells and tumor angiogenesis, Int J Cancer, 18:703 (1976).
4. M. E. Gottlieb, Modelling blood vessels: a deterministic method with fractal structure based on physiological rules, Proc Annu Int Conf IEEE Eng Med Biol Soc, 12:1386 (1990).
5. M. E. Gottlieb, Vascular networks: fractal anatomies from non-linear physiologies, Proc Annu Int Conf IEEE Eng Med Biol Soc, 13:2196 (1991).
6. M. E. Gottlieb, The VT model: a deterministic model of angiogenesis, IEEE Trans Biomed Eng, In press.
7. M. F. Barnsley, "Fractals Everywhere," Academic Press, San Diego (1988).
8. J. D. Murray, A pre-pattern formation mechanism for animal coat markings, J Theor Biol, 88:161 (1981).
9. H. Meinhardt, M. Klingler, A model for pattern formation on the shells of molluscs, J Theor Biol, 126:63 (1987).
10. K. D. Miller, J. B. Keller, M. P. Stryker, Ocular dominance column development: analysis and simulation, Science, 245:605 (1989).

Illustrations reproduced with permission: (3a) MJW Rees, GI Taylor, A simplified lead oxide cadaver injection technique, Plast Reconstr Surg, 77:141 (1986), p. 143. (3b) GI Taylor, et al, The venous territories (venosomes) of the human body: experimental study and clinical implications, Plast Reconstr Surg, 86:185 (1990), p. 190. (3c) RC Truex, MB Carpenter, "Human Neuroanatomy," 6th ed, The Williams & Wilkins Company, Baltimore (1969), p. 78. (3d) HM Duvernoy, Cortical veins of the human brain, in "The Cerebral Veins, An Experimental and Clinical Update," LM Auer and F Loew, eds., Springer-Verlag, Wein (1983), p. 14. (3g) JH Barker, et al, The hairless mouse ear for in vivo studies of skin microcirculation, Plast Reconstr Surg, 83:948 (1989), p. 952. (3i) W Waugh, The ossification and vascularisation of the tarsal navicular and their relation to Köhler's disease, J Bone Joint Surg, 40B:765 (1958), p.769. (10a) JCB Grant, "An Atlas of Anatomy," 6th ed., The Williams & Wilkins Co., Baltimore (1972), fig. 157. (10b) W Bloom, DW Fawcett, "A Textbook of Histology," 9th ed., W. B. Saunders Co., Philadelphia (1968), p. 338. (10c) "Catalog 61", Carolina Biological Supply, Burlington, NC (1990-1991), p. 341. (10d) V Pearse, et al, "Living Invertebrates," Blackwell Scientific Publications, Boston (1987), p. 226. (10e1) TJ Sejnowski, C Koch, PS Churchland, Computational neuroscience, Science, 241:1299 (1988), p. 1303. (10e2, left) JB West, "Respiratory Physiology," The Williams & Wilkins Company, Baltimore (1979), p. 5. (10e2, right) RMH McMinn, RT Hutchings, "Color Atlas of Human Anatomy," Yearbook Medical Publishers, Chicago (1977), p. 231.

MECHANISMS OF BIOLOGICAL PATTERN FORMATION AND CONSTRAINTS IMPOSED BY GROWTH

Hans Meinhardt

Max-Planck-Institut für Entwicklungsbiologie
Spemannstr. 35, D-74 Tübingen, Germany

Complexity and reproducibility in biological pattern formation

The development of a higher organism starts, as the rule, with a single fertilized egg. The result of the development is a complex arrangement of differentiated cells. As shown by many contributions in this book, relatively simple mechanisms allow the generation of very complex structures. In contrast to most biological pattern forming processes, most non-biological patterns of the same type are very different in details. No two lightning strokes are identical. Such similar but non-identical structures are also found in biological systems. The veins of the leaves of a tree are different from each other despite that they emerge under control of the same genetic information. However, the elements of the basic body pattern are very reproducible, there is one head, two arms etc., and all these structures are laid down at precise positions.

How development has to proceed must be encoded by the genes. The similarity of identical twins provides some intuition about how precisely the final pattern is determined. The reference to the genes, however, does not provide an explanation of how the spatial pattern of an organism is generated since, as the rule, during cell division both daughter cells obtain the same genetic information.

Pattern formation causes that originally identical cells become different from each other. The reliable reproduction of complex patterns require hierarchically ordered cascades of pattern forming events. By cell division new material becomes available on which pattern forming mechanisms can act.

In his pioneering paper, Turing (1952) has discovered that the interaction of two substances with different rates of diffusion can lead to pattern formation. Since many systems grow during development, control by diffusion imposes special opportunities and problems. Different systems react very different on growth. More and more leaves are added during growth of a plant but the number of legs in a higher animals remain constant. In some insects (such as *Drosophila*) the final number of segments is present as soon as segments can be recognized while in others (for instance in Locusts) further segments are added during outgrowth at the posterior end until completeness.

We have proposed several models for different developmental situations. This article contains a brief overview with emphasis on the effect of growth. A more detailed account can be found elsewhere (Meinhardt 1982).

Generation of a primary pattern by autocatalysis and lateral inhibition

As shown by many contributions in this volume, pattern formation from almost homogeneous initial conditions is by no means a privilege of living systems. We have proposed that primary pattern formation results from a short ranging self-enhancing (autocatalytic) process that is coupled with a long-ranging antagonistic reaction (Gierer and Meinhardt, 1972; Gierer, 1981; Meinhardt, 1982). A simple molecular realization consists of an "activator" $a(x)$ whose autocatalysis is antagonized by a long-range inhibitor $h(x)$. An example for the production rates of both substances that allow the generation of a stable pattern is given by equation 1.

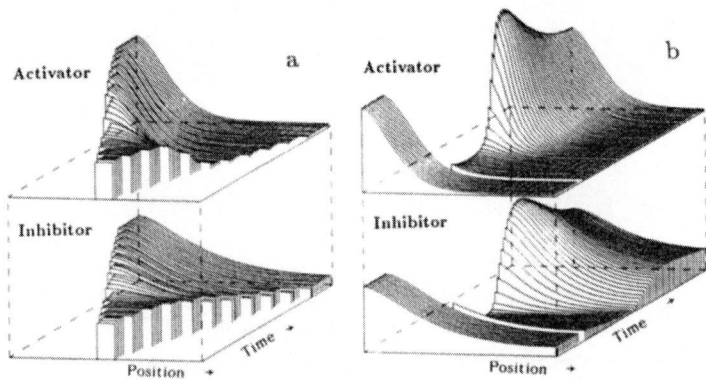

Fig. 1. Formation (a) and regeneration (b) of a graded concentration profile. Assumed are an autocatalytic substance, the activator (top) and its highly diffusing antagonist, the inhibitor (calculated with Eq.1a,b). Both concentrations are plotted as functions of position and time. Assumed further is a linear array of cells (to enable a space-time plot) that grows at both margins. If a critical size (the range of the activator) is exceeded, random fluctuations are sufficient to initiate pattern formation. A high concentration appears at a marginal position of the field since a central maximum would require space for two activator slopes which is not available below the critical size. A polar pattern results that can be maintained upon further growth. (b) After separation into an activated and a non-activated half, the activator maximum and thus the gradient regenerates in the non-activated part after the decay of the remnant inhibitor.

$$\frac{\partial a}{\partial t} = \frac{\rho a^2}{h} - \mu a + D_a \frac{\partial^2 a}{\partial x^2} + \rho_0 \qquad (1a)$$

$$\frac{\partial h}{\partial t} = \rho a^2 - \nu h + D_h \frac{\partial^2 h}{\partial x^2} + \rho_1 \qquad (1b)$$

where D_a and D_h are the diffusion coefficients, μ and ν the decay rates of a and h. The source density ρ describes the ability of the cells to perform the autocatalysis. A small activator-independent activator production ρ_0 can initiate the system at low activator concentrations. The total time corse can be calculated by an integration of these equations, starting from certain initial conditions.

The simulations shown in Fig. 1 demonstrate that the interaction (1) has properties that are basic for the explanation of biological pattern formation. A pattern emerges whenever the size of the field exceeds the range of the activator. A high activator concentration is formed at one end of the field. Thus, the system described by (1) is able to generate "positional information" (Wolpert, 1969). The graded activator and/or the graded inhibitor distribution can activate different genetic information in different parts of the tissue in an ordered sequence (see Fig. 3).

An important feature of many developing systems is their ability to regenerate. The removal of parts are compensated by pattern regulation. This is a property of the activator-inhibitor system since, for instance, after removal of the activated region, the remnant inhibitor decays until a new maximum is formed via autocatalysis (Fig. 1).

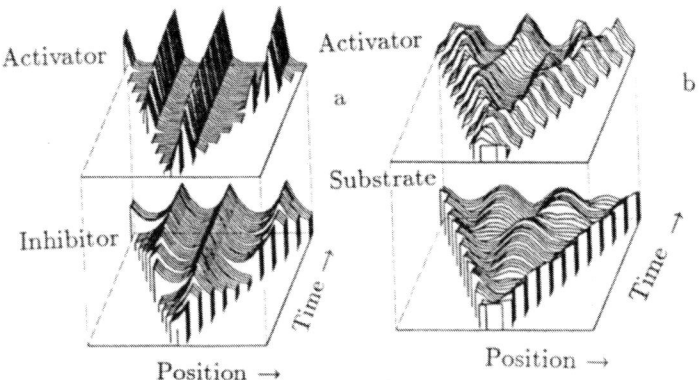

Fig. 2. Different behaviour of (a) an activator-inhibitor (equation 1) and (b) an activator - depleted substrate model (equation 2) in growing fields. If the antagonistic reaction results from an inhibitor, new maxima appear at a distance to the existing maxima since there the inhibitor is low enough to allow the onset of autocatalysis. If the antagonistic reaction results from the depletion of a substrate, new maxima are formed by a splitting and shifting of existing maxima.

The antagonistic effect can also result from the depletion of a substrate $s(x)$ which is consumed during the autocatalytic activator production.

$$\frac{\partial a}{\partial t} = \rho s a^2 - \mu a + D_a \frac{\partial^2 a}{\partial x^2} + \rho_0 \tag{2a}$$

$$\frac{\partial s}{\partial t} = \sigma - \rho s a^2 \; (-\nu s) + D_s \frac{\partial^2 s}{\partial x^2} \tag{2b}$$

This reaction has similarities with the so-called Brusselator reaction (Prigogine and Lefever, 1968; Lefever, 1968) but is somewhat simpler. It has some properties distinctly different of the activator-inhibitor mechanism. In growing systems new maxima are formed preferentially by a split and movement of existing maxima (Fig. 2b). With growth, the substrate concentration increases in the enlarging space between the activated regions since the substrate is not used up there. This can lead to a higher activator production at the side than in the centre of an existing maximum, i.e. the maximum begins to wander towards higher substrate concentrations until a new optimum position is obtained. A split of an activated region would indicate a mechanism of the activator-substrate type. Based on this feature, Lacalli (1981) has simulated pattern formation during growth of unicellular algae by such a depletion mechanism. In contrast, in the activator-inhibitor system new maxima appear at a distance to existing maxima (Fig. 2a). The appearance of new buds in the fresh water polyp hydra (new activated regions) at a distance from the head indicate that a real inhibitior spreads out from the head. In hydra, an inhibitor has been partially purified (Berking 1979).

Saturation of autocatalysis introduced, for instance, by a substitution of a^2 by $a^2/(1 + \kappa a^2)$ in equation 1 or 2 can lead to a regulation of the activated region in relation to the total size of the field (Gierer and Meinhardt, 1972) a well as to stripe-like patterns (Meinhardt, 1988)

Pattern formation by long range activation of mutual exclusive cell states

A modification of the lateral inhibition mechanism consist of a long range activation of cell states that locally exclude each other. If two or more feedback-loops locally compete with each other, it can be achieved that in a particular cell only one of these feedback loops remain active. A cell has to make a choice. If on long range such cell states activate each other and, for instance, the feedback loop a_1 has won the competition in one region, the long range help will cause that the second loop a_2 will win at a more distant position (Meinhardt and Gierer, 1980).

The chain of feedback loops that locally exclude but on long range activate each other in a symbiotic way can contain more than two elements. I have shown that many experimental observations on insect segmentation can be understood under the assumption that each segment contain at least three such each other activating cell states. Reiterated sequences of at least three cell states always have a polarity $(.../ABC/ABC/...)$. The borders between two cell states can be used to generate particular structures, the A/C border, for instance to initiate a segment border or the B/C border to initiate legs and wings (see below)

The wavelength problem: maintenance of a graded concentration profile during growth by a feedback on the source density

As a rule, the size of a morphogenetic field increases during growth of the embryo. In a usual reaction-diffusion system a graded concentration profile can be maintained only over a range of about a factor two. With increasing field size, the tendency exist to change from a monotonic distribution into a symmetric and ultimately into a periodic distribution (Fig. 2). This creates a severe problem if the graded concentration profile should be used in the growing embryo as positional information for the determination of the primary body axes.

The dynamic range over which a monotonic gradient can be maintained is increased by an order of magnitude by a feedback of the activator on its source density (ρ in equation 1 or 2) (Meinhardt and Gierer, 1974). Usually the sources of the activator and inhibitor synthesis are assumed to be homogeneously distributed, except of small random fluctuations (Fig. 1). If, however, an increased activator concentration leads to an increase of the source density, the source density obtains a graded distribution too. A possible addition to equation 1 or 2 for the change of the source density ρ is given by (3).

$$\frac{\partial \rho}{\partial t} = \gamma(a - \rho) + D_\rho \frac{\partial^2 \rho}{\partial x^2} \tag{3}$$

At a distance from the activated region, the source density becomes lower. Consequently it is less likely that the inhibition emanating from the existing maximum can be overcome. Thus, the dominance of an existing maximum is enlarged and a polar pattern can be maintained although substantial growth takes place (Fig. 4). The graded source density provides the long-lasting information about the polarity of the system. A small fragment regenerates a pattern according to the original polarity. The regeneration is very fast since no symmetry breaking is required.

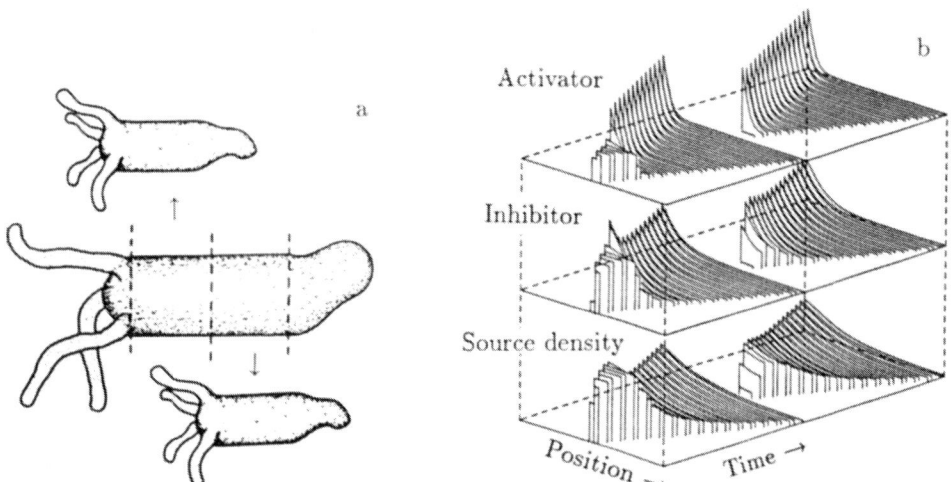

Fig. 3. Maintenance of a polar distribution in a growing field. (a) Biological observation: A hydra maintains its polar structure over a large range of sizes. Small fragments regenerate. The original polarity is maintained. Depending on the position of the cuts, the cells form either a head or a food. (b) Model: the tendency to form periodic structures upon growth can be reduced if the activator has a feedback on the source density. The source density becomes graded which, in turn, stabilizes the polar distribution. At a region of low source density, the initiation of secondary maxima is unlikely.

Since the wavelength of pattern forming system is no longer determined by the activator diffusion, the activator diffusion can be very small. Therefore, small fragments also can regenerate. The graded source density distribution provides information about the polarity of the tissue. A small fragment regenerates an activator maximum in the region of relatively highest source density, i.e. according to the original polarity. This requires that the source density has a longer time constant in comparison with the activator since the source density should not change significantly during regeneration. The simulation Fig. 3 shows the rapid regeneration of a fragment according to the original polarity. Since after regeneration the ρ-gradient will be restored in the course of time, this gradient will not be diluted out in repeating rounds of regenerations. In hydra the source density is presumably the density of a certain cell type to which the activator and inhibitor synthesis is restricted.

Cell determination and region-specific gene activation

In higher organisms, the pattern generated by a reaction-diffusion mechanism is necessarily transient since, due to growth, the polar pattern cannot be maintained over the whole expansion of the growing organism. This requires that, at an appropriate stage, the cells make use of position- specific signals, i.e. that they become determined for a particular pathway by activating particular genes. Afterwards the cells maintain this determination whether or not the evoking signal is still present.

To see which types of molecular interactions are required for selective activation of few alternative genes, it is helpful to realize that the activation of a particular gene has many formal similarities to the formation of a pattern. In pattern formation, a

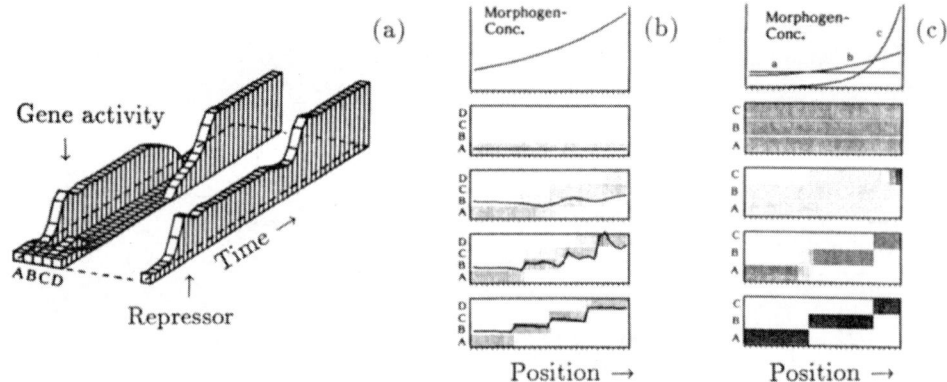

Fig. 4. Formation of cell states and space-dependent gene activation. (a) A set of genes (1,2...5) is assumed whose products feed back on the activation of the corresponding genes. In addition, all genes compete with each other by a repressor (Equation 4a,b). This has the consequence that only one of the genes can be active within one cell. In this simulation, random fluctuations have been decisive that gene A initially wins this competition. By an external signal, gene D becomes activated. Full activation results from autoregulation and is accompanied by the repression of the previously active gene. (b, c) Odered activation of several genes under control of graded morphogen distribution. A single gradient can lead to a promotion from gene A to gene B and so on, depending on the local level of the morphogen (b). (c) Alternatively, different gradients of different steepness activated different genes. Gene A is activated in a position-independent manner (line a) while B and C are activated by the gradients b and c, respectively. Shown are the initial, two intermediate and the final pattern of gene activities. Sharply confined regions of gene activities emerge despite the fact that the initiating signals are graded.

particular substance is produced at a particular location but this production is suppressed at other locations. Correspondingly, determination requires the activation of a particular gene and the suppression of the other alternative genes of a set. Gene activation may thus be regarded as a pattern formation in the gene space.

In analogy to the activator-inhibitor system for pattern formation, one can formulate the following set of equations for activation of particular genes via their gene products g_i ($i = 1...n, n$ is the number of alternative pathways (Meinhardt, 1978, 1982).

$$\frac{\partial g_i}{\partial t} = \frac{c_i g_i{}^2}{r} - \mu g_i + \frac{\delta m g_{i-1}}{r} \qquad (4a)$$

$$\frac{\partial r}{\partial t} = \sum_i c_i g_i{}^2 - \nu r \qquad (4b)$$

Each gene product g_i is autocatalytic, but also produces and reacts upon the repressor r. The last term in Eq. 4a describes a possible influence of the graded morphogen concentration m which provides the positional information. This could be, for instance, an activator or inhibitor gradient (Eq. 1; Fig. 1). Under the driving force of the morphogen, the cells switch from one activated gene to the next (Fig. 4b). The number of steps depends on the morphogen concentration. Due to the competition via a common repressor (or via a direct negative influence of the alternative genes), only one gene of the set can be active in a particular cell. The result of this "interpretation" of positional information is that in groups of neighbouring cells a particular gene is active. An abrupt transition from one activated gene to the subsequent one takes place between neighbouring cells despite the smooth distribution of the morphogen. The determinations form an ordered sequence in space.

Due to the self-activation, a gene remains active even after the signal that has caused its activation is gone. However, after an increase of the signal to higher levels, cells may be "promoted" further and obtain a differentiation corresponding to a structure closer to the high point of the gradient. Such behaviour is essential for the maintenance of the determination under normal condition and re-specification of cells during regeneration. Due to an overall growth, cells in the middle of a field become exposed to lower and lower morphogen concentration due the increasing distance from the source. (The morphogen profile depends on diffusion and is not influenced by growth). This decrease of morphogen concentration is without effect on the gene activation since a once obtained activation is stable due to the autoregulation. However, after removal of the source and its regeneration (see Fig. 1), due to the proximity of the source the cells become exposed to a morphogen concentration that is higher than the cells have seen before. These cells become re-specified to form structures corresponding to a position close to the source. The complete structure can be regenerated. A biological example is the regeneration of an insect leg (Meinhardt, 1983b)

Meanwhile, many genes with autocatalytic properties (autoregulation) have been found (for review see Serfling, 1989), so that the predicted principle, the maintenance of the determined state by feedback of a gene on its own activity, appears to be a more general process.

The structure of a higher organism is much more complex than can be achieved by the interpretation of a single gradient or two orthogonal gradients. A reliable finer and finer subdivision can be achieved in hierarchical way if the borders generated by one process provide a scaffolding for a subsequent process. A hierarchical model for pattern formation in *Drosophila* is discussed in detail elsewhere (Meinhardt, 1986, 1988). The initiation of substructures such as legs and wings at intersections of borders will be discussed below.

How to determine legs and wings?

Primary pattern formation such as described above must also control the position at which substructures such as legs or wings are to be formed. On the other hand, many experiments indicate that, after initiation, the formation of these structures is a more or less autonomous, self-regulating process. The question is then how a particular group of cells can be determined to form, for example a limb. Moreover, how is it achieved that a limb has a particular handedness (left or right) and a particular orientation with respect to the main axes of the developing embryo?

The determination of appendages clearly requires not only a preceding determination along the antero-posterior axis but also a pattern formation perpendicular to the first, along the dorsoventral axis. One could imagine that in this way a particular group of cells - specified like a particular field of a checkerboard - obtains the signal to produce, for instance, a leg, and that the further pattern formation proceeds as described above for the primary pattern formation: by autocatalysis and lateral inhibition. However, an analysis of the experimental data indicate another mechanism. Not the homogeneous region that has obtained a particular specification forms the future substructure but the region that surrounds the border between different determinations (Meinhardt, 1980, 1983 a,b). A border can become the source region for a new morphogen if cells of the two types cooperate to produce a new morphogen m. For instance, if the cells of the type A produce a precursor and the P-cells the final product m, the m-production is restricted to the A-P border, since at a distance from the border, one of the necessary components is missing - either no precursor molecules or no cells which can produce the final morphogen are available. No autocatalysis is required for that process (autocatalysis may, however be required for the region-specific gene activation that generates the borders).

The A-P border has a long extension and surrounds the embryo in a belt-like manner. To determine the position of a limb, not only a single border but the intersection of two borders is necessary. The intersection of two borders defines unique points which becomes, due to the required cooperation of the different cell types, a local source region of the morphogen.

How does a limb obtain its handedness

The hypothesis that the intersection of two borders is the necessary condition for the formation of legs or wings solves the problem of handedness. The requirement of two intersecting borders is equivalent with the requirement that four quadrants or three sectors are close to each other (depending on whether the intersection of the two borders is of the X- or of the T-type). Such an arrangement necessarily has a handedness since it is different from its mirror image. Assuming a cylindrical shape of the embryo a particular intersection is always present twice, one is located on the left, the other on the right side of the body. Both intersections have the opposite handedness. For instance, anterior-dorsal, anterior-ventral and posterior are arranged in counterclockwise fashion in a left leg. The boundary-model for limb development is supported by many experimental observations with insect (Meinhardt, 1980,1983b) and vertebrate limbs (Meinhardt, 1983a).

Formation of net-like structures

A very common structure in almost all higher organisms are filament-like branching structures. The venation of leaves, the tracheae of insects, the blood or lymph vessels as well as neurons are examples. Growing colonies of bacteria can lead to very similar patterns (see Matsushita, this volume; Matsushita and Fujikawa, 1990). A model for the formation of such structures has been proposed (Meinhardt, 1976, 1982). The simulation in Fig. 5 is based on the interaction of four substances. An activator-inhibitor system generates local high concentrations. Cells exposed to a high concentration differentiate. The differentiation is implemented in the model by the assumption of a gene whose gene product has a saturating positive feedback on its own production. Exposure of a cell to a sufficiently high activator concentration leads to the activation of this gene. Due to the feedback, the gene remains in the ON-state even if the activator concentration becomes low again. Further it is assumed that all cells produce a substrate, that the differentiated cell remove the substrate and that the activator-inhibitor system depends on this substrate. Due to the depletion of the substrate around differentiated cells, the maximum of the activator concentration becomes shifted into neighbouring cells that will differentiate too, and so on. Since the shift is directed away from other differentiated cells, in the absence of other constrains, the elongation of a filament will be straight. Branches are formed whenever activator maxima (the points of tip elongation) become sufficiently remote from each other. Then, the inhibitor concentration can become locally so low that a new activator maximum is triggered along an existing filament.

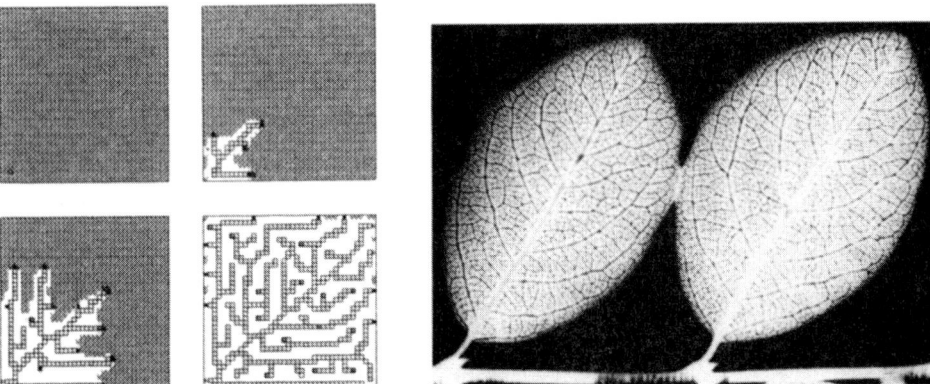

Fig. 5. Stages in the formation of a net-like structure. An interaction of four substances can create a very complex net-like structure. Two substances are used to form local activator maxima (⊔) (see Fig. 2). The corresponding cells differentiate (⊔) and remove a substance (wavy lines) from the surrounding cells. The activator maximum moves towards higher substance concentration, i.e. away from cells already differentiated, leaving a trail of newly differentiated cells behind. At right, for comparison, venation of two leaves of the same tree.

This model of vein formation is also able to account for pattern regeneration. After removal of some filaments, the substrate is no longer removed there. An increase of the substrate concentration results that attracts activator maxima on filaments near the border of the injured region. New filaments grow into the damaged region. The resulting pattern is similar but not identical when compared with the original one.

Although the it may be surprising that the interaction of only four substances can generate such complicated structures, it is not the simplest possible model. There are two inhibitions involved, one that keeps the activator signal localized and another that causes its shift. Both inhibitions can be mediated by one substance, but this has the consequence that an existing filament is so inhibitory that no lateral branches can be formed. Branches can only occur by a split of the activator maximum at the tip of a growing filament, i.e. by bifurcation of the growing filaments. The venation of leaves of the evolutionary older Ginkgo tree as well as those of some lower ferns show this pattern. Thus, the model suggest a separation of the two inhibitons as a particular step in leaf evolution.

Leaf venation is frequently regarded as an example for a fractal process in which similar structures are formed at very different scales. According to the model, there is only one process involved that works only at a single scale. Since the leaves grow, also the spaces between the veins enlarge and from a certain size onwards, new filaments grow into these regions, in a similar way as in a regeneration such as mentioned above.

Conclusion

Relatively simple molecular interactions are able to account for pattern formation during the development of higher organisms. The postulated main steps include the generation of positional information by a system of short range autocatalysis and long range inhibition and, under its control, the regional activation of different genes at particular locations. Self-regulating sequences of cell states emerge if two or more cell states activate each other on long range but exclude each other locally. Borders between regions of different gene activities act as a scaffold to generate new positional information systems. The intersection of borders between regions of different determination act as organizing regions for the formation of new structures such as legs or wings. Therefore, the new structures have necessarily the correct location and orientation in relation to the already determined cells. Thus, a cascade of basically simple molecular interactions may be responsible for the pattern formation during embryogenesis.

References

Berking, S. (1979). Analysis of head and hoot hormation in Hydra by means of an endogenous inhibitor. *Wilhelm Roux's Archives* 186: 189-210.

Gierer, A. (1981). Generation of biological patterns and form: Some physical, mathematical, and logical aspects. *Prog. Biophys. molec. Biol.* 37: 1-47.

Gierer, A. and Meinhardt, H. (1972). A theory of biological pattern formation. *Kybernetik* 12: 30-39.

Lacalli, T.C. (1981). Dissipative structures and morphogenetic pattern in unicellular algae. *Phil. Trans. R. Soc. Lond. B* 294: 547-588.

Lefever, R. (1968). Dissipative structures in chemical systems. *J.Chem.Phys.* 49: 4977-4978.

Matsushita, M. and Fujikawa, H. (1990). Diffusion-limited growth in bacterial colony formation. *Pysica A* 169: 489-506.

Meinhardt, H. (1976). Morphogenesis of lines and nets. *Differentiation* 6: 117-123.

Meinhardt, H. (1978). Space-dependent Cell Determination under the control of a morphogen gradient. *J. theor. Biol.* 74: 307-321.

Meinhardt, H. (1980). Cooperation of Compartments for the Generation of Positional Information. *Z. Naturforsch.* 35c: 1086-1091.

Meinhardt, H. (1982). Models of biological pattern formation. Academic Press, London

Meinhardt, H. (1983a). A boundary model for pattern formation in vertebrate limbs. *J. Embryol exp. Morph.* 76: 115-137.

Meinhardt, H. (1983b). Cell determination boundaries as organizing regions for secondary embryonic fields. *Devl. Biol* 96: 375-385.

Meinhardt, H. (1986). Hierarchical inductions of cell states: a model for segmentation in *Drosophila*. *J. Cell Sci. Suppl.* 4: 357-381.

Meinhardt, H. (1988). Models for maternally supplied positional information and the activation of segmentation genes in *Drosophila* embryogenesis. *Development* 104 (Supplement): 95-110.

Meinhardt, H. and Gierer, A. (1974). Applications of a theory of biological pattern formation based on lateral inhibition. *J. Cell Sci.* 15: 321-346.

Meinhardt, H. and Gierer, A. (1980). Generation and regeneration of sequences of structures during morphogenesis. *J. theor. Biol.* 85: 429-450.

Prigogine, I. and Lefever, R. (1968). Symmetry breaking instabilities in dissipative systems. *II. J. chem. Phys.* 48: 1695-1700.

Serfling, E. (1989). Autoregulation, a common property of eucariotic transcription factors?. *Trend Genetics* 5: 131-133.

Turing, A. (1952). The chemical basis of morphogenesis. *Phil. Trans. B.* 237: 37-72.

Wolpert, L. (1969). Positional information and the spatial pattern of cellular differentiation. *J. theoret. Biol.* 25: 1-47.

GROWTH PATTERNS IN FRACTURE

Stéphane Roux

CERAM, Ecole Nationale des Ponts et Chaussées
1 Av. Montaigne, 93167 Noisy-le-Grand Cedex, France
LPMMH, URA CNRS 857, ESPCI
10 rue Vauquelin, 75231 Paris Cedex 05, France

INTRODUCTION

In a number of occasions, one is interested in characterizing and understanding the behavior of heterogeneous materials on a scale much larger than that of the elementary constituents. When the local behavior is linear (elasticity, conductivity, permeability, ...) the macroscopic behavior will also be linear, and with very few restrictions, the macroscopic properties will converge under coarse graining toward a well defined limit, justifying the powerful concept of equivalent homogeneous medium. However, when the local properties are non-linear, it may happen that the macroscopic behavior will be controlled by the heterogeneities of the medium *at all scales*. This turns out to be the case for instance in brittle fracture. Other examples may also be found in various frameworks: critical currents in disordered superconductors, clogging of porous filters by deposition of particles in pores, flow of threshold fluids in a porous medium, plastic behavior of heterogeneous solids, *etc* ... In many a case, it is possible to distinguish different regimes for the local behavior, and thus different "phases", and the geometrical arrangement of these phases produces naturally patterns which reveal some information about the development of the structure, and thus the complex interactions at play at the local scale.

We will concentrate in this chapter on the particular case of brittle fracture of disordered media. In this case, the two "phases" are trivially the broken and the intact elements. Thus the pattern which emerges in the fracture process is nothing but the geometry of the set of cracks. We will see below that the structure of these cracks is non trivial and that scaling concepts are needed to account for their sample-size dependence.

The problem of incorporating disorder at the local scale in brittle fracture is attracting a lot of effort as can be judged from the recent studies on this subject, Refs.[1-11]. The first two references are reviews of these approaches. Despites the variety of models considered, and the basic representation of the local level (networks, finite-element methods, fiber bundles, ...) most of these approaches aim at exploring the macroscopic consequences of fluctuating local properties, in order to study the coarse-grained behavior law of such heterogeneous solids, and the systematic effects associated with a change in scale. In most cases, analytic works are limited to very simple geometries or very early stages of damage and thus most information originates from numerical simulations. Thus although these recent results are interesting, one should be careful in extrapolating their limit of validity. Despite this warning, we will tentatively suggest asymptotic scaling properties.

The paper is organized as follows: We first introduce one simple breakdown model, chosen to be as simple as possible, while describing accurately a disordered system and the interaction between micro-cracks as they develop. Then we report a basic fundamental property of this model, and extract a few observable properties which result from this key observation.

MODEL OF BREAKDOWN

We consider a simple discrete model which captures the essential ingredients of a brittle behavior of a disordered material, and simplify all features which are not expected to play a key role. Let us imagine a regular lattice whose bonds are the basic constituents of the model. Each bond is assumed to have a linear behavior up to a threshold where it fails irreversibly. These elements can either be elastic (linear brittle springs or beams) or electrical circuits (linear fuse). From numerical simulations performed in both frameworks, it turns out that the scaling properties are insensitive to this character [7,13-15]. The bonds are thus characterized by two numbers: their stiffness (or conductance), and their threshold stress (or current). Since the elastic modulus of an heterogeneous solid is a property which converges rather fast toward a well defined value, we neglect at the local scale the fluctuations of this property, and set the local stiffness to a single value chosen to be one. On the contrary, we wish to take into account the effect of a random distribution of thresholds. We thus introduce the probability distribution, $p(t)$, of thresholds t. This function, together with the lattice size, are the only input of the model.

The fracture of the lattice is monitored in the following way. At each stage of the fracture process, the behavior of the model is linear from the origin (zero displacement and zero force applied onto the lattice) to a threshold external force. The model makes use of this property to follow the development of the fracture. At each stage, the force applied on the lattice is reset to zero and progressively increased up to the breaking of one single bond. The number of broken bonds (geometric damage) is therefore the control parameter of the problem.

With this definition of the model, many properties can be studied [7,13-15]. We will simply underline one of them and derive some consequences.

ONE BASIC PROPERTY

One of the basic properties observed at the very final stage of fracture (*i.e.* when a single bond remains to be broken) concerns the distribution of local stresses. It has first been reported by de Arcangelis and Herrmann [13] that this distribution had a *multifractal* character [16]. This means that this distribution exhibits a size dependence which can be accounted for by the introduction of reduced variables. Let $N(\phi, L)$ be the logarithmically binned histogram of local force ϕ in a lattice of size L subjected to a unit external force applied on the lattice. The use of the scaling index

$$\alpha = \log(\phi)/\log(L),$$

and of the fractal dimension

$$f = \log(N(\phi, L))/\log(L)$$

allows to characterize the histogram $N(\phi, L)$ through the function $f(\alpha)$, the so-called *multifractal spectrum*, which is *size independent*. The variable α gives the scaling of the local force with the system size $\phi \propto L^{\alpha}$, whereas f is the fractal dimension of the support of those forces since $N(\phi, L) \propto L^{f}$. One basic consequence is that the correlation length defined on the distribution of forces is equal to the size of the lattice at the final stage. Figure 1 illustrates this size independence for five system sizes ($L = 4, 8, 16, 32$ and 64).

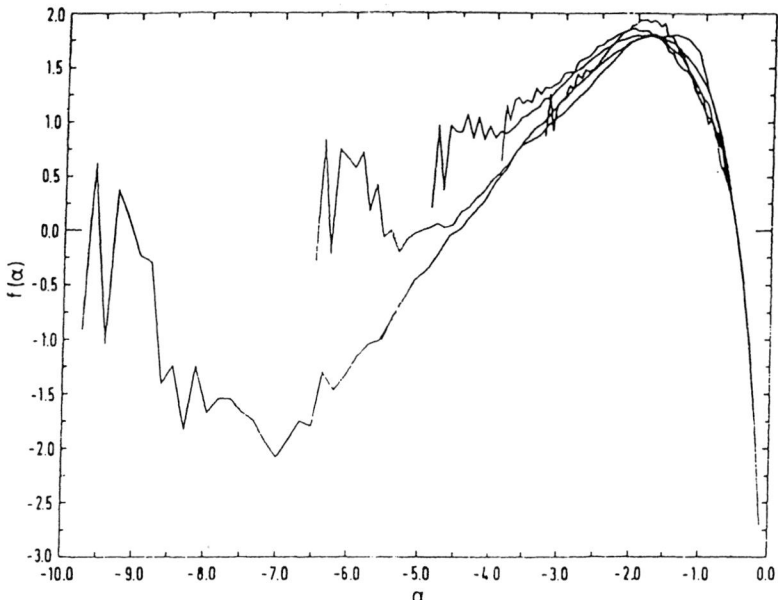

Figure 1. Rescaled distribution of local forces in a beam network under shear with a uniform distribution of threshold stress. These distributions collapse onto a single curve which is the multifractal spectrum.(from Ref.[14])

Two extreme cases can be mentioned where this multifractal character is obviously encountered. The first case is a homogeneous system. In this case a simple straight crack will develop throughout the medium, and thus at the very late stage of fracture a single bond remains to be broken. The distribution of force in this case will display a very simple multifractal spectrum consisting in a few points: one describing the bulk of the medium (f equal to the space dimension d) and a trivial α index resulting from the equilibrium $\alpha = 1 - d$. Another isolated point will describe the singularity of forces at the tip of the crack $f = 0$, and $\alpha = 0$, and finally depending on the precise geometry, another "cold stress" singularity may appear (the latter playing in any case a negligible role).

Another extreme case is to be found when the distribution of thresholds is extremely broad, in the limit where it dominates completely the distribution of local forces in the criterion for fracture. This case can be described as a percolation type problem (with the additional requirement that the complete screening of bonds has to be taken into account [17]). It is well known that at the final stage of fracture, i.e. at the percolation threshold, the distribution of currents in a random resistor network (or the mechanical equivalent) has a multifractal character [18].

We also note that this multifractal character of the current distribution is a very common property which appears in a number of other models. In the case of the Dielectric Breakdown Model (DBM) introduced by Niemeyer et al [4], the distribution of the current at the surface of the growing crack is also multifractal. When this model is changed in such a way that breaking may occur at any place in the lattice, with a probability proportional to the local current raised to some power η [19], a similar property is observed.

The above mentioned multifractal character of the stress distribution at the fracture point may appear as a very academic property, which is difficult to observe experimentally and whose consequences are not obvious. It is the purpose of the next two sections to show that this property is useful in predicting observable features.

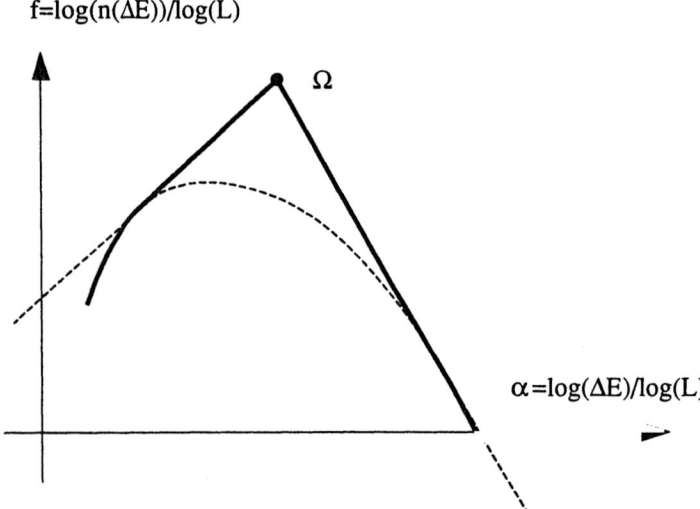

Figure 2. Schematic plot of the construction of the distribution of jumps recorded during the entire fracture process. If the location of the point Ω lies above the spectrum obtained at the end of fracture (shown in dashed line) the histogram will contain two power-laws as shown by the bold line. Otherwise, it is not possible to distinguish the two spectra.

ACOUSTIC EMISSION

Having noted that the force distribution is multifractal at the end of the process, it is straightforward to realize that the distribution of macroscopic elastic modulus jumps ΔE or compliance jumps ΔS that would result from the breaking of one single bond is also multifractal at the late stage of fracture.

Before reaching the final stage of the process, it is reasonable to assume that there exists a correlation length ξ which reaches the system size only at the final stage. This correlation length should be used instead of the system size in order to characterize the distribution of local current at any intermediate stage, in the definition of the variables α and f, in order to recover a size independent distribution. The initial stage, where the correlation length is equal to the mesh size of the lattice is characterized by a multifractal spectrum reduced to a single point. When one uses the usual definition of the spectrum (using L and not ξ) it is straightforward to see that the spectrum can be obtained from that at the late stage rescaled by a factor $\log(\xi)/\log(L)$ from the point which corresponds to the initial stage [21].

It is thus possible to use this information to investigate the distribution of a local quantity which is recorded as the fracture process progresses from the initial to the final stage. In particular the stiffness jumps which occur during fracture originate from various stages, which can be ordered by the correlation length. If we are interested in the series of jumps which take place during each individual breaking, we can use the previously described contruction, as well as the correct weighting specifying how many events are recorded as a function of the correlation length. Under very general conditions, the entire distribution of these jumps regardless of the precise state of the lattice when they were generated can be obtained using a simple geometric construction: One should consider the convex enveloppe of the multifractal spectra at the final stage dilated and of one point Ω in the $f - \alpha$ plane — which depends on the way the correlation length reaches L and on the spectrum in the initial stage. Thus if

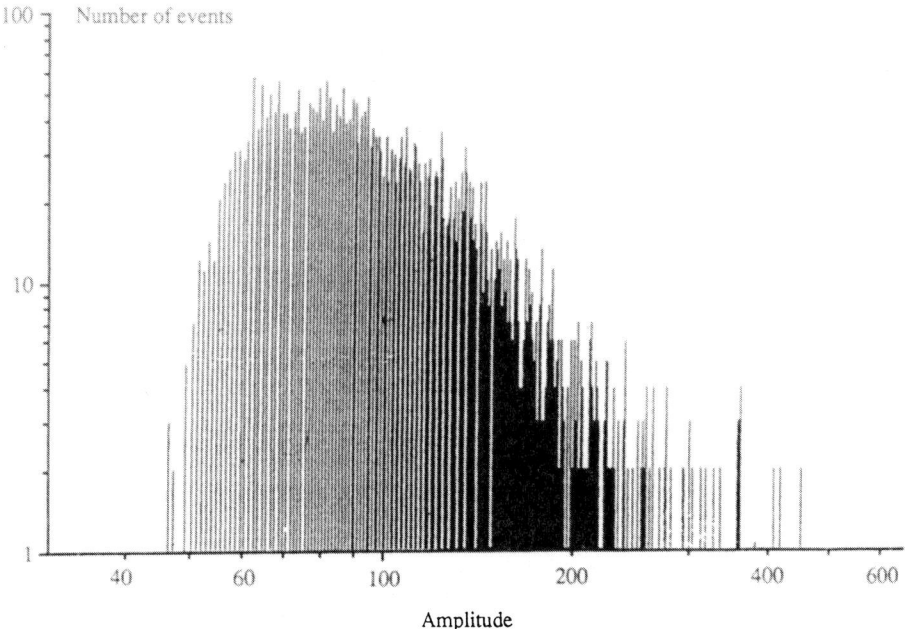

Figure 3. Log-log plot of the histogram of acoustic emission amplitudes recorded during the fracture of a single block of concrete under tension. More than 3700 events have been identified. (Courtesy of Y. Berthaud)

the point Ω lies below the final spectrum, the envelope is simply equal to that of the final stage. On the contrary, if Ω is above the spectrum as shown schematically on figure 2, a cone shape whose summit is Ω, and which is tangent to the spectrum has to be added. Since, we are dealing with a log-log plot, this means in practice that for the latter case the histogram will contain two power-laws.

Figure 3 shows a log-log plot of the histogram of acoustic emission amplitudes of events which occured during the fracture of one sample of concrete under a tensile test. For each of these events, the location of the source has been computed so as to exclude spurious events. A total of about 3700 events were recorded and identified. The experiment has been performed by Y. Berthaud and J. L. Robert. More details about the experiment can be found in Ref.20. The right hand part of the curve displays a very clear power-law behavior, which suggest that acoustic emission may provide a very direct way of investigating the scaling of the local stress field during fracture. These data will be analysed in more details in a future work.

CLASSIFICATION OF SCALING BEHAVIORS

Apart from the predictions on the distribution of acoustic emission events, the property of having a multifractal distribution of stress or current at the final stage of fracture has some theoretical implications, on the classification of scaling behaviors which can be expected depending on the distribution of local thresholds [23]. The distribution of local thresholds $p(t)$ is an input of the model. Obviously, this distribution should be set according to some experimental data analysis. It corresponds to an infinite number of degrees of freedom.

However, we have seen that the "correct" size independent variable of the problem was not the stress but the scaling index $\alpha = \log(\phi)/\log(L)$. In the model, the criterion for choosing the next bond to break is to find the minimum of the ratio t/ϕ where t is the local threshold. Since α should be used, naturally, we conclude that the "correct" variable which represents the threshold is $\alpha_t = \log(t)/\log(L)$, *i.e.*

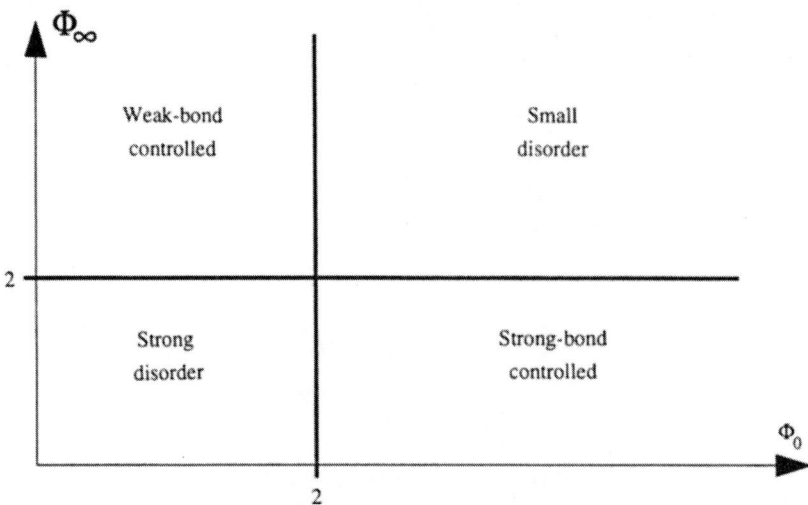

Figure 4. Schematic phase diagram of the scaling behaviors in the fracture process. Depending on the parameters Φ_0 and Φ_∞, different regimes are expected as discussed in the text.

the scaling index of the threshold. Thus, the relevant representation of the threshold distribution is its multifractal spectrum.

This statement seems odd at first sight. Indeed $p(t)$ is not expected to be size dependent, since it is a local material property. Nevertheless, the multifractal spectrum of $p(t)$ is not simply reduced to a single point. Generally, it contains three points. One which correponds to the "typical" threshold: $f = d$, $\alpha = 0$, and two additional points which indicate how the distribution reaches zero and infinity. More precisely, let us call

$$\Phi_{0/\infty} = \lim_{t \to 0/\infty} \left(\frac{log(t\, p(t))}{log(t)} \right)$$

These two numbers, Φ_0 and Φ_∞, are enough to characterize the scaling of the smallest and the largest element in a sampling of a given size. The spectrum of the distribution also contains two points which indicate this size-effect: $(\alpha = -\Phi_0,\ f = 0)$ and $(\alpha = \Phi_\infty,\ f = 0)$. If the distribution does not reach zero (resp. infinity), then the limit is not defined, and we take $\Phi_0 = \infty$ ($\Phi_\infty = \infty$ respectively) as a convention.

It is obvious that the spectrum of a distribution contains much less information than the distribution itself (two numbers are enough to characterize it), however, the identification of the proper "thermodynamic" variables allows to expect that the scaling properties of the fracture process are uniquely determined by the two scalar numbers defined above. Thus one should be able to draw a "phase-diagram" giving the scaling behavior in the Φ_0–Φ_∞ plane [23].

Let us note that there exist a few simple models where an analytic solution can be obtained (one dimensional case, hierarchical lattices, ..). In all those cases, the above prediction applies. However, in two- or three-dimensional euclidian lattices, it is not possible to map exactly this phase diagram. Thus, we have to rely on numerical simulations in order to identify the scaling properties at any given point. We may also use some perturbation techniques so as to extract the domain of validity of a regime

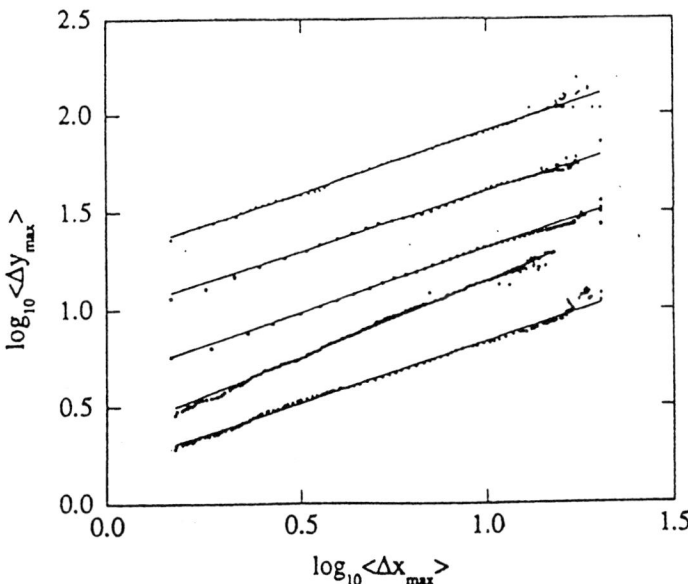

Figure 5. Mean width of the largest crack as a function of its extension during the fracture of a fuse network in log-log scale for five different threshold distributions (see Ref.[25] for details).

which is well-known (no-disorder case for instance) [24].

Figure 4 shows such an attempt to identify various phases, in terms of scaling properties. When Φ_0 and Φ_∞ are large enough, we are close to the situation where no disorder is present (all bonds would have the same threshold). Reducing progressively those parameters allows to obtain an estimate of the validity of this regime. Working along those lines, we have obtained the limits $\Phi_0 = 2$ and $\Phi_\infty = 2$. For lower Φ_0, the frature process is controlled by the weakest bonds in the system, which induce a sort of "diffuse damage" rather than the initiation of a single dominant crack. For lower Φ_∞, then very strong bonds are numerous enough to stop any propagating crack, and thus diffuse damage will also result. When both Φ_0 and Φ_∞ are smaller than 2, then it is extremely difficult to extract any reliable law.

The results of numerical simulations [7,13-15] seem to indicate that in particular for the weak-bond controlled regime, the scaling properties are insensitive to the value of Φ_0 and Φ_∞, provided they still lie in the same domain.

FRACTURE GEOMETRY

Let us finally discuss the geometry of cracks generated in the fracture process. It is obvious that in the lattice models, the crack have not a simple straight geometry. They always display some roughness. It is thus tempting to measure the width of the cracks as they grow and merge together. Figure 5 shows an example of this for a simple fuse network with a few examples of distributions $p(t)$ [24]. At each time step, the largest crack was extracted, and the width of the projection of this crack on a line perpendicular to the mean orientation was computed. Then all widths obtained for a given extension of the crack were averaged (irrespective from the stage of the fracture where they were generated). Figure 5 shows a log-log plot of this variation with a mean slope of about 0.65 (with one exception where the slope has been measured to be 0.78 — second set of data from bottom).

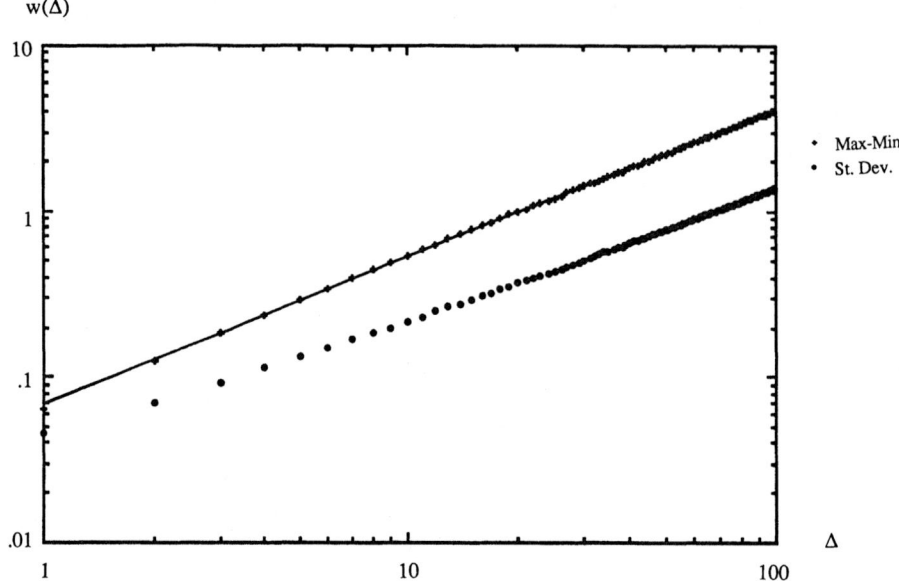

Figure 6. Log-log plot of the roughness of a granite fracture as a function of the size over which this roughness is estimated. A straight line of slope 0.88 is indicated on the graph. (Courtesy of S. Gentier [29]).

From an extensive study of many distribution which belonged to various domains of the phase diagram (Fig.4), it turned out that the same type of property was obtained: all cracks displayed a self-affine geometry even for very small extensions. Moreover, the roughness exponent ζ relating the width w to the extension ℓ of the cracks : $w \propto \ell^\zeta$ turned out to be fairly robust compared to the typical fluctuations of values found for a single distribution. In all cases for this two-dimensional fuse model, ζ was found to lie in the range $\zeta \approx 0.70 \pm 0.10$. A similar property was also observed after breaking, where the roughness of the main crack was dependent on the size of the domain over which it was estimated.

Such a result is rather easy to test experimentally provided one can have access to a good description of the geometry of the fracture surface. Indeed such a study had already been performed in the past by Mandelbrot et al [26] through the analysis of the fractal dimension of the intersection between the fracture surface and a plane parallel to it. The fractal dimension obtained this way was reported to vary with the toughness of the material. Recently similar experiments were performed by Bouchaud et al [27] for ductile fracture, and contrary to the first reference, the fractal dimension of the intersection was observed to be independent of various treatments performed on the Al alloy used.

Måløy et al have recently studied systematically very different materials (ranging from plaster to bakelite) and recorded the profile of a cut through the fracture surface $z(x)$. Different tests were performed in order to reveal a self-affine structure. The first is a generalization of the intersection method developed in the three dimensional case: For each point x along the profile, we studied the distance $\delta(x)$ at which the profile reaches again and for the first time the same height $z(x)$. The statistical distribution $N(\delta)$ of the distances δ displays a power-law distribution with an exponent $2 - \zeta$ for a self-affine surface. The second method is to investigate the pair correlation function of the height through its Fourier transform. The modulus of the Fourier transform of the profile varies on average as the wavenumber to the power $0.5 + \zeta$. These two criteria gave for five different materials a similar estimate of the roughness exponent: $\zeta \approx 0.87$ [28].

We have recently complemented our initial approach with an estimate of the roughness $w(\Delta)$ obtained over an interval $[x, x + \Delta]$, and averaged over x [30]. The roughness could be estimated using various definitions. One obvious way is to use the standard deviation of the profile with respect to the mean position. A second way is to define w as the difference between the maximum and the minimum of $z(x)$ over the interval. For a self-affine surface, $w(\Delta)$ is expected to vary as Δ^ς. Figure 6 shows such an analysis with the two definitions of the roughness. Surprisingly, the difference between maximum and minimum gives a power-law with extremely small deviation, whereas the quality of the plot of the standard deviation of the height gives a somewhat poorer fit (although both are consistent). The latter example is an analysis of a data set provided to us by S. Gentier [29], from an in-situ measurement of a granite fracture over length scales ranging from half a millimeter to one meter.

CONCLUSION

We have reported a few properties of brittle fracture of disordered media. We would like to point out the fact that the use of tools and concepts of statistical physics provides a very natural and powerful guideline to the analysis of such material properties in the case of heterogeneous materials. We have focused our presentation on very general properties, trying to be as distant as possible from a specific model. Some properties reported above have to be checked experimentally very carefully, since it is really through this confrontation that scaling analysis may become a useful tool. Moreover, the numerical modelling is always restricted to a rather small range of sizes.

Let us conclude that the analysis of fracture patterns through the self-affine character of the surface generated, may provide a good framework for analysing other properties which may be affected by this geometry, such as the friction and punching of cracks, or flow of fluids in a cracked solid.

ACKNOWLEDGEMENTS

We acknowledge the collaboration of L. de Arcangelis, A. Hansen, H. J. Herrmann, E. L. Hinrichsen, K. J. Måløy and J. Schmittbuhl with whom most of the material reported above has been obtained. We are also grateful to Y. Berthaud for providing us with the acoustic emission data, and S. Gentier for the data concerning the granite fracture surface. This work is supported by the GRECO "Géomatériaux".

REFERENCES

[1] H. J. Herrmann and S. Roux eds., *Statistical models for the fracture of disordered media*, North-Holland, (Amsterdam, 1990)

[2] J. C. Charmet, S. Roux and E. Guyon eds., *Disorder and Fracture*, Plenum, (New York, 1990)

[3] P. M. Duxbury, P. D. Beale and P. L. Leath, Phys. Rev. Lett. **57**, 1052, (1986); P. M. Duxbury, P. L. Leath and P. D. Beale, Phys. Rev. B **36**, 367, (1987)

[4] L. Niemeyer, L. Pietronero, H. J. Weissman, Phys. Rev. Lett. **52**, 1033, (1984)

[5] M. Sahimi and J. D. Goddard, Phys. Rev. B **33**, 7848, (1986)

[6] E. Louis and F. Guinea, Europhys. Lett. **3**, 871, (1987)

[7] L. de Arcangelis, A. Hansen, H. J. Herrmann and S. Roux, Phys. Rev. B **40**, 877, (1989)

[8] R. L. Smith and S. L. Phœnix, J. App. Mech. **48**, 75, (1981)

[9] W. A. Curtin and H. Scher, J. Mater. Res. **5**, 535, (1990)

[10] P. Rossi and S. Richer, Mat. and Struct. **20**, 334, (1987)

[11] P. Meakin, Cryst. Prop. and Prep. **17**, 1, (1988)

[12] P. D. Beale and D. J. Srolovitz, Phys. Rev. B **37**, 5500, (1988)

[13] L. de Arcangelis and H. J. Herrmann, Phys. Rev. B **39**, 2678, (1989)

[14] H. J. Herrmann, A. Hansen and S. Roux, Phys. Rev. B **39**, 637, (1989)

[15] A. Hansen, S. Roux and H. J. Herrmann, J. Physique (Paris) **50**, 733, (1989)

[16] T. C. Halsey, M. H. Jenssen, L. P. Kadanoff, I. Procaccia and B. I. Shraiman, Phys. Rev. A **33**, 1141, (1986)

[17] S. Roux, A. Hansen, H. J. Herrmann and E. Guyon, J. Stat. Phys. **52**, 237, (1988)

[18] L. de Arcangelis, S. Redner and A. Coniglio, Phys. Rev. B **31**, 4725, (1985)

[19] A. Hansen, E. L. Hinrichsen and S. Roux, Europhys. Lett. **13**, 517, (1990)

[20] Y. Berthaud, E. Ringot and D. Fokwa, Cement and Concrete Research, **21**, 928, (1991)

[21] S. Roux and A. Hansen, Europhys. Lett. **8**, 729, (1989)

[22] S. Roux and A. Hansen, in Ref.[2] p.17;
S. Roux, A. Hansen and E. L. Hinrichsen, preprint.

[23] A. Hansen, E. L. Hinrichsen and S. Roux, Phys. Rev. B **43**, 665, (1991)

[24] S. Roux and A. Hansen, Europhys. Lett. **11**, 37, (1990)

[25] A. Hansen, E. L. Hinrichsen and S. Roux, Phys. Rev. Lett. **66**, 2476, (1991)

[26] B. B. Mandelbrot, D. E. Passoja and A. J. Paullay Nature **308**, 721, (1984)

[27] E. Bouchaud, G. Lapasset and J. Planés, Europhys. Lett. **13**, 73, (1990)

[28] K. J. Måløy, A. Hansen, E.L. Hinrichsen and S. Roux, preprint

[29] S. Gentier, Private communication.

[30] J. Schmittbuhl, Y. Berthaud and S. Roux, Compte-Rendu du GRECO Géomatériaux, (1991)

ON THE STABILITY OF GROWTH WITH A THRESHOLD

H. J. Herrmann

HLRZ, KFA Jülich
Postfach 1913, D-5170 Jülich, Germany

Abstract

Moving boundary problems with a threshold in the growth velocity occur in dielectric breakdown or fracture. A linear stability analysis at exactly the threshold shows differences between Dirichlet and von Neumann boundary conditions. For example, a circular hole in the stretched membrane of a drum is stable, while a circular hole with pressure inside is unstable.

INTRODUCTION

Many growth situations can be cast into moving boundary problems. So dielectric breakdown or viscous fingering have been described by a Laplace equation with the velocity of the boundary proportional to the gradient in the field, eventually raised to some power. Various instabilities (Saffman-Taylor, tip splitting, side branching) are then responsible for the complex morphologies encountered. Similarly one can write the fracture of solids in terms of the Lamé equations with the moving crack surface as boundary condition.

It has been pointed out in several occasions that the growth velocities in many realistic cases do not increase smoothly with the gradient at small values of the gradient but that a threshold in the gradient must be overcome before growth can set in. For dielectric breakdown a critical electric breakdown field E_c has been introduced and its role in experiments has been discussed[1]. Also viscous fingering in porous media has a threshold capillary pressure p_c[2]. To study these effects a threshold was introduced into the probabilistic "Dielectric Breakdown Model" (DBM) on a lattice[3] and a crossover was found from usual DBM patterns to more spiky structures after a certain characteristic size.

The propagation of cracks in elastic media is similar in this respect. The scalar field of the Laplace problem is replaced here by a vector field, the displacement **u**, which obeys Lamé's equation. In this case a threshold is given by the cohesion stress σ_c which must be overcome locally to tear the material apart[4].

In this contribution I want to present the linear stability analysis of the two dimensional scalar and vectorial moving boundary problems with both Dirichlet and von Neumann conditions. It turns out that the Laplacian case, presented in the next section, is unstable in most cases except when for von Neumann boundary conditions the system is small enough and the perturbation has a long wavelength. The section after treats the case of fracture which has been presented in ref. 5. There the effect of the nature of the boundary condition is more dramatic. In the last section the results are summarized.

LAPLACIAN GROWTH WITH A THRESHOLD

Information about the intrinsic instability leading to scalar Laplacian pattern formation can be obtained from linear stability analysis[6]. This is a consequence of the moving boundary condition which is attached to the set of linear field equations. The linear stability analysis is carried out in the following way. First, the solution with a symmetric boundary is taken. Then a small amplitude periodic perturbation (with wavenumber k) of the boundary is superimposed and the time dependence of the amplitude is calculated by taking into account the equation of motion of the interface and allowing deviations from the symmetric solution up to first order. Due to the linear character of the resulting equations, the solution for the amplitude has the form $e^{\omega(k)t}$ where $\omega < 0$ (> 0) means that mode k is stable (unstable).

Let us consider the pressure field p governed by the Laplace equation

$$\Delta p = 0 \tag{1}$$

in a two dimensional radial geometry as shown in Fig. 1a. Inside we have a circular hole of radius R_1 and on its boundary Γ a constant pressure p_1 is imposed (for instant by a penetrating fluid). On the outer boundary at radius R_0 we will consider two cases A: constant pressure p_0 $(< p_1)$ and B: constant flux I_0. So we have the boundary conditions:

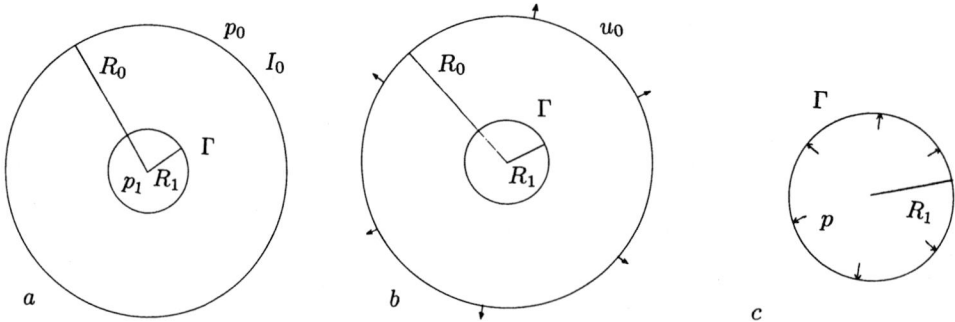

Figure 1 Schematic plot of considered situations: (a) Laplacian case, (b) stretched membrane, (c) hole with pressure inside.

$$p(\Gamma) = p_1 \quad \text{and} \quad \text{A}: \quad p(R_0) = p_0 \quad \text{or} \quad \text{B}: \quad \nabla_n p(R_0) = -I_0. \tag{2}$$

The normal growth velocity of the moving inner boundary shall be given by

$$v_n = C[-\nabla p(\Gamma) - \frac{p_c}{\ell}] \tag{3}$$

where p_c is the capillary pressure, ℓ the length of the capillary and $[..]$ denotes the positive part, since negative velocities are unphysical in most cases.

It is practical to use polar coordinates r and ϕ and one finds for the pressure field $\bar{p}(r)$ in the unperturbed case, i.e. for $\Gamma = R_1$, the solution

$$\text{A}: \quad \bar{p}(r) = p_1 + \frac{\Delta p(\ln R_1 - \ln r)}{\ln(R_0/R_1)} \tag{4a}$$

$$\text{B}: \quad \bar{p}(r) = p_1 + I_0 R_0(\ln R_1 - \ln r) \tag{4b}$$

where $\Delta p = p_1 - p_0$.

Let us now place ourselves exactly at the threshold: $\nabla \bar{p}(R_1) = -p_c/\ell$ which means in case A: $p_c = \ell \Delta p/(R_1 \ln(R_0/R_1))$ and in case B: $p_c = \ell I_0 R_0/R_1$. We analyse the stability of the shape of the inner boundary Γ by considering a perturbation of the form

$$\Gamma = R_1 + \varepsilon e^{ik\phi + \omega t} \tag{5}$$

where the wavenumber k is a positive integer and ε is infinitesimally small. We expect then the solution to have a form

$$p(r, \phi) = \bar{p}(r) + \varepsilon P(r) e^{ik\phi + \omega t}. \tag{6}$$

Making the Ansatz $P(r) = ar^x$ and inserting eq. (6) into the Laplace equation in polar coordinates one finds $x = \pm k$. Using the boundary conditions of eq. (2) one then obtains the solution

$$P(r) = ar^x + br^{-x} \tag{7}$$

with

$$\text{A}: \quad a = \frac{R_1^k p_c/\ell}{R_1^{2k} - R_0^{2k}} \quad \text{and} \quad b = -a R_0^{2k} \tag{8a}$$

$$\text{B}: \quad a = \frac{R_1^k p_c/\ell}{R_1^{2k} + R_0^{2k}} \quad \text{and} \quad b = a R_0^{2k}. \tag{8b}$$

The velocity of the moving boundary is given by eq. (5)

$$v_n = \frac{d\Gamma}{dt} = \omega \varepsilon e^{ik\phi + \omega t}. \tag{9}$$

Using eq. (3) to order ε, i.e. considering the unperturbed solution on the perturbed boundary and the perturbed solution on the unperturbed boundary one finally obtains

$$\text{A}: \quad \omega = [\frac{Cp_c}{R_1 \ell}(k \frac{R_0^{2k} + R_1^{2k}}{R_0^{2k} - R_1^{2k}} - 1)] \tag{10a}$$

$$\text{B}: \quad \omega = [\frac{Cp_c}{R_1\ell}(k\frac{R_0^{2k} - R_1^{2k}}{R_0^{2k} + R_1^{2k}} - 1)]. \tag{10b}$$

We see that case A, i.e. keeping the pressure constant, is unstable, i.e. $\omega > 0$ for all real perturbations, i.e. for any $k > 1$. The case B, i.e. keeping the flow constant, is more complicated: The perturbation is unstable if the wavevector k is larger than a value k_c that can be determined (numerically) from the equation

$$(\frac{R_0}{R_1})^{2k_c} = \frac{k_c + 1}{k_c - 1} \tag{11}$$

and the perturbation is stable if $k < k_c$. So, if R_1 is large enough with respect to R_0 only short wavelength perturbations are unstable. This is consistent with the more spiky patterns observed in ref. 3. We see however also that this is a finite size effect since when $R_0 \to \infty$ all perturbations will be unstable.

STABILITY ANALYSIS FOR THE PROPAGATION OF FRACTURE

The propagation of cracks is a problem of both technological and scientific interest It turned out that the propagation of a single crack in an elastic medium can be considered as a vector analogue problem to Laplacian pattern formation[4]. Let us restrict our attention here to slow "brittle" fracture, i.e. to elastic solids. In this case the medium is described on a mesoscopic level by the equations of motion of the displacement field **u**. In order to study how a void in the material becomes a crack and how this crack grows it is useful to formulate a moving boundary problem in **u**. The equation of motion of the elastic solid is the Lamé equation

$$\nabla(\nabla \cdot \mathbf{u}) + (1 - 2\nu)\Delta\mathbf{u} = 0 \tag{12}$$

which in two dimensions consists of two coupled equations where ν is the "Poisson ratio". Again two cases can be considered: A: a stretched membrane with a central hole (Fig. 1b) and B: pressure p applied inside a hole (Fig. 1c).

In case A we radially stretch the membrane at radius R_0 by an amount u_0. Then on the internal boundary, i.e. the void surface, the condition is that the stress $\sigma_\perp$ (force) perpendicular to the surface be zero:

$$\sigma_\perp = 0. \tag{13a}$$

In case B let us for simplicity put the outer boundary at infinity, i.e. a vanishing displacement there. On the surface of the hole one has:

$$\sigma_\perp = -p. \tag{13b}$$

The normal growth velocity v_n of the crack surface Γ will be a function of the difference between the stress $\sigma_\parallel$ parallel to the crack surface and a material dependent cohesion strength σ_c. Since no first principles law is known for this function one assumes the general behaviour

$$v_n = \frac{d\Gamma}{dt} = C[\sigma_{||} - \sigma_c]^\eta \tag{14}$$

where C and η are constants. Let us consider here for simplicity only $\eta = 1$. The cohesion stress σ_c that the material must overcome to break represents again a threshold and we will present here the stability analysis of ref. 5 of a a circular hole of radius R_1 when the stress at a boundary has just reached the value of σ_c.

Again we use polar coordinates r and ϕ. Denoting the radial and angular components of **u** by u_r and u_ϕ respectively and derivatives with respect to r by a prime and derivatives with respect to ϕ by a dot, the Lamé equations can be written as[8]:

$$2(1 - \nu)\left(u_r'' + \frac{u_r'}{r} - \frac{u_r}{r^2} + \frac{\dot{u}_\phi'}{r} - \frac{\dot{u}_\phi}{r^2}\right) + (1 - 2\nu)\left(\frac{\ddot{u}_r}{r^2} - \frac{\dot{u}_\phi}{r^2} - \frac{\dot{u}_\phi'}{r}\right) = 0 \tag{15a}$$

$$2(1 - \nu)\left(\frac{\dot{u}_r'}{r} + \frac{\dot{u}_r}{r^2} + \frac{\ddot{u}_\phi}{r^2}\right) + (1 - 2\nu)\left(u_\phi'' + \frac{u_\phi'}{r} - \frac{\dot{u}_r'}{r} - \frac{\dot{u}_r}{r^2} - \frac{u_\phi}{r^2}\right) = 0. \tag{15b}$$

Hooke's law relates the stress and the strain tensors:

$$\sigma_{rr} = 2(1 - \nu)e_{rr} + 2\nu e_{\phi\phi}, \quad \sigma_{\phi\phi} = 2\nu e_{rr} + 2(1 - \nu)e_{\phi\phi}$$

$$\text{and} \quad \sigma_{r\phi} = \sigma_{\phi r} = (1 - 2\nu)e_{r\phi} \tag{16}$$

and the strain tensor is defined as:

$$e_{rr} = u_r', \quad e_{\phi\phi} = \frac{\dot{u}_\phi}{r} + \frac{u_r}{r} \quad \text{and} \quad e_{r\phi} = e_{\phi r} = \frac{1}{r}(\dot{u}_r - u_\phi + ru_\phi'). \tag{17}$$

Using $\sigma_{||} = \sigma_{\phi\phi}$ one finds from eq. (14) since $\eta = 1$:

$$v_n = C[2\nu u_r' + 2(1 - \nu)(\frac{\dot{u}_\phi}{r} + \frac{u_r}{r}) - \sigma_c] \quad . \tag{18}$$

Let us next consider cases A and B separately.

Case A: Stretched membrane

The symmetric solution is

$$\bar{u}_r(r) = \alpha[(1 - 2\nu)r + R_1^2 r^{-1}] \tag{19}$$

with $\alpha = u_0/[(1 - 2\nu)R_0 + R_1^2 R_0^{-1}]$ while due to symmetry $\bar{u}_\phi = 0$. In this case the stress parallel to the surface is $\bar{\sigma}_{||} = 4\alpha(1 - 2\nu)$. We assume that the stretching displacement u_0 at the outer boundary R_0 is exactly such that this parallel stress just equals the cohesion strength: $\bar{\sigma}_{||} = \sigma_c$. So we place ourselves again exactly at the threshold and impose now again a perturbation of the boundary Γ of the form of eq. (5). The solutions are then expected to have the form

$$u_r(r, \phi) = \bar{u}_r(r) + \varepsilon U_r(r)e^{ik\phi + \omega t}, \tag{20a}$$

$$u_\phi(r, \phi) = i\varepsilon U_\phi(r)e^{ik\phi+\omega t} \tag{20b}$$

In polar coordinates the boundary condition (13a) takes the form

$$(1 - \nu)u_r' + \nu\left(\frac{\dot{u}_\phi}{r} + \frac{u_r}{r}\right) = 0 \tag{21a}$$

$$\dot{u}_r - u_\phi + ru_\phi' = 0. \tag{21b}$$

For the amplitudes we make the ansatz $U_r = Ar^x$ and $U_\phi = Br^x$ which leads to the general solution:

$$x = \pm(k \pm 1), \quad \text{and} \quad B = \frac{k^2 - x^2 + 2}{3k - xk}A \ . \tag{22}$$

Since the effect of the perturbation should decay with increasing distance from Γ, the radial solutions take the form

$$U_r = ar^{1-k} + br^{-1-k} \tag{23a}$$

$$U_\phi = \gamma ar^{1-k} - br^{-1-k} \tag{23b}$$

with $\gamma = [4(1 - \nu) - k]/[k - 2(2\nu - 1)]$. The coefficients a and b are then determined from the boundary conditions (21) to be:

$$a = \alpha\frac{k + 2(1 - 2\nu)}{k - 1}, \quad b = \alpha\frac{-k}{1 + k}. \tag{24}$$

Knowing the solution we have to consider the condition for the moving boundary of eq. (18). The zeroth order term in ε, i.e. the unperturbed solution on the unperturbed boundary by definition just cancels with the cohesion strength σ_c. Thus we have to gather the first order terms: Inserting into eq. (18) the unperturbed solution on the perturbed boundary and the perturbed solution on the unperturbed boundary gives:

$$\frac{\omega}{2C} = \nu\bar{u}_r''(R_1) + (1 - \nu)\left(\frac{\bar{u}_r'}{r} - \frac{\bar{u}_r}{r^2}\right)_{R_1} + \nu U_r'(R_1) + (1 - \nu)\left(\frac{U_r}{r} - k\frac{U_\phi}{r}\right)_{R_1} \tag{25}$$

which leads to

$$\omega = -8C(1 - 2\nu)\alpha R_1.$$

Since $\nu \leq 1/2$, the sign of ω can never be positive, i.e. there is no instability. Note that the k-dependence cancels out in the calculation so that this result holds for any perturbation.

Case B: Cracking due to pressure

We now turn to the stability analysis of the geometry illustrated in Fig. 1c. We assume that at infinity the outer boundary is free. A constant pressure p is applied at the inner boundary. Similarly to the case A we start from the unperturbed solution:

$$\ddot{u}_r(r) = \frac{pR_1^2 r^{-1}}{2(1 - 2\nu)} \quad . \tag{26}$$

Again, we assume that the cohesion strength is just reached: $\bar{\sigma}_{\phi\phi}(R_1) = \sigma_c = p$.

The perturbation and the perturbed solution have the same form as in case A. (eqs. (5) and (20), respectively). The boundary condition (13b) in polar coordinates becomes:

$$2(1 - \nu)u_r' + \frac{2\nu}{r}(\dot{u}_\phi + u_r) = -p \tag{27a}$$

$$\frac{1 - 2\nu}{r}(\dot{u}_r - u_\phi + ru_\phi') = 2i\varepsilon kp \tag{27b}$$

where it has been taken into account (up to first order in ε) that the pressure acts perpendicular to the perturbed boundary. Again, eq. (18) completes the definition of the problem.

The solution goes parallel to the case A and the coefficients which are calculated from the boundary conditions turn out to be

$$a = -p\frac{k - 4\nu + 2}{2(1 - 2\nu)}, \qquad b = p\frac{k}{2(1 - 2\nu)}. \tag{28}$$

From eq. (25) we now have

$$\frac{\omega}{2C} = -\frac{p}{R_1} + \nu U_r'(R_1) + (1 - \nu)\left(\frac{U_r}{r} - k\frac{U_\phi}{r}\right)_{R_1} \tag{29}$$

leading to

$$\omega = \frac{4Cp}{R_1}(k - 1). \tag{30}$$

So we find that all real perturbation modes ($k > 1$) are unstable, irrespective of the value of the Poisson ratio ν.

SUMMARY

In the Laplacian case having a threshold, as for instance a capillary pressure p_c, we considered the linear stability of the interface when the highest pressure gradient at the interface just equals the threshold gradient. When the pressure is imposed on the outer boundary any perturbation is unstable. When a flux is imposed only perturbations of wavevectors larger than k_c are unstable where k_c goes to one when the ratio of the outer radius to the inner radius goes to infinity.

For the case if an elastic solid[5] at the point when the highest stress in the medium just reaches the value of the cohesion force we discussed two cases: On one hand a circular hole in a membrane that is stretched at the outer boundary is stable against small perturbations of the shape of the hole and the damping rate ω is independent on the wavevector k of the perturbation. When on the other hand when the driving force is given by the pressure in the hole any perturbation causes propagating cracks with a growth rate that is independent of the Poisson ratio ν. The stability or instability of the model comes from the fact that in one case a displacement is imposed whereas in the second case, the stress (or equivalently the strain) is imposed.

The range of validity of the linear stability analysis is of course limited since it does not take into the strong non-linearities arising as time evolves. If the growth rule has a threshold the von Neumann boundary condition tends to stabilize the growth interface with respect to the Dirichlet conditions and the stabilization of long range perturbations in the Laplacian case are consistent with the spiky patterns observed in ref. 3.

I thank my collaborator Janos Kertész and Stephanefor many Roux for many elightening discussions.

REFERENCES

1. H.J. Wiesmann and H.R. Zeller, J. Appl. Phys. **65**, 1770 (1986); L. Pietronero and H.J. Wiesmann, Z. Phys. B **70**, 87 (1988)
2. R. Lenormand and C. Zarcone, Phys. Rev. Lett. **54**, 2226 (1985); R. Lenormand, Physica A **140**, 114 (1986); R. Lenormand, E. Taboul and C. Zarcone, J. Fluid. Mech. **189**, 165 (1986)
3. E. Arian, P. Alstrøm, A. Aharony and H.E. Stanley, Phys. Rev. Lett. **63**, 2005 (1989)
4. H.J. Herrmann and S. Roux (eds.), *Statistical Models for the Fracture of Disordered Media* (Elsevier, Amsterdam, 1990)
5. H.J. Herrmann and J. Kertész, Physica A **178**, (1991)
6. W.W. Mullins and R.F. Sekerka, J. Appl. Phys. **34**, 323 (1963); P.G. Saffmann and G.I. Taylor, Proc. R. Soc. London A. **255**, 312 (1957)
7. H.J. Herrmann in *Random Fluctuations in Pattern Growth*, eds. H.E. Stanley and N. Ostrowski, (NATO ASI Series Vol.157, Kluwer, Dordrecht, 1988), p. 149
8. Y.C. Fung, *Foundations of Solid Mechanics* (Englewood Cliffs, Prentice Hall, 1965)

EVIDENCE FOR UNIVERSALITY IN TRANSIENTS

J. V. Maher and H. Zhao

Department of Physics and Astronomy
University of Pittsburgh
Pittsburgh, PA

Throughout the extensive work of recent years on pattern formation and evolution in viscous fingering[1], most theoretical attention has been given to steady states and asymptotic behavior, with the little attention given to transient stages of evolution centering on the linear regime at the onset of pattern evolution. This neglect of nontrivial, nonlinear stages of pattern dynamics arises partly because the mathematics is so much more difficult than that for onset and asymptotic regimes and partly because of a prejudice that the transient is less important (i.e., less universal) than the asymptotic behavior.

Any universality in the transient pattern dynamics is certainly less obvious than in the case of asymptotic behavior. The dynamics are generally nontrivial with no transparent universality, but recently some universal arguments have been advanced for other evolving systems under the name of intermediate asymptotics [2]. In such a situation experimental work can be especially helpful in pointing the way with empirically determined regularities which can lead to helpful insights into useful simplifying assumptions. If there are indeed universal aspects to the transient stages of pattern formation, their very elucidation would greatly expand our understanding of pattern dynamics.

In addition to the value for deep theoretical understanding in detecting any universal features of transient dynamics, transients are important to understand for experimental and technological reasons. That is, most laboratory systems do not have a steady state, and even when there is theoretical reason to believe in the existence of a statistical steady state at asymptotic times, most experiments cannot be controlled long enough to reach that state convincingly. Thus most experimental data are measuring transients, and the connection with theory requires care of interpretation. Similarly, most technologically interesting procedures in materials processing normally operate in transient regimes, and any empirical evidence for universal behavior would expand our ability to predict the effects of untested changes in procedure.

In this paper we review the evidence for universality in transient patterns found in our study of various versions of viscous fingering in Hele-Shaw flows [3,4,5,6,7]. We begin with extremely simple examples which are little more than consequences of dimensional analysis of the Hele-Shaw equations, but then we progress through non-trivial collapses of the simplest Hele-Shaw flows to striking regularities in very complicated patterns found in non-Newtonian flows.

Growth Patterns in Physical Sciences and Biology, Edited
by J. M. Garcia-Ruiz *et al.*, Plenum Press, New York, 1993

The simplest case involves flow in a channel of width W. Both the steady state and the linear regime have been extensively studied and reviewed [1,4,5,6,7]. While three dimensional effects have been shown to modify the experimentally attainable states [8], agreement between experiment and theory is very good. Much less is known about the transient domain between the initial breakup of the flat interface and the attainment of the steady state single finger, but calculations by Tryggvason and Aref [9] bear close resemblance to laboratory observations [4,5,6,7]. Figure 1 is taken from work done with M.W. DiFrancesco [5] and shows the collapse of data from many different flows in this transient regime. In the figure, L′ is the reduced length of the interface between two phases of a critical binary liquid mixture (L′=L/Lo-1, where L is the instantaneous interface length and Lo is the initial interface length). Time is shown in seconds on the left side of the figure for many different values of the capillary number (B), and the same data are shown as a function of dimensionless time on the right side of the figure. While the dimensionless units are suggested by the Hele-Shaw equations and the collapse of the data might thus be considered to be a trivial example of scaling, it is important to realize that the dynamical regime is not well understood and that nothing in our understanding of the equations tells us that L′ must collapse. A less trivial example from the same set of flows is shown in

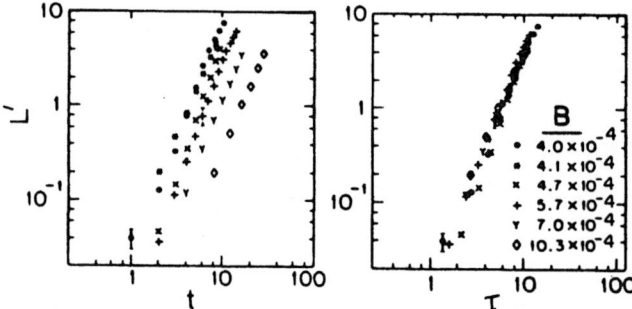

Fig. 1   Reduced length of the interface in low-contrast Saffman- Taylor flow vs time (in seconds on the left and in dimensionless units on the right) for indicated values of the capillary number B. (see reference 5)

Figure 2. On the left side of this figure, the average value of the interfacial curvature is plotted as a function of time for flow realizations at several different values of the capillary number. On the right side of the figure the same data are seen to collapse

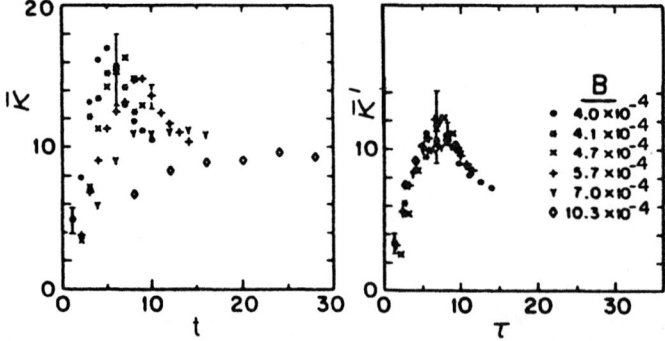

Fig. 2   Average interfacial curvature vs time in low-contrast Saffman-Taylor flow for indicated values of the capillary number, B. Data are shown in cgs units on the left and in dimensionless units on the right. (see reference 5)

when the dimensionless average curvature is shown as a function of dimensionless time.

While both Figures 1 and 2 compare cases which differ in capillary number, similar results were obtained when the other control parameter, the viscosity aspect ratio was varied, suggesting that there is indeed something universal about the dynamical path to the Saffman-Taylor steady state. This universality is not easy to study, however, because there appears to be a non-trivial slowing down of the approach to the steady state (and possibly a shrinkage of its zone of attraction) when viscosity aspect ratio is reduced [4]. In work done with S.A. Curtis [6], an attempt was made to study the full range of the competition leading to the steady state; this work involved setting up a "racetrack" in which a large number of steady state fingers were prepared in narrow channels and then introduced into a wide channel simultaneously, using large values of the viscosity contrast to hurry progress (nitrogen was used to displace paraffin oil). It was not possible to predict which finger would win the competition in any one case, nor was it possible to predict which finger would place in other arbitrary positions in the "race". Nevertheless, great regularity was observed in the dynamics of the process. On the left side of Figure 3 a typical set of finger trajectories are shown (position in the Hele-Shaw cell in cm vs time in seconds). While there is no well motivated way to define dimensionless variables, an arbitrary choice was found to collapse the finger trajectories. That is, a characteristic length and time were selected arbitrarily by choosing the position and time at which the fourth-place finger's trajectory departed from that of the winner. All positions and times were then made dimensionless by dividing them by the characteristic position or time. This procedure resulted in apparently universal trajectories for all fingers; that is, we cannot predict which finger will finish a given race in any one position, but we can predict that the nth place finger's trajectory will fall on the same universal trajectory as those of the nth place fingers for all the other races, regardless of the rather large overall variation in the flow conditions. The right side of Figure 3 shows several cases each of the trajectories for the fingers which finish fourth and fourteenth.

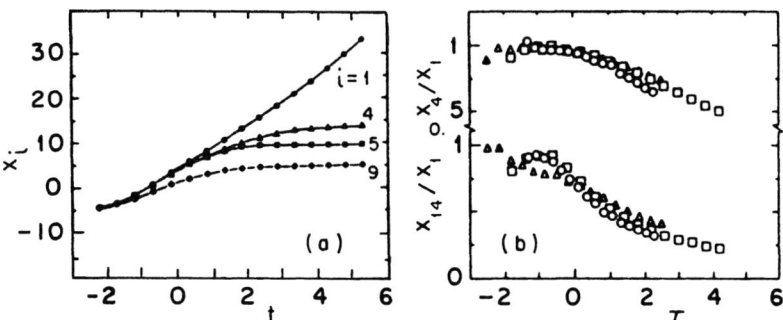

Fig. 3 Trajectories of "racing" Saffman-Taylor fingers. Left side: finger positions in cm vs time in sec; i is final position in the "race". Right side: trajectories of fourth (top) and fourteenth (bottom) place finishers for several "races" in dimensionless units. (see reference 6)

When viscous fingering is initiated in the radial geometry, there is no steady-state toward which the system evolves. Instead, the finger-tips keep splitting as the interface length continues to grow. Much has been written on the nature of the statistical steady-state toward which such a system may or may not evolve asymptotically [1]. In work done with Barnes and Rauseo [3], we showed that there are many features of the transient flow which collapse in an apparently universal fashion, even though the dynamical problem is too complicated to predict universality with any great confidence. An example is shown in Figure 4 where the mean wavenumber averaged over all curvature values is plotted, in dimensionless units, vs the dimensionless area of the mixing zone for many different flow realizations. Sarkar and Jasnow [10] have predicted that this function should asymptotically approach a power law with exponent –1, which the data of Figure 4 may well be doing, but for present purposes the interesting feature is that the data collapse onto an apparently universal curve at much smaller dimensionless mixing zones than are needed to begin to show asymptotic behavior.

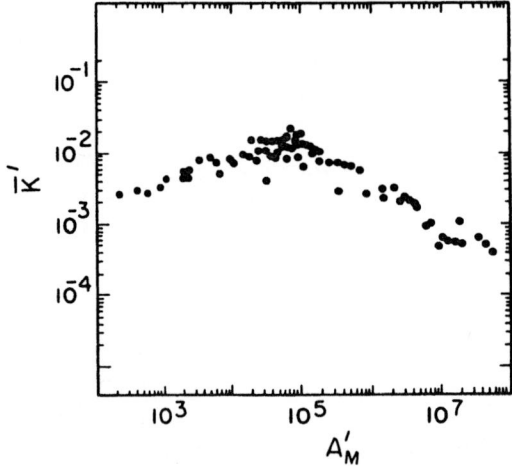

Fig. 4   Mean wavenumber, in dimensionless units, of modal analysis of curvature distributions for a wide variety of radial Hele-Shaw fingering flows vs dimensionless area of the mixing zone. (see reference 3)

Having seen evidence of unpredicted simplicity in the simplest of the pattern-forming flows, let us now turn our attention to a much more complicated case. We have studied the displacement of an aqueous polymer solution by water in a radial Hele-Shaw cell, and we have succeeded in collapsing two measures of the lengths involved in the very complicated patterns by using the known power-law behavior of polymer solutions to construct a dimensionless number for the flow. In this experiment we have four control parameters, the molecular weight, M, of the polymer, the concentration, c, of the polymer, the injection rate, Q, of the water, and the gap, b, between the plates of the Hele-Shaw cell. For the polymer we used polyethylene oxide of molecular weights 300,000 and 5,000,000. We measured viscosity as a function of shear rate, molecular weight and concentration, and we measured viscous fingering only at concentrations where the low-shear viscosity increased as molecular weight to the power 3.4. In this same concentrated regime, the polymer solution's plateau modulus is expected to vary as concentration to the power 2.2. Thus, as is discussed by DeGennes [11], it is possible to construct a characteristic time for the relaxation of polymer entanglements, $\tau_p \sim M^{3.4}/c^{2.2}$. A characteristic time can also be constructed from the shear rates imposed on the fluids during injection, $\tau_f \sim b^2/Q$. The ratio of τ_f to τ_p yields a dimensionless number, $A \equiv b^2 c^{2.2}/QM^{3.4}$ which is expected to be a power law in our four control parameters. While other schemes were tried in various attempts to collapse the data presented below, the dimensionless number A provided the only successful approach.

By variation of the four control parameters discussed above, a very rich variety of patterns was observed. Figure 5 provides a representative sample of these patterns. The mass-fractal dimension of the patterns varied from 1.2 to 1.8. In general, high values of Q produced smoother fingers with more branching at lower values of Q. High values of c produced strong screening of the interior regions of the patterns, while low c led to continual growth of the cores left behind by the advancing fingers. High molecular weight produced fewer fingers than did lower M, and the only uninteresting parameter was b which changed the patterns only by making the smallest length scale courser or finer in a linear way. In our search for regularity in this rich variety of

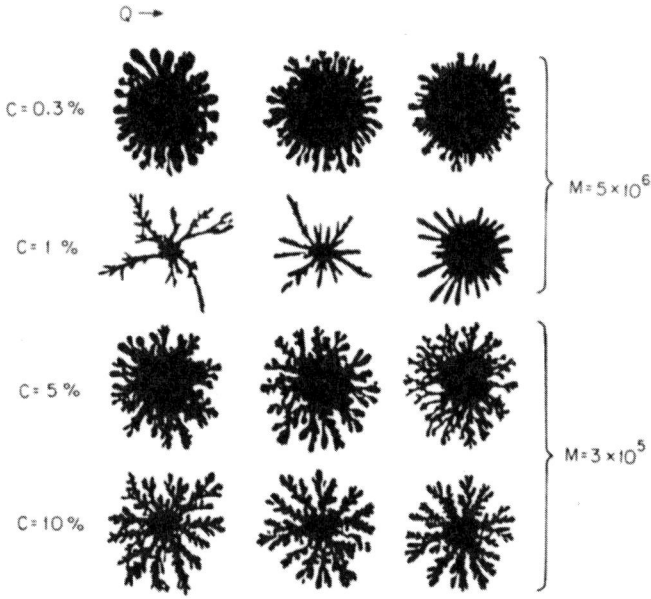

Fig. 5 Typical patterns as water (dark) advances against an acqueous solution of polyethylene oxide in a radial Hele-Shaw cell. In each row, the injection rate, Q, increases from left to right.

patterns, we defined two length scales and formulated one modal analysis. The length scales were ℓ_1, the twig width observed at the onset of the instability and ℓ_2, the twig width late in the flow. For the modal analysis we defined a density as shown schematically in Figure 6: circles were drawn at a series of radii for a given pattern, and then on each circle the density at a given polar angle was defined to be one if the water (dark region in the pattern) was present and zero if the polymer solution was present. Densities were then averaged as functions of polar angle and the averaged densities were Fourier transformed to give a modal distribution (essentially the same modal distributions were obtained if the Fourier transforms were performed on the unaveraged densities and then the Fourier spectra were averaged).

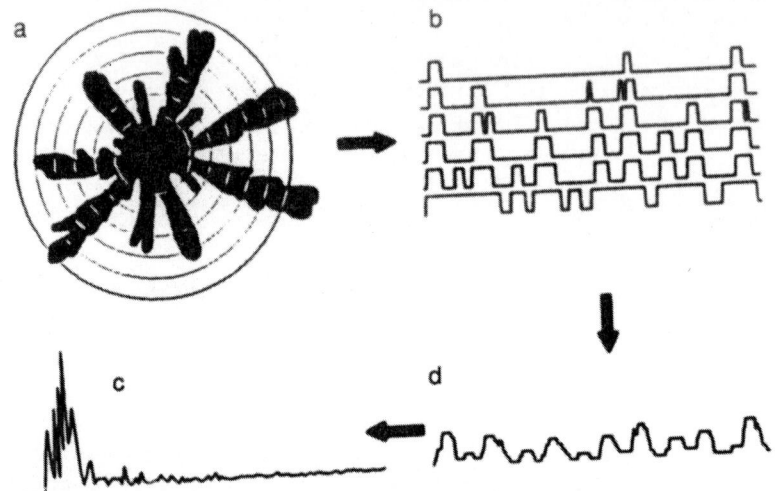

Fig. 6 Chart of stages in modal analysis of patterns, as discussed above.

Figure 7 shows ℓ_1, ℓ_2 and several properties of the averaged Fourier distributions as a function of the dimensionless number A. The initial twig width, ℓ_1 (in units of gap, b), increases from about 1.5 units of b to approximately 4.5 units of b as A varies through the whole range covered by our $M=5,000,000$ data. Less regularity is seen for the low M data. The peak wavenumber, the mean wavenumber and the standard deviation in wavenumber of the modal analysis are all seen to be very smooth functions of A. (Unfortunately, the strong dependence on M prevented us from varying the other parameters enough to span the gap between high and low M data.) The peak value of the modal analysis should give information about the average spacing of major branches of the patterns, while the mean value should average all length scales, not just the major branches. The standard deviation should be related to the width of the length scale distribution. The regularity of all these measures of the modal analysis along with the less-striking but clear regularity of the initial twig width, ℓ_1, represents the only case of which we are aware in which truly complicated patterns have been demonstrated to vary in an understandably regular way. The failure of the late-stage twig width, ℓ_2 to collapse may be related to the fact that we do not yet know how to define a characteristic time for this problem; thus "late" means something different for each different flow realization.

In summary, pattern evolution data show great regularity in the transient regime. The regularity is present and easily demonstrable when the experiment is designed to constrain the patterns to be very simple and to approach a well-understood steady-state very rapidly. Regularity is perhaps more surprising when the patterns are less constrained, but it is nevertheless demonstrable. Our knowledge of pattern-forming processes in nature will have taken a giant step forward when this empirically determined regularity can be explained in terms of systematic properties of the dynamical equations and their boundary conditions. This work was supported by the USDOE under grant #DE-FG02-84ER45131.

References

1. D. Bensimon, L. P. Kadanoff, S. Liang, B. I. Shraiman, and C. Tang, Rev. Mod. Phys. **58**, 977 (1986); P. G. Saffman, J. Fluid Mech. **173**, 73 (1986); G. M. Homsy, Ann. Rev. Fluid Mech. **19**, 271 (1987); and references therein.

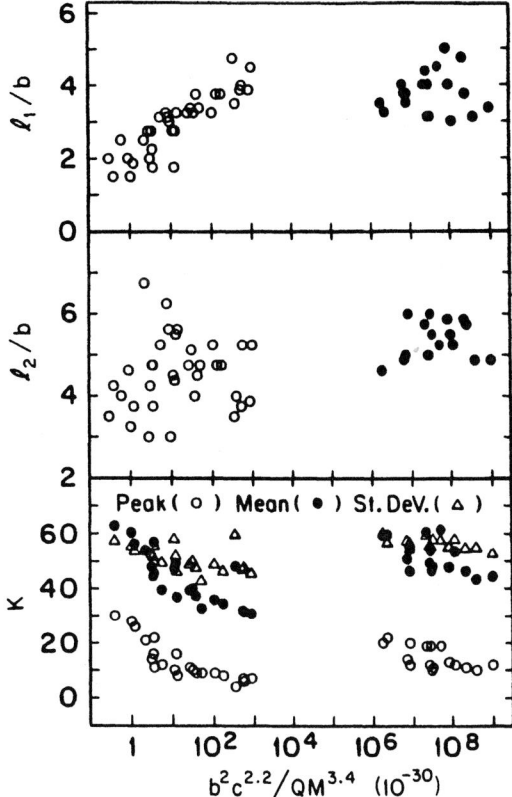

Fig. 7 Top: ℓ_1/b vs A; middle: ℓ_2/b vs A; Bottom: the peak position, the mean, and the standard deviation of the Fourier spectra vs a.

2. N. Goldenfeld, O. Martin, and Y. Onno, J. Sci. Comput. **4**, 355 (1989).

3. S. N. Rauseo, P. D. Rauseo, P. D. Barnes Jr., and J. V. Maher, Phys. Rev. A **35**, 1245 (1987).

4. M. W. DiFrancesco and J. V. Maher, Phys. Rev. A **39**, 4709 (1989).

5. M. W. DiFrancesco and J. V. Maher, Phys. Rev. A **40**, 295 (1989).

6. S. A. Curtis and J. V. Maher, Phys. Rev. Lett. **63**, 2729 (1989).

7. H. Zhao and J. V. Maher, Phys. Rev. A **42**, 5893 (1990).

8. C. W. Park and G. M. Homsy, J. Fluid Mech. **139**, 291 (1984); P. Tabeling and A. Libchaber, Phys. Rev. A **33**, 794 (1986).

9. G. Tryggvason and H. Aref, J. Fluid mech. **136**, 1 (1983); G. Tryggvason and H. Aref, *ibid.* **154**, 287 (1985).

10. S. Sarkar and D. Jasnow, Phys. Rev. A. **35**, 4900 (1987).

11. P. G. de Gennes, Europhys. Lett. **3 (2)**, 195 (1987).

MICELLES AND FOAMS: 2-D MANIFOLDS ARISING FROM LOCAL INTERACTIONS

Humberto Terrones and Alan L. Mackay

Department of Crystallography
Birkbeck College, (University of London)
Malet Street, London WC1E 7HX
England, U.K.

Abstract

Surfaces as 2-D manifolds play an important role in the description of structures, from inorganic materials to biological systems. These surfaces which can be planar, spherical or hyperbolic (saddle-shaped), arise as a consequence of interatomic forces. We are concerned with the generation and application of 2-D manifolds, in particular, periodic minimal surfaces. We show that surfaces can be decorated with atoms to obtain structures with different curvatures, related to the mean coordination number CN. When $CN = 6$ a planar surface or a cylinder can be obtained, if $CN < 6$ we get a closed spherical surface as in Buckminsterfullerenes, and if $CN > 6$, an infinite structure, periodic or otherwise, can be generated. Regarding the case $CN > 6$, we have found that the existence of ordered graphite foams with topologies similar to periodic minimal surfaces is quite possible. various transformations of surfaces, such as the Bonnet transformation, the Goursat transformation and a new combination of both, are analysed, since they might be useful in the description of physical and biological processes.

1 Introduction

An area which is gaining importance is the study of surfaces or 2-D manifolds. Surfaces are present in different natural structures from inorganic materials to biological systems. Mackay and Andersson [2], [3], [37], [38] have proposed to describe atomic structures such as zeolites by using periodic surfaces. The idea of representing arrangements of atoms with surfaces is the next step. The first approach has been to localize atomic postions in 3-D space with coordinates (x, y, z). The second approach consists of joining these points (atomic coordinates) by straight lines obtaining in this way the representation of bonds. Therefore, we have gone from points with zero dimension to lines with dimension one. Many methods for characterising structures use points and lines, but in some cases the information obtained it is not accurate or representative, so we have to go further by using 2-dimensional surfaces. It is

Growth Patterns in Physical Sciences and Biology, Edited
by J. M. Garcia-Ruiz *et al.*, Plenum Press, New York, 1993

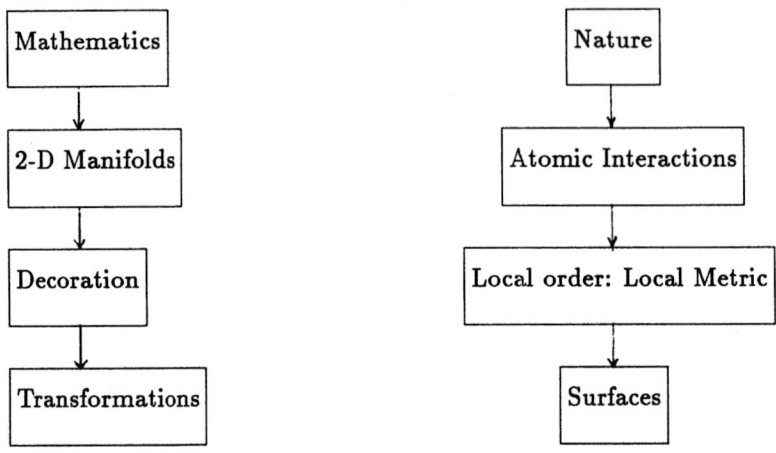

Figure 1. Connection between mathematical and real surfaces.

important to emphasize that a surface gives more information than a point or a line. Moreover, this way of thinking can be extended to volumes or hypersurfaces.

It turns out that Triply Periodic Minimal Susrfaces (TPMS) fit very nicely in some inorganic structures [2], [23] [24]. A minimal surface has zero mean curvature at every point and the *Gaussian* curvature is everywhere non positive, so the surface bends equally to both sides. A good example of a minimal surface is a soap film over a wire loop. Von Schnering and Nesper [46] have shown that some zero equipotential surfaces (ZEPS) have topologies similar to TPMS. Fermi and nodal surfaces are other 2-D manifolds which look like TPMS [40].

In water-surfactant-lipid systems, planar, spherical and hyperbolic surfaces can be observed. Planar bilayers are called *lamellae*; spherical arrangements are *micelles* or *vesicles*; and hyperbolic structures with cubic symmetry have been named *cubic phases* [36],[50]. The cubic phases present the same symmetries and topologies as TPMS, but this does not mean that they are exact TPMS [47], [48], [1],[22].

Other systems in which 2-manifolds are important involve liquid crystals [8]-[13], [44], [43], cellular membranes [6], [7], skeletal nets of echinoderms [42], [14] and graphitic structures [39]. Regarding these graphitic structures we have found that introducing octagonal rings of carbon, periodic surfaces with the same topology as TPMS can be constructed. It becomes important to analyse the possible patterns that can be built with patches of 2-D manifolds.

The formation of surface patterns with curvature is the product of local interactions which are very complicated and yet we do not know very much about it. The number of particles involved in such systems as cubic phases for example it is very high, so computer simulations are difficult and time consuming. Therefore, the strategy should be to generate, with the aid of geometry and mathematics, 2-D manifolds which can be connected with real structures (see figure 1).

2   Description of 2-D Manifolds

In 2-D manifolds the *Gaussian* curvature K (the product of the two principal cur-

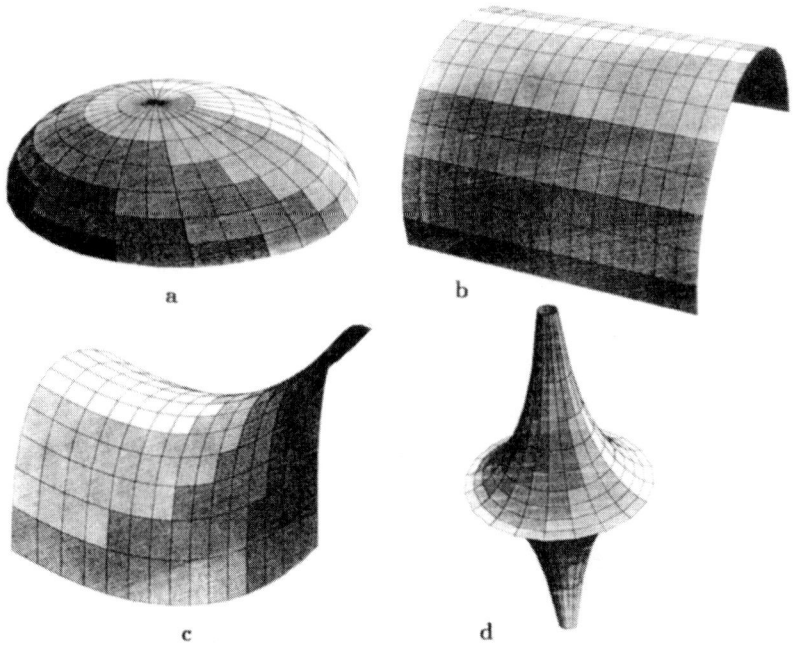

Figure 2. (a) Elliptic points. (b) Parabolic points. (c) Hyperbolic points. (d) The pseudo-sphere, a surface with negative constant Gaussian curvature.

vatures) at each point can be positive, negative or zero. When $K > 0$ the point is called *elliptic*, if $K < 0$ *hyperbolic* and if $K = 0$ *parabolic* (see figure 2). In the case of a sphere, every point on it is an elliptic point with all principal curvatures equal, so the surface has constant positive Gaussian curvature. As a consequence of this, the sum of the interior angles of a spherical triangle is greater than $180°$. The plane is composed of parabolic points with both principal curvatures equal to zero giving a surface of zero Gaussian curvature (The sum of the interior angles of a triangle is $180°$). In a cylinder the Gaussian curvature is also zero, but with one non-zero principal curvature. A point which has the two principal curvatures zero is called a *flat* point. On the other hand, there are surfaces formed with points having negative curvature everywhere such as the *pseudo-sphere* (see figure 2 d.). In the pseudo-sphere the sum of the inerior angles of a triangle on the surface is less than $180°$.

Besides the Gaussian curvature, another important descriptor is the *mean curvature* which is the average of the two principal curvatures ($H = \frac{k_1 + k_2}{2}$). Surfaces with $H = 0$ everywhere are called minimal surfaces. A minimal surface has non positive Gaussian curvature, so each point can be an hyperbolic or a flat point. It turns out that minimal surfaces can have periodicity and can be singly, doubly or triply periodic (see figure 3).

3 Lipid-Water-Surfactant Systems

Surfactants (surface active agents) are composed of molecules called *amphiphiles* which have an hydrophilic head and an hydrophobic tail (phospholipids have more

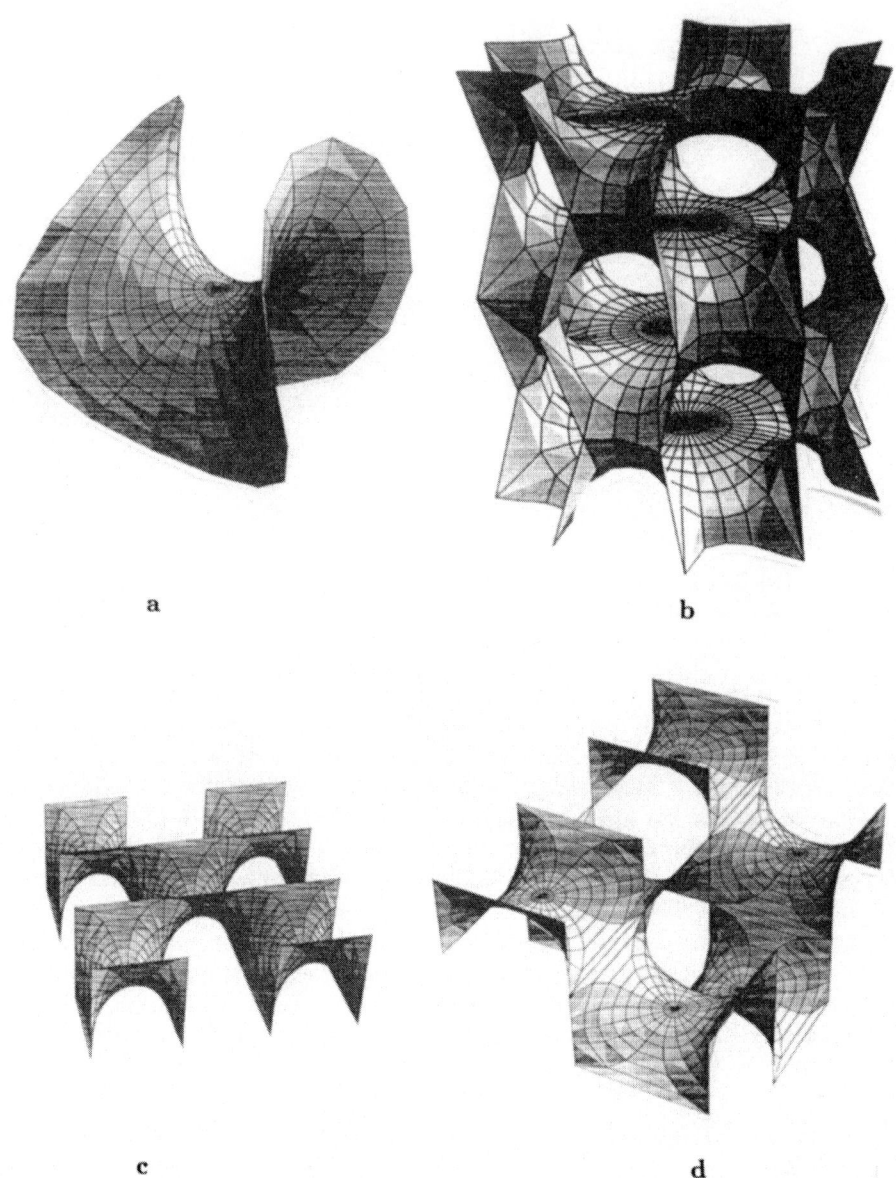

a

b

c

d

Figure 3. Minimal surfaces. (a) Enneper's Surface (non-periodic). (b) Scherk's tower with k=4 (singly periodic). (c) Adjoint of Scherk's k=1 (doubly periodic). (d) Tetragonal-surface (triply periodic).

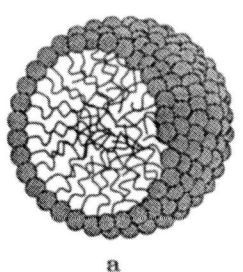
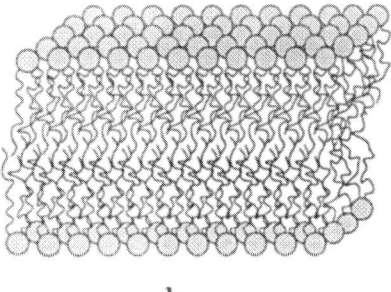

a   b

Figure 4. (a) A micelle where $K > 0$. (b) A bilayer where $K = 0$. (courtesy of N.A. Warrender).

than one tail). When a surfactant is mixed with oil and water, the hydrophilic heads tend to be exposed towards the water while the hydrophobic tails are directed towards the oil. Depending on the concentration and teperature, different shapes with distinct curvatures can be recognised, such as micelles, vesicles ($K > 0$) [34], [35], layers and bilayers ($K = 0$) [47], [48] or cubic phases ($K < 0$) [36] (see figure 4). According to experimental data, cubic phases have topologies and symmetries similar to triply periodic minimal surfaces (TPMS) [1],[48]. We do not yet how far these phases are from being minimal and for this reason, other periodic surfaces, although not minimal, should be studied; among these are: The Fermi surfaces [28], zero equipotential surfaces [41], [46] and nodal surfaces [40]. Quantitative studies on the mean and Gaussian curvatures of these surfaces reveal important differences which allow the discrimination among themselves and therefore, essential characteristics that can be presented in real structures are possible of being identified [4], [52].

Periodic surfaces divide space in two regions which can be congruent, where the inside is the same as the outside, or non-congruent. A congruent space partitioner is called *homogeneous* and a non-congruent *heterogeneous* [46]. Examples of homogeneous space partitioners are the D or F, the P, the G and the Neovius TPMS. In the I-WP TPMS the inside is different from the outside, so it is an heterogeneous space partitioner. (see figure 5)

4   Periodic Graphite Foams

The discovery of the third form of carbon known as *fullerenes* has brought the possibility of having new materials with very interesting properties. The most common fullerene is the C_{60} which has the shape of a truncated icosahedron (soccer ball) having 20 hexagons and 12 pentagons (see figure 6). C_{70} is also common and has an ellipsoidal shape which looks like a rugby ball [30], [31]. It has been found that doping C_{60} with potassium an $18K$ superconductor is obtained [21]. If the doping is with rubidium and caesium the superconducting temperature increases to $33K$ [49], [27].

C_{60} can be considered as a structure with positive Gaussian curvature since the atoms rest on the surface of a sphere. This curvature is produced by the presence of pentagonal rings [20]. In the absence of these rings the structure can be planar as in

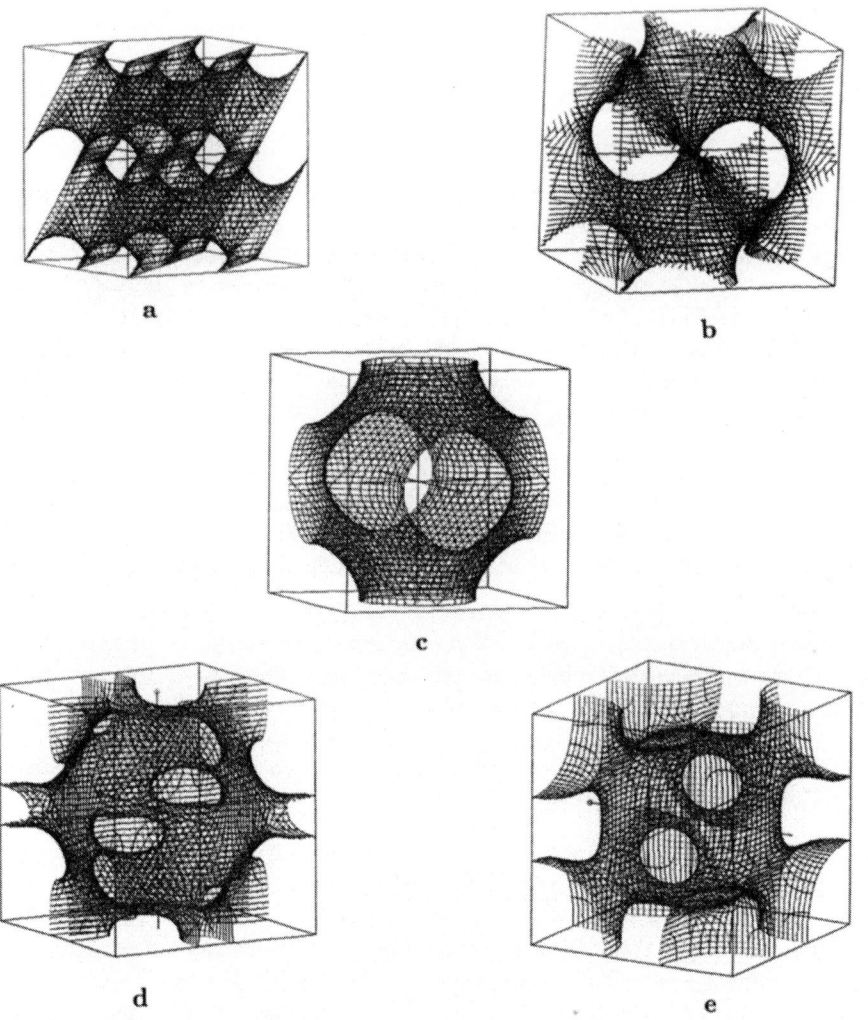

Figure 5. Unit cells of periodic surfaces (a) D Surface. (b) G surface. (c) P surface. (d) Neovius surface. (e) I-WP surface.

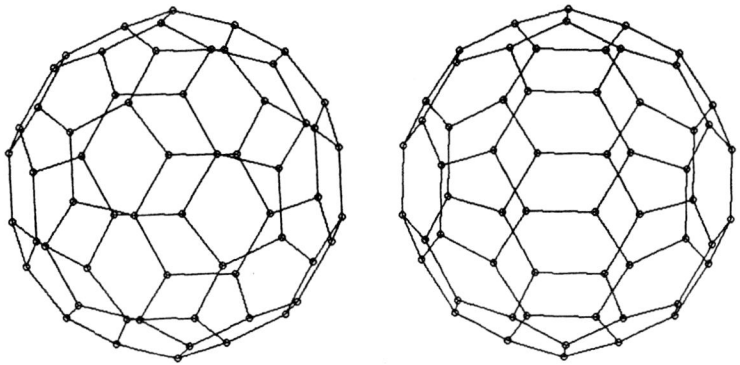

Figure 6. Stereo picture of C_{60}.

ordinary graphite (atoms are arranged in hexagonal layers), or cylindrical as in the graphite filaments obtained by IIjima [26]; In both cases the Gaussian curvature is zero.

We have found that by introducing octagonal rings into an hexagonal layer, periodic graphite structures with topologies similar to those observed in TPMS can be constructed [39]. The octagons produce the negative Gaussian curvature needed for periodicity. Models for the D, P and H TPMS have been built. A remarkable point in these structures is that the 120° bond angles of ordinary graphite are preserved with very little strain while in the C_{60} some angles change to 108°. Although in the cubic cell of the P graphite there are 192 atoms and in the D graphite 768, the number of atoms in the primitive cells of both is 192. These two surfaces have genus three and present the same topology as a sphere with three handles (see figure 7).

In ordinary graphite the mean coordination number CN of rings is six (the structure is made of hexagons), in C_{60} is 5.625 and in the D and P graphite is 6.26. Therefore, in a 2-Dimensional manifold with positive Gaussian curvature ($K > 0$) CN is less than six, if $K = 0$ then CN is equal to six and if $K < 0$, CN is greater than six.

Lenosky *et al.* [32] have proposed periodic graphite structures with heptagonal rings of carbon. In this case the surfaces do not divide space into two congruent regions as in the ones constructed with octagons. However, due to the difference in the volumes separated, the structures with heptagons can be put inside themselves obtaining in this way a bilayer. Unfortunately, the distance between the two layers is too short compared with graphite. Nevertheless, seems possible to get the right spacing for graphite layers, if the whole structure is scaled up. Finally, periodic graphite foams look quite possible, but the problem is how to synthesize them.

5 Transformation of Surfaces

5.1 Bonnet and Goursat Transformations

Transformations of 2-D manifolds can be relevant in the description of physical and biological processes. A transformation of minimal surfaces studied by Bonnet [5]

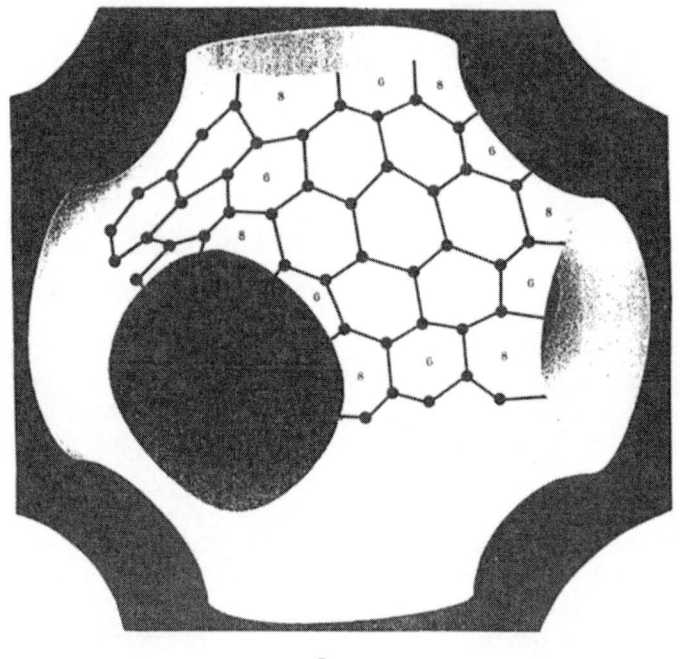

a

b

Figure 7. (a) P surface decorated with graphite. (b) Sphere with three handles having the same topology as the P surface.

allow us to bend a surface preserving the Gaussian and mean curvatures, so the surface is not stretched. The Bonnet transformation is achieved by multiplying the Weierstrass equations for minimal surfaces by a factor $e^{i\beta}$ [51], [33]; these equations can be written in different forms, but the most common are

$$x = \Re\left[e^{i\beta}\int_0^{\omega_0}(1-\omega^2)\,R(\omega)\,d\omega\right]$$

$$y = \Re\left[e^{i\beta}\int_0^{\omega_0}i\,(1+\omega^2)\,R(\omega)\,d\omega\right] \qquad (1)$$

$$z = \Re\left[e^{i\beta}\int_0^{\omega_0}2\,\omega\,R(\omega)\,d\omega\right]$$

$$x = \Re\left[e^{i\beta}\int_0^{\rho_0}\frac{1}{2}F(\rho)\,(1-G(\rho)^2)\,d\rho\right]$$

$$y = \Re\left[e^{i\beta}\int_0^{\rho_0}\frac{1}{2}i\,F(\rho)\,(1+G(\rho)^2)\,d\rho\right] \qquad (2)$$

$$z = \Re\left[e^{i\beta}\int_0^{\rho_0}F(\rho)\,G(\rho)\,d\rho\right]$$

Where (x,y,z) are the coordinates of the surface in real space, $\omega = u + iv$, $\rho = \sigma + i\tau$ and $i = \sqrt{-1}$.

By knowing the suitable complex functions of the Weierstrass equations called Weierstrass functions ($R(\omega)$, $F(\rho)$ and $G(\rho)$), the metric, the Gaussian curvature and the unit normal vectors to the surface can be found [51]. Until know, not all the Triply Periodic Minimal Surfaces (TPMS) have a Weierstrass representation since some of these were found experimentally by dipping closed wires into soapy water [15]-[18]. On the other hand, just about 50 TPMS have been discovered and there is no reason to believe that these are all TPMS, so should be others waiting to be found.

Aplying the Bonnet transformation to the tetrahedral patch for constructing the D TPMS, it is possible to get the surface patches for the G (gyroid) and P TPMS. For the G surface $\beta = 38.0147°$ and for the P surface $\beta = 90°$ (see fig 8). Hyde and Andersson have proposed that the mechanism of the martensite transition is through the Bonnet transformation [25]. Also by using this transformation we can change a catenoid into a helicoid. A transformation analogous to the Bonnet has been implemented on surfaces of constant mean curvature where $H = c$ [52].

Another transformation of minimal surfaces is the Goursat transformation [19]. During this transformation the metric and the Gaussian curvature change, so the surface is bent and stretched. This transformation is obtained by introducing a term k_G (Goursat Parameter) in the Weierstrass functions. The coordinates of a Goursat transformed surface are obtained from:

$$x = \Re\int_0^{\omega_0}\left(k_G - \frac{\omega^2}{k_G}\right)R(\omega)\,d\omega$$

$$y = \Re\int_0^{\omega_0}i\left(k_G + \frac{\omega^2}{k_G}\right)R(\omega)\,d\omega \qquad (3)$$

$$z = \Re\int_0^{\omega_0}2\,\omega\,R(\omega)\,d\omega$$

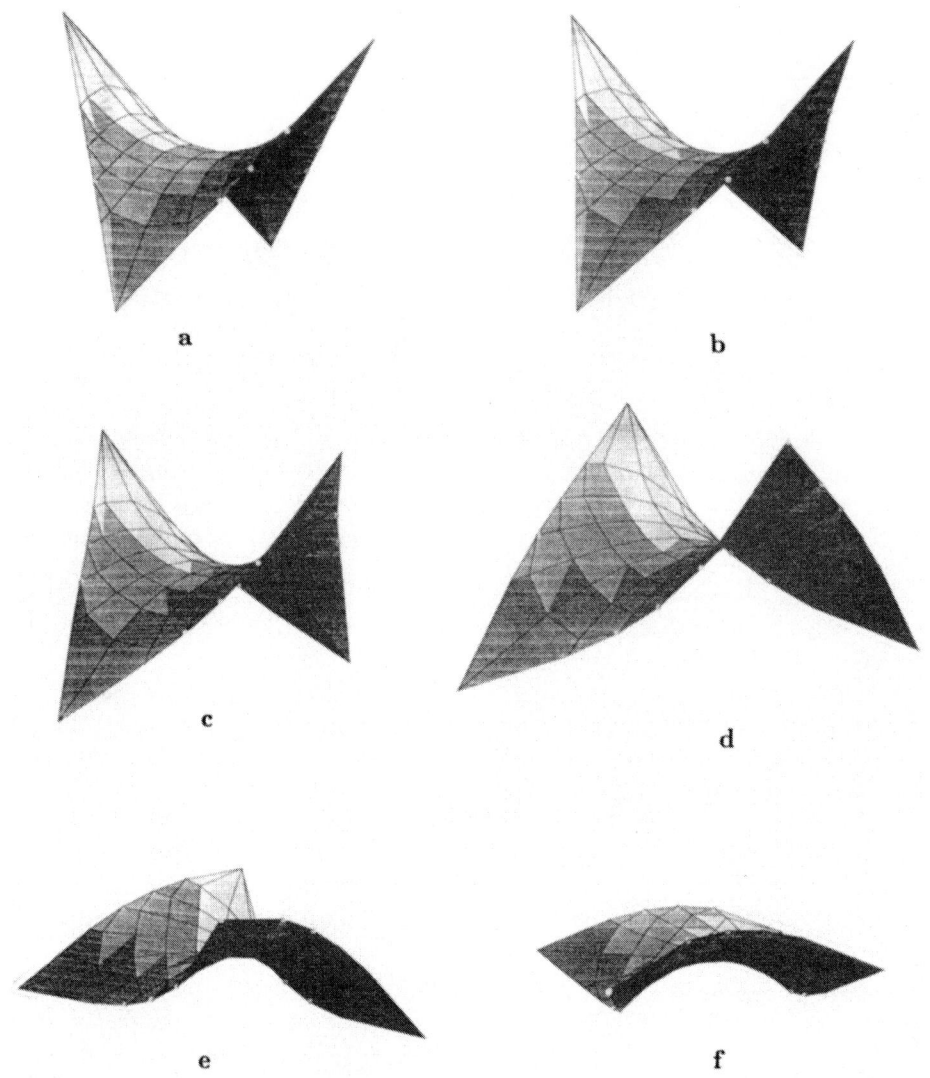

Figure 8. The Bonnet transformation and the tetrahedral saddle of the D surface.
(a) $\beta = 0$. (b) $\beta = 10°$. (c) $\beta = 20°$. (d) $\beta = 38.015°$ (piece of the G surface).
(e) $\beta = 60°$. (f) $\beta = 90°$ (piece of the P surface).

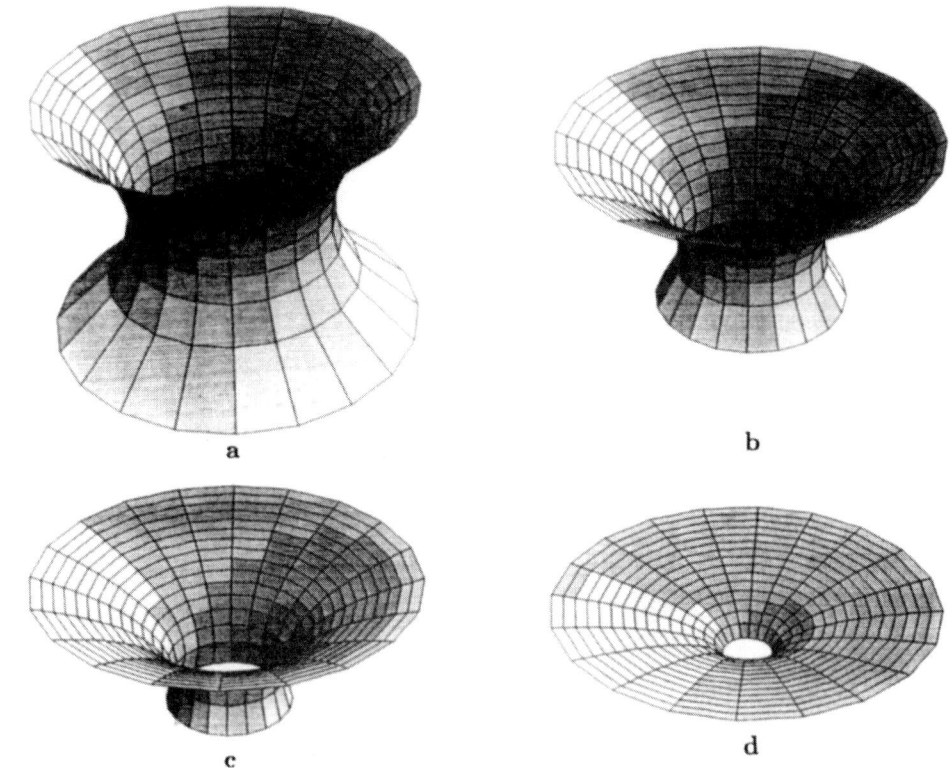

Figure 9. The Goursat transformation on a catenoid. (a) $k_g = 1$. (b) $k_g = 1.5$. (c) $k_g = 2$. (d) $k_g = 5$.

$$x = \Re \int_0^{\rho_0} \frac{1}{2} F(\rho)(k_G - \frac{G(\rho)^2}{k_G})\, d\rho$$

$$y = \Re \int_0^{\rho_0} i\frac{1}{2} F(\rho)(k_G + \frac{G(\rho)^2}{k_G})\, d\rho \qquad (4)$$

$$z = \Re \int_0^{\rho_0} F(\rho)\, G(\rho)\, d\rho$$

Different values of k_G give different surfaces (see fig. 9)

5.2 Combination of Bonnet and Goursat Transformations

A new transformation which combines the stretching of the Goursat transformation and the bending of the Bonnet transformation is obtained when the Weierstrass functions F and G of a known minimal surface are changed to the functions $F^\dagger$ and $G^\dagger$ given by

$$F^\dagger = F(2 + \cos\gamma)\, e^{i\gamma} \quad , \quad G^\dagger = \frac{G}{2 + \cos\gamma} \qquad (5)$$

Where γ can take any value from zero to 2π. Note that in equation 5 we have a stretching Goursat term $2 + \cos\gamma$ and a Bonnet term $e^{i\gamma}$. If the Weierstrass functions of the catenoid are used [51], then the stretching term that we have chosen allow us to obtain a perfect catenoid when $\gamma = \pi$. Let us describe in more detail this transformation: if $\gamma = 0$, a stretched catenoid is generated; by gradually increasing the value of γ, we start to get helicoidal surfaces with different degrees of stretching and bending until we reach $\gamma = \frac{\pi}{2}$. From $\gamma = \frac{\pi}{2}$ to $\gamma = \frac{3\pi}{2}$ the surface compresses itself and changes to a perfect catenoid at $\gamma = \frac{3\pi}{2}$. In the interval from $\frac{3\pi}{2}$ to 2π the surface stretches again until gets the same state as in $\gamma = 0$. Considering a Goursat term equal to $2 + \sin \gamma$, it is possible to get, during the transformation, a perfect helicoid. It is important to emphasize that all the surfaces generated by this transformation are minimal.

6 Conclusion

2-D manifolds or surfaces play an important role in the description of structures which, due to the complexity of the atomic interactions, have to be analysed with geometry and mathematics. The idea of allowing atoms to lie in curved 2-D manifolds (flexi-crystallography) is introduced. Here, a model for periodic graphite foams with negative Gaussian curvature is given. In addition, transformations of surfaces which can be important in real processes have been studied.

References

[1] Anderson D. M., *"Studies in the Microstructure of Microemulsions"*, Ph.D. Thesis, University of Minnesota, 1986.

[2] Andersson, S., Hyde, S.T. and von Schnering, H.G., (1984). "The intrinsic curvature of solids", Z. f. Krist., **168**, 1-17.

[3] Andersson, S., Hyde, S.T., Larsson, K. and Lidin, S., (1988). "Minimal Surfaces and structures: from inorganic and metal crystals to cell membranes and biopolymers", Chem. Rev. **88**, 221-242, 1988.

[4] Barnes, I., Hyde S.T., Ninham, B.W., "The Caesium Chloride Zero Equipotential is not the Schwarz P Surface", Colloque de Physique, C 7, No.23, 19-24, 1990.

[5] Bonnet, O., "Note sur la théorie générale des surfaces". C.R. Acad. Sci. Paris, **37**, 529-532, 1853.

[6] Bouligand, Y., "Comparative Geometry of Cytomembranes and Water-Lipid Systems, Colloque de Physique, C 7, No.23, 35-51, 1990.

[7] Bouligand, Y., "Geometry and Topology of Cell Membranes", Chapter IV of *"Geometry in Condensed Matter Physics"*, Edited by J.F. Sadoc, World Scientific, (1990).

[8] Charvolin, J., "Polymorphism of interfaces", J. de Chimie Physique, 80, No.1, 15-23, (1983).

[9] Charvolin, J., (1984). "From micelles to liquid crystals", Molec. Crystals, **113**, 1-11.

[10] Charvolin, J., "Crystals of Interfaces: The cubic phases of amphiphile/water systems", Journal de Physique, Colloque C3, supp. au no. 3, Tome 46, (1985).

[11] Charvolin, J. and Sadoc, J. F., "Periodic systems of frustrated films and cubic liquid crystalline structures". J. de Physique, **48**, 1559-1569 (1987).

[12] Charvolin, J. and Sadoc, J.F., "Films of Amphiphiles: Packing Constraints and Phase Diagrams", Jour. Phys. Chem., **92**, No. 20, 5787-5792, (1988).

[13] Charvolin J., Sadoc J.F., "Morphology of Stratified Fluids", Chapter II of "Geometry in Condensed Matter Physics", Edited by J.F. Sadoc, World Scientific, (1990).

[14] Donnay G., Pawson D.L., "X-ray Diffraction Studies of Echinoderm Plates", Science, **166**, 1147-1150, 1969.

[15] Fisher, W., Koch, E., "On 3-periodic minimal surfaces". Z. Kristallogr. **179**, 31-52, (1987).

[16] Fisher, W., Koch, E., "Topological Characterisation of 3-periodic minimal balance surfaces", Z.Kristallogr., **185**, 293, (1988).

[17] Fisher, W., and Koch, E., "New Surface Patches for Minimal Balance Surfaces. III. Infinite Strips", Acta Cryst. **A45**, 485-490, (1989)

[18] Fisher, W., and Koch, E., "Genera of Minimal Balance Surfaces", Acta Cryst. **A45**, 726-732, (1989)

[19] Goursat, E., "Sur un mode de transformation des surfaces minima", Acta Mathematica, **11**, 135-186, (11 Feb. 1888)

[20] Harris, W.F., "Disclinations", Sci. Amer., 130-145, (Dec. 1977).

[21] Hebard, A.F., Rosseinsky, M.J., Haddon, R.C., Murphy, D.W., Glarum, S.H., Palstra, T.T.M., Ramirez, A.P., and Kortan A.R., "Superconductivity at 18 K in Potassium-doped C_{60}", Nature, **350**, 600-601, 1991.

[22] Hoffman D., "Some Basic Facts, Old and New, About Triply Periodic Embedded Minimal Surfaces", Colloque de Physique C7, **51**, 197-208, 1990.

[23] Hyde, S.T., Andersson, S., "A Systematic net description of saddle polyhedra and periodic minimal surfaces", Z. f. Kristallogr. **168** 221-254, (1984)

[24] Hyde, S.T., Andersson, S., "Differential geometry of crystal structure descriptions, relationships and phase transformation", Z. Kristallogr., **170**, 225-239, (1985)

[25] Hyde, S.T., Andersson, S., "The martensite transition and differential geometry", Z. f. Kristallogr., **174**, 225-236, (1986)

[26] Iijima S., " Helical Microtubules of Graphitic Carbon", Nature, **354**, 56-58, 1991.

[27] Kelty, S.P., Chen Chia-Chun and Leiber, C.M., "Superconductivity at 30 K in Caesium-doped C_{60}", Nature, **352**, 223-224, 1991.

[28] Kittel C., *"Introduction to Solid State Physics"*, John Wiley & Sons, Fourth Edition, (1971).

[29] Krätschmer W., Lamb L.D., Fostiropoulos K., Huffman D.R., "Solid C_{60}: a new form of carbon", Nature, **347**, 354, 1990.

[30] Kroto, H.W., Heath, J.R., O'Brien, S.C., Curl, R.F., and Smalley, R.E., "C_{60}: Buckminsterfullerene", Nature, **318**, 162-163, 1985.

[31] Kroto, H.W., "The Stability of the Fullerenes C_n, With $n = 24, 28, 32, 36, 50, 60$ and 70", Nature, **239**, 529-531, 1987.

[32] Lenosky T., Gonze X., Teter M., and Elser V., " Energetics of Negatively Curved Graphitic Carbon", Nature, in press 1991.

[33] Lidin, S. and Larsson, S., "Bonnet transformation of infinite periodic minimal surfaces with hexagonal symmetry ", J. Chem. Soc. Faraday Trans, **86**, (5), 769-775, (1990).

[34] Lipowsky R., "Critical Behaviour of Vesicles and Membranes", Colloque de Physique, C 7, No.23, 243-248, 1990.

[35] Lipowsky R., "The Conformation of Membranes", Nature, **349**, 475-481, 1991.

[36] Luzzati, V., Mariani, P. and Gulik-Kryzywicki,T., (1987). "The cubic phases of lipid-containing systems: Physical structure and biological implications", Workshop "Physics of Amphiphilic Layers", Les Houches, 10-19 Feb. 1987. (Springer-Verlag)

[37] Mackay, A.L., "Periodic minimal surfaces", Physica, **131B**, 300-305,(1985)

[38] Mackay, A.L., "Periodic Minimal Surfaces", Nature, **314**,604-606, (18 April 1985)

[39] Mackay, A.L., and Terrones H., "Diamond from Graphite", Nature, **352**, 762, (Aug. 1991).

[40] Mackay, A.L., "Lonsdale Invariants – Periodic Nodal Surfaces", Per. Mineralogia di Roma, (in press) 1991.

[41] Nesper, R. and von Schnering, H.G., "Periodic potential surfaces in crystal structures", Angew. Chem. int. Ed. Engl.,**25**, 110-112, (January 1986)

[42] Nissen, H., "Crystal Orientation and Plate Structure in Echinoid Skeletal Units", Science, **166**, 1150-1152, 1969.

[43] Pansu B., Dubois-Violette E., "Blue Phases: Experimental Survey and Geometrical approach", Colloque de Physique, C 7, No. 23, 281-296, 1990.

[44] Sadoc, J.F., and Charvolin, J., "Infinite periodic minimal surfaces and their crystallography in the hyperbolic plane", Acta Cryst. A**45**,10-20,(1989).

[45] Schoen, A.H., "Infinite Periodic Minimal Surfaces Without Self-Intersections", NASA Technical Note D-5541 (1970).

[46] von Schnering, H.G. and Nesper, R., "How Nature adapts chemical structure to curved surfaces". Angew. Chem., **26**, No. 11, 1059-1080, 1987.

[47] Seddon, J.M., "Structure of the Inverted Hexagonal (H_{II}) Phase, and Non-Lamellar Phase Transitions of Lipids", Biochim. Biophys. Acta. **1031**, 1-69 (1990).

[48] Seddon, J.M. and Templer, R., "Liquid Crystals in the Living Cell", New Scientist, **130**, No. 1769, 45-49, (1991).

[49] Tanigaki, K., Ebbesen, T.W., Saito, S., Mizuki, J., Tsai, J.S., Kubo, Y., and Kuroshima, S., " Superconductivity at 33 K in $Cs_xRb_yC_{60}$", Nature, **352**, 222-223, 1991.

[50] Tardieu, A., and Luzzati, V., "Polymorphism of lipids a novel cubic phase–a cage-like network of rods with enclosed spherical micelles",Biochim. Biophys. Acta, **219**, 11-17, (1970).

[51] Terrones, H., "Computation of Minimal Surfaces", Colloque de Physique C7, **51**, 345-361 (1990).

[52] Terrones H., "Bending Transformation of Constant Mean Curvature Surfaces", (in preparation)

[53] Wolfram S., *Mathematica: a System for Doing Mathematics by Computer*, Addison-Wesley Publishing Company, Inc., Second Edition, 1991.

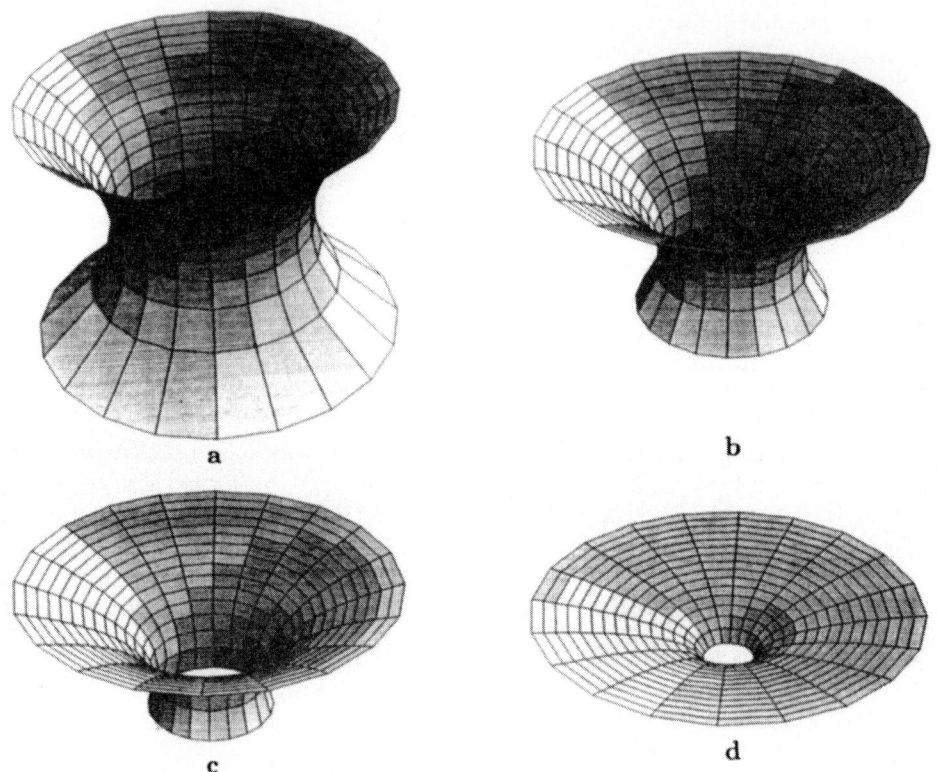

Figure 9. The Goursat transformation on a catenoid. (a) $k_g = 1$. (b) $k_g = 1.5$. (c) $k_g = 2$. (d) $k_g = 5$.

$$x = \Re \int_0^{\rho_0} \frac{1}{2} F(\rho) (k_G - \frac{G(\rho)^2}{k_G}) \, d\rho$$

$$y = \Re \int_0^{\rho_0} i \frac{1}{2} F(\rho) (k_G + \frac{G(\rho)^2}{k_G}) \, d\rho \qquad (4)$$

$$z = \Re \int_0^{\rho_0} F(\rho) \, G(\rho) \, d\rho$$

Different values of k_G give different surfaces (see fig. 9)

5.2   Combination of Bonnet and Goursat Transformations

A new transformation which combines the stretching of the Goursat transformation and the bending of the Bonnet transformation is obtained when the Weierstrass functions F and G of a known minimal surface are changed to the functions $F^\dagger$ and $G^\dagger$ given by

$$F^\dagger = F \, (2 + \cos\gamma) \, e^{i\gamma} \quad , \quad G^\dagger = \frac{G}{2 + \cos\gamma} \qquad (5)$$

Where γ can take any value from zero to 2π. Note that in equation 5 we have a stretching Goursat term $2 + \cos\gamma$ and a Bonnet term $e^{i\gamma}$. If the Weierstrass functions of the catenoid are used [51], then the stretching term that we have chosen allow us to obtain a perfect catenoid when $\gamma = \pi$. Let us describe in more detail this

SIMULATING RADIATE ACCRETIVE GROWTH USING ITERATIVE GEOMETRIC CONSTRUCTIONS

J.A. Kaandorp

Department of Computer Science / Institute of Taxonomic Zoology
University of Amsterdam, Kruislaan 403, 1098 SJ Amsterdam
The Netherlands

INTRODUCTION

Laplacian models[1] (Diffusion Limited Aggregation-models[2]) can be generalized to describe many fractal growth phenomena[3,4]. An attractive application in biology of these Laplacian models is to model the process of morphogenesis. Several examples have been developed to model growth processes in biology by applying Laplacian models, where the physical environment of the growing object consists of a diffusing nutrient[5]. Especially in case-studies, where the growth of a bacterium colony on a flat substrate in a petri dish is simulated, this model delivers good results[6,7]. In these colonies, the building elements can be considered as loose particles, which do not exhibit structural coherence. For most organisms the building elements (spicules, corallites, zooids, cells etc.) will be consolidated in a mesh with a certain architecture. A morphological model for these organisms based on a cellular automaton will be quite artificial. For these purposes a geometric model seems to be the most appropriate.

In this paper a model will be presented for radiate accretive[8] growth based on an iterative geometric construction, in combination with a model of the physical environment. The model of the physical environment consists of a nutrient concentration model, based on the DLA model and a light model. The idea is to make a combination of iterative geometric constructions and cellular automata techniques such as the DLA model, in order to take benefit of both approaches. For the purpose of demonstrating the generic character of the model examples will be shown of organisms where the nutrient concentration in the environment is the key-factor in the growth process and of organisms where light is the main environmental parameter.

Radiate accretive growth can be defined as an iterative growth process in which layers of material are added externally to the tip of a preceding growth step, which remains unchanged in the next growth steps. In this process the thickness of the layers is highest at a minimal angle between a tangential element and an axis of growth (see Figure 1) and decreases towards the sides of the tip. In this way a typical radiate architecture is formed, where the longitudinal elements are set perpendicular to the preceding tangential elements. This type

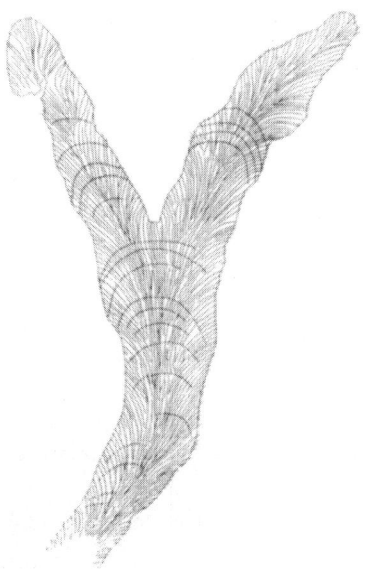

Figure 1. Drawing of a longitudinal section the sponge *Haliclona oculata* (A) and a longitudinal section through a column shaped colony of the stony-coral *Montastrea annularis* (B).

of growth is found among members of various groups of modular[9] marine organisms: e.g. stony-corals[10], sponges[8,11], coralline algae[12] and a symbiotic octocoral / sponge association[13]. Among these a distinction can be made between organisms where the growth process is mainly light-dependent and a group where growth is mainly determined by the supply of nutrients, suspended material in the surrounding water. Mixtures of these autotrophic and heterotrophic sources of carbon for the metabolic synthesis are also possible. In many stony-corals photosynthesis of the symbiont zooxanthellae is the most important part of the energy source, while zooplankton feeding represents only an insignificant component of the energy intake[14-16]. As an example of a heterotrophic organism with radiate accretive growth, the sponge *Haliclona oculata* will be used, whereas as an example of an (mainly) autotrophic organism the scleractinian coral *Montastrea annularis*[10] in this paper will be referred to.

A typical characteristic of many modular organisms is that they often exhibit a wide range of growth forms, caused by differences in the physical environment. Basically the growth forms of *Haliclona oculata* range from quite regular, thin-branching to irregular plate-like forms (see Figure 2). The thin-branching form is typical for sheltered conditions while the plate-like form is found under conditions with more exposure to water movement[11]. For a mainly light-dependent organism with radiate accretive growth, such as *Montastrea annularis*, the range of growth forms varies with the light intensity. This species exhibits a hemi-spherical colony form under circumstances with a maximum light intensity, i.e. when the colony grows close to the water surface. The colony gradually transforms from hemi-spherical, column-shaped, tapered forms to a substrate-covering plate; when the light intensity decreases[10].

As a consequence of the radiate accretive growth process both organisms have the potentiality to develop erect growth forms. In, for example sponges, where the elements are not arranged in a specific way and are oriented randomly (halichondrid skeleton[8]) usually quite irregular (often encrusting) and seldom erect forms emerge in the growth process.

A   B

Figure 2. Two examples of extremes of growth forms of the sponge *Haliclona oculata*. Form A is typically for a sheltered growth site, B for a more exposed one.

In the case of a filter feeding organism (or an organism with any other mechanism for absorbing nutrients from its environment) with erect growth forms, the form emerges in a gradient of water velocities, with a corresponding gradient of supply in suspended material.

METHODS

Iterative geometric constructions

The radiate accretive growth process in two dimensions can be modelled as an iteration process[17] (see Figure 3), in which the formation of the skeleton, as seen in a longitudinal section (see Figure 1), is simulated. In each iteration step a layer consisting of new longitudinal elements (perpendicular to the preceding layer of tangential elements) and new tangential elements (the new surface of the object) is constructed.

The length l of a new longitudinal element is determined by a product of four functions, $f(\alpha)$, $h(rad\_curv)$, $l(\theta)$ and $k(c)$ (see formula 1). The values s and $inhibition\_level$ in formula 1) are constants. s is the size of a tangential element and $inhibition\_level$ is the minimal length of a longitudinal element.

$$(1)$$
$$l \;=\; \begin{cases} s.f(\alpha).h(rad\_curv).l(\theta).k(c) \;\; for \; f(\alpha).h(rad\_curv).l(\theta).k(c) > inhibition\_level \\ 0.0 \;\; for \; f(\alpha).h(rad\_curv).l(\theta).k(c) \leq inhibition\_level \end{cases}$$

All functions return values in the range 0.0 - 1.0.

The argument α in function f (formula 2) represents the angle between an axis of growth (see dotted lines in Figure 3) and a tangential element.

$$f(\alpha) \;=\; \begin{cases} 1.0 \; for \; \pi/2 \le \alpha \le (\pi/2 + \pi/n) \\ sin((\pi/2)/(\pi/2 - \pi/n).(\pi - \alpha)) \; for \; (\pi/2 + \pi/n) < \alpha \le \pi \\ n > 2 \end{cases} \qquad (2)$$

The function f attains a maximum value when α equals $\pi/2$ and a minimum for $\alpha = \pi$.

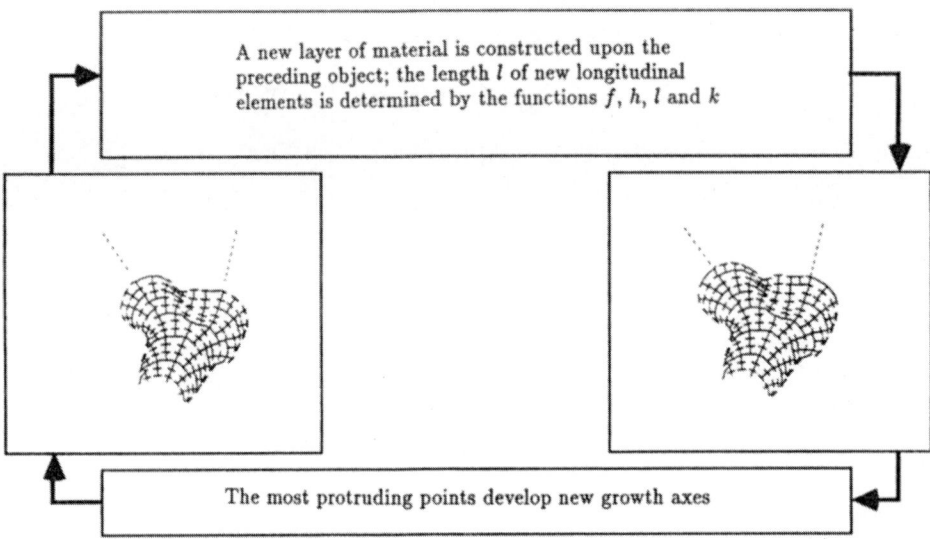

Figure 3. Iteration process in which the radiate accretive growth process is modelled.

Function h (formula 3) returns a normalized version of the local curvature, computed from the radius of curvature rad_curv, which is in turn estimated from a set of points situated on neighbouring tangential elements.

$$h(rad\_curv) \;=\; 1.0 - \qquad (3)$$
$$(rad\_curv - min\_curv)/(max\_curv - min\_curv) \; for$$
$$min\_curv \le rad\_curv \le max\_curv$$
$$h(rad\_curv) \;=\; 1.0 \; for \; rad\_curv < min\_curv$$
$$h(rad\_curv) \;=\; 0.0 \; for \; rad\_curv > max\_curv$$

The values min_curv and max_curv in formula 3 are constants, the value max_curv is the maximum allowed radius of curvature.

The functions $l(\theta)$ and $k(c)$, representing the influence of the physical environment on the growth process, will be discussed in the next section. In the iteration process a rule is applied which specifies that the most protruding points of the object develop new growth axes.

The physical environment

In the model the light intensity I on a surface is determined by the cosine of θ, the angle of incidence of the light beam with respect to the surface normal, and the intensity I_S of the light source (Lambert's law, formula 4). The light source is assumed to be infinitely far away and the light beam direction corresponds to the vertical. The light intensity I_S, which varies with the depth, is assumed to remain constant in this light model.

$$I = I_S cos(\theta) \tag{4}$$

The light model can be extended by including reflection from the environment. When the light is reflected the angle between the normal of a surface and the vertical, where I becomes zero, may vary between $\pi/2$ and π. For autotrophic organisms, the relative decrease in light intensity ($l(\theta) = I/I_S$) was used in the model Figure 3 to mimic the influence of the light intensity on the growth process.

With the Laplace equation the nutrient concentration c in a diffusion process can be described. It is possible to determine c in a field where an object grows and consumes the nutrient[5], when the diffusion proces is assumed to be fast compared to the growth process. The growth of the object can be simulated in a two dimensional lattice. In this paper a high-resolution lattice of 1000 X 1000 sites was used. An exponent η was assumed in formula 5 to describe the relation between the local field and the concentration[1,5].

$$k(c) = c^\eta \tag{5}$$

After each iteration step (Figure 3) the circumference of the object was mapped on the lattice by drawing the edges with the Bresenham algorithm[18]. The region within the edges, consisting of occupied sites in the lattice, was filled by applying a scan-line algorithm[18]. The growing object, represented by occupied sites in the lattice, was set to the concentration value zero. A linear source of nutrient[5] was used by setting the top of the lattice to the value 1.0 (the maximum concentration) and the bottom row to zero. After the computation of the concentration distribution, for each tangential element in the geometric model an estimation was made of the local concentration gradient. In this estimation an exponent η was used (formula 5).

RESULTS

Two results of the iteration process (Figure 3) are shown in Figure 4. In form A l was determined by the light model $l(\theta)$ (no reflection from the environment), the other functions $f(\alpha)$, $h(rad\_curv)$ and $k(c)$ remained constant (were set to the value 1.0). In Figure 4B the same construction was applied with reflection from the environment.

In Figure 5 a branching form is simulated. In this construction the function $l(\theta)$ in the determination of l remained constant (was set to the value 1.0). In Figure 6 the preset range within the quantity rad_curv was allowed to vary in the function $h(rad\_curv)$ (formula 3) was chosen larger than in Figure 5. In Figure 5 η (formula 5) was set to 1.0 and in Figure 6 to 0.5. The surrounding basins of equal ranges of concentration are visualized in alternating black and white[19]. The concentration is zero on the object itself, which is displayed in black, the concentration increases in the successive surrounding basins and is maximal (1.0) at the top of the picture.

The effect of the concentration gradient on the form of the object was tested by rotating the object in the iteration process. In Figure 7 the object was positioned after the 50th iteration step horizontally in the lattice. For this object the same parameter setting as in object Figure 5 was used.

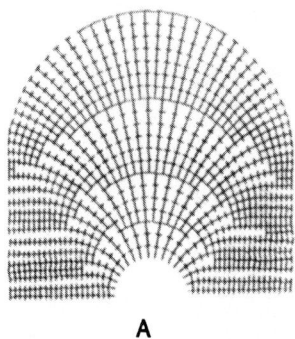

A	B

Figure 4. Two simulated forms of the radiate accretive growth process, by applying the iteration process shown in Figure 3. All functions, except the light model $l(\theta)$, remained constant. In form B the reflection from the environment was included.

DISCUSSION

The tangential and longitudinal elements in the iteration process represent the basic building elements of the simulated biological objects. For *Montastrea annularis* this basic building element is the corallite. In *Haliclona oculata* (see Figure 1A) the skeleton elements (the spicules) are the basic building elements of the sponge.

In a longitudinal section (see Figure 1A) of the heterotrophic example, the highest growth velocities occur in parts of the sponge where the surface and the axis of growth make an angle of 90 degrees, while the velocities decrease to zero at an angle of about 180 degrees. The spicule secreting cells, exhibit the highest activity close to the tip, where they are situated in an area with the highest supply of material suspended in the water[11]. Towards the sides this access to the suspended material as well as the secretion decreases. In the heterotrophic example, with internal secretion of elements, growth of the longitudinal elements is related to the angle α between a tangential element and an axis of growth. This dependence is modelled by the function $f(\alpha)$ (formula 2).

In the autotrophic *Montastrea annularis* (see Figure 1B) growth velocities are highest at the tip of the of the column shaped colony, the corallites are secreted superficially. Growth of the corallites is related to the angle θ of the corallite with respect to the light source[10]. In this case the vertical is the axis of growth and the length l of new longitudinal elements is determined by the light model, the function $l(\theta)$. When there is enough reflection from the bottom a hemi-spherical form emerges (comparable with the simulated form Figure 4B). When the light intensity and reflection decrease, the form transforms in Figure 4A which can be used as a simulation of the actual object shown in Figure 1B.

In the section of Figure 1A the tip has split into two new tips the moment it widened too much. The aquiferous system of the sponge *Haliclona oculata* is relatively poor developed. In the case the sponge assumes a plate-like form food supply can decrease locally in parts of the tissue of the sponge which are not in short distance with the environment. The result is that a local minimum in growth velocity emerges and two (or more) local maxima are formed[11], from which two new axes of growth emerge. The formation of new growth axes is modelled by the rule in the iteration process (Figure 3). In general the tip-splitting for an organism with radiate accretive growth can be modelled with the function $h(rad\_curv)$ (formula 3), which expresses the limitations of the transport mechanism of nutrients through

Figure 5. Simulated form generated with a combination of the geometric and the DLA model. The parameter η (formula 5) was set to 1.0.

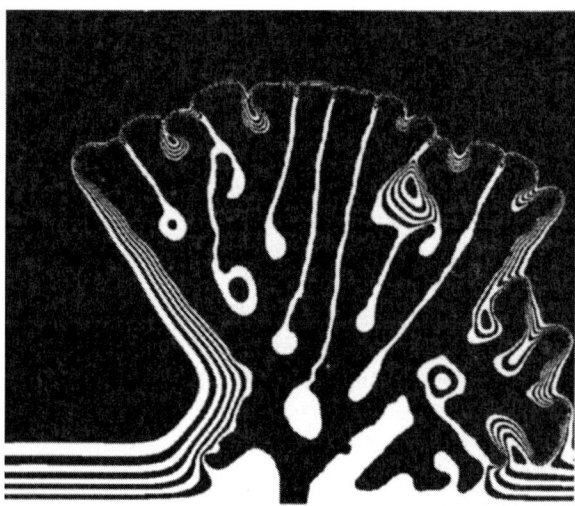

Figure 6. Simulated form generated with a combination of the geometric and the DLA model. In this form the preset range in which rad_curv (function h) is allowed to vary is extended, when compared to the form in Figure 5. The parameter η in (formula 5) was set the value 0.5.

the tissue. The preset range within rad_curv is allowed to vary, depends on external water movements as well: in an environment with a higher rate of water movement, the aquiferous system is supported by these external movements and a relatively larger range is applied. The effect of changing this range is demonstrated in Figure 6: the simulated form exhibits more plate-like ends. For *Montastrea annularis* this limitation as well as the formation of new axes of growth is not relevant, since there is only one axis of growth (the vertical).

Figure 7. Simulated form generated with a combination of the geometric and the DLA model. The object was positioned horizontally after the 50th iteration step.

Besides the secretion of new layers of growth proportional to the angle of the axis of growth and the tip-splitting, two more aspects in the growth process of the heterotrophic example can be observed: negative substrate-trophy and collision of branches. In sheltered conditions the vertical gradient in water velocity (together with the food supply and resulting growth velocity) will be less steep than in exposed conditions, resulting in a larger minimal angle between an axis of growth and the horizontal, when compared to the exposed situation. This effect can be seen in Figure 2 and in the simulated objects in Figures 5 and 6. The gradient in food supply can be modelled with the function $k(c)$ (formula 5), where for each tangential element an estimation is done of the (gradient of) the local nutrient concentration. The steepness of the gradient can be modelled with the exponent η in $k(c)$. For the autotrophic example $k(c)$ is not relevant and remains constant in the iteration process.

Especially on a sheltered site food supply as well as growth velocity decreases when branches collide. This effect can also be modelled with the function $k(c)$. In Figures 5 and 6 it can be seen that the nutrient is locally depleted at sites which are enclosed by branches, resulting in suppressed growth of screened branches.

Conclusions

Growth of *Montastrea annularis* can be modelled with a subset of the rules applied in the model (Figure 3), where only the light model $l(\theta)$ is relevant. Longitudinal sections of ecotypes of this coral can be simulated, by varying the amount of reflection from the environment.

The radiate structure in *Haliclona oculata* can be modelled with $f(\alpha)$, this function describes the influence of the food supply to the secreting cells. This function is only applicable to organisms with internal secretion, it is not relevant for organisms where new tissue is deposited superficially upon a dead core. The branching process itself is modelled by $h(rad\_curv)$, which describes the limitations of the transport system of nutrients. In order to model growth, which exhibits negative substrate-trophy, suppression of growth in the lower branches, suppression of growth of branches which collide; it is neccesary to add a model of the physical environment $k(c)$ which describes the distribution of nutrients. The simulated forms shown in Figures 5 and 6, exhibit negative substrate-trophy and suppression of growth by the canopy of branches of the screened branches. The negative substrate-tropy of the model is demonstrated in Figure 7, where the object grows from the substrate (the bottom line) towards the nutrient source. This effect shows that the model can be used to predict forms which are found in experiments where a sponge was positioned horizontally.

The diffusion model can describe the situation under sheltered conditions, but under exposed conditions laminar and turbulent flows will disturb this pattern of nutrient distribution. In order to describe this situation accurately it will be necessary to replace the Laplace equation by the Navier- Stokes equations. A simplified simulation of the difference between the growth process under sheltered and exposed conditions is shown in Figure 6 where a steeper nutrient gradient was assumed in order to simulate a more exposed environment. Of course another important simplification is that the growth process is modelled in 2D. In reality the situation that the nutrient is depleted between the branches (see Figures 5 and 6) will occur less frequently because nutrient will be supplied from more directions.

It is useful to create a model for radiate accretive growth which unifies autotrophic and heterotrophic organisms, since with such a generic model it is possible to model growth of organisms where light as well as a heterotrophic nutrient source are the main environmental parameters determining the growth form, a situation which occurs among many Scleractinia[20] and some Porifera[21]. The formation of branched growth forms can indicate a significant contribution of the heterotrophic component. For autotrophic organisms with radiate accretive growth and the formation of branches, as for example found in coralline algae[12], the formation of branches may be induced by local nutrient concentrations (non-heterotrophic nutrients essential for building the skeleton). In a radiate growth process, not signicantly influenced by local nutrient concentrations and mainly controlled by local light intensities, spherical or columnar forms as shown in Figure 4 (for lower light intensities, tapered columns and substrate covering plates) are expected. For organisms which exhibit a combination, the simple light model will not be sufficient in many cases. In the case where branching appears, cast shadows will suppress growth in the lower over-shadowed branches. This can only be modelled adequately with a 3D geometric model and a light model which takes cast shadows into account. A model, based on ray-tracing techniques, could be a good reflection of the actual environment.

ACKNOWLEDGEMENTS

The author would like to thank Dr. E.H. Dooijes and Prof. Dr. ir F.C.A. Groen (Department of Computer Science, University of Amsterdam); Drs. M.J. de Kluijver, Dr. R.W.M. van Soest, Prof. Dr. J.H. Stock (Institute of Taxonomic Zoology, University of Amsterdam) for their comments on the first drafts of the manuscript. Mr. L.A. van der Laan (Institute of Taxonomic Zoology, University of Amsterdam) is thanked for preparing the photographs and Drs. E. Meesters (Carribean Marine Biological Institute, Curaçao) for providing a section of *Montastrea annularis*.

REFERENCES

1. L. Niemeyer, L. Pietronero, and H.J. Wiesmann, Fractal dimension of dielectric breakdown, *Phys. Rev. Lett.*, 52(12):1033–1036 (1984).
2. T.A. Witten and L.M. Sander, Diffusion-limited aggregation, a kinetic critical phenomenon, *Phys. Rev. Lett.*, 47(19):1400–1403 (1981).

3. L.M. Sander, Fractal growth processes, *Nature*, 322:789–793 (1986).

4. H.E. Stanley and N. Ostrowsky, On Growth and Form: Fractal and Non-Fractal Patterns in Physics, Martinus Nijhoff, Boston (1987).

5. P. Meakin, A new model for biological pattern formation, *J. theor. Biol.*, 118:101–113 (1986).

6. H. Fujikawa and M. Matsushita, Fractal growth of *Bacillus subtilis* on agar plates, *J. phys. Soc. Japan*, 58(11):3875–3878 (1989).

7. T. Matsuyama, M. Sogawa, and Y. Nakagawa, Fractal spreading growth of *Serratia marcescens* which produces surface active exolipids, *FEMS Microbiology Letters*, 61:243–246 (1989).

8. F. Wiedenmayer. Shallow-Water Sponges of the western Bahamas. Birkhauser Verlag, Basel (1977).

9. J. L. Harper, B. R. Rosen, and J. White, The Growth and Form of modular Organisms, The Royal Society, London, (1986).

10. R.R. Graus and I.G. Macintyre, Variation in growth forms of the reef coral *Montastrea annularis* (Ellis and Solander): a quantitative evaluation of growth response to light distribution using computer simulation, *Smithson. Contr. mar. Sci.*, 12:441–464 (1982).

11. J.A. Kaandorp, Modelling growth forms of the sponge *Haliclona oculata* (Porifera; Demospongiae) using fractal techniques, *Mar. Biol.*, 110:203–215 (1991).

12. D.W.J. Bosence, Ecological studies on two unattached coralline algae from western Ireland. *Paleontology*, 19(2):365–395 (1976).

13. R.W.M. van Soest and J. Verseveldt, Unique symbiotic octocoral-sponge association from Komodo, *Indo-Malayan Zoology*, 4:27–32 (1987).

14. J.W. Porter, Zoo-plankton feeding by the Caribbean reef-building coral *Montastrea cavernosa*, *Proceedings of the 2th International Coral Reef Symposium vol. I*, pages 111–125 (1974).

15. P. Spencer Davies, The role of zooxanthellae in the nutritional energy requirements of *Pocillopora eydouxi*, *Coral Reefs*, 2:181–186 (1984).

16. P.J. Edmunds and P. Spencer Davies, An energy budget for *Porites porites* (Scleractinia) growing in a stressed environment, *Coral Reefs*, 8:37–43 (1989).

17. J.A. Kaandorp, Modelling growth forms of sponges with fractal techniques, In Fractals and Chaos, pages 71–88, A.J. Crilly, R.A. Earnshaw, and H. Jones, Eds, New York, Springer Verlag (1991).

18. J.D. Foley, A. van Dam, S.K. Feiner, and J.F. Hughes, Computer graphics: principles and practice, Addison-Wesley, New York (1990).

19. B.B. Mandelbrot and C.J.G. Evertsz, The potential distribution around growing fractal clusters, *Nature*, 348:143–145 (1990).

20. J.C. Bythell, A total nitrogen and carbon budget for the elkhorn coral *Acropora palmata* (Lamarck), *Proceedings of the 6th International Coral Reef Symposium vol. II*, pages 535–540 (1988).

21. C.R. Wilkinson, A.C. Cheshire, D.W. Klumpp, and A.D. McKinnon. Nutritional spectrum of animals with photosynthetic symbionts-corals and sponges. *Proceedings of the 6th International Coral Reef Symposium vol. III*, pages 27–30 (1988).

PHYLLOTAXIS AS A SELF-ORGANIZED GROWTH PROCESS

Stéphane Douady * and Yves Couder

Laboratoire de Physique Statistique
24 rue Lhomond, 75231 Paris Cedex 05, France.
* also at : Laboratoire de Physique, E.N.S. Lyon
46 allée d'Italie, 69364 Lyon Cedex 07, France

Abstract

The large morphological differences through the botanical world seem contradictory with the universality of the botanical arrangements of leaves, florets, etc. This universality comes from an identical growth process which can be translated into simple assumptions for an iterative process. As we obtain the phyllotactic patterns in a physics experiment and in a simulation, this iterative self-organization proves to be sufficient. The analysis of our results permits an understanding of the patterns selection and of the Fibonnaci series predominance.

Introduction

To observe easily the most common arrangements take for example a pine cone or a sunflower head (Fig. 1). The elements (scales or florets) form regular lattices on the surface.

Figure 1. A sunflower head (*Helianthus annuus* L.), is constituted of a large number of small flowers, the florets. Their regular arrangement, recalling crystalline order, is clearly visible. Here 21 and 34 spirals in the two opposite directions are counted at the periphery of the inflorescence.

Growth Patterns in Physical Sciences and Biology, Edited
by J. M. Garcia-Ruiz *et al.*, Plenum Press, New York, 1993

The eye is attracted to conspicuous spirals *(the parastichies)*, linking each element with its nearest neighbours. The entire surface is covered with a given number i of parallel spirals running in one direction, and j in the other. The most striking feature is that *(i, j)* are nearly always two consecutive numbers of the Fibonnaci series : $\{F_k\}=\{1, 1, 2, 3, 5, 8, 13, 21, 34 ...\}$ where each new number is the sum of the two preceding ones. These regular arrangements (with low parastichies numbers) are also commonly visible in the disposition of leaves around a stem (Phyllo-Taxis in greek). In the most common case where the elements appear one after the other, early works[1-3] introduced *the generative spiral*, the unique tightly wound spiral joining the elements in their order of appearance. The basic quantity is then *the divergence* φ which is the angle between the radial directions of two consecutive elements. By convention φ is taken between $-\pi$ and π. Its sign indicates whether the generative spiral is clockwise or anti-clockwise. The Bravais brothers[3] carefully measured the divergences on mature plants and found them surprisingly close to $\Phi=2\pi(1-\tau) \approx 137.5°$, where $\tau = (-1+\sqrt{5})/2$ is the golden mean.

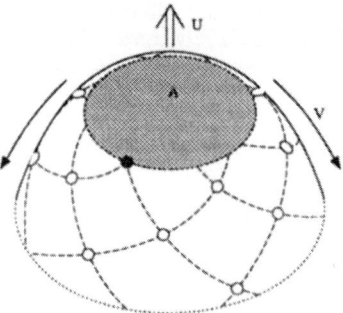

Figure 2. Sketch of the growth in the vascular plants. At the extremity of the shoot is situated a region of undifferentiated tissues: the apex (A, in grey). Around it, a specific mitotic activity creates small protrusions: the primordia (only the centers of the primordia base (o) have been drawn). They evolve later into various types of botanical elements: leaves, bracteae, sepals, petals, stamens, florets... The primordia already form conspicuous parastichies which are drawn here. Because the shoot grows with velocity U, in the reference frame of the apex the primordia seem to be advected away from the tip with a velocity V. The appearance and position of the new primordium (●) around the apex can be related to the position of the older ones.

A basic hypothesis[4] is that these patterns result from the conditions of appearance of the primordia near the tip of the growing shoots. This hypothesis relates the universality of these arrangements in the vascular plants to the universality of their growth caracteristics (reviews can be found in Ref. 5 - 7). The stem tips (the apical meristems) show various axisymetric profiles, from very sharp to flat (Fig. 2). The summit (or the center), is always occupied by a relatively quiet region: the apex. The primordia (which will evolve into leaves, stamens, florets…) are first visible as small protrusions around the apex. As the shoot grows, the distance between the apex and the existing primordia increases. In the reference frame of the tip, they seem to be advected away radially, while new primordia continue to be formed near the apex[8]. Note that after the very beginning of their growth the primordia expand but do not move tangentially ; they become linked to the vascular system of the shoot. As a result the divergence remains practically unchanged through all the growth and maturation of the plant. The generic quantities characterising the growth are : the mean apex radius R_0 (typically 100μm), the advection velocity $V(r)$ (V_0 at R_0), and the mean time interval T between the formation of two successive primordia (called the plastochrone[9], typically 1 day). An adimensionnal quantity, the plastochrone ratio has been introduced by Richards[10], it is the ratio of the distance to the center of two consecutive leaves $r_m/r_{m-1} = \exp[V_0 T/R_0]$ (around the apex the growth is approximately exponential).

Botanical observations have long tried to find the position and the time of formation of a new primordium in relation to the preceding ones. In 1868, Hofmeister[4] suggested that the new primordium appears periodically at a regular plastochrone T, at the apex boundary, in the largest gap left by the preceding primordia. Several models based on this hypothesis obtained partial results (for a review see Ref 11).The primordia interaction was simulated with hard disks packing[12-14] or the diffusion of an inhibitor [15-18]. After experimenting on shoot apices Snow & Snow[19] proposed a variant of Hofmeister's hypothesis : the plastochrone T is free to vary, but a new primordium appears only when the elder ones have left enough space for its formation.

Physical model

We now want to demonstrate that the phyllotaxic patterns can result from a purely physical mechanism for the primordia organisation during the plant's growth. For this purpose we implemented a laboratory experiment and a numerical simulation. The basic physical characteristics we retained to obtain phyllotactic growth are the following: identical elements are generated in a plane surface[20] at a given radius R_0 from a center with a constant plastochrone T . These elements are radially advected at velocity V, and there is a repulsive interaction between them (so that the new element will appear as far as possible from the preceding ones, i.e. in the largest available place). As control parameter we chose $G=V_0T/R_0$. It is similar but more general than the plastochrone ratio which depends on the velocity profile (for a linear growth $R_1/R_0= 1+G$).

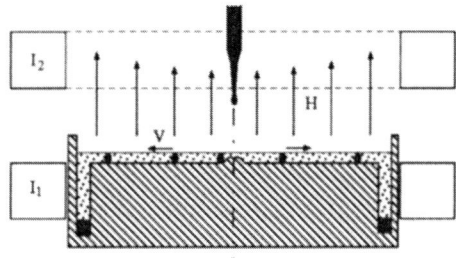

Figure 3. Sketch of the experimental apparatus. Drops of ferrofluid are used to simulate the primordia. The drops (of volume v ≈ 10mm³) fall with a tunable periodicity T at the center of a horizontal teflon dish, of diameter 8cm filled of silicon oil. This dish is in the vertical magnetic field H created by two coils in the Helmholtz position. The magnetic field polarizes the ferrofluid drops so that they form small magnetic dipoles repelling each other. In order to simulate the apex, the drops fall at the center of the dish onto a small truncated cone. Due to the interaction with the previous drops, the new drop falls from this cone swiftly in the direction of the point of minimum energy. These dipoles are radially advected by the magnetic field gradient (controlled by the currents I_1 and I_2 in the two coils). To prevent accumulation the drops ultimately fall into the deep ditch at the periphery.

The experimental system (Fig. 3) consists of a horizontal dish filled with silicone oil and placed in a vertical magnetic field H(r) created by two coils. Drops of ferrofluid[21] of equal volume fall with a tunable periodicity T at the center of the cell. The drops are polarized by the field and form small vertical magnetic dipoles. They thus repell each other with a force proportional to d^{-4} (where d is their distance). These dipoles are advected by a radial gradient of the magnetic field, their velocity V(r) being limited by the viscous friction of the oil. In order to modelize the apex, the dish has a small truncated cone at its center, so that the drop introduced at its tip quickly falls to its periphery toward the place of minimum repulsive energy. R_0 is then related to the radius of the cone base plus the drop's radius. G can be tuned either by changing the periodicity T or by changing the gradient of H. In most of our experiments H varied from $2.4 \ 10^4$ A/m at the center to $2.48 \ 10^4$ A/m at the border of the dish. We used an ionic ferrofluid in nitric acid with a volumic concentration of 8%. For the values of H we used the magnetization was $M = \chi H$ with $\chi = 0.493$.

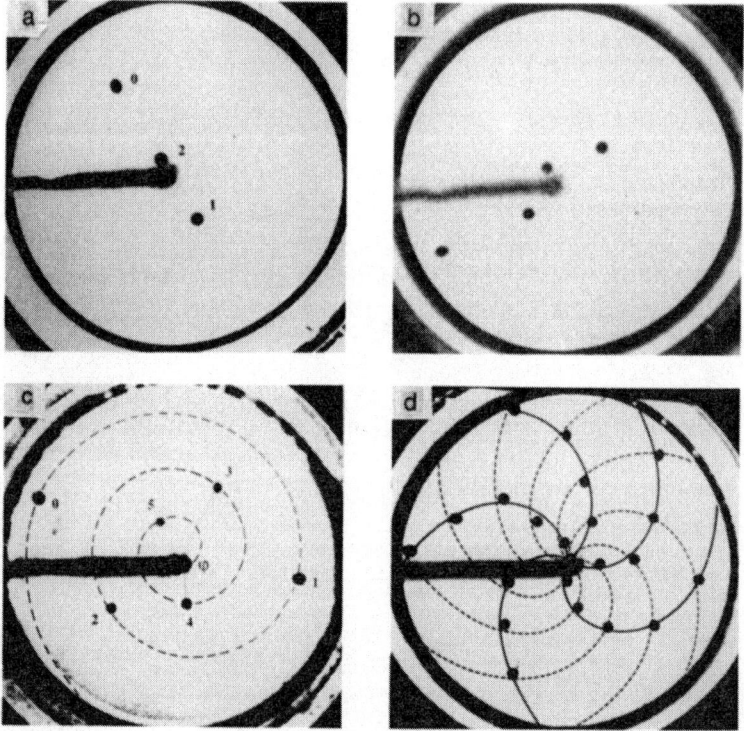

Figure 4. Four photographs (seen from above) of typical phyllotactic patterns (The tube for the ferrofluid supply hides the central truncated cone).

(a) For strong advection, $G \approx 1$, each new drop is repelled only by the previous one and a distichous mode is obtained, $\varphi = 180°$.

(b) Transient below the first symmetry-breaking bifurcation (at $G \approx 0.7$): while the first two drops have moved in opposite directions the third one has broken the symmetry and determined the direction of the generative spiral. The divergence angle is $\varphi \approx 170°$.

(c) The generative spiral obtained at $G \approx 0.6$ where $\varphi \approx 150°$.

(d) For $G \approx 0.15$, $\varphi \approx 139°$ and the parastichies numbers are (5, 8). The friction of the drops on the bottom during their advection is responsible for slight irregularities of the pattern.

The experiment shows that the final steady pattern depends crucially on G (Fig. 4). For a strong advection (Fig. 4a, G=1), a drop which has just fallen onto the dish is only repelled by the previous one. Successive drops then move away in opposite directions, $\varphi = 180°$. This mode is called the alternate or distichous mode in Botany. As G is decreased below a threshold G_c, the new drop becomes sensitive to the repulsion of both of the two previous ones and cannot remain on the straight line formed by these drops. It slides to one of either side of this line. This is a symmetry-breaking into a chiral mode (Fig. 4b)which selects once for all the direction of rotation of the generative spiral: the following drops wind along the same direction . Later a steady regime is reached with a constant divergence φ (150° on Figure 4c). For smaller G, the new drop becomes sensitive to the interaction of three or more

344

previous ones, and the divergence gets nearer to Φ. In Fig. 4d, φ=139° for G=0.15. These experiments produce very common botanical patterns, with Fibonnaci numbers of conspicuous parastichies. The trajectories of each drop were reconstructed from the videotape recordings. Except very near the center, no reorganization occurs and each particle is simply advected on a radial trajectory : this is in good agreement with botanical observations, where no tangential motion of the primordia were ever reported, and with the simplifying assumption of our model, where these motions are not allowed.

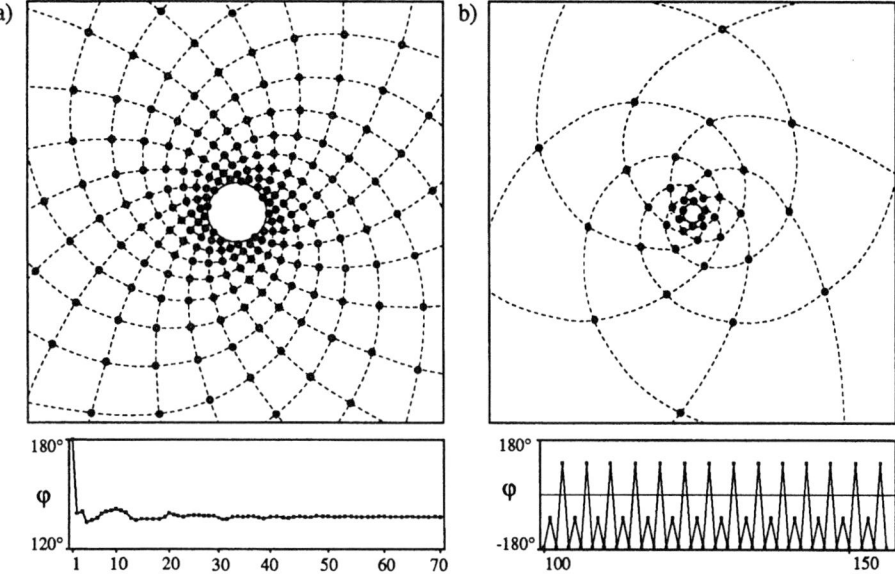

Figure 5. Two typical patterns obtained in the numerical simulation (the parastichies and the apex radius were drawn for an easier visualisation of the structure).

(a) Pattern obtained after a transient from G=1 (distichous mode), to a constant small value G=0.01 , in a characteristic time of 6 plastochrones. The final steady divergence angle is φ≈137.47° and the parastichies numbers (13, 21). At the bottom of the figure a plot gives the evolution of φ with the number of deposited particles. The transient is clearly visible: the first two particles are opposite to each other then the divergence converges quickly towards 137.47°.

(b) Example of pattern obtained when the simulation is started directly with a small G (here 0.077). The divergence does not reach a stable value and periodically oscillates in time (φ(t) is shown at the bottom). The structure is however regular and the drawing of the parastichies shows that this pattern corresponds to a bijugate mode: the parastichies numbers are (4, 6).

Our numerical simulations are based on the same physical hypotheses but even more simplified. Also performed in a plane radial configuration, the locus of appearance of elements is a circle C of radius R_o centered at the origin. These elements will be punctual particles, each generating a repulsive energy E(d) where d is the distance to the particle. Several energy laws were used, 1/d, $1/d^3$ (which is the interaction between the ferrofluid drops), and exp(-d/l) with various l : the results were qualitatively the same. To decide the place of birth of a particle, we compute in each point of the circle C the value of the total energy due to all the previous particles, and place the new element at the point of minimum energy. All particles after their appearance are given the same radial motion with a velocity V(r), thus neglecting any later reorganization due to the interaction of particles. In this case the results are qualitatively independant of the chosen radial motion. In order to remain close to botany[22] we chose an exponential growth (V(r)= $V_0 r/R_0$).

The initial conditions chosen to start running these simulations are very important. There is well established [12,18,23] botanical evidence that during the growth, for instance from a seedling to a plant, G decreases from an initially large value to a small one. The best way to reach a botanically relevant pattern at a fixed value of G is thus to start the simulation at a larger value of G (e.g. 1) and decrease it progressively (e.g. during 6 T) to its constant value. With such transients, we obtain very easily the high order phyllotactic patterns: Figure 5a shows a typical steady pattern at G=0.01 where the divergence (φ= 137.47°) is very close to Φ, and parastichies numbers are successive elements of the Fibonnaci series, (i, j)=(13, 21). This steady regime is reached after a transient lasting less than 20 T.

To understand this result, i.e. the selection of the Fibonnaci patterns after a transient, we performed a systematic investigation of the patterns that could be obtained with various types of initial conditions and we plotted the resulting divergences as a function of G (Fig. 6).

A brutal way to run the simulation is to impose immediately at the initial time the given value of G and to see which pattern is spontaneously obtained. For large G we observe, as in the experiment, a typical symmetry-breaking bifurcation, the threshold of which G_{11}, depends on the energy profile. For smaller values of G, the same divergence angles as in the experiment are obtained and φ starts to converge oscillatingly towards Φ corresponding to the first Fibonacci modes. But for very small G, the system undergoes long transients, and either converges on a steady regime, with several possible values of φ, or stabilizes on more or less complicated periodic regimes. The possible steady values of φ form several new curves in the diagram (Fig. 6). In order to investigate the limits of existence of each solution, we also used initial conditions in which we initially forced an artificial pattern and observed whether it could keep growing.

With all the results obtained with the various types of initial conditions we get a complete diagram shown on Figure 6. In this diagram there is one main curve along which φ converges oscillatingly from 180° towards 137.508°. Each change of sign of the slope of the curve φ(G) occurs at a value G_{ij}. Comparison of the patterns obtained just above and just below G_{ij} shows that a transition from parastichies numbers (i, j) to (j, i+j) has occured at this value. A remarkable asset of the diagram (Fig. 6) is that near each G_{ij}, a new, different possible value of φ also appears. It corresponds to a regular spiral pattern with parastichies numbers (i, i+j). With a further decrease of G this new value of φ will evolve continuously and thus form in the φ(G) diagram a new curve disconnected from the main one. Following this new curve φ(G) towards G=0, φ tends oscillatingly towards an irrational angle and a Fibonacci type of series builds up with the same transition rule (i, j) -> (j, i+j). Each of the curves φ(G) thus corresponds to a different pair of initial terms for the series and converges for G=0, towards the related irrational angle[3] (99.502°, 77.955°, 151.135°, etc). This structure of the diagram φ(G) is independant of the energy profile and of the chosen geometry (plane, cone, cylinder, etc.). If G undegoes too abrupt a change during growth, it can induce a jump to the other curves φ(G) and result into the patterns with secundary Fibonnaci series. They indeed are found in plants as rare anomalies[3].

Finally let us examin the only regimes that were not represented on Figure 6 in which φ oscillates in time. They correspond either to the normal patterns in which pairs of particles have their appearance order inverted, or to a rarer family of phyllotactic patterns (fig.5b): the bijugate arrangements, with "two generative spirals" and parastichy numbers double those of the Fibonnaci series[3].

Interpretation

In terms of iterative dynamical systems, all the transitions at $G_{i,j}$ are bifurcations. We find that near the threshold of the first bifurcation G_{11}, the divergence φ varies as : (180-φ)$\propto$ $(G_{11}-G)^{1/2}$, and the number of iterations (i.e. of particles) necessary to reach a steady regime diverges. These are the characteristics of a direct symmetry-breaking bifurcation. Here it leads from an alternate pattern to a chiral spiral pattern. All the other bifurcations,

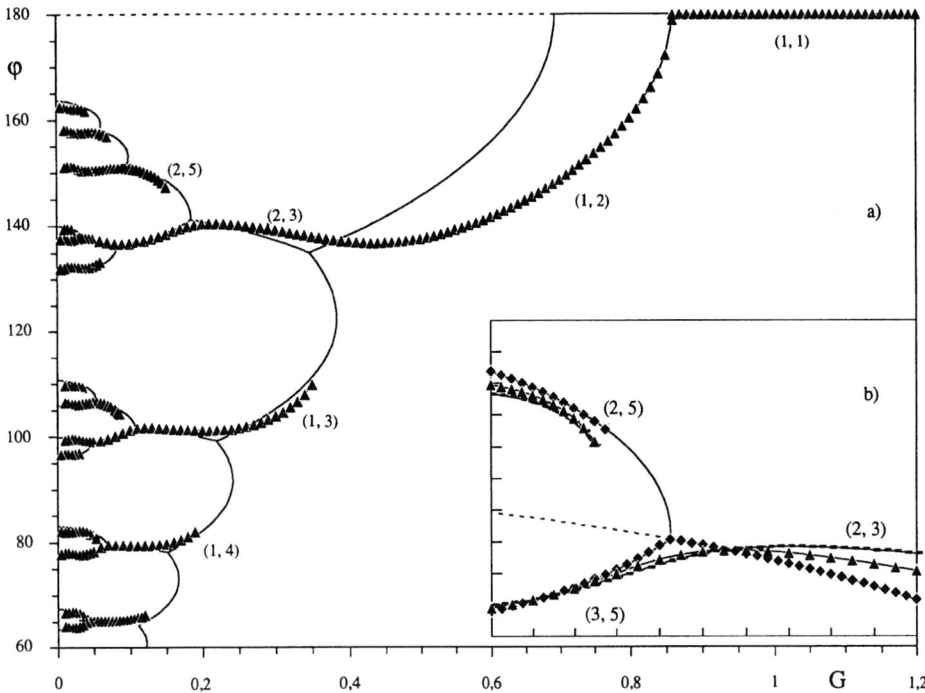

Figure 6. (a) Diagram of the values of the steady divergences φ obtained as a function of G (energy profiles $1/d^3$). Only the positive values of φ have been drawn (the diagram being symmetrical). This diagram can be approximated using a geometrical condition that the new particle appears exactly equidistant from two older ones (thin lines). Comparison of the structure of the diagram with the geometrical approximation shows the imperfect nature of the bifurcations in our model.

Inset (b) Detail of the relation between the modes with parastichies numbers (2, 3), (3, 5) and (2, 5). Three energy profiles were used (1/d, dashes, 1/d3, triangles, and exp[-d/l] with l=0.01, squares). All the simulations' results are similar. Depending on the chosen profile they can be as close as desired to the geometrical solution, but the gap is unchanged. This gap only comes from a dynamical constraint, and can be interpreted as the fact that the symmetry is already broken at the transition. This asymmetry is directly visible on the diagram : at the transition the slope of the curve (j, i+j) is always smaller than that of (i, i+j).

because the symmetry is already broken, are imperfect (see below): at each of them only one curve is continuous and a new curve appears disconnected (Fig. 6).

This is the key point of our result, and the main difference with previous works[12-14] obtaining a type of diagram first derived by Van Iterson[12]. He reproduced the phyllotactic arrangements by looking to the regular patterns formed by hard disks paving a cylinder (or a cone) and found a relation between φ and the ratio of the radius of the disks to that of the cylinder. In our case, a diagram of this type can also be obtained by stating that a new particle appears exactly equidistant from two previous ones, n-i and n-j (cf Fig. 7). This approximation transforms our dynamical problem into a geometrical one. If a geometrical constant H of the assumed regular spiral[24] is set equivalent to G it is possible to compare the geometrical relation between φ and H (for each (i,j)) and our simulations results (Fig. 6). As in all geometrical models[12-14], the curve corresponding to parastichies (i, j) (with j>i) is connected at G_{ij} with the two curves (j, i+j) and (i, i+j). The geometrical models fail to obtain the selection because they do not take into account the fact that the condition on the new element is not only to be located between two previous ones, but also that its place should be the best (e.g. the largest space). To our knowledge only one previous work[18] had used such a criterium in the simulation of the diffusion of an inhibitor in a cylindrical

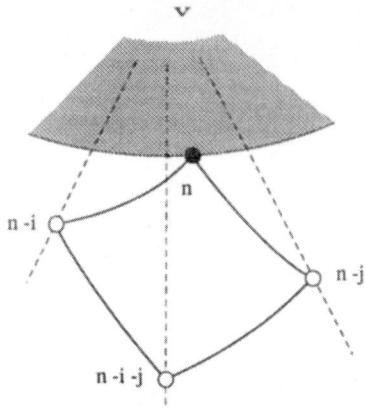

Figure 7. Sketch of the position of the new primordia. The growth is assumed to be already in a steady regular regime. The primordia then form a regular lattice. If the primordia are numbered by order of appearance, n being the new one, the appearance numbers of its two neighbours, n-i and n-j, give the two numbers of parastichies: (i,j). From the regularity, we then know for instance that when (n-i) appeared, its two neighbours were (n-i)-i and (n-i)-j. So the primordium between n-i and n-j is n-(i+j). As n-j is older than n-i (i<j), it is located further form the apex boundary, and the rhomb [n,n-i,n-i+j,n-j] is inclined. When the advection velocity is reduced the two neighbours of the new primordium n thus naturally become n-j and n-(i+j). So by decreasing the advection velocity the parastichies numbers are increased from (i, j) to (j, i+j). This explain the bifurcations and their imperfections. These bifurcations would be perfect only if the rhomb was not inclined, i.e. if the symmetry not already broken before the transition. This implies that n-i = n-j, which is only possible for i=j=1, and corresponds exactly to the first (perfect) transition. This also shows that in Phyllotaxis the Fibonnaci rule (the new term is the sum of the two preceeding one) only derives from the periodicity rules of a regular lattice.

geometry. A convergence of φ towards Φ had then effectively been obtained, though the overall structure of the bifurcations diagram had not been described.

All the bifurcations and their imperfections can be simply interpreted (Fig. 7). For parastichies numbers (i, j), the new particle n is repelled essentially by the j previous ones, and is precisely between n-i and n-j [24]. If G was decreased and no other particles than the j-latest were taken into account, the divergence would tend toward a rational : n would appear at the same angular position as a previous particle, which, from the regularity of the pattern would be n-(i+j). But near G_{ij}, the repulsion due to n-(i+j) becomes no longer negligible and n slides to avoid the proximity of this particle. As n-j is older than n-i, the situation is not symmetrical and n is always angularly between n-(i+j) and n-j (Fig. 7). Below G_{ij}, n thus slides between n-j and n-(i+j) selecting the transition with the Fibonacci rule. When G is decreased, the system thus avoids all the successive possible periodic arrangements and φ converges towards the simplest irrational numbers.

Our results can be compared to those of Levitov[26] published during the course of our work. In a cylindrical geometry he assumed a regular helical lattice with repelling elements, and sought the lattice slope for which the interaction energy is minimum. He showed that the compression of the whole lattice produces a similar diagram of imperfect bifurcations. This theoretical work appears far from the problem of botanical growth. The relation to our work, however, is similar to that between the investigation of the energy of periodic lattices and the search for the growth mechanism of crystals. The convergence of both results shows that the dynamics of appearance of the new primordia at the place of lowest repulsive energy, creates a final structure of minimum global interaction energy. It also confirms the fact that after the primordia appearance, no later reorganization is needed.

Conclusion : Relevance to botany

Botany provided the basic hypotheses we used, and we can return to it to judge the relevance of our results. We can first investigate how they are affected when the criterium for the primordium appearance is changed. In a variant of our simulation we replaced Hofmeister's criterium by the Snows'[19]. We left free the time of appearance of a new element but added the condition that a particle can only appear when the minimum energy becomes smaller than a chosen threshold E_s. The same steady regimes are reached with a regular plastochrone T selected by E_s, and the same diagram shown on Fig.6 is obtained[27]. But with this criterium other phyllotactic patterns can also appear. For exemple around G=0.27, two particles can appear practically simultaneously in opposite positions of the circle[28]. The two consecutive particles will appear 2 T later at right angles to the previous ones (Fig. 8). This corresponds to the very common opposite decussate phyllotaxy. Note that for the same simulation, with only different initial conditions, a spiral (2, 3) mode is obtained.

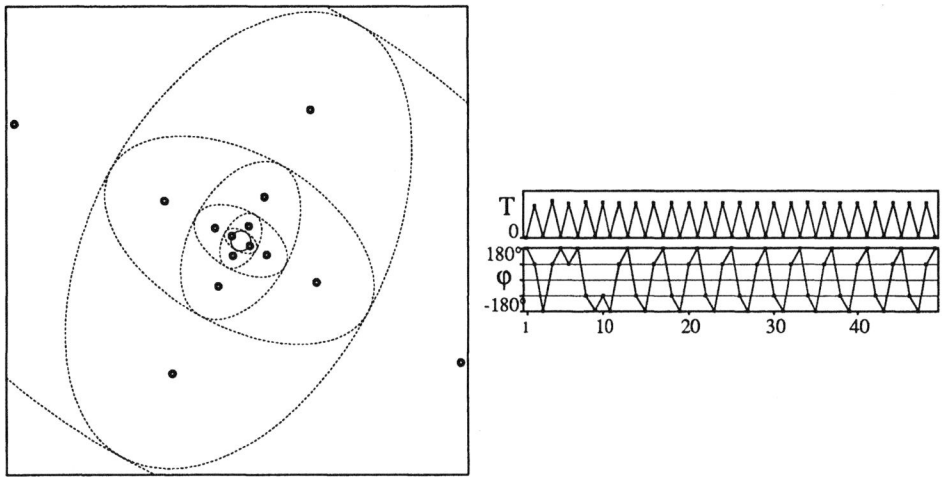

Figure 8. A pattern obtained when the criterium for the primordia appearance of Snow & Snow is used (see text). For the imposed value of the energy threshold, the resulting mean plastochrone is G= 0.27. The oscillations of both the divergence and the plastochrone (bottom) show that two particles appear almost at the same time in opposite positions, and that the two consecutive ones appear at right angles. This corresponds to the decussate mode. Arbitrary primordia bases, typical of a decussate apical meristem, have been drawn centered on each particle for an easier visualisation of the structure.

With the Snows' criterium, we are thus able to obtain not only the spiral Phyllotaxis, where each element appear one after another, but also the other familly of phyllotactic arrangements, where several elements appear regularly at the same time. The possibility of observing opposite decussate phyllotaxy in the same situation as spiral (2, 3) phyllotaxy with only different initial conditions, is also in agreement with an important botanical experiment of Snow & Snow[29]. They showed that by breaking the symmetry of an apex with a diagonal cut, a usually decussate plant could form a spiral pattern, even though the plastochrone ratio was unchanged[10,30].

The continuity of the phyllotaxy when a meristem goes from a vegetative growth to flowering had been remarked early[3]. Meicenheimer recently showed that the corresponding phyllotactic evolution was essentially linked to a decrease of the plastochrone ratio[23]. The same result was obtained about the effects induced by growth hormones[31,32] or about those due to light exposition[33]. Such botanical experiments show that the older parts create the conditions for the later growth through the iterative appearance of the primordia ; and that the phyllotactic pattern depends on the whole history of its growth. On the other hand, we showed that the self organisation of elements based on Hoffmeister's or Snow & Snow's hypothesis leads to all of the various phyllotactic types, and our main result, the structure of the diagram with imperfect bifurcations, permits an explanation of the pattern selection.

We can finally consider the spectacular ordering of the sunflower heads (Fig. 1). Inspection of the whole plant shows a continuity in the spirals formed by the leaves along the stem, the bracteae, and then the florets. This evolution corresponds to a continuous decrease of G which is due to the slowing down of the growth and to the widening of the apex. It explains the astonishingly high orders of the Fibonnacci numbers that can be reached at the periphery of sunflower heads (e.g. i= 144, j= 233). Furthemore the detailed structure of the head can also be obtained. With the completion of the flower head the growth stops; in this process the apex shrinks and G increases again[34]. We performed a simulation of this growth by performing a simulation with G first decreasing and then increasing back. The outer points show the angular positions of the leaves[35] and bracteae. The central region reproduces the main characteristics of the flower head, in particular the highest phyllotactic order obtained at the periphery; going inwards successive transitions between parastichies orders ((i,j)-> (i,j-i)) are observed. This is clearly seen by comparing Figure 9 with Figure 1.

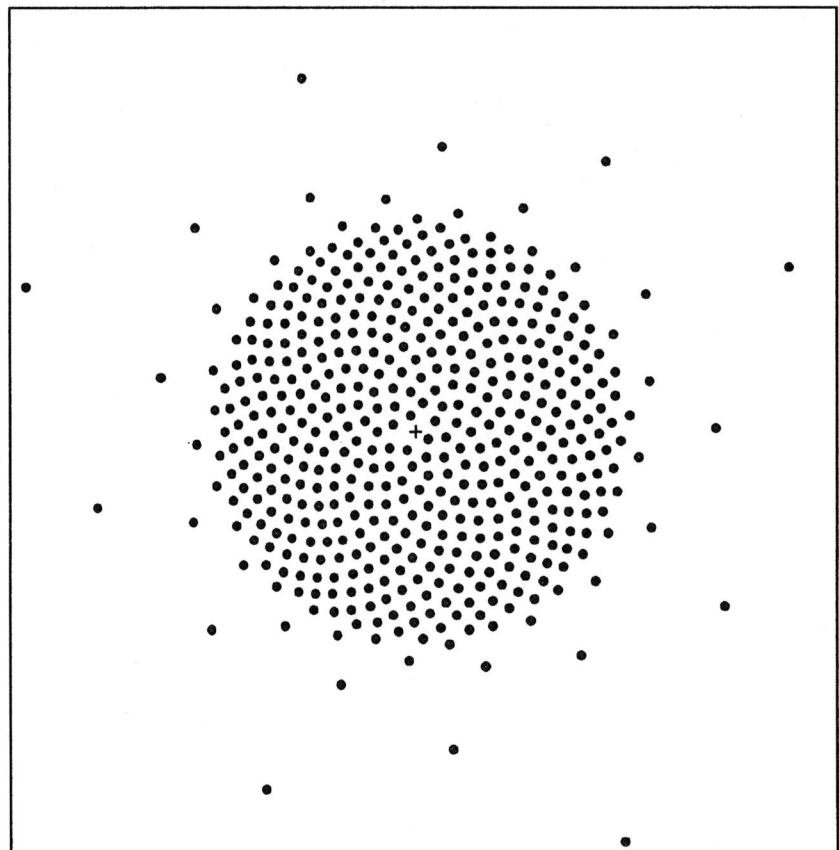

Figure 9. Pattern resulting from a simulation reproducing the growth of a sunflower : G is first decreased from 1 to 0.005 and then increased back to 0.2. The outer points show the positions of some leaves (from seedling). They give a divergence already close to Φ. As G then decreases, the primordia now evolves into bracteae. When G is at its minimum, the primordia evolve into florets. This corresponds to the higher parastichies numbers, here (34,55), and to the border of the flower head. The final increase of G corresponds to the decrease of the parastichies numbers from the border to the center of the flower.

The robustness and the simplicity of our results, practically independant of the interaction law and of the geometry, is the reason why we think they can be botanically relevant. They are also coherent with botanical observations, and using the Snow &Snow criterium, all the phyllotactic patterns can be reproduced. In particular our results show that, with only an assumed (genetically) prefixed growth mechanism, the observation of the most comon arrangements is explained, but also the possibility to observe some abnormal cases.

This leads to think that the same iterative self-organizing process occurs in most of the vascular plants, as they have similar genetically determined apical meristem structure and primordia interactions[36]. But for a given plant shoot, our results (together with the botanical experiments[23,29,31-33]) suggest that the final organisation results only from the effective growth, i.e. from one particular sequence of iterations. This sequence is determined by *the initial conditions* and by *the time evolution of the factor G*, which depend not only of the genetics, but also of the environnement and of the whole plant history. In general, if some direct relations between genes and a final growth result are well known, the way these genes effectively control this result is still rather unkonwn[7]. This is perhaps the reason why there is a recurrent trend towards thinking that the final growth result is *a priori* fixed by the genes, i.e. directly encoded. To the contrary we think, at least concerning the plant's shoot organisation, that the final result is not prefixed and that the genes only determin the possible evolutions.

Acknowledgements: Throughout the setting-up of the experiment we benefitted from very helpful advice from J.C. Bacri and R. Perzynski, who also gave us the ferrofluid we used. We also thank members of the "Biomembranes Végétales E.N.S." laboratory, A.M. Catesson, J.C. Thomas, C. Berkaloff, and D. Douady for useful discussions.

References

(1) Schimper, K.F. (1830) Geiger's Mag. für Pharm. 29, 1.

(2) Braun, A. (1831) Nova Acta Acad. Caesar Leop. Carol. 15, 197. and (1835), Flora, Iena, 18, 145.

(3) Bravais, L. and Bravais, A.
 (1837), Ann. Sci. Nat. second series 7, 42-110, 193-221, and 291-348 ;
 (1837) Ann. Sci. Nat. second series 8, 11-42 ;
 (1839) Ann. Sci. Nat. second series 12, 5-14, and 65-77.

(4) Hofmeister, W. (1868). Allgemeine Morphologie der Gewachse, Handbuch der Physiologishen Botanik, 1 Engelman, Leipzig, 405-664.

(5) Steeves, T.A. and Sussex, I. M. (1989) Patterns in Plant Development, Cambridge University Press, Cambridge.

(6) Lyndon, R.F. (1990) Plant Development : the Cellular Basis, Unwin Hyman, London

(7) Sachs, T. (1991) Pattern Formation in Plant Tissues, Cambridge University Press, Cambridge.

(8) This growth is somewhat similar to that of crystalline dendrites. However, in this case, there is not only anisotropy in the growth's direction but also across it, and this fixes the direction of the side branches. So that in this case the phyllotactic modes cannot show up.

(9) Church, A.H. (1904) On the Relation of Phyllotaxis to Mechanical Laws. London: Williams and Norgate.

(10) Richards, F.J. (1948) Symp. Soc. Exp. Biol. 2, 217.

(11) Jean, R.V. (1984) Mathematical Approach to Patterns and Form in Plant Growth. Wiley & Sons.

(12) Van Iterson, G. (1907) Mathematische und Microscopisch-Anatomische Studien über Blattstellungen, nebst Betrschungen über den Schalenbau der Miliolinen, Gustav Fisher-Verlag, Iena.

(13) Adler, I. (1974) J. Theor. Biol. 45, 1-79.

(14) Rothen, F., Koch, A.J., (1989) J. Physique 50, 633-657 and 1603-1621.

(15) Schoute, J.C. (1913) Recl. Trav. Bot. Néerl. 10, 153.

(16) Turing, A.M. (1952) Phil. Trans. R. Soc. of London, 237B, 37-72.

(17) Veen, A.H. and Lindenmayer, A. (1977) Plant Physiol. 60, 127-139.

(18) Mitchison, G.H. (1977) Science, 196, 270-275.

(19) Snow, M. and Snow, R. (1962) Phil. Trans. Roy. Soc. London Ser. B 244, 483.

(20) The exact geometry should not be an important parameter, provided it has axisymmetry. In botany, similar phyllotactic modes are obtained with various apical meristem profiles.

(21) Massart R.(1981) I.E.E.E. Trans. Magn. 17,1247

(22) Richards, F.J. (1951) Phil. Trans. Roy. Soc. B 225, 509-564.

(23) Meicenheimer, R.D. (1979) Am. J. Botany 66, 557-569.

(24) In the geometrical model the positions of the particles are a priori fixed on an artificial generative spiral : $\theta(n)=n\phi$, $R(n)=R_0 exp(nH)$.

(25) It is equivalent to say (Ref. 3) that the appearance order of the two neighbours of the new particle n are n-i and n-j, and that the parastichies numbers are (i, j).

(26) Levitov, L.S. (1991) Europhys. Lett. 14, 533-539.

(27) Note that the same dynamical selection of the bifurcations is obtained. The condition on the new place is now not to be the largest one but the first available one.

(28) The resulting G can be compared with the geometrical prediction for the value of G around which this mode would be observed : $Gd = 1/4 \ln[3] \approx 0.275$.

(29) Snow, M. and Snow, R. (1935) Phil. Trans. Roy. Soc. London Ser. B 225, 63-94

(30) The diference between the plastochrone ratio measured by Richards, and the value of exp(G) for which we observe the decussate mode comes from the fact that, for a quantitative comparition between theory and botany, the exact form of the primordia must be taken into account (see ref 12 and 32).

(31) Schwabe, W. W. (1971) Soc. Exp. Biol. Symp. 25, 301,322.

(32) Maksymowych, R. and Erickson, R.O. (1977) Amer. J. Bot. 64, 33-44.

(33) Erickson, R.O. and Meicenheimer, R.D.(1977) Amer. J. Bot. 64, 981-988.

(34) Charles-Edwards, D.A. Cookshull, K.E. Horridge, J.S. and Thornley (1979) Ann. Bot. 44, 557-566

(35) Which in reality would have been advected on the approximately cylindrical stem.

(36) They also share the possibility of an abnormal apical meristem evolution, called the fasciation process : the apex becomes a long macroscopic line, instead of a staying a microscopic point. See for instance *celosia cristata* (the cockscomb) or some cacti. In this case of course the usual arrengements are not observed, and the growth forms a flattened shoot undergoing successive buckling instabilities.

MULTIPLICATIVE NOISE IN DOMAIN GROWTH: STOCHASTIC

GINZBURG-LANDAU EQUATIONS

A. Hernández-Machado[1], L. Ramírez-Piscina[2], J.M. Sancho[1]

[1] Departament d'Estructura i Constituents de la Matèria,
Universitat de Barcelona, Av. Diagonal 647,
E-08028 Barcelona, Spain.
[2] Departament de Física Aplicada,
Universitat Politècnica de Catalunya,
Jordi Girona Salgado 31, E-08034 Barcelona, Spain.

INTRODUCTION

Ginzburg-Landau equations with multiplicative noise are proposed to study the effects of fluctuations in domain growth. These equations are derived from a coarse-grained methodology. Multiplicative noise gives new contributions to the Cahn-Hilliard linear stability analysis. We also derive numerical algorithms for the computer simulation of these equations. The numerical results corroborate the analytical predictions of the linear analysis. In particular, multiplicative noise introduces a delay in the domain growth dynamics.

A MODEL OF DOMAIN GROWTH

Recent experiments indicate that noise seems to play an important role in dynamic processes like pattern formation and growth. Examples of such effects have been observed in the generation of sidebranching in dendritic growth[1], cells in Rayleigh-Benard convection[2] and William domains in the electrohydrodynamic instability of nematic liquid crystals.[3] Ref. 4 gives an overview of this field. In some of these experiments[1,2], the origin of the noise is not known, and a thermal additive noise asumption gives rise to discrepancies between the theoretical and experimental results[1,2]. In these cases, a new type of modeling of fluctuations may be needed to explain them. In other experiments in liquid crystals, the noise has been deliberatly superimposed to the AC voltage[3]. The results imply a strong effect on the response of the system, like changes in the threshold of the instability points.[3] A modeling of the last situation is given by a Langevin equation with multiplicative noise, for which the noise appears multiplying a function of the relevant variables. Then, the effects of the noise depend on the state of the system and, due to the coupling, they are, in general, more important than those induced by simple additive noise.

Growth Patterns in Physical Sciences and Biology, Edited
by J. M. Garcia-Ruiz *et al.*, Plenum Press, New York, 1993

An internal noise could also appear in a multiplicative way. This is the situation that we consider in this paper. We start by obtaining our model from a mesoscopic derivation using a coarse-grained procedure[5]. In this way, it is easier to give an interpretation of the multiplicative noise equations. In general, the study of these type of equations would have relevance in domain growth in phase separation dynamics,[6] pattern formation,[7] polymers,[8] etc.

Here, we consider a system of two components, like a binary liquid or alloy, which could undergo phase separation.[6] The system is suddenly quenched from a one-phase region inside its coexistence region. Then, the homogeneous phase becomes unstable and domains of the new stable phases start growing. This mechanism is called spinodal decomposition. In a previous paper[9], the deterministic evolution of such a system was studied when a variable dependent diffusion coefficient was taken into account. This assumption has been considered to model deep quenching[10] or to take into account the presence of an external field, like gravity.[7] We find that the assumption of a concentration dependent diffusion coefficient implies multiplicative thermal fluctuations. The field model that we have derived is given by the following Ginzburg-Landau type equation with multiplicative noise :

$$\frac{\partial c(r,\tau)}{\partial \tau} = \nabla M \nabla \frac{\delta F}{\delta c} + \frac{\beta^{-1}}{2} \nabla (\nabla \frac{\delta}{\delta c})M + \nabla^i m \, \xi^i(r,\tau) \tag{1}$$

where $c(r,t)$ is the concentration variable, $F[c]$ is the Ginzburg-Landau free energy functional:

$$F(\{c\}) = \frac{1}{2} \int dr(- \frac{c^2}{2} + \frac{c^4}{4} + \frac{(\nabla c)^2}{2}) \tag{2}$$

and $M(c) = m^2(c)$ is the concentration dependent diffusion coefficient. The noise is a d-dimensional vector with a correlation:

$$<\xi^i(r,\tau) \, \xi^j(r',\tau')> = 2\beta^{-1}\delta_{i,j} \, \delta(r - r')\delta(\tau - \tau') \tag{3}$$

β^{-1} is the intensity of the gaussian white noise. A common assumption regarding the dependence of M on the concentration has been obtained by phenomenological arguments.[11] That is:

$$M(c) = 1 - ac^2 \tag{4}$$

where a is a parameter related to temperature. For $a=0$ we obtain the usual model B of phase separation dynamics with additive noise.[6] For $a \neq 0$, apart from the multiplicative term, we find a spurious term, the second term on the right hand side (r.h.s.) of Eq.(1), of stochastic origin. This spurious term ensures the evolution of the system to the correct equilibrium solution for long times. We show that both terms of stochastic origin give new relevant contributions even in a linear stability analysis, that is to the standard Cahn-Hilliard theory. However, nonlinear effects of such models are difficult to study analytically. In order to get some insight into the main effects of the multiplicative noise, we present an algorithm for the numerical integration of general equations of the type of Eqs.(1-3). The numerical results indicate that the multiplicative noise induces a delay in the short time behavior of the domain growth dynamics, in accordance with the linear analysis.

A COARSE-GRAINED DERIVATION

In the standard coarse-grained procedure[5], one divides the lattice into regular cells of volume Δx^d containing N sites and defines the concentration of the binary mixture at the cell α, c_α, by:

$$c_\alpha = \frac{1}{N} \sum_{k \in \alpha} \sigma_k \qquad (5)$$

where $\sigma_k = 1, -1$ indicates a lattice site occupied by a particle A and B, respectively. Then, it is assumed that a markovian master equation is obeyed by the probability $P(\{c\}, t)$ of the configuration of cells, $\{c\} = \{c_1, c_2, \dots\}$, :

$$\partial_t P(\{c\}, t) = \sum_{\alpha i} \sum_{\in} (W(\{c\}^{\alpha i} \to \{c\}) P(\{c\}^{\alpha i}, t) - W(\{c\} \to \{c\}^{\alpha i}) P(\{c\}, t) \qquad (6)$$

where the indexes α and i numerate the cells and their nearest neighbors in the positive direction, respectively. $W(\{c\}^{\alpha i} \to \{c\})$ is the transition probability between the initial configuration $\{c\}^{\alpha i} = \{c_1, c_2, \dots, c_\alpha - \in, c_{\alpha+i} + \in, \dots\}$ and the final one $\{c\} = \{c_1, c_2, \dots, c_\alpha, c_{\alpha+i}, \dots\}$. $\in$ is the concentration interchanged in an elementary step of the evolution.

We consider situations for which the system evolves to an equilibrium state and the steady state distribution, $P_{st}(\{c\})$, is proportional to the Boltzmann factor. Then, we write for the transition probabilities, W:

$$W(\{c\}^{\alpha i} \to \{c\}) = M(\{c\}^{\alpha i}, \{c\}) e^{\beta \frac{\Delta F}{2}} \qquad (7)$$

where $F(\{c\})$ is a coarse-grained free energy. The detailed balance condition is fulfilled provided M is symmetric by interchange of the initial and final states. In the usual derivation of the field model, with constant diffusion coefficient and additive noise, no dependence of M on the configurations is considered, and it is assumed that $M(\{c\}^{\alpha i}, \{c\}) = P(\in)$, where $P(\in)$ is a sharp function around $\in = 0$. The generalization that we present here will give us a model with a variable dependent diffusion coefficient and a multiplicative noise. By assuming that $\in$ is a small quantity, we can expand the different terms of the r.h.s. of Eq.(6) in power series of $\in$ and we get to the lowest order:

$$\frac{\partial P}{\partial t} = - \Delta x^2 \Gamma \frac{\partial}{\partial c_\alpha} (\nabla_L^i)_{\alpha\beta} M_{\beta i} (\nabla_R^i)_{\beta\sigma} \left\{ \frac{\partial F}{\partial c_\sigma} + \beta^{-1} \frac{\partial}{\partial c_\sigma} \right\} P \qquad (8)$$

where $\Gamma = <\in^2> \beta/2$, being $<\in^2>$ the second moment of $P(\in)$, and ∇_L^i and ∇_R^i are the left and right discrete versions of the gradient operators. Summation over indexes is understood.

The Langevin equation, in the Stratonovich interpretation, associated with the Fokker-Planck Equation (8) is given by:

$$\dot{c}_\alpha = \Gamma \Delta x^2 (\nabla_L^i)_{\alpha\beta} M_{\beta i} (\nabla_R^i)_{\beta\sigma} \frac{\partial F}{\partial c_\sigma}$$

$$- \frac{\Gamma}{2} \beta^{-1} \Delta x^2 (\nabla_L^i)_{\alpha\beta} (\nabla_R^i)_{\beta\sigma} \frac{\partial M_{\beta i}}{\partial c_\sigma} + (\nabla_L^i)_{\alpha\beta} m_{\beta i} \xi_\beta^i(t) \qquad (9)$$

where $m_{\alpha i}^2 = M_{\alpha i}$, and $\xi_\beta^i(t)$ is a gaussian white noise of zero mean and correlation

$$\langle \xi_\alpha^i(t) \xi_\beta^j(t') \rangle = 2\Delta x^2 \Gamma \beta^{-1} \delta_{ij} \delta_{\alpha\beta} \delta(t - t') \qquad (10)$$

The way to prove that the Langevin Eq.(9) corresponds to the Fokker-Planck Eq.(8) is to derive the latter from the former. This can be done, by means of Novikov theorem. See Ref.12 for details of such derivation.

At this point, Eqs.(8-9) are formal and general equations in which the expressions of $M_{\alpha i}(\{c\})$, or equivalently $m_{\alpha i}(\{c\})$, need to be specified for each particular model. Furthermore, these equations are given in terms of the cell variables and we are also interested in finding the corresponding equations in the continuous spatial limit. Then, we need to make some assumptions on the form of $M_{\alpha i}(\{c\})$. In general, $M_{\alpha i}(\{c\})$ is a function that depends on the concentration values of all the cells of a given configuration. In the continuous spatial limit, this gives rise to a functional expression of $M[c]$. In order to obtain a local mobility function $M(c)$ like the one considered in the macroscopic model[10-11], Eq.(4), we take into account that the transition probabilities, Eq.(7), only involve interchanges of matter between nearest neighbour cells α, $\alpha+i$ at each elementary step. Then, we restrict ourselves to functions $M_{\alpha i}(\{c\})$ that only depend on the concentration values of the cells α, $\alpha+i$ and on a limited number, **n**, of cells in the vicinity of α and $\alpha+i$. In terms of the function $m_{\alpha i}(\{c\})$, we write:

$$m_{\alpha i} = \sum_\beta Q_{\alpha\beta}^i f(c_\beta) \qquad (11)$$

where $f(c_\beta)$ is a function of only one variable c_β and the matrix elements $Q_{\alpha\beta}^i$ are different from zero only when the indexes α and β correspond to the lattice points α, $\alpha+i$ or the **n** lattice points in the vicinity of this couple. A normalization condition on Eq.(11) insures that, in the continuous limit, $m(c) = f(c)$. A characteristic mesoscopic lenght is present in the family of models described by Eq.(11). It gives the size of the region which include all the cells which appear in the definition of $m_{\alpha i}(\{c\})$, Eq.(11), and that are involved in the interchange of matter in an elementary step. Now, we can write the Fokker-Planck Eq.(8) in the continuous limit:

$$\frac{\partial P}{\partial \tau} = - \int dr \frac{\delta}{\delta c(r)} \nabla M \nabla \left[\frac{\delta F}{\delta c(r)} + \beta^{-1} \frac{\delta}{\delta c(r)} \right] P \qquad (12)$$

where the new time scale is

$$\tau = t \Gamma \Delta x^{2+d} \qquad (13)$$

Analogously, Eq. (1) is the continuous Langevin equation corresponding to Eq.(9). One of the advantages of the family of models given by Eq.(11) is that the continuous and discrete versions of the dynamics obtained from them are equivalent term by term by using standard calculus[12]. An explicit and simple example of $m_{\alpha i}(\{c\})$, which correspond to Eq.(3) in the continuous limit is given by:

$$m_{\alpha i}(\{c\}) = \frac{1}{2} \left(f(c_\alpha) + f(c_{\alpha+i}) \right) \tag{14}$$

where $f(c)=(1+ac^2)^{1/2}$. For this example, only a dependence on the concentration of the couple α, $\alpha+i$ is considered and the characteristic mesoscopic lenght is $R=\sqrt{2}\Delta x$.[12]

NEW CONTRIBUTIONS TO THE LINEAR STABILITY ANALYSIS

A simple way to analyse some of the effects of the multiplicative noise is by means of the linear approximation of the equation of motion of the concentration variable. Certainly, this analysis is limited to short times after the quench, but it will give results useful to understand the evolution of the long wavelength instability.

In a linear approach, the equation of motion for the first moment of the Fourier transformed variable $c_k(t)$ takes the form:

$$\langle \dot{c}_k \rangle = - \omega(k)\langle c_k \rangle \tag{15}$$

By studying the behavior of $\omega(k)$ as a function of k, one could obtain which are the modes that grow or decay in the early stages of evolution. The dispersion relation is:[12]

$$\omega(k) = \frac{1}{2}k^2(k^2 - 1 + \frac{4a\beta^{-1}}{R^d}) \tag{16}$$

From this result, one can conclude that, for the early stage of the evolution, those modes with $k < k_c = 1 - 4a\beta^{-1}/R^d$ are unstable and grow with time. In contrast, the modes with $k > k_c$ relax. For the case $a = 0$ (the case studied until now in the literature), $k_c = 1$. Hence, the presence of the multiplicative noise reduces the domain of the unstable modes in the k-space. This implies a delay in the domain growth dynamics at an early stage. This is an explicit prediction of our theory which is confirmed by computer simulation, as we will obtain below.

THE ALGORITHM

Although computer simulations of field equations with additive noise are standard, by using, for example, first order Euler algorithms, there is no systematic methods to deal with multiplicative noise. The algorithm for the one-variable simulation[13] can be easily generalized to multivariable systems with non-conserved order parameters. However, the case of conserved order parameters could not be easily implemented following the standard procedure. This is mainly due to the fact that, in this last case, the standard algorithm involves not only the simulation of simple gaussian processes but also the appearance of nongaussian processes that could not be simulated in an exact way. To solve this problem, we will take a different point of view. It is well known that the presence of a multiplicative noise makes necessary to choose a prescription in order to interpret the stochastic integrals that appear in the formal integration of the stochastic differential equations.[14] Here, our attitude is to use, in the derivation of the algorithm, the prescription of the stochastic integrals that was employed in the formulation of the model. In this way, and in the corresponding approximation, we will get a closed algorithm in terms of only gaussian processes.

We write a general Langevin field equation with multiplicative noise in the following discrete form:

$$\dot{\psi}_\mu(t) = v_\mu(\{\psi\}) + g_{\mu\alpha}(\{\psi\})\xi_\alpha(t) \tag{17}$$

where we have used now the notation $\psi_\mu(t) = c(\vec{r}_\mu,t)$. Eq.(17) is a set of coupled stochastic differential equations for the variables $\psi_\mu(t)$. The noise $\xi_\alpha(t)$ has the gaussian white-noise correlation

$$\langle\xi_\alpha(t)\xi_\beta(t')\rangle = 2D\,\delta_{\alpha\beta}\,\delta(t-t') \tag{18}$$

where the intensity of the noise is $D = \beta^{-1}\Delta x^{-d}$. $g_{\mu\nu}(\{\psi\})$ is the multiplicative function that couples the variable to the noise. Eq.(17) is defined in a d-dimensional lattice of total volume V and cubic cells of volume $\Delta V=(\Delta x)^d$. This scheme avoids ultraviolet and infrared divergences as far as the relevant lengths of our system are between Δx and $V^{1/d}$. Then, the problem is the integration of a finite number of coupled multiplicative Langevin equations.

The numerical algorithm is obtained from the formal integration of Eq.(17) during a time step Δ:

$$\psi_\mu(t+\Delta) = \psi_\mu(t) + \int_t^{t+\Delta}\{v_\mu(\psi(t'))+g_{\mu\alpha}(\psi(t'))\xi_\alpha(t')\}\,dt' \tag{19}$$

The Stratonovich prescription stablishes that a stochastic integral, like the one in Eq.(17), should be interpret as[14]

$$\int_t^{t+\Delta}dt'\,g_{\mu\alpha}(\psi(t'))\xi_\alpha(t') = g_{\mu\alpha}\left(\tfrac{1}{2}(\psi(t)+\psi(t+\Delta))\right)\int_t^{t+\Delta}dt'\xi_\alpha(t') + o(\Delta^{3/2}) \tag{20}$$

and now $g_{\mu\alpha}$ is expanded as

$$g_{\mu\alpha}\left(\tfrac{1}{2}(\psi(t) + \psi(t+\Delta))\right) = g_{\mu\alpha}(\psi(t)) + \frac{1}{2}\frac{\partial g_{\mu\alpha}(\psi(t))}{\partial\psi_\beta(t)}X_\beta(t)+o(\Delta) \tag{21}$$

Substituting Eqs.(20-21) into Eq.(19), we get up to first order in Δ:

$$\psi_\mu(t+\Delta) = \psi_\mu(t) + v_\mu(\psi(t))\Delta + g_{\mu\alpha}(\psi(t))X_\alpha(t)$$
$$+ \frac{1}{2}\frac{\partial g_{\mu\alpha}(\psi(t))}{\partial\psi_\nu}g_{\nu\beta}X_\alpha(t)X_\beta(t) + o(\Delta^{3/2}) \tag{22}$$

where

$$X_\alpha(t) = \int_t^{t+\Delta}\xi_\alpha(t')\,dt' \tag{23}$$

$X_\alpha(t)$ is a gaussian process of order $\Delta^{1/2}$ with zero mean and variance $2D\Delta$. It can be simulated easily by using gaussian random numbers η_α of zero mean and variance equal to one:

$$X_\alpha(t) = \sqrt{2D\Delta}\,\eta_\alpha \qquad (24)$$

The algorithm (22) can present some technical difficulties when one trays to make the corresponding computer program so it would be worth to simplify it. The idea is to substitute the expression (22) by other algorithm with the same statistical properties. These properties are contained in the probability density, $P[\psi]$, so this substitution has to maintain the same dynamics for $P[\psi]$. It can be obtained [15] that the equation of $P[\psi]$ is also preserved if Eq.(22) is substituted by the more simple algorithm:

$$\psi_\mu(t+\Delta) = \psi_\mu(t) + \left(v_\mu + D\frac{\partial g_{\mu\alpha}}{\partial \psi_\nu}g_{\nu\alpha}\right)\Delta + g_{\mu\alpha}(\psi(t))X_\alpha(t) \qquad (25)$$

where we have replaced the last term of Eq.(22) by its mean value. Then, both algorithms are stochastically equivalent to first order in Δ. We call Eq.(25) the Minimum Algorithm for Multiplicative Langevin Equations (MAMLE). As a general rule, we suggest to use the intermediate algorithm (22), but in some circumstances the simplified version (25) would be more appropriate saving valuable computing time.

APPLICATION TO DOMAIN GROWTH

Now, we aply the MAMLE to the model given by Eqs.(9,10,14). The result is:

$$c_\mu(\tau+\Delta) = c_\mu(\tau) + \left(-\frac{1}{2}(\nabla_L^i)_{\mu\nu}\,m_{\nu i}^2\,(\nabla_R^i)_{\nu\sigma}(c-c^3+(\nabla)^2 c)_\sigma \right.$$

$$\left. -2\beta^{-1}(\nabla_L^i)_{\mu\nu}m_{\nu i}(\nabla_R^i)_{\nu\sigma}f'(c_\sigma)\right)\Delta + (\nabla_L^i)_{\alpha\beta}\,m_\beta\,X_\beta^i(\tau) \qquad (26)$$

A simulation of this system with different values of the parameters has been performed in order to confirm the effects of the multiplicative noise predicted by the linear theory. We will also find how the intermediate times far from the linear regime become affected. We have taken different values of the intensity of the noise below the critical point, and two values of the constant a. We start from an homogeneous initial state $c = 0$, and the system is let to evolve until a time $\tau = 3000$. The lattice spacing used in the simulation has been $\Delta x = 1$, and the size of the system was 120×120. Each data results from the statistical average of ten independent runs.

It can be seen that in the multiplicative noise case the pattern has more diffuse interfaces than in the additive one.[15] In order to have a quantitative characterization of the pattern, we have studied the evolution of the structure function. In Figs.(1.a-b), we show the evolution of the structure function for a fixed time and two values of a. We observe the effects of an increase in the intensity β^{-1}. Fig.(1.a) corresponds to additive noise and it is plotted as a reference. First, the peaks are located at the same position but the heights depend on β^{-1}. Then, the patterns have the same characteristic length but they have more diffuse interfaces by increasing β^{-1}, as one should expect.[16] In Fig.(1.b), for the same values of the parameters than in Fig.(1.a) but with $a = 0.8$, we observe important differences, specially for larger values of β^{-1}. In particular, the position of the peaks depend strongly on β^{-1}. They are located at smaller values of k and have reduced drastically their height. This effect can be understood from our analysis of Eq.(16). The fact that more modes are now stable makes the peak of $S(k,t)$ to grow at smaller values of k.

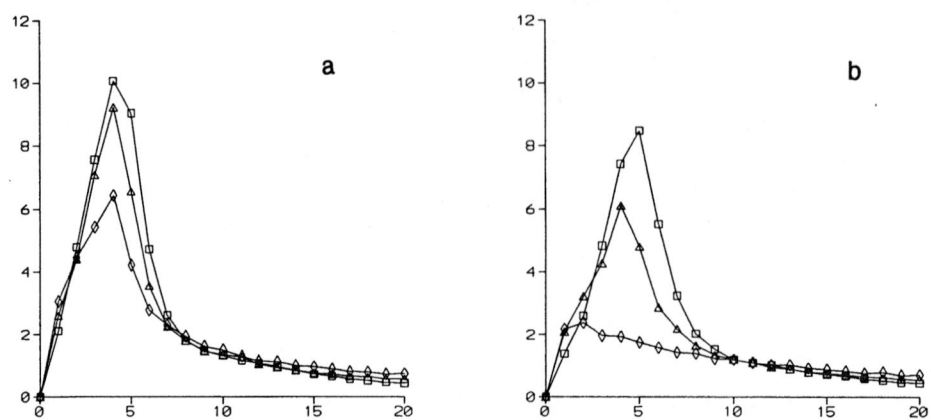

Fig. 1. Structure function corresponding to the results of simulation of Eq.(26) at $\tau=1500$ for a) $a=0$ and b) $a=0.8$. Squares, $\beta^{-1}=0.1$; triangles, $\beta^{-1}=0.2$ and rombous, $\beta^{-1}=0.3$.

The possibility of scaling properties on the late stage and the characterization of this regime by means of the structure function and the time behavior of the charcteristic domain size is in progress.

ACKNOWLEDGEMENTS

We acknowledge financial support of the Direccion General de Investigacion Cientifica y Tecnica (Spain) Pro. No. PB90-0030. A.H.M. and J.M.S. thank NATO for partial support under the Collaborative Research Grant No. 900328.

REFERENCES

1. A. Dougherty, P.D. Kaplan, J.P. Gollub, Phys. Rev. Lett. **58**, 1652 (1987).
2. C.W. Meyer, G. Ahlers, D. Cannell, Phys.Rev. Lett. **59**, 1577 (1987); C. W. Meyer, G. Ahlers and D.S. Cannell, Phys. Rev A **44**, 2514 (1991).
3. T. Kawakubo, A. Yanagita, S. Kabashima, J. Phys. Soc. Japan, **50**, 1451 (1981).
4. Proceedings of the Workshop on "External noise and spatial degrees of freedom". Eds. C.R. Doering, H.R. Brand, R.E. Ecke. J. Stat. Phys. **54**, Nos. 5/6 (1989).
5. J.S. Langer, Annals of Physics 65,53 (1971).
6. J.D. Gunton, M. San Miguel and P.S. Shani, in "Phase Transitions and Critical Phenomena", Vol.8, 267. C. Domb and J.L. Lebowitz, Eds. (Academic Press, 1983).
7. D. Jasnow, in "Far from Equilibrium" Lectures Notes in Physics, Vol. 319. Ed. L. Garrido. Springer-Verlag, (1988).8. M. Doi and S.F. Edwards, "The Theory of Polymer Dynamics", International series of monographs on Physics, Vol.73. Oxford University Press (1989).
9. A. Lacasta, A. Hernández-Machado, J.M. Sancho and R. Toral, Phys. Rev. B (1992).
10. J.S. Langer, M. Bar-on and H.D. Miller, Phys. Rev. A **11**, 1417 (1975).
11. K. Kitahara, Y. Oono andd. Jasnow, Mod. Phys. Lett. B **2**, 765 (1988).
12. L. Ramírez-Piscina, A. Hernández-Machado and J.M. Sancho, Preprint (1991).

13. J. Sancho, M. San Miguel S. Katz and J.D. Gunton, Phys. Rev. A **26**, 1589 (1982)
14. R. Stratonovich, in "Noise in Nonlinear Dynamical Systems", Vol.1, Eds. F. Moss and P.V.E. McClintock. Cambridge University Press (1989).
15. L. Ramírez-Piscina, J.M. Sancho and A. Hernández-Machado, Preprint (1991).
16. T.M. Rogers, K.R. Elder and R.C. Desai, Phys. Rev. B **37**, 9638 (1988).

BURSTING INTERMITTENCY AND MICROWAVE POPCORN: COMMENTS ON

THE "REPORTING OUT" OF NEURON-LIKE FIRING BEHAVIOR

Karen A. Selz and Arnold J. Mandell

Laboratory of Experimental and Constructive Mathematics
Departments of Psychology and Mathematics
Florida Atlantic University
Boca Raton, FL 33431

Patterns of intermittency dominate information transport mechanisms in neural and neuroendocrine systems. The statistical coding and realizations of these patterns, therefore, continue to challenge theoretical and mathematical biology. The intermittent bursting pattern of "pop"-like sounds heard during microelectrode studies of sequences of action potentials in single neurons (intracellularly recorded) and clusters of neurons (extracellularly recorded) when they are auditorily monitored are similar to those made by microwave popcorn as it is being popped. This inspired a study of the potential information transport properties, the "reporting out" of experimental conditions by microwave popcorn as an intermittent system. Salt concentration was of particular interest.

WHY A MICROWAVE POPCORN SIMULATION MODEL FOR NEURONAL ACTIVITY?

Neurophysiologists study the discharge rates and rhythms of single neurons and groups of neurons in living experimental animals by placing micron-diameter, finely drawn electrolyte containing glass electrodes through skull and *among* or *into* nerve cells, and then monitor fluctuations in transmembrane microvoltage, using amplifier-oscilloscope techniques. The major observables are "spikes" indicating the all-or-none reversal of the resting membrane potential of the neuron associated with its "firing." A characteristic pattern of neuronal discharge sequences is bursting intermittency[31,36,38].

A now classic study demonstrated that for a local neuronal network, the time series of an intracellularly recorded single neuron or of a cluster of extracellularly recorded neurons were statistically similar [13]. Except for the problem of a *nonuniform distribution of the phase points* in intermittency dynamics, this observation might lead naturally to a kind of *local neuronal ergodic hypothesis* concerning patterns of activity in microelectrode recordings of neuronal events. In other words we may restate the *redundancy as reliability* hypothesis proposed by neurophysiological theorists in the past[17]. We justify our use of a population simulation for single neuronal events as Gibbs might, as a microcanonical ensemble of copies of single neurons. It is even more realistic to make this ergodic assumption when comparing a single neuron's behavior over time with that of the statistical aggregate of its local network, due to the dense connectivity among them. The intracellularly recorded single cell can be perceived as a living antenna, nonlinearly monitoring the activity of its brethren.

The signals generated by large sudden shifts in membrane resistance resulting in changes in the direction and the amount of dominant electrolyte transport, the cell's "action potential," is often transformed into an auditory output, as well as a running oscilloscope record. This allows the physiologist to monitor his preparations. The changes in the patterns of pitch and rhythm of the action potentials are used by experienced "single unit" electrophysiologist to predict the location of the electrode relative to the cell (inside or out, near or far), the type of cell, and its "health," since cells tend to deteriorate when studied intracellularly. The information content in these sequences of sounds can also include the "reporting out by the cell about the presence of a variety of infused neurotransmitters, drugs, ions and local pH levels, as biological changes induce characteristic patterns in the interspike intervals, isi's[40,2].

The neuronal sound sequences are composed of "pops" signifying the neuron's action potentials, with varying interpop intervals, ipi's (i.e. an observable result of the cell's isi's). Runs of the smallest length ipi's ("shorts") are near-periodic, bounded from below by the absolute refractory period—the minimal time required to repolarize the membrane sufficiently as to be responsive to excitation again. For this reason short ipi's bunch up into temporal regions of fast regular popping, embedded in and interspersed with the variable "longs" of the neuron's "spoken" history. The series of longs and shorts can be analyzed from the standpoint of run-lengths in a random series with fair or unfair "coin flipping" tasks. We recall the Erdos-Renyi "new law of large numbers" which, intuitively stated, suggests that in random variable, binary tasks, the largest run of consecutively similar outcomes grows as $c \, log_{1/p} \, n$ (where $p \equiv$ fairness of the coin and c is a constant). For example, the base of the logarithm is 2 if $p = .5$ (i.e. a fair coin). We assume $c = 1$.

While an intermittent series of ipi's can be modeled using a series of random "coin flips" with variable outcome probabilities, another model of the same kind of ipi behavior involves the deterministic intermittency of nonlinear dynamical systems in parameter regions such as those engendering inverse saddle node bifurcations[25], saddle node bifurcations, tangent bifurcations in one dimensional maps [33] and pitchfork bifurcations, as well as homoclinic tangencies in dimension two[27], behavior characteristic of orbits at the "fractal basin boundaries" of complex attractors[8] , or around "chaotic band mergings" in low dimensional discrete dynamical systems[34]. "Bursts" of nearly regular "shorts" – trapped by either minimal cellular mechanisms or limits on the techniques for the temporal resolution of microevents – interspersed with irregularly longer ipi's constitute the neurophysiological temporal intermittency referred to here.

Listening to the patterns of sound made by popping microwave popcorn and using the popcorn ipi's as observables, we were struck by the similarities of this system to neuronal systems, particularly in the use of a variety of patterns of intermittency to "report out" the current state of the system. The microwave popcorn popping $epoch$ manifests three distinct phases: (1) $kindling$--the increasing rate of popping up to transiently stable state; (2) the transiently stable state of $intermittent \ popping$, and (3) $exhaustion$--the decrease in the finite number of available kernels, including residue composed of those kernels with the resistance to popping at even high internal temperatures. Physical measures made on heat and microwave processes in popping corn demonstrate that extra-kernel conditions such as oil content, salt concentrations, breed and harvest pattern of the corn alter the pattern of discontinuities in the temperature curve, expansion volume, and residue number[21,12,20].

That finer levels of discrimination within the intermittency regime might be possible was suggested by the evidence that there were low order correlations in our data, showing that decreases in maximal entropy, decreases in the normalized difference between the maximal and minimum entropy, increases in normalized maximal run length, decreases in the kindling exponent, and decreases in residues (number of unpopped kernels) were associated with the increases in the amount of salt in the popcorn preparation. Salt content varied across brands and across products within brands of the commercially available microwave popcorns used.

Due to the assumption of uniform hyperbolicity in the use of the quasi-isomorphisms of the maximal and minimal entropies[5] , and the yet unresolved problems in the nonuniform case[15], which is more representative of the non-constant curvatures of biological manifolds, we sought an equivalent measure. Furthermore, working in the realm of the finite, nonasymptotic orbit lengths of applied mathematics, we sought an equivalence measure which did not presuppose large finite or infinite series (e.g. FFT via power spectra analyses). We conjectured that using the variable logarithmic base $1/p$ of the Erdos-Renyi new law of large numbers as an order parameter, that the maximal run length and $c \log_{1/p} n$ would be *ergodically equivalent*. We further supposed that for the popcorn preparations (or for members of families of deterministic maps with intermittency), the maximum run lengths normalized for n are *entropically equivalent, quasi-isomorphic* to those of the random coin tossing system with a related *specifiable* unfair p.

Some aspects of the cooperative physics of the microwave popcorn system can be analogized to those of the neuron embedded in a network. The variety of short-range connectivities created by axosomatic, axodendridic, and autorecurrent axons are simulated by the variety of *packing arrangements* among the pre-popped kernels, which release latent heat locally in the form of steam at the time of their popping, leading naturally to spatio-temporal bunching in kernel pops. In the same way that very small rocks increase the connectivities between large ones in piles, we predicted that salt would fill gaps (relative thermal sinks) between popcorn kernels, and when compared with the nonsalt containing condition, salted preparations would demonstrate decreases in kindling times (i.e. increases in kindling slopes) and/or changes in the measures of irregularity in the *ipi*'s. The diffusive heat gradient "field" of the microwave oven is analogous to the "nonspecific" activating systems of brain (the reticular formation and the biogenic amine neurons) which serve as the field of long-range connectivities, globally altering the excitability of the neurons.

ENTROPIC QUASI-ISOMORPHISM AND THE INFORMATION CONTENT OF INTERMITTENCY

On theoretical and experimental grounds, using studies of the dynamics of heart, neuroendocrine systems, single neurons, electroencephalic recordings, and drug-perturbed animal behavior, we have concluded that the complexity required in biological information transport, as entropy generated per discrete unit time with respect to the partition, ξ, and the system generator of times, $f^{n}(\tau)$, characteristically occupied the interval between zero and one in healthy systems $[H_0, H_{max}]$[22,29,30]. A fixed point or limit cycle exists in an entropy minimum, bounded from below by H_{min}, and carries no potential for entropy reduction through encoding. A system at maximum entropy, bounded from above by H_{max}, though having maximal potential for information transport, can be said to be without a sufficient "in place" code, in the form of systematically reduced entropy, to receive messages about the "readiness" for the reception and transmission of any functionally meaningful message. Entropy of this sort is called arousal and/or attention in neurobiology.

Somewhere between the biological entropy maximum and minimum there is room for adaptive regulation[24] via changes in the nonuniformity of the system's expansion, in the direction of both increases and decreases in H.

In an abstract ergodic dynamical system, the actions of a transformation are those that preserve the structure of the probability space (i.e. while transformations may alter specific orbits' characteristics, the probability characteristics are unchanged through the transformation). The system measure, μ, is therefore invariant over time evolution. However, without the uniform expanding and folding and maximum mixing of horseshoe-like processes (e.g. the idealized uniform maximally entropic Axiom-A or Anosov processes), the ergodic, measure theoretic description of these motions, *without uniformity in their sensitivity to initial conditions*, become more difficult to characterize.

Hadamard[10], Hedlund[11], and others showed that *geodesic* (i.e. minimal length) flows on surfaces of constant and non-constant negative curvature were ergodic. These are analogous to expansive flows in real space. In other words, the probability spaces of fully covered surfaces and surfaces not fully covered by the behavior of their occupying systems are equally preserved under transformation.

The non-uniformity of our system's expansion is, then, no longer a problem. Uniform and maximal expansion in the tangent space of the unstable manifold, and contraction in that of the stable manifold (i.e. so called "resonance"), generates a point set in the Poincare section with maximum entropy. Then we may say that in the modeling of biological systems, when the vector field of the unstable manifold is *maximally* and *uniformly* expanding (i.e. positive non-zero eigenvalue), and that of the stable manifold is contracting (i.e. non-positive eigenvalue), our system is not healthy. In fact, it may lack an internally consistent language with which to interpret itself and its environment.

Though the existence of bounds on the entropies of nonuniform hyperbolic systems (as geodesic flows on manifolds of nonpositive curvature) have been proven[26] , little in the way of knowledge about the generic internal structure of such systems is available. Differential systems, in contrast to the derivative form, are those represented by equations in which the δt term is treated as a dependent variable, existing on the right hand side of the equation. Theorems of this form involve conditions on what Dennis Sullivan calls "*bounded variation*" of the logarithmic derivative of the first derivative (i.e. f''/f'). This "normalization" of the second derivative by the first implies that the systems *stability* is of interest.

With respect to manifolds of negative curvature, early theorems of Kuiper (personal communication with second author) demonstrate that such manifolds can be extended but not continuously deformed, suggesting the existence of some universal structural features. Modern dynamical systems theorists have referred to these topological constraints on metric spaces as *rigidity*. The very recent (unpublished) work by Gromov (1989-91) is also moving in the direction of universalities in the hyperbolic "parameters" of path metric spaces in what he calls near- or δ-hyperbolic systems.

We here treat our nonuniformly hyperbolic popcorn systems as measurable on experimental partitions (i.e. a finite set) of the one dimensional space of the actions of $f^n(\tau)$. Measures of maximal and minimal entropy and their difference (nonuniformity) serve as group theoretic quasi-isomorphisms with reflexive, symmetric, and transitive properties preserved[28]. Quasi-isomorphic equivalence invokes these properties on equal entropy spaces (except for sets of measure zero). It is here that Ornstein's 1989 result plays an important role in the spaces of dynamical systems. Take, for example, two coin-flipping processes with variously unfair coins but with equal entropies. These processes are spatially isomorphic. It is this equivalence relation of entropy quasi-isomorphism which may serve as the coding scheme for the "reporting out" of internal conditions by neural mechanisms and microwave popcorn alike. We explore the possibility of using this $log_{1/p} n$ equivalence similarly.

In the case of the log base equivalence relationships, the slope of the log of the growth rate of the longest run of "shorts" (i.e. the longest burst) over n is compared with the slopes of bounding p values for similar $log_{1/p} n$ slopes.

We define the *topological entropy* (H_{max}) of a global system popping function, f^n (the manifold of the *ipi*'s, where L, $f^n:L \longrightarrow L$), as a real number, H_{max}, which indicates the number of different orbits f^n is generating in the real space L, over events, n. This one dimensional space can be analogized to the two dimensional idealized case of Axiom-A diffeomorphic systems with a proven generating partition, ξ , where the orbit can be traced by a symbol sequence, S_i, a one-sided Markov subshift of finite type, $\sigma:S_i \longrightarrow S_{i+1}$ [0,1]. That is, the orbits of the system can be nicely mapped into phase space.

An incidence matrix is created through the transformation of the transition matrix constructed on ξ. We rely on the Frobenius-Perron theorem [9] which guarantees the existence a largest eigenvalue, λ_m, of this non-negative, [0,1] incidence matrix, such that the asymptotic growth rate of $\text{Tr}(M)^n$ as $n \rightarrow \infty$, serves as an easily implemented computational method for $\log(\lambda_m) \leq H_{max}$. This is the case because the Frobenius-Perron theorem states that any entirely non-negative matrix has a distinct largest eigenvalue, λ_m, which is equal to the log of the asymptotic growth rate of the major diagonal of the incidence matrix, and $\lambda_m \leq H_{max}$. This works in part because expanding (hyperbolic) dynamical systems, in which unstable periodic points are dense, collect periodic orbits which are equidistributed with respect the measure, μ,[4] . We note that the use of symbolic dynamics on a generating partition coarse grains the orbit and n bounds the "time observation" such that H_{max} is a finite real number.

The lower bound on H, $H_{min}(f^n)$ represents the *metric entropy* of the system, and reflects the statistical structure of the time (event)-averaged distribution of most points with respect to the orbital complexity of the system, H_{max}, on the partition ξ. The computational result for H_{min} from the densities on ξ is a real number. It is calculated as $-(\Sigma(p_i\log(p_i)))$ on the asymptotic composition of the Markov matrix (transformed from the transition matrix on ξ), $\underline{M}^n$, which converges geometrically, as n $\longrightarrow \infty$ [16], to the eigenvalues of the stochastic matrix.

H_{max} and H_{min}, and other invariants of the dynamical system, $f^n:L\longrightarrow$ L, are characteristically computed as the supremum (i.e. highest value) taken over all measurable partitions,

$\xi(N)$, or are dependent on the condition that ξ is a generating partition on the space L. Essentially, the latter requires that the partition ξ is such that no more than one point is contained in any component of the partition. In the entropy formalism, it is this condition that is required for H_{max} and H_{min} to be invariants of both ξ and f^n. In effect, this defines bounds on the arbitrariness with which ξ can be manipulated to generate unique entropies. It is noted, however, that in the context of a generating partition, the variability of the partition component sizes (i.e. bin sizes), describes the stretching and folding of the system's behavioral space.

In applied work, a non-generating, but still appropriate partition, ξ, can be derived from intrinsic and observable properties of the real system (e.g. available measurement resolution and/or characteristic behaviors of the system under study). This "meaningful" partition[30] can then be used in an empirical way, for example, it can be held constant in comparisons among treatment conditions. In this way, ξ can be nonarbitrarily *fixed* without being generating. H_{max} and H_{min} in this context will be used with an implicit reference to the *experimentally determined* ξ rather than with respect to some limiting value of the partition in abstract ergodic theory.

With respect to information transport, if we assume two recursivefunctions, f_1^n and f_2^n, that have finitely (i.e. experimentally) determined partitions which are held constant, and if $H(f_1^n) = H(f_2^n)$, then f_1^n and f_2^n are metrically isomorphic (modified from Ornstein). Information transport in the brain functions largely on global behaviors such as "avoid" and "approach." These nomothetic brain states are, in fact, experimentally discriminable through a measure of complexity on the electrophysiological signals, from electrodes in the hippocampus and other limbic sites in cats, for instance, and are conjectured to be encoded in metric equivalencies of ergodic redundancies in the behavior of local neural networks. Bowen[5] has commented that since the dynamical (i.e. topological and metric) entropies were "crude invariants" it likely that they could serve some applied purposes.

More generally, it is well known that experienced brain scientists can discriminate definable states of consciousness (e.g. alert, relaxed, drowsy, dreaming, and non-dreaming sleep) either from the qualitative aspects of time series of electrovoltage variations in skull recordings (EEG), or from the spectral dimension of the power spectra of the same series, or from the capacity dimension of the suitably embedded signal. Thus, electrical signals from aggregate neuronal activities may "report out" on a finite set of global neurophysiological and behavioral states.

It is the log base equivalence classes, along with these metric isomorphisms in H_{max}, H_{min}, and their difference (as an index of nonuniformity) ΔH, that we regard as the dependent variables of interest from the neuronal-popcorn packing network. The range of local connectivities, rather than being mediated by axons and dendrites, is governed by influences on the local volume-densities (i.e. kernel packing).

The consistency of the "reported out" representation of a state up to some limiting equivalence relations (the isomorphism problem) is required for the intermittent processes to serve as a "look-up table." The representations can be treated as isomorphic up to the bound of allowable dissimilarity, beyond which the signals may no longer be treated as decodable information of dynamic transport. Our $c \log_{1/p} n$ equivalences serve just such a purpose.

As noted above, uniformly hyperbolic systems can be partitioned and studied using symbolic dynamics. Also, Ornstein proved that any two Bernoulli automorphisms with the same entropy are metrically isomorphic. The classic, uniformly random Bernoulli systems, mapped onto themselves, provide us with a distribution of entropies which we can divide into a meaningful partition of fractional occurrences. We extend these classical results to conditions of experimentally-determined partitions and nonuniform expansion in the intermittency dynamics of neuron-like microwave popcorn discharge patterns. Here we divide our time series elements into the categories of bursting and non-bursting elements, based on their *ipi* values relative to our criterion. The results reported are based on the definition of a bursting, "short" element as any element with a value 10% or less of the maximum *ipi* of the series.

In considering such transformed signals, we might recall the Morse code alphabet, in which most codes for the more probable letters are simpler than those for letters that are rare. This serves as intuitive model of how a relational statistical property such as $H_{min} < H_i < H_{max}$ can provide a continuous representation of even discrete, global dynamical codes. We then compare our entropy results with the $c \log_{1/p} n$ equivalence.

Common forced dissipative biological oscillators, such as the neural membranes and cardiac models, using the reduced Hodgkin-Huxley-like Cartwright-Littlewood-Levinson-Levy-Van-der-Pol system, operate in the biologically relevant parametric regions of homoclinic tangency and intersection[19,23] (i.e. where stable and unstable manifolds touch or cross, resulting in intermittent or chaotic behavior, respectively). This parametric preference in biology is manifested by the nonuniformity of the inter-event intervals, most notably irregular bursting discharges.

Using the real numbers H_{max}, H_{min}, and $H\Delta$ as invariants of the expanding dynamical system, f $^n(\tau)$, we want controls upon both the expansion rate and the uniformity of the expansion in order to model a popcorn system with entropic isomorphisms. We remind ourselves of Kolmogorov's 1959 proof that all coin-tossing (Bernoulli shift) processes are not isomorphic. He accomplished this proof by demonstrating that the H_{min} (i.e. $-(\Sigma(p_i log(p_i)))$) of a three partition random system (a three-sided coin) was higher than that of a comparable two partition system (a more standard two-sided coin). Adler and Weiss[1] and Ornstein[28] proved that sequences generated by coins of the same unfairness were entropically isomorphic *up to a renormalization of time*. Because order (*n*) is a dynamic variable in our non-asymptotic series, sequences of unequal lengths cannot be considered as entropically isomorphic in the above consideration (e.g. f^n and $f^{n/2}$ are not isomorphic). A renormalization (e.g. truncation) of the number of events a series (i.e. time) may be necessary to reinstate the equivalence relationship.

Some co-dimensional control of rate and its variation is required in the modeling of isomorphisms of f^n.

We model the *ipi* $\equiv \tau$ intermittency phenomenologically using a piece-wise, variably convex endomorphism, on which the convexity of either of its two component manifolds is controlled by the *r* parameter. The map is limited to the unit interval ($\tau \in [0,1]$), with a repelling fixed point at $\tau = 0$. This is an overall expanding map, (i.e. $f' > 1$), with the measure $\mu(f)$ being asymptotically stable. In addition, it is a smooth function, at least twice differentiable, C^2, with a nonzero nonlinear parameter, ($r \neq 0$). The endomorphism {1} is also nonuniformly expanding (i.e. f'' is non-constant across all $\tau \in [0,1]$). Renyi[35] was able to prove the existence of a unique invariant measure, μ, for such convex piece-wise maps.

$$f^n(\tau) = \tau_{n+1} = (1 + r)\tau_n + (\beta - r)\tau_n^2 \ (mod \ 1 \ ) \qquad \{1\}$$

Endomorphism {1} is in codimension two, such that β and r jointly regulate f'. At the $\tau = 0$ repelling fixed point, $f' = 2\tau(b - r) + r \ 1)$ and $f'' = 2(b - r)$.

We note in passing that intermittent "bursting" (i.e. "runs" of "shorts" among "longs" with frequency and length inversely related in both) of popcorn and neurons need not be modeled by a deterministic map such as {1}. Since coin-flipping randomness (Bernoulliness) is a *factor* of any unpredictable orbit[39], entropic isomorphism with r,β-dependent changes in {1} can be modeled by random variable processes using a value for the parameter p, different from the fair coin $p = 0.5$. For example, in the case of the *Quickbasic 4.5* pseudorandom number generator (Heinrich Niederhausen, personal communication with authors), $X_{i+1} = \alpha X_i + k \ (B)$, partition, p, $\equiv X_i \leq .5 = 0$; $X_i > .5 = 1$, in which k and B are relative prime, $\alpha = 16,598,013$, $k = 2293 \circ 5591$, and $B = 2^{24}$, clearly as parametrically dependent, deterministic a process as {1}. With respect to the entropies, an H_i of pseudorandomness can be found isomorphic with any deterministic $H_i(r,\beta)$.

EXPERIMENTAL METHODS

Twenty examples of two commercially available brands of standard bagged microwave popcorn preparations were used in these studies: Orville Redenbacher and Pop-Secret. Five "types" of packaged popcorn were used. They varied in groups of four or six with respect to their levels of salt in milligrams: 0, 125-160, and 260-380; and the presence or absence butter; 1, 4, 6, or 7 grams of fat (as butter and/or oil). All packages had net weight of 3.5 oz. each.

A Sharp RK-6C, 120 VA.C, 0.95KW, 60Hz, single phase microwave oven which was surge protected with a GSC-Powermate, with maximum output 115 VAC, 15A, a1875 W was used in these studies. All preparations were run in the microwave oven for four minutes (± error of 5 sec. maximum difference in timers) at the highest power setting, and without preheating. Interexperimental intervals of 30 minutes were employed to allow cooling of the microwave chamber.

Inter-pop intervals were recorded through a flat, directional microphone placed in a fixed position, 0.5 cm. from the microwave door, by a Sony TCS-40 cassette recorder using Maxell XLII, high bias, high resonance proof magnetic tape. They were then played through a sound transducer into a Grass Model 70 polygraph with 7P511 electroencephalographic amplifiers which transformed the sounds of popcorn discharges onto an oscillograph, via an event recorder with minimal *ipi* resolution of 0.05 seconds. Pops were seen as negative pen deflections on this paper record.

The kindling component of the *ipi's* was composed of those events (*ipi's*) that occurred up to the first incidence of four consecutive *ipi's* each $\leq$ 0.5 seconds. The slope of the log-log plot of these *ipi's* over time is what we called the *kindling exponent*. The series of *ipi's* that followed, up to the last incident of four consecutive *ipi's* each $\leq$ 0.5 seconds is the *regime of intermittency* that was studied in this paper. The final *ipi's*, following the last burst and constituting the *phase of exhaustion* were not subjected to analysis.

Although a pre-experimental count of the available popping corn kernels was not possible due to the potential for disturbing the microwave bag, a post-popping epoch count of the *residue*, the number of unpopped kernels, was recorded. The levels of salt in each preparation and the presence or absence butter were recorded for each experiment as noted above.

Seven real numbered dependent variables were available from each experiment: the slope of the growth rate of the log of the longest burst length, the kindling exponent, H_{max}, H_{min}, ΔH, the longest run length of consecutive *ipi's* each < 0.5 seconds and the residue number. The conditions of the microwave popcorn preparation and the brand name were the source of the independent variables.

In addition to the microwave popcorn experiments, six simulations of system {1} generated time series of *ipi's* in which the slope of the log of the growth rate of the longest burst, H_{max}, H_{min}, ΔH were determined.

TWO CO-VARYING QUASI-ISOMORPHISMS

H_{max} as a global invariant of an expanding dynamical system characterized the total exponential complexity of the orbit structure. We have calculated here after the Artin and Mazur[3] zeta function which is a power series, $\zeta(t) = exp\Sigma(1/m) N_{max} t^m$ which counts m fixed points, N_m of f^m as the incidence matrix t is exponentiated (theoretically m—>∞ times). N_m is seen to be countably infinite. H_{max} correlates well with the growth rate of a variety of stochastic behaviors[38], and we conjecture here that maximal run length (of zero-one random process--though easily extendable to higher dimensions) normalized by the length of the series of observations (i.e. *n*) is one of these.

Numerically, we know that whereas $\zeta(t)$ converges as m —>∞, as the Erdos-Renyi law implies, the maximal run length as $c \ log_{1/p} n$ does not. However, applied work on neurons and microwave popcorn involve finite *n*'s such that, practically, the relationship between the mixing properties of H_{max} and the loss of normalized run length is intuitively clear. Discriminating random and deterministic systems, such as the one represented by the endomorphism {1}, appears not possible due to the intrinsic limitations in sample size: a low bounded number of kernels in a microwave popcorn preparation; neuronal spike trains with a relatively short time stationarity. We did, however, find an inverse relationship between H_{max} and the longest run length in the intermittent regime of the popcorn recordings.

THE KINDLING EXPONENT, RESIDUE NUMBER AND SALT CONCENTRATION

Though "far from equilibrium," and therefore not clearly relevant to the theory, the fluctuation-dissipation theorem relating the characteristic times fluctuation to those of relaxation of "macrovariables"[18] suggested the possibility that the negative power-law kindling slope (relaxation to a transient, metastable equilibrium) could be related to the fluctuations, as in an equilibrium or near equilibrium state. We, however, failed to find other than a small positive correlation between the kindling exponent and H_{max}. As one might expect, the theorem breaks down in intermittent systems. Generally, those systems like intermittent ones, without a computable correlation function (i.e. a characteristic time), are not subject to either the perturbation-relaxation or fluctuation-spectral approaches to system characterization.

The same issue of the *nonuniformity of hyperbolic systems* that impairs the straight forward application of entropic isomorphism also weakens and/or abolishs most of the other measure techniques. Area measures, like that of Lesbegue and Haar, of course, cannot be used on a patchy or curdled terrain.

It is for this reason that we are working on theorems relevant to $log_{1/p}$ equivalence which, rather than suffering from the nonuniformity, supply a metric relative to a random process for its measurement. We conjecture that it must supply the parameter-sensitive continuity in a measure of complexity under circumstances unmet by the uniform hyperbolicity requirement for the metric isomorphism of the entropies.

STUDIES OF SYSTEM {1} H(MAX) $\cong$ LOG(1/p)n

Studies of {1} at r parameter values ranging from 0.02 to .99 ($\tau(0) = 0.1$) demonstrated a nonsignificant correlation between changes in the normalized longest run length and H_{max}. However, near linear decreases in longest run length with increasing r were observed. For example, in n-normalized run length pairs: (0.05-0.113), (0.1-0.087), (0.3-0.033), (0.5-0.30), (0.8-0.023), a(0.99-0.020).

The latter finding is consistent with the possibility of using the maximum run length equivalence to the r parameter of map {1} as a way of mapping empirical observations of real data onto a generic, discrete, deterministic intermittency system as well as run lengths in binary random variables.

Of course, since the Erdos-Renyi law yields a p for any maximum run length for a given n, we can compute the *informational (Shannon) entropy* via this without the assumption of uniform hyperbolicity in the underlying system[16,37].

ACKNOWLEDGEMENTS

This work is supported by the Office of Naval Research sections on Cognitive and Neurosciences and Systems Biophysics. We extend additional thanks to Alan Nash for the use of a human EEG laboratory, in which the odor of buttered popcorn may confound the work of serious brain scientists for months to come.

REFERENCES

1.   Adler, R.L. & Weiss, B., "Entropy is a complete metric invariant for automorphisms of the forms", *Proc. Nat. Acad.*
      Sci.,   57, 1573-1576 (1967).
2.   Aghajarian, G.K., Foote, W.E. & Sheard, M.H., "Sensative neuronal units in the midbrain raphe", *Science*, 161, 706-709 (1968).
3.   Artin, E. & Mazur, B., "On periodic points", *Ann. Math.*, 81, 82-90 (1965).
4.   Bowen, R., "Periodic points and measures for axiom A diffeomorphisims", *Tran. Am. Math. Soc*, 154, 377-397 (1971).
5.   Bowen, R., "Entropy versus homology for certain diffeomorphisms", *Topology*, 13, 61-67 (1974).
6.   Bowen, R., "A model for   flow", *lect. Notes Math.*, 615, 117-134 (1977).
7.   Cowan, J., "The problem of organismic reliability", in *Cybernetics of the Nervous System*, eds. Weiner, N. & Schade,
      J.P., Elsevier, N.Y. (1965), pp. 9-63.
8.   Ditto, W.L., Rauseo, S., Cawley, R., Grebogi, C., Hsu, G.H., Kostelich, E., Ott, E., Savage, H.T., Segnam, R., Spano,
      M.L., & Yorke, J.A., "Experimental   observation   of crisis-induced intermittency and its critical exponent", *Phys. Rev. Lett.*, 63, 923-926 (1989).
9.   Gantmacher, F.R., *The Theory of Matrices: Vol. I*, Chelsea, N.Y. (1959).
10.   Hadamard, J., "Les surfaces a courbures opposees et leurs lignes geodesiques", *J. Math. Pures et Appl.*, 4, 27-73 (1898).

11. Hedlund, G., "The dynamics of geodesic flows", *Bull. Am. Math. Soc.*, 45, 241-246 (1939).

12. Hoseney, R.C., Zeleznak, K., & Ahdelrahman, A., "Mechanism of popcorn popping", *J. Cereal Sci.*, 1, 43-54 (1983).

13. John, E.R., *National Science Foundation Report*, (1976).

14. Katok, A., "Lyapounov exponents, entropy and periodic orbits for diffeomorphisms", *Publ. Math. Inst. Hautes E'tudes Sci.*, 51, 137-173 (1980).

15. Katok, A., "Nonuniform hyperbolicity and the structure of smooth dynamical systems", *Proc. Int. Cong. Math.*, 2, 1245-1254 (1983), Warsaw.

16. Katok, A., Kneiper, G., Pollicott, M., & Weiss, H., "Differentiability of entropy for anosov and geodesic flows", *Bull. Am. Math. Soc.*, 22, 285-294 (1990).

17. Kolmogorov, A.N., "On the entropy per time unit as a metric invariant of automorphisms", *Dokl. Akad. Nauk.*, 124, 754-755 (1959).

18. Kubo, R., Matsuo, K., & Kitahara, K., "Fluctuation and relaxation of macrovariables", *J. Stat. Phys.*, 9, 51-96 (1973).

19. Levi, M., "Qualitative analysis of the periodically forced relaxation oscillations", *Mem AMS*, 214, 1-147 (1981).

20. Lin, Y.E., & Anantheswaran, R.C., "Studies of popping of popcorn in a microwave oven", *J. Food Sci.*, 53, 1746-1749 (1988).

21. Lyerly, P.J., "Some genetic and morphilogical characters affecting the popping expansion of popcorn", *J. Am. Soc. Agron.*, 34, 986-997 (1942).

22. Mandell, A.J., "Dynamical complexity and pathological order in the cardiac monitoring problem", *Physica*, 27D, 235-242 (1987).

23. Mandell, A.J., Russo, P.V., & Blonigren, B.W., "Geometric universality in brain allosteric protein dynamics: Complex hyperbolic transformation predicts mutual recognition by polypeptides and proteins", *Ann. N.Y. Acad. Sci.*, 504, 88-117 (1987).

24. Mandell, A.J., & Kelso, J.A.S., "Neurobiological coding in nonuniform times", in *Essays on Classical and Quantum Dynamics*, eds. Ellison, J.A., & Uberall, H., Gordon and Beach, N.Y. (1990).

25. Manneville, P., "Intermittency in dissipative dynamical systems", *Phys. Lett.*, 79A, 33-37 (1980).

26. Manning, A., "Curvature bounds for the entropy of the geodesic flow on a surface", *J. London Math. Soc.*, 24, 351-357 (1981).

27. Newhouse, S., Palis, J. & Takens, F., "Bifurcation and stability of families of diffeomorphisms", *Publ. Math. IHES*, 57, 5-72 (1983).

28. Ornstein, D.S., "Ergodic theory, randomness and chaos", *Science*, 243, 182-187 (1989).

29. Paulus, M.P., Gass, S.F., & Mandell, A.J., "A realistic, minimal 'middle layer' for neural networks", *Physica*, 40D, 135-155 (1989).

30. Paulus, M.P., Geyer, M.A., Gold, L.H. & Mandell, A.J., "Ergodic measures of complexity in rat exploratory behavior", *Proc. Nat. Acad. Sci.*, 87, 723-727 (1990).

31. Plant, R.E. & Kim, M., "On the mechanism underlying bursting in the Aplysia abdominal ganglion R-15 cell", *Math. Biosci.*, 26, 357-375 (1975).

32. Plant, R.E. & Kim, M., "Mathematical description of a bursting pacemaker neuron by a modification of the Hodgkin-Huxley equations", *Biophys. J.*, 16, 227-244 (1978).

33. Pomeau, Y. & Manneville, P., "Intermittent transiyions to turbulence in dissipative dynamical systems", *Commun. Math. Phys.*, 74, 189-196 (1980).

34. Post, T., Capel, H.W. & van der Weele, "Short-phase anomalies in intermittent band switching", *Phys. Lett. A*, 133, 373-377 (1983).

35. Renyi, A., "Representation for real numbers and their ergodic properties", *Acta Math. Acad. Sci. Hung.*, 8, 477-493 (1957).

36.   Rinzel, J., "Bursting oscillations in an excitable membrane model", *Lect. Notes Math.*, Vol 1151 (1985).
37.   Rinzel, J., "A formal classification of bursting mechanisms in excitable systems", *Lecture Notes in Biomathematics*, 71,
267-281 (1987).
38.   Selz, K. A. & Mandell, A. J. "Bernoulli partition-equivalence of intermittent neuronal discharge patterns", *Intnl. J. Bifurcation & Chaos*, 1 (3), 234-237 (1992).
39.   Sinai, J., "On weak isomorphism of transformations with invariant measures", *Math. USSR*, 63, 23-42 (1964).
40.   Stein, R.B., "Some models of neuronal variability", *Biophys. J.*, 7, 37-68 (1967).

ORDER, PATTERN SELECTION AND NOISE IN LOW DIMENSIONAL SYSTEMS

M. San Miguel

Departament de Física
Universitat de les Illes Balears
E-07071 Palma de Mallorca, Spain

INTRODUCTION

In this paper I will address some aspects of the role of noise in pattern formation problems. Pattern formation is associated with spatial symmetry breaking, but broken symmetries might be restored by noise. Equilibrium phase transitions is a well known class of problems associated with symmetry breaking. A phase transition takes place when, in the thermodynamic limit, thermal fluctuations are not able to mix states with different symmetries. This only occurs for a large enough spatial dimensionality, which depends on the type of broken symmetry. An example of the opposite situation of symmetry restoring by noise is given by the ordinary laser instability. The proper description of this system is zero dimensional and spontaneous emission noise restores the phase symmetry of the lasing field after a time of the order of the coherence time. Pattern formation usually occurs in open systems of low dimensionality which correspond to situations somehow intermediate between the two discussed above.

Generally speaking, in the emergence and growth of spatial pattern a given system selects a state among a set of possible stable states. Relevant questions on the role of noise in this process are: Is symmetry restored by noise destroying long range order in the pattern? How does noise influence the dynamical process of pattern selection from a given initial condition? Does noise drive the system through different patterns towards a preferred one? These rather general questions are here considered in two specific examples. In a first example I consider the one-dimensional Swift-Hohenberg equation [1] as a prototype model displaying spatially periodic patterns. It is shown that long range order is destroyed by noise[2]. However, the dynamical mechanism of pattern selection is by the mode of fastest growth and this is robust against moderate fluctuations, so that noise does not drive the system towards the pattern minimizing the Lyapunov functional of the problem (except at extremely long times) [3]. The second example considers the breaking of the continuous cylindrical symmetry in the transverse profile of an optically pumped laser [4, 5]. It is shown that symmetry is restored by a phase diffusion mechanism induced by noise.

PATTERN SELECTION IN THE STOCHASTIC SWIFT-HOHENBERG EQUATION

The Swiff-Hohenberg equation was originally introduced to model the onset of a convective instability in simple fluids [1]. It describes the dynamical evolution of a scalar real variable $y(x, t)$,

$$\partial_t \, y(x, t) = \left[\, \gamma^2 - (k_0^2 + \partial_x^2)^2 \right] y(x, t) - y^3(x, t) + \xi(x, t) \tag{1}$$

where $\xi(x, t)$ models thermal fluctuations and it is taken to be a Gaussian process that satisfies

$$\langle \xi \rangle = 0 \, , \, \langle \xi(x, \, t) \, \xi(x', \, t') \rangle = 2\varepsilon\delta(x - x') \, \delta(t - t') \tag{2}$$

Thermal fluctuations had been commonly neglected because its smallness for simple fluids [6]. However, in several recent experiments [7], in particular in the electrohydrodynamic instability in nematic liquid crystals [8], these fluctuations have been either directly or indirectly observed.

In the absence of noise ($\varepsilon = 0$) and for $\gamma^2 > 0$, eq. (1) admits stationary periodic solutions $y_q(x)$ characterized by a wavenumber q [9]

$$y_q(x) = \sum_{i=0}^{\infty} A_i(q) \, \sin\left[(2i + 1) \, qx \right] \tag{3}$$

These solutions exist for a band of wavenumbers $q \in (q_{-L}, q_L)$ centered around $q = k_0$, which are linearly unstable around $y(x) = 0$. Stable solutions only exist in a restricted band $q \in (q_{-E}, q_E)$. Stationary solutions with wavenumbers $q \in (q_E, q_L)$ or $q \in (q_{-L}, q_{-E})$ are unstable with respect to a long wavelength instability known as the Eckhaus instability. Such instability is easily described [10] in terms of the amplitude equation for the complex amplitude $A(x,t)$ which characterizes the slowly varying envelope of a solution $y_q(x)$:

$$y(x, t) = A(x, t) \, e^{ik_0x} + cc. \tag{4}$$

The issue of pattern selection in this problem is to decide (for a bulk instability) which of the stable solutions within the Eckhaus stable band is chosen from a given initial condition. Eq.

(1) admits a Lyapunov functional L,

$$\partial_t \, y\,(x,t) = -\frac{\delta L\,(\{y\})}{\delta \, y\,(x,t)} + \xi\,(x,t) \tag{5}$$

so that a preferred solution is the one minimizing L, which for small γ is characterized by q_{min} = $k_0 - \gamma^4 / 1024 q_0^7$. Another special solution is the one associated with a mode of fastest growth. For the initial condition $y = 0$ this is the wavenumber k_0 which is extremely close to q_{min}. However, for an initial condition of the type (3) with $q \in (q_E, q_L)$ the mode of fastest growth can be clearly differentiated from q_{min} and k_0.

It should be noted that, in the presence of noise, the problem of pattern selection as stated above is ill-posed. This is so because noise destroys long range order, and typical stationary configurations are not characterized by a single wavenumber q. Pattern coherence is lost in d=1 because fluctuations of the phase of A(x,t) scale as $<\varphi^2_q> \sim q^{-2}$. A numerical simulation of the stochastic equation (1)[2] makes explicit the absence of long range order in several ways: Starting from any initial condition, the long time dynamics is characterized by mode competition with no single dominant mode. Such nonlinear mode competition is not well described in conventional theories of transient dynamics in which some effective mode decoupling of the nonlinear contribution around a dominat mode is introduced [11]. The absence of long range order is also clearly displayed in the normalized correlation function of y (x,t). This function oscillates in space but its envelope decays rather fast with a correlation length r_0 small compared with the system size L. For $\gamma=0.5$ and $\varepsilon = 10^{-2}$ it is found that L/r_0 ≈ 50. Alternatively one can look at the stationary structure factor

$$\lim_{t \to \infty} P\,(q,t) = \frac{1}{N} \left| \sum_{i=1}^{N} \, y_i\,(t)\, e^{-iqx_i} \right|^2 \tag{6}$$

where $y_i(t)$ is the discretization of y(x,t) in a grid of N points. The structure factor is rather broad with a width W that gives idea of the range of coherence. For a grid of N=8192 points with $(L/N) = \Delta x = 2\pi/32$ and $\gamma=0.5$, $k_0=1$, $\varepsilon = 5.\,10^{-3}$ it is found that $W \approx 0.065$, which corresponds to 5 convective rolls in a system of 512 rolls. A final clear way of displaying the absence of a selected wavenumber is by looking at a local structure factor in which $y_i(t)$ in (6) is replaced by a filtered value around $x_0 : \tilde{y}_i(x_0,t) = e^{-(x_i - x_0)^2/\beta^2} y_i(t)$. Such local structure factor is shown in Fig.1 where a local wavenumber can be identified.

Although in the presence of noise a periodic solution of wavenumber q is not selected, the number of changes of sign (NCS) of y(x,t) is essentialy a constant in any typical

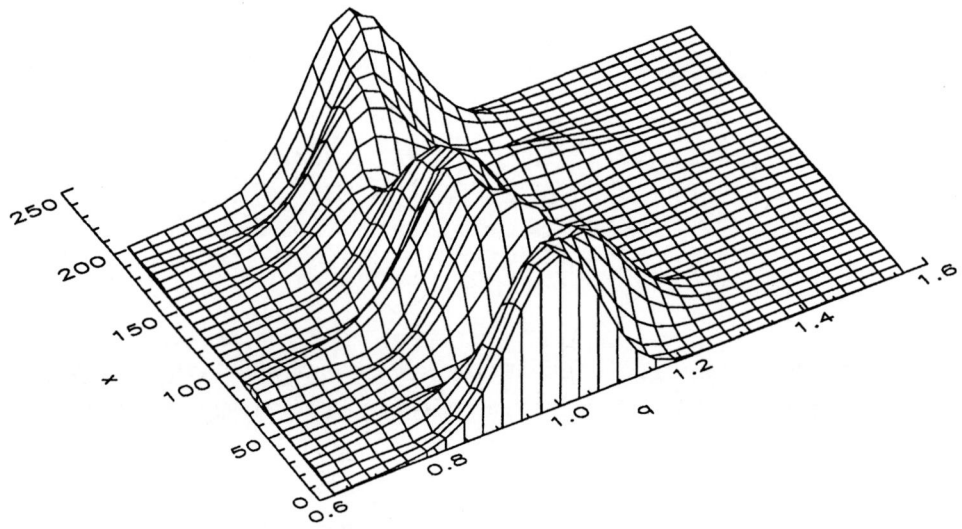

Fig. 1. Local stationary structure factor P_{xo} (q), as defined in the text, for a grid of $N = 1024$ and $\Delta x = 2\pi / 32$. Parameter values: $\gamma = 0.5$, $k_0 = 1$, $\varepsilon = 2.10^{-2}$ and $\beta = 10$.

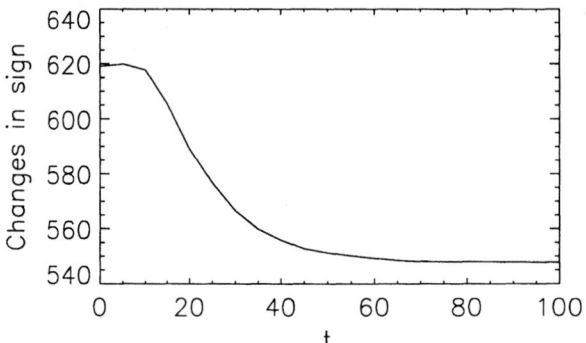

Fig 2. Time dependence of the NCS in the decay of an Eckhaus unstable configuration for a grid of $N = 8192$ and $\Delta x = 2\pi / 32$. Parameter values: $\gamma = 0.75$, $k_0 = 1$, $\varepsilon = 10^{-2}$ and initial wavenumber $q_i = 1.21$ ($q_E \approx 1.186$).

stationary configuration. Thus, a selection of a NCS occurs, corresponding to a selection of a number of convective rolls. The selected value of NCS can be identified by looking at the transient dynamics from an initial periodic configuration with wavenumber in the Eckhaus unstable regime. The time dependence of the NCS averaged over 20 realizations of the noise is shown in Fig.2. It is seen that the evolution gets trapped in a metastable state after the disappearence of 72 rolls. The final observed NCS is consistent with the global wavenumber associated with fastest growth and it differs from the one minimizing L in another 36 rolls. Therefore, noise of moderate strength along the dynamical path is not able to drive the system, in the explored time scale, to the configuration which according to eq.(5) is the most stable. Eventually, and for extremely long time scales, noise will induce jumps to configurations with a lower value of L.

SYMMETRY RESTORING IN TRANSVERSE LASER PATTERNS.

The study of spatial and spatiotemporal phenomena in nonlinear optics has emerged [12] as an intersting alternative to more conventional pattern formation studies in fluids. Following a theoretical prediction [4], it was shown experimentally [5] that when the pump bias level of an optically pumped laser is increased above a critical value, a spontaneous breaking of cylindrical symmetry in the transverse radiation profile occurs. The emerging state displays a spatial pattern associated with phase locking of three appropiate spatial modes. The angular position of the pattern was seen to fluctuate strongly over time intervals larger than a few milliseconds. Further instabilities of such pattern (not considered here) have also been analyzed [13].

The slowly varying envelope F of the electric field inside a ring laser with spherical mirrors can be expandend in terms of Gauss-Laguerre cavity modes $A_{p,l}(\rho,\varphi)$. They describe the transverse profile [4] with ρ and φ being the polar transverse coordinates and p and l the radial and angular index respectively. In conditions in which the atomic line is resonant with cavity modes p=1, l=0; p=0, l=2 cosine profile and p=0, l=2 sine profile, the electric field can be described in terms of the three corresponding complex amplitudes f_1, f_2, f_3:

$$F(\rho, \varphi, t) = \sum_{i=1}^{3} f_i(t) \, A_i(\rho, \varphi) \qquad (7)$$

In this description the system becomes effectively zero dimensional with 3 relevant degrees of freedom. Under standard approximations detailed elsewhere [14], the dynamics of the system is given by

$$d_t f_i = -f_i(t) + 2C \left[M_i f_i - \sum_{j,k,l=1}^{3} A_{ijkl} f_j f_k f_l^* \right] + \eta_i(t) \tag{8}$$

where C is the pump parameter, M_i and A_{ijkl} given coefficients and η_i model spontaneous emission noise as Gaussian stochastic processes satisfying

$$\langle \eta_i(t) \rangle = 0, \ \langle \eta_i(t) \eta_j^*(t') \rangle = 2\epsilon \, \delta_{ij} \, \delta(t-t') \tag{9}$$

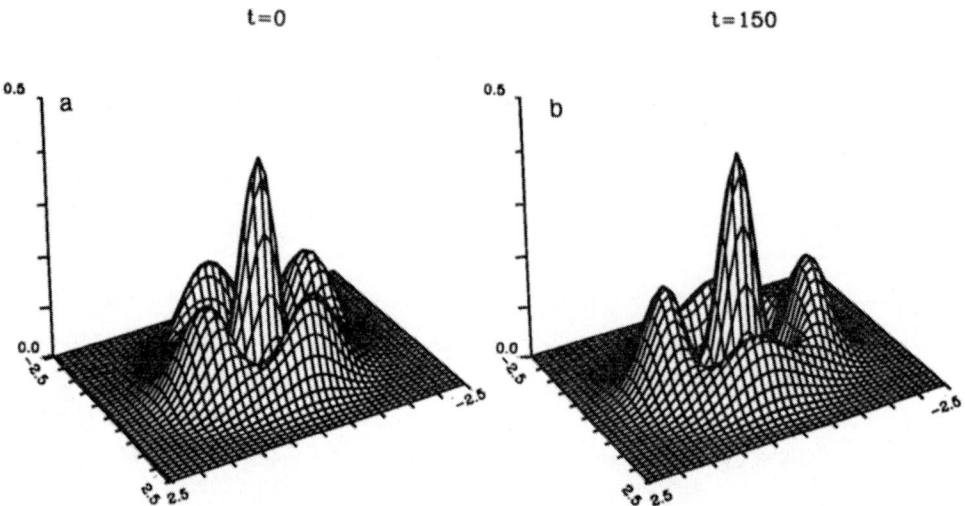

Fig 3. (a). Noise free stationary pattern for C = 1. 1029. (b) Configuration at time t = 150 obtained by stochastic evolution with $\epsilon = 10^{-3}$

In a noise-free framework, the laser instability occurs at $C=(2M_1)^{-1}$, leading to a cylindrically symmetrical pattern with $f_1 \neq 0$ and $f_2=f_3=0$. Increasing further the pump parameter to $C = (2M_2)^{-1} = (2M_3)^{-1}$ the modes f_2, f_3 become unstable and a new pattern with $f_i \neq 0$, i=1,2,3 emerges. This pattern shown in Fig 3a breaks the cylindrical symmetry. The side peaks disappear for $f_2 = f_3 = 0$. A continuous set of solutions with any angular orientation exists.

Introducing new mode variables $g_i = |g_i| e^{i\beta_i}$ by $g_1 = f_1$, $g_{2,3} = (\sqrt{2})^{-1} (f_2 \pm if_3)$, the electric field becomes

$$F(\rho, \varphi, t) =$$

$$\left(\frac{2}{\pi}\right)^{1/2} e^{-\rho^2} e^{i\beta_1} \left\{ (1 - 2\rho^2) |g_1| + \sqrt{2}\, \rho^2 \left[e^{2i(\varphi-\xi)} |g_2| + e^{-2i(\varphi-\xi)} |g_3| \right] e^{2i\delta} \right\} \tag{10}$$

where $\delta = ((\beta_2+\beta_3)/4) - (\beta_1/2)$ and $\xi = (\beta_3-\beta_2)/4$. This expression makes clear that the variable ξ characterizes the global orientation of the pattern and that this orientation is unchanged by simultaneous changes of δ and ξ in $\pm \pi/2$. The effect of noise in this problem is to restore the cylindrical symmetry by driving the system through different values of ξ. This effect of noise can be visualized by a numerical integration of the stochastic equations for the modal amplitudes [14]. Taking as the initial condition a deterministic solution with arbitrary orientation (Fig. 3a), fluctuations of the modal amplitudes induced by noise lead to a changing pattern in time with global rotation and distortion (Fig 3b). A long time average, or an ensemble average over many realizations, results in an averaged cylindrically symmetric pattern.

The steady state properties of the stochastic system are described by the stationary solution of the Fokker-Planck equation for the probability distribution of the complex amplitudes. It can be seen that this distribution has the form [14].

$$P_{st}(g_1, g_2, g_3) = N |g_1| |g_2| |g_3| \exp \{ \frac{-2}{\varepsilon} V(|g_1|, |g_2|, |g_3|, \delta) \} \tag{11}$$

where N is a normalization factor. Symmetry restoring manifests itself in the fact that P_{st} is independent of the angular orientation variable ξ. The deterministic stationary solutions are given by the minima of the potential located at $|g_2|^2 = |g_3|^2$ and $\delta \pm (\pi/4), \pm (3\pi/4)$. The discrete values of δ evidentiate phase locking. In the stochastic evolution obtained by numerical integration it is possible to observe that ξ diffuses freely while δ is fixed in one of its possible values. Noise also induces from time to time jumps among the possible values of δ. These are jumps in $\pi/2$ accompanied by the same jumps of ξ so that the orientation of the pattern remains unchanged. Such evolution is described by the stochastic equation for ξ [14]:

$$d_t\xi = (C/2 M_2) |g_1|^2 \frac{|g_2|^2 - |g_3|^2}{|g_1| |g_3|} \sin 4\delta + \frac{1}{4} (|g_3|^{-1} \overline{\eta}_3 - |g_2|^{-1} \overline{\eta}_2) \tag{12}$$

where $\overline{\eta}_3$ and $\overline{\eta}_2$ are real and independent Gaussian random processe of intensity ε. The drift part of (12) vanishes in the deterministic steady states. Linearizing around these states, (12) describes a diffusive motion of ξ. The pattern shown in Fig. 3 has four phase singularities in which $F(\rho, \varphi)=0$. These singular points also exhibit a diffusive motion.

ACKNOWLEDGMENT

Most of the ideas and results summarized in this paper are the result of collaborations with M. Brambilla, P. Colet, E. Hernández-García, L. Lugiato, R. Toral and J. Viñals. Financial support from the Dirección General de Investigación Científica y Técnica, Project PB89-0424 is acknowledged.

REFERENCES

1- J. Swift and P. C. Hohenberg, Phys. Rev. A **15**, 319 (1977).

2- J. Viñals, E. Hernández-García, M. San Miguel and R. Toral, Phys. Rev. A **44**, 1123 (1991).

3- J. Viñals, E. Hernández-García, M. San Miguel and R. Toral, (unpublished).

4- L. Lugiato, G. L. Oppo, J. R. Tredicce and L. M. Narducci, Opt. Commun. **69**, 387 (1989).

5- C. Tamm and C. O. Weiss, Opt. Commun. **78**, 253 (1990).

6- G. Ahlers, M. C. Cross, P. C. Hohenberg and S. Sfran, J. Fluid Mech. **110**, 297 (1981).

7- G. Ahlers, Physica D **51**, 421 (1991).

8- I. Rehberg, S. Rasenat, M. de la Torre, W. Schöpf, F. Horner, G. Ahlers, and H. R. Brand, Phys. Rev. Lett. **67**, 596 (1991).

9- Y. Pomeau and P. Manneville, J. Physique **40**, L609 (1979).

10- S. Fauve in "Instabilities and Nonequilibrium Structures", E. Tirapegui and D. Villarroel, eds., Reidel Publ. Co., Dordrecht (1987).

11- F. de Pasquale, P. Tartaglia and P. Tombesi, Phys. Rev. A 31, 2447 (1985); K. Elder and M. Grant, J. Phys. A **23**, L803 (1990).

12- J. Opt. Soc. Am. B **7** (6-7)(1990), Special issues on transverse effects in nonlinear optical systems.

13- M. Brambilla, F. Battipede, L. Lugiato, V. Penna, F. Prati, C. Tamm and C. O. Weiss, Phys. Rev. A **43**, 5090 (1991).

14- P. Colet, M. San Miguel, M. Brambilla and L. Lugiato, Phys. Rev. A 43, 3862 (1991).

PATTERN FORMATION IN EXTENDED CONTINUOUS SYSTEMS

Michael Bestehorn[1,2] and Carlos Pérez-García[2]

[1]Institut für Theoretische Physik und Synergetik
Universität Stuttgart, Pfaffenwaldring 57/4
7000 Stuttgart 80, Germany

[2]Dpto. Física y Matemática Aplicada
Universidad de Navarra
31080 Pamplona, Navarra, Spain

INTRODUCTION

Pattern formation as a self organized process in non-equilibrium systems is surely one of the most fascinating fields of research in our time. A unifying treatment developed in Synergetics [1,2,3] allows the mathematical description of the formation of spatio-temporal structures in quite different subjects, such as physics, chemistry, and biology. The behavior of a system in the vicinity of a non-equilibrium phase transition can be very often described by a few state variables which in analogy to Ginzburg-Landau theory of equilibrium thermodynamics are called *order parameters*. The large (in continuous systems even infinite) number of state variables assigned to the dynamics of the complete system can be expressed by the order parameters in a unique way; the order parameters enslave the many degrees of freedom that are linearly damped. By elimination of these enslaved modes [4]. a drastically reduced description in terms of few order parameter equations (OPE) is obtained. It thereby turns out that these equations have a similar form as the Ginzburg-Landau equations derived phenomenologically for phase transitions in thermal equilibrium. Beyond that, these *generalized Ginzburg-Landau equations* (GGLE) for phase transitions in open systems can be derived from basic physical laws, such as the Navier-Stokes equations [5] or the Maxwell equations [6] for hydrodynamic problems and for laser instabilities, respectively.

Using the convection instability as a paradigm, the purpose of this article is to outline the tools leading to the description of pattern formation by means of order parameter equations. It thereby turns out that the systems under consideration may be divided into two groups, according to their temporal behavior: 1.) The temporal evolution is relaxational and after a transient phase the system reaches a final stable state. 2.) The pattern stays time dependent, after initial transients a periodic, quasi-periodic, or even chaotic state emerges.

The underlying basic hydrodynamic equations are in general not variational, i.e.

may not be found by an extremum principle. On the other hand, deriving the OPE it turns out that part of them may originate from a potential, at least under some approximations. If these approximations are violated or if non-potential forces determine the spatio-temporal evolution of the system, we may find time dependent solutions. As examples we shall refer to the case where the instability of the conducting state is oscillatory (binary mixtures [7]) or to the occurence of a horizontal mean flow in convection of fluids with a low Prandtl number [8]. Both effects violate the variational character of the governing OPE and additional, non-variational terms become necessary. It thereby turns out that spatial defects and grain boundaries of the patterns play a crucial role and can mediate the non-relaxational behavior.

THE CONVECTION INSTABILITY

Pattern formation in the field of hydrodynamic instabilities is investigated since the pioneering experimental work of H.Bénard [9] at the beginning of the century. (For a review see e.g. [5,10,11,12] and references therein.) The Bénard instability of a pure fluid (see fig.1) is concerned with a homogeneous fluid layer contained between two horizontal plates and a uniform vertical temperature gradient. The external heating that drives the system away from thermal equilibrium induces an unstable density distribution of the fluid. At a certain critical temperature gradient, convection sets in in various forms of ordered regular patterns.

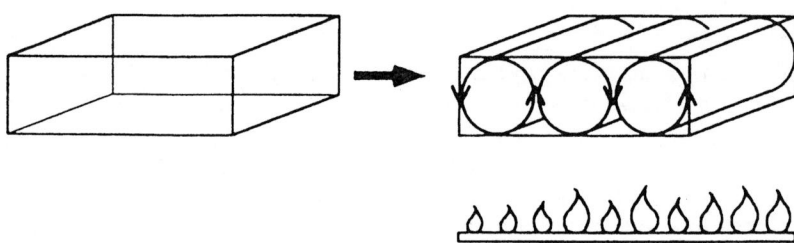

Figure 1. The convection instability. A fluid in a rectangular container is heated from below. At a certain temperature gradient the fluid starts to move and a regular pattern of up and down welling fluid particles emerges.

A fluid with density ρ, viscosity η, and thermal conductivity κ is described by the velocity field $\mathbf{v}(\mathbf{r},t)$, the temperature field $T(\mathbf{r},t)$, the pressure $p(\mathbf{r},t)$ as well as a state equation for the density. The conservation laws for an incompressible fluid under the influence of an externally applied gravitational acceleration g read:

$$\rho(\mathbf{r},t)\{\partial_t\mathbf{v}(\mathbf{r},t) + \mathbf{v}(\mathbf{r},t)\bullet\nabla\mathbf{v}(\mathbf{r},t)\} = \rho(\mathbf{r},t)g\mathbf{z}_0 - \nabla p(\mathbf{r},t) + \eta\Delta\mathbf{v}(\mathbf{r},t)$$
$$\nabla\bullet\mathbf{v}(\mathbf{r},t) = 0 \tag{1}$$
$$\partial_tT(\mathbf{r},t) + \mathbf{v}(\mathbf{r},t)\bullet\nabla T(\mathbf{r},t) = \kappa\Delta T(\mathbf{r},t)$$

where $\mathbf{z}_0$ is the unit vector in vertical direction. In the Boussinesq approximation the variation of the density ρ as a function of the temperature is neglected except for the external force term, where it results in buoyancy effects. The transport coefficients are assumed to be constant.

For the velocity we make the usual decomposition into two scalar fields:

$$\mathbf{v}(\mathbf{r},t) = \nabla\times\{\phi(\mathbf{r},t)\mathbf{z}_0\} + \nabla\times\nabla\times\{\psi(\mathbf{r},t)\mathbf{z}_0\} \tag{2}$$

Introducing the variation $\Theta(\mathbf{r}, t)$ of the temperature from the basic linear temperature profile and eliminating the pressure by forming the curl and twice the curl of the Navier-Stokes-equations we get:

$$\{\Delta - \frac{1}{Pr}\partial_t\}\Delta\Delta_2\psi(\mathbf{r},t) = -R\Delta_2\Theta(\mathbf{r},t) - \frac{1}{Pr}\{\nabla \times \nabla \times (\mathbf{v}(\mathbf{r},t) \bullet \nabla\mathbf{v}(\mathbf{r},t))\}_z$$

$$\{\Delta - \frac{1}{Pr}\partial_t\}\Delta_2\phi(\mathbf{r},t) = -\frac{1}{Pr}\{\nabla \times (\mathbf{v}(\mathbf{r},t) \bullet \nabla\mathbf{v}(\mathbf{r},t))\}_z \qquad (3)$$

$$\{\Delta - \partial_t\}\Theta(\mathbf{r},t) = \Delta_2\psi(\mathbf{r},t) + \mathbf{v}(\mathbf{r},t) \bullet \nabla\Theta(\mathbf{r},t)$$

where $Pr = \eta/\rho_0\kappa$ is the Prandtl-number and $\Delta_2 = \partial_{xx} + \partial_{yy}$ the horizontal Laplacian. The control parameter R is the Rayleigh-number and given by $R = -\rho_0 g \alpha \beta d^4/\kappa\eta$ where α is the thermal expansion coefficient of the fluid. R is proportional to the temperature gradient β. Assuming vanishing velocity components on the boundaries (rigid boundary conditions), (2) leads to

$$\phi(\mathbf{r},t) = \psi(\mathbf{r},t) = \partial_\mathbf{n}\psi(\mathbf{r},t) = 0 \qquad (4)$$

for $\mathbf{r}$ on the horizontal and vertical walls and $\mathbf{n}$ perpendicular. On the lateral walls, we have in addition

$$\partial_\mathbf{n}\phi(\mathbf{r},t) = 0, \qquad \Delta_2\psi(\mathbf{r},t) = 0 \qquad (5)$$

General boundary conditions for the temperature field can be expressed in the form

$$\partial_n\Theta(\mathbf{r},t) = \pm Bi\Theta(\mathbf{r},t) \qquad (6)$$

where Bi is another dimensionless parameter, the so-called Biot-number, standing for the ratio of the thermal conductivity of the corresponding wall and to that of the fluid.

For pure fluids between rigid, conducting plates, patterns of quite regular rolls are usually obtained. Squares are obtained if the fluid is enclosed between poorly conducting plates or in binary mixtures in concentration dominated convection. Hexagonal cells may be obtained in fluids with so-called non-Boussinesqian material properties. These properties can become important if one uses a very thin layer, and, as a consequence, a large temperature gradient along the cell. Then the viscosity and thermal conductivity are also varying with the temperature (see e.g. [11]). Hexagons are also formed if the instability is not dominated by bouyancy but by the temperature dependence of the surface tension of the fluid at a free surface (Marangoni-effect) [10].

VARIATIONAL MODELS

In this and in the next section we wish to give an overview on recent theoretical development on pattern formation in the frame of generalized Ginzburg-Landau equations. This section is devoted to pattern formation showing a relaxational behavior in time and we shall see that the order parameter equations are potential, at least under certain approximations.

The Order Parameter Equation in Real Space

Relaxational time evolution towards an ordered steady state is obtained in a pure fluid with a moderate or large Prandtl number near the onset of convection. In the limit of $1/Pr = 0$, the number of dependent variables in (3) is reduced. We get:

$$\Delta\Delta_2\phi(\mathbf{r},t) = 0 \qquad (7)$$

and, together with (4,5):

$$\phi(\mathbf{r}, t) = 0 \tag{8}$$

in the whole layer. We project the remaining variables ψ, Θ onto a Galerkin basis formed by the eigenvectors $\mathbf{q}_\ell$ of the linearized version of (3):

$$\begin{bmatrix} \psi(\mathbf{r}, t) \\ \Theta(\mathbf{r}, t) \end{bmatrix} = \sum_\ell \mathbf{q}_\ell(\Delta, z) \Psi_\ell(\mathbf{x}, t) \tag{9}$$

with $\mathbf{x} = (x, y)$. The wave functions Ψ_ℓ contain the fully horizontal and temporal dependence of the non-equilibrium structures. Along the lines of Synergetics, we classify the eigenvectors $\mathbf{q}_\ell$ into two groups, according to their eigenvalues λ_ℓ:

$$\lambda_u \approx 0, \Longrightarrow \Psi_u(\mathbf{x}, t), \ \text{u=unstable, order parameter}$$

$$\lambda_\ell$$

$$\lambda_s \ll 0, \Longrightarrow \Psi_s(\mathbf{x}, t), \ \text{s=stable, enslaved modes}$$

The slaving principle allows us to express the amplitudes of the enslaved modes Ψ_s as a functional of those of the order parameters Ψ_u in a unique way. Substituting (9) into (3) we may completely eliminate the enslaved modes from the equations of motion. The simplest way to do this is invoking the adiabatic elimination [1,2]. In this case, the dynamics of the enslaved modes is neglected, they follow instantaneously the order parameters. The remaining 2D-equation for the order parameters Ψ_u read (here and in the following we supress the index "u") formally:

$$\partial_t \Psi(\mathbf{x}, t) = F[\Psi, \nabla^n \Psi] \tag{10}$$

where F denotes a nonlinear functional that depends on spatial derivatives of Ψ.

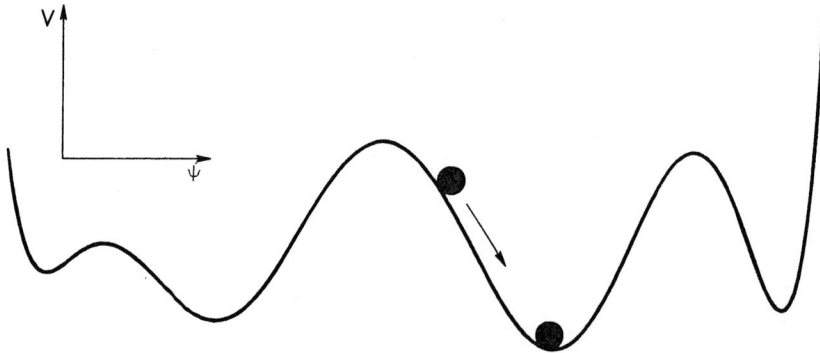

Figure 2. In the variational case, eq.(10) has its analogy in classical mechanics. A particle moves downwards (transient phase) in a landscape corresponding to the potential and stops at a minimum (stationary solution). All local minima correspond to stationary, stable solutions of (10).

On this level of description, we may distinguish naturally between variational and non-variational dynamics. If we split F into:

$$F[\Psi] = \frac{\delta V[\Psi]}{\delta \Psi} + W[\Psi] \tag{11}$$

then V denotes the potential and W the non-variational effects. Boundary conditions may be directly derived from (9),(4), and (6). For perfect thermal conductors, they read:

$$\Psi = \partial_\mathbf{n} \Psi = 0 \tag{12}$$

For variational models ($W = 0$), eq.(10) with (11) are analogous to the overdamped motion of a particle moving in a potential landscape downwards until a local minimum of V is reached (fig.2).

Extended Swift-Hohenberg Models

We wish to outline the particular form of (10) derived from (3). In the linear part of F we allow for a variation of the wave vector in a band centered at k_c, where k_c denotes the wave vector of the pattern for the smallest Rayleigh number R_c that leads to instability. This can be approximated by the first terms of a Taylor series:

$$\lambda(\Delta_2) \approx \varepsilon - (1 + \Delta_2)^2 \tag{13}$$

where $\varepsilon = (R - R_c)/R_c$ is the separation from threshold. (Here and in the following we rescale the spatial coordinates to have $k_c^2 = 1$.) The third order term in F is nonlocal and may be approximated by spatial derivatives, or, with regard to a rotationally invariant formulation of the problem, by powers of the 2D-Laplacian. If we consider the general expression:

$$\Psi \sum_{n=0}^{N/2} A_{2n} \left(\frac{\Delta_2}{2} + 1 \right)^{2n} \Psi^2 \tag{14}$$

then the cubic interaction between two amplitudes $\xi(\varphi)$ of plane waves having the critical wave length 2π and the orientation φ can be written as:

$$\int_0^{2\pi} d\varphi' f(\varphi - \varphi')\xi(\varphi)|\xi(\varphi')|^2 \tag{15}$$

If only variational couplings are considered, f may only depend on scalar products of the wave vectors, i.e. it may be developed in a Taylor series with respect to $\cos(\varphi - \varphi')$:

$$f(\beta) = \sum_{n=0}^{N/2} a_{2n}(\cos \beta)^{2n} \tag{16}$$

Because of the symmetry with respect to $\varphi' \mapsto \varphi' + \pi$ in the expression under the integral (15), odd powers of $\cos \beta$ cancel and only even powers have to be taken into account. The relations between A_n and a_n of (16) and (14) may be established as follows:

$$\begin{aligned} a_0 &= 3A_0 + \sum_{n=1}^{N/2} A_{2n} \\ a_{2n} &= 2A_{2n} \qquad n = 1, 2, 3 ... N/2 \end{aligned} \tag{17}$$

(17) enables us to express the fully angular dependence intrinsic in (10) of the mode coupling between plane waves with given orientation via the local form (14) including only rotationally invariant expressions of spatial derivatives.

Numerical Results: Rolls, Squares, Quasi-Periodic Structures

Now we discuss briefly the properties of several truncations of (13) and (14) with respect to N. The crudest approximation corresponds to $N = 0$ and is the well-known Swift-Hohenberg equation [13] that was numerically treated in two spatial dimensions in [14] as a model of the convection instability:

$$\partial_t \Psi(\mathbf{x}, t) = [\varepsilon - (1 + \Delta_2)^2]\Psi(\mathbf{x}, t) + A_0 \Psi^3(\mathbf{x}, t) \tag{18}$$

and $A_0 < 0$. In terms of (16), no angular dependence of the coupling coefficients can be expressed. This leads to a pattern of parallel rolls that corresponds quite well to that observed in experiments with a a high Prandtl number fluid having perfectly heat conducting top and bottom plates [15]. A numerical solution is presented in fig.3.

The next level is $N = 2$. Now we may describe a coupling with one extremum in the interval $[0, 2\pi]$ of coupling angles. A simplified equation reads:

$$\partial_t \Psi(\mathbf{x}, t) = [\varepsilon - (1 + \Delta_2)^2]\Psi(\mathbf{x}, t) + A\Psi^3(\mathbf{x}, t) + B\Psi(\mathbf{x}, t)\Delta_2\Psi^2(\mathbf{x}, t) \tag{19}$$

and $B < 0$. If A changes from negative to positive values, rolls loose stability and give way to squares. Fig.3 shows a stable, regular square planform. The coefficients in (19) were calculated numerically for poor thermal conductors on the bottom and the top of the layer. Experimental evidence may be found in [16] and for a binary mixture in [17].

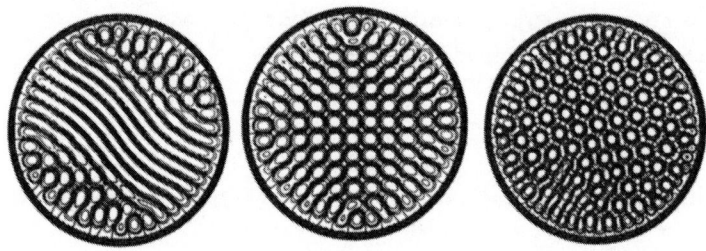

Figure 3. Stable states of extended Swift-Hohenberg equations. All states where obtained by integration of a completely random initial state. Left: Rolls for the simple model (18), $\varepsilon = 0.1, A_0 = -1$. Middle: Squares for the extension (19), $\varepsilon = 0.1, A = 1.16, B = -1$. Right: Hexagons if a quadratic term is included in (18), $\varepsilon = 0.1, A_0 = -1, \delta = 0.4$.

In contrast to (18), (19) cannot longer be derived from a potential. Nevertheless we know from (15) that all sorts of coupling coefficients are potential, if they fulfil $f(\beta) = f(-\beta)$ and if the patterns consist mainly in modes with the same wave length. We conjecture that this is the reason why all numerical investigations of equations having the general form (13) and (14) as well as most experiments using simple fluids near threshold present a relaxational behavior.

Including even higher derivatives, one can model hypothetic couplings with more minima. In that way a hexagonal pattern can be stable with $N = 6$. Here we wish to show the formation and stabilization of a pattern having 10-fold symmetry in a very large aspect ratio system. For f we take ($N = 10$):

$$\begin{aligned}
f(\beta) = \;& -3 + 56\cos^2\beta - 392\cos^4\beta + 1120\cos^6\beta \\
& -1280\cos^8\beta + 512\cos^{10}\beta
\end{aligned} \tag{20}$$

This fixes the values of A_n according to (17). Fig.4 shows two stages of pattern formation obtained in this case starting with an initial condition that was chosen randomly. Notice that these patterns ahve a pentagonal structure typical for quasi crystals (see e.g. [18]).

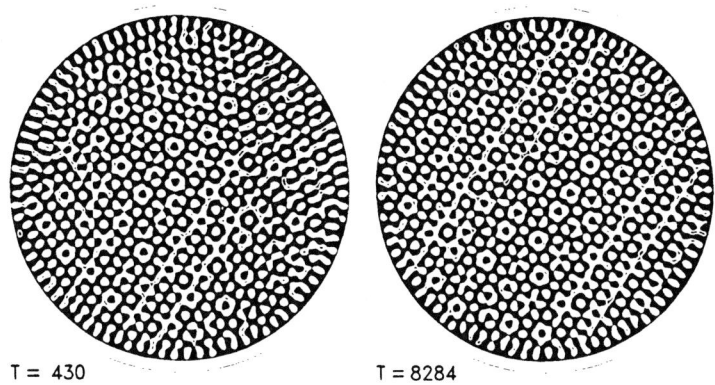

Figure 4. Formation of a pattern having five-fold symmetry with an angular coupling according to (20). Times in units of (13) with (14), $\varepsilon = 0.1$.

Even Nonlinearities − Hexagons

If the mirror symmetry with respect to the vertical midplane is violated (Marangoni or non-Boussinesqian effects), terms with even powers in the order parameter occur. The simplest one is quadratic, and, in the form without spatial derivatives, it keeps the variational property of (18).

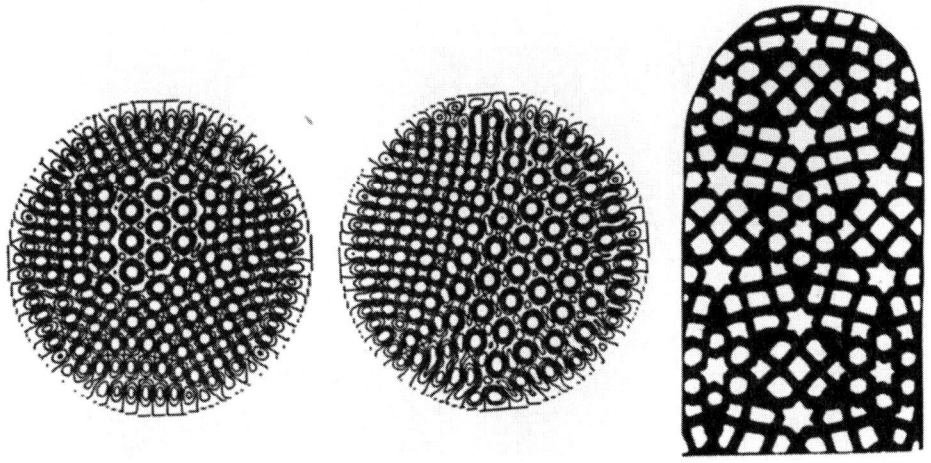

Figure 5. Left ($\varepsilon = 0.1, \delta = 0.01, A = 1.07, B = -1$) and middle ($\varepsilon = 0.1, \delta = 0.07, A = 1.17, B = -1$): Adding a quadratic (hexagon) term to (19) may lead to the formation of mixed states with stable fronts between hexagons and squares. Right: A mixed state of square and hexagonal symmetry invented by the Moors some hundred years ago and found in the Alhambra palace, Granada, Spain.

Adding the expression

$$\delta\Psi^2(\mathbf{x},t)$$

to the right-hand-side of (18), we arrive at an equation first derived by Haken [2] and numerically investigated in [19]. Stability analysis shows that hexagons are stable if

$$\varepsilon < -\frac{4}{3A_0}\delta^2$$

(see figs.3,5). δ corresponds directly to non-Boussinesqian or surface tension effects and can be calculated as a function of the Marangoni-number. A detailed study of the stability of rolls and hexagons in the case of Bénard-Marangoni convection is given in [20]. Recent experiments can be found in [21].

NON-VARIATIONAL MODELS

Now we shall discuss two outstanding examples of convection that lead to pattern formation exhibiting a non-variational dynamics. One is the thermal instability in a mixture of two miscible fluids, the other one is concerned with convective patterns in a liquid with low viscosity.

Oscillatory Instabilities – Binary Mixtures, Reaction-Diffusion Systems

A convective instability with an oscillatory character can be observed in the case of a fluid consisting of two miscible components like for instance a water-ethanol mixture. Oscillatory behavior of the instability is expected if instead of a simple increase of an initial displacement a temporal oscillation around the inital position of the fluid particle sets in. Linear theory shows that this may happen in a binary mixture due to the Soret-effect [22]. Experiments on the onset of convection in binary fluid mixtures reveal an astonishingly complex spatio-temporal behavior of the flow already close to onset [23]. The fluid motion consists of convection rolls which move in horizontal direction forming traveling waves. Usually there exist several wave trains which interact and behave chaotically in time and simultaneously exhibit irregular spatial patterns [24].

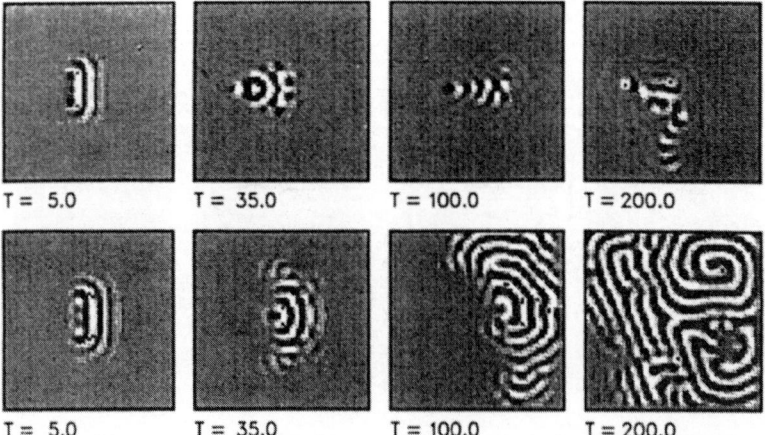

| T = 5.0 | T = 35.0 | T = 100.0 | T = 200.0 |

| T = 5.0 | T = 35.0 | T = 100.0 | T = 200.0 |

Figure 6. The formation of pulses as solution of (21). Two time series are shown for the same initial condition of a wave front. $\varepsilon = -0.1, A = 3+i, B = 0, C = -2.75+i, \gamma = -0.3$ (top), $\gamma = 0.3$ (bottom).

The order parameter equation for the oscillatory instability is the complex generalization of (10). The order parameter is now a complex wave function whose solutions consist of traveling or standing wave trains. The linear expression (13) has to be completed by dispersion. The simplest nonlinear model that may differentiate between the interaction of left and right traveling waves reads:

$$\dot{\Psi}(\mathbf{x},t) = [\varepsilon - (1+\Delta)^2 + i(\omega_c - \gamma(1+\Delta))]\Psi(\mathbf{x},t) \tag{21}$$
$$+ A\Psi(\mathbf{x},t)|\Psi(\mathbf{x},t)|^2 - B\Psi(\mathbf{x},t)|\nabla\Psi(\mathbf{x},t)|^2 + C|\Psi(\mathbf{x},t)|^4\Psi(\mathbf{x},t)$$

Here we included 5th order terms to have the possibility to describe a subcritical bifurcation, as it is the usual case for a binary mixture. In one spatial dimension, traveling waves are selected if $\mathrm{Re}(A-B) > 2\mathrm{Re}(A)$ if $C = 0$, $A, B < 0$. We performed a numerical treatment of (21) in a circular layer using a pseudo-spectral method and a semi-implicit time integration scheme.

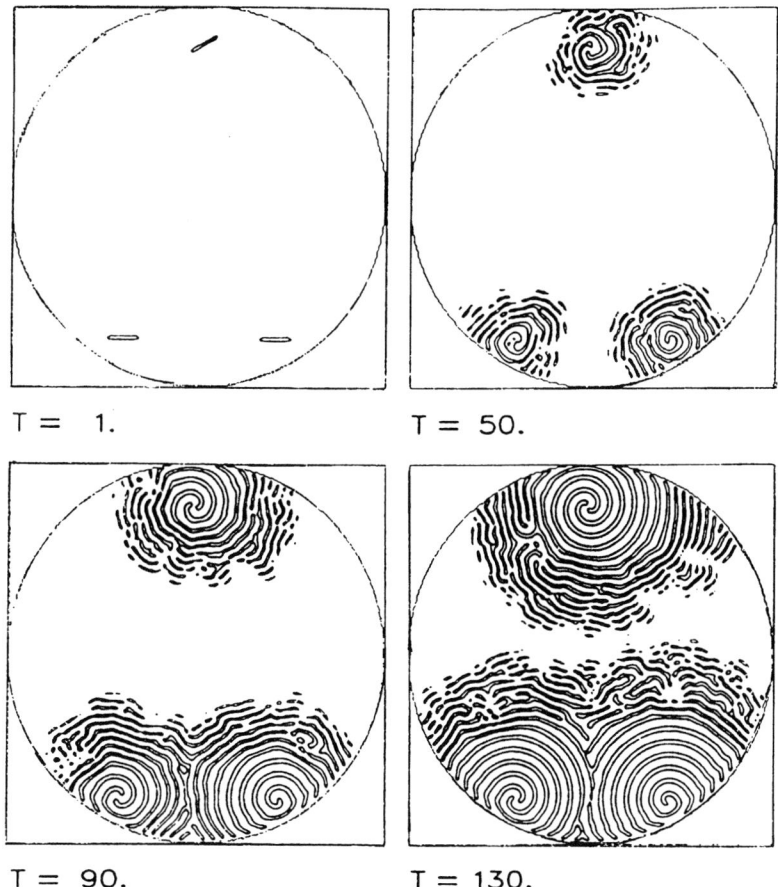

T = 1. T = 50.

T = 90. T = 130.

Figure 7. Starting with three localized fronts, the evolution according to (21) shows the formation of spirals known from reaction diffusion equations. The fronts separating different spirals are stable. Parameter values as in fig.6, bottom.

Results for a binary mixture having perfect vertical boundary conditions are presented in [24]. Here we show the subcritical evolution of a spatial pulse for several parameters of (21) (fig.6). A qualitatively comparable evolution was found in recent experiments in binary mixtures [27]. The envelope shapes change on a slow time scale compared to the phase velocity of the underlying traveling wave trains. For other parameter values, the formation and stabilization of soliton like pulse solutions could be found [25].

Another time dependent solution is shown in fig.7. The initially chosen singularities spread out in form of spirals and reach eventually stable envelopes. Nevertheless they continue to turn clock or counterclockwise, depending on their charge. Sinks are formed where the borders of two spirals meet. We note that the same structures are encountered in chemical reactions far from thermal equilibrium in an excitable (subcritical) medium [26].

Fluids With Low Viscosity – Mean Flow Effects

Experiments in fluids having a low Prandtl number show time dependent behavior just at onset. This is theoretically understandable since the work of Siggia et al. [8]. They considered the horizontal mean flow field ϕ (see eq.(2)) as a second order parameter.

In a recent paper Bodenschatz et al. [28] report on convection in a cylindrical large aspect ratio system with pressurized CO_2 under non-Boussinesquian conditions. They discovered n-armed spiral patterns formed by convective rolls. These spirals are stationary in a rotating frame of references.

Based on the models discussed in [29], we examined two coupled evolution equations, one for the order parameter Ψ, the other for the stream function ϕ of the horizontal mean flow [30]:

$$\begin{aligned}
\partial_t \Psi(\mathbf{x},t) &= [\varepsilon - (1+\Delta_2)^2]\Psi(\mathbf{x},t) - \Psi^3(\mathbf{x},t) - g\mathbf{V}_H(\mathbf{x},t) \bullet \nabla_2 \Psi(\mathbf{x},t) \\
\partial_t \Delta_2 \Phi(\mathbf{x},t) &= Pr(\Delta_2 - c^2)\Delta_2 \Phi(\mathbf{x},t) + [\nabla_2 \Psi(\mathbf{x},t) \times \nabla_2 \Delta_2 \Psi(\mathbf{x},t)]_z \qquad (22) \\
\mathbf{V}_H(\mathbf{x},t) &= (-\partial_y \Phi(\mathbf{x},t), \partial_x \Phi(\mathbf{x},t))
\end{aligned}$$

The constant c depends on the vertical boundary conditions (for rigid-rigid conditions, the horizontal drift may be approximated as a Poiseuille flow leading to $c^2 = 10$; for free-free conditions, there is no z-dependence, $c^2 = 0$). The boundary conditions at the circular lateral wall read:

$$\begin{aligned}
\Psi(\mathbf{x},t) &= d, & \partial_n \Psi(\mathbf{x},t) &= 0, \qquad (23) \\
\phi(\mathbf{x},t) &= 0, & \partial_n \phi(\mathbf{x},t) &= 0
\end{aligned}$$

The parameter d parametrizes lateral heating on the sidewall that plays a major role in the experiments in [28,27]. Patterns with stable n-armed spirals can be obtained already for the Swift-Hohenberg equation forced by lateral heating ($d \neq 0$). To this end we started from an initial condition with a one-armed spiral with its tip located at the center of the cylindrical container. The pattern evolves towards a stable one-armed spiral that matches the concentric rolls near the sidewall due to the lateral forcing. In the matching zone, a defect is created (see fig.8).

If non-variational terms are included ($g \neq 0$) the spiral begins to rotate rigidly. It is seen that the mean flow is essentially created by the defect, where the horizontal drift velocity is maximal. In turn, the mean flow acts on this dislocation and rotates the whole spiral.

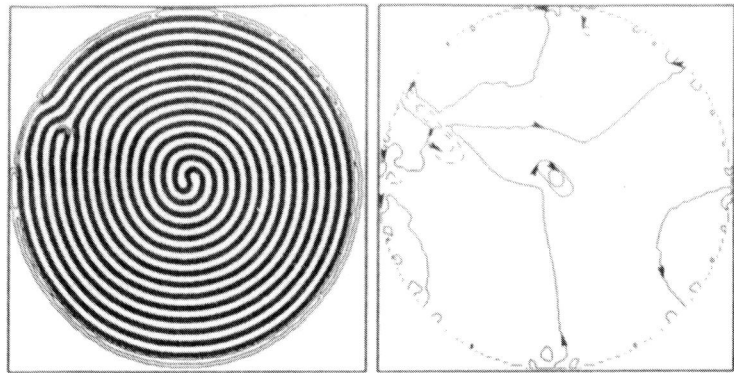

Figure 8. Rigidly rotating pattern found by numerical integration of (22). The spiral rotates counter-clockwise (left) due to the mean flow (right). The mean flow is a horizontal drift that is created mainly by the defects. In turn the mean flow acts on these defects and rotates the spiral with a constant angular velocity. $\varepsilon = 0.7, g = 20, Pr = 1, c^2 = 10, d = 0.15$.

We note that the angular velocity is proportional to the strength of non-potential effects. There is no threshold value for Pr for the onset of rotation and the occurence of non-variational behavior. This is due to the fact that the continuous rotation symmetry in the cylindrical layer is broken by the convection pattern. The disturbances due to any pattern that represent infinite rotation have zero growth rate. These disturbances are excited by the vorticity field. In contrast, the numerical results for rectangular cells [29] showed a significant influence of the mean flow on pattern formation only for very small Pr.

Acknowledgments

It is a pleasure to thank R.Friedrich and M.Fantz for fruitful discussions. We thank H.Herrero Sanz for leaving us the first two frames of figure 5. We acknowledge financial support from the EEC project SC311, and from a german-spanish Integrated Action (No26,1991).

References

[1] H.Haken, *Synergetics. An Introduction*, Springer Berlin 3rd Ed. (1983)

[2] H.Haken, *Advanced Synergetics*, Springer Berlin, 2nd print (1987)

[3] H.Haken, Rep. Prog. Phys. **52**, 515 (1989)

[4] A.Wunderlin and H.Haken, Z. Phys. **44**, 135 (1981)

[5] R.Friedrich, M.Bestehorn, and H.Haken, Int. J. Mod. Phys. **B4**, 365 (1990)

[6] H.Haken in *Encyclopedia of Physics*, Vol XXV/2c, Springer Berlin (1970)

[7] D.Gutkowicz-Krusin, M.A.Collins, and J.Ross, Phys. Fluids **22**, 1443, 1457 (1979)

[8] D.Siggia and A.Zippelius, Phys. Rev. Lett. **47**, 835 (1981)

[9] H.Bénard, Rev. Gen. Sci. Pur. Appl. **11**, 1261 (1900)

[10] C.Normand, Y.Pomeau, and M.Velarde, Rev. Mod. Phys. **49**, 581 (1977)

[11] F.H.Busse, Rep. Prog. Phys. **41/II**, 1931 (1978)

[12] R.P.Behringer, Rev. Mod. Phys. **57**, 657 (1985)

[13] J.Swift and P.C.Hohenberg, Phys. Rev. **A15**, 319 (1977)

[14] H.S.Greenside, W.M.Coughran Jr., and N.L.Schreyer, Phys. Rev. Lett. **49**, 729 (1982);
H.S.Greenside and W.M.Coughran Jr., Phys. Rev. **A30**,398 (1984)

[15] P.Bergé and M.Dubois, Contemp. Phys. **25**, 535 (1984)

[16] P.Le Gal, A.Pocheau, and V.Croquette, Phys. Rev. Lett. **54**, 2501 (1985)

[17] E.Moses and V.Steinberg, Phys. Rev. **A43**, 707 (1991); Pys. Rev. Lett. **57**, 2018 (1986)

[18] G.M.Zaslavsky, R.Z.Sagdeev, D.A.Usikov, and A.A.Chernikov *Weak chaos and quasi-regular patterns*, Cambridge University Press (1991)

[19] M.Bestehorn and H.Haken, Phys. Lett. **A99**, 265 (1983)

[20] M.Bestehorn and C.Pérez-García, Europhys. Lett. **4**, 1365 (1987)

[21] S.Ciliberto, E.Pampaloni, and C.Pérez-García, J. Stat. Phys. **64**, 1045 (1991)

[22] J.K.Platten and J.C.Legros, *Convection in Liquids*, Springer Berlin (1984)

[23] V.Steinberg, E.Moses, and J.Fineberg, Nucl. Phys (proc. sup.) **B2**, 109 (1987);
D.Bensimon, P.Kolodner, C.M.Surko, H.Williams, and V.Croquette, J. Fluid Mech. **217**, 441 (1990)

[24] M.Bestehorn, R.Friedrich, and H.Haken, Z. Phys. **B75**, 265 (1989);
M.Bestehorn, R.Friedrich, and H.Haken, Z. Phys. **B77**, 151 (1989);
M.Bestehorn, R.Friedrich, and H.Haken, Physica **D37**, 295 (1989)

[25] M.Bestehorn and H.Haken, Phys. Rev. **A42**, 7195 (1990)

[26] A.T.Winfree, Scient. Americ. **6**, 82 (1974);
A.N.Zaikin and A.M.Zhabotinsky, Nature **225**, 535 (1970);
Z.Nagy-Ungvarai, S.C.Müller: In *Propagation in Systems Far from Equilibrium*,
J.E.Wesfreid, H.R.Brand, P.Manneville, G.Albinet, N.Boccara (eds.), Springer Series in Synergetics **Vol 41**, Springer Berlin (1988)

[27] G.Ahlers, at the NATO-ARW in Estella, Spain, September 1991, to appear in *New Trends in Nonlinear Dynamics: Non-Variational Aspects*, Physica D

[28] E.Bodenschatz, J.R. de Bruyn, G.Ahlers and D.S.Cannell, Preprint (1991)

[29] P.Manneville, J. Phys. (Paris) **44**, 759, L-903 (1983)

[30] M.Bestehorn, M.Fantz, R.Friedrich, H.Haken, and C.Pérez-García, preprint

ROLE OF CATALYSIS ON THE EVOLUTION OF

ERROR-PRONE SELF-REPLICATIVE MOLECULES

F. Montero[+] & J.C. Nuño[*]

[+] Dpto. de Bioquímica y Biología Molecular I, Fac. CC. Químicas
Universidad Complutense. 28040 Madrid, Spain

[*] Dpto. de Matemáticas. E.T.S.I. de Montes
Universidad Politécnica. 28040 Madrid, Spain

1 INTRODUCTION

Nucleic acids are biological molecules that have intrinsic template activity by themselves, without the aid of any other molecular species. The behavior of populations of this kind of molecules is the main subject of a paper published by Eigen in the early seventies[1]. At that time, some experiments were carried out in order to find out what the dynamics of these systems was like[2]. Just at the beginning, the mathematical model agrees with biological experiments in relation to the general behavior of these systems. This fact caused Eigen's model to be seriously taken into account and deeply analyzed.

In the same article[1], the author also proposed a model where different selfreproductive molecules were linked by means of an specific polymerase catalytic activity. This first attempt to describe the dynamics of catalytic networks followed the central dogma of Molecular Biology: the replication of nucleic acids is catalyzed by the action of proteins, which had been previously sintethized following the code written in the primary structure of nucleic acids.

At this time, the existence of catalytic organization formed by nucleic acids (contrary to the central dogma) was extensively discussed, although experimental results about these systems were not available[3,4]. After confirming the catalytic activity of special kinds of RNAs molecules[5,6] (the so called ribozymes), this process should be introduced as a basic assumption in models describing the behavior of populations formed by these RNA-like molecules.

In this paper, we briefly outline some ideas on the dynamics of catalytic networks, summarizing our own contribution on this subject.

2 MODELLING EVOLUTION OF CATALYTIC NETWORKS

The process of evolution of self-reproductive molecules may be discretized in steps, in each of them an element of the system being able to undergo one of the following kinetics reactions:

$$(\mu^*) + I_k \overset{Q_{kk}\mathcal{R}_k}{\rightarrow} 2I_k \tag{1}$$

$$(\mu^*) + I_k \overset{Q_{kj}\mathcal{R}_k}{\rightarrow} I_k + I_j \tag{2}$$

$$I_k \overset{\mathcal{D}_k}{\rightarrow} (\mu) \tag{3}$$

$$I_k \overset{\phi}{\rightarrow} \tag{4}$$

where I_k is one of the n possible molecular network species (replicators or information carriers). Equation (1) represents the free-error self-replication of I_k, i.e., the process by means of which this molecule makes an exact copy of itself. Equation (2) takes into account the error-prone replication. In both reactions μ^* are rich energy monomers, from which I_k is built. Equation (3) drives the degradation of replicators, i.e. the process of breaking the replicator molecular structure yielding low energy monomers μ. In addition, equation (4) account for the output flux of information carriers from the system.

The replication function $\mathcal{R}_k$ is a reaction-rate parameter that can be expanding in a polymonical way as a consequence of mass action kinetics[7]:

$$\mathcal{R}_k(\vec{y}) = R_k + \sum_{j=1}^{n} R_{kj} y_j + \sum_{i=1}^{n} \sum_{j=1}^{n} R_{kij} y_j y_i + \dots \quad k = 1, 2, \dots, n \tag{5}$$

$\vec{y}$ being the vector species whose components are the concentration of each replicator ($y_i \ i = 1, \dots, n$). This function takes into account all the possible mechanisms involved in the replication process: the first term governs the uncatalyzed replication; the second one drives the bimolecular-catalyzed self-reproduction. The rest of the terms of this expansion represent the catalyzed self-replication by means of the simultaneous action of more than one network species. These last processes seem highly unlikely and can be neglected. In every case, the catalysis intensity is measured by the catalytic constants $R_{i,\dots,j}$.

As in the replication process, the degradation function $\mathcal{D}_k$ represents a generalized reaction-rate parameter, so it can be also expanded as a polynomial

$$\mathcal{D}_k(\vec{y}) = D_k + \sum_{j=1}^{n} D_{kj} y_j + \sum_{i=1}^{n} \sum_{j=1}^{n} D_{kij} y_j y_i + \dots \quad k = 1, 2, \dots, n \tag{6}$$

Each term of this equation has similar meaning to the corresponding term in the replication function expansion (Eq. (5)). Most models only consider the first term, although catalyzed degradation may exist.

As described in reaction (2), the process of replication is error-prone: the primary structure can be changed during the process. This possibility is introduced in the model by means of the mutation matrix $\mathcal{Q}$. Q_{kj} are the elements of this matrix, determining the rate of formation of the species I_j as a consequence of replication of the species I_k. The diagonal elements of $\mathcal{Q}$ (Q_{kk}) are the free-error rate parameters (eqn. (1)).

In order to get competition some kind of constraint on the system must be imposed. Either a constant population constraint CP (number of replicators fixed) or constant flux constraint CF (input flux of monomers constant) is usually assumed. It can be demonstrated that the asymptotic behaviors obtained under CP constraint is equivalent to those obtained under CF constraint[8].

It's worth noting that the dimension of this system is astronomically large (if we consider replicators as chains of digits, the number of different primary strutures of replicators of lenght 100 that can be formed from four different digits is around 10^{30}).

3 UNCATALIZED DYNAMICS: QUASILINEAR MODEL

If a very low catalytic activity of the replicators is assumed, both replication function $\mathcal{R}$ and degradation funtion $\mathcal{D}$ become linear, i.e. only the first terms in the expansion may be considered. This assumption would imply that any species is independent each other from a kinetic point of view, but any kind of constraint (e.g. CP or CF) leads to a kinetic link among the network species. In addition, all the network species are related via mutation matrix. Thus, the evolution of each species depends directly on the existence of other species.

Due to the enormous complexity of this system only in very special situations a complete analysis has been carried out[9,10]. From these simplified models, it has been understood the general behavior of populations composed of uncatalyzed self-replicative species, although some important characteristics remain hidden.

One relevant result derived from the quasilinear model is that coexistence among different networks species is not possible. In special fitness landscapes (one peak or two) and depending on the mutation parameter value, the system can reach either a metastationary state, where only the fitest species remains, or a distribution in which the mean life time for every species is relatively short. In the first case the fitest species is surrounded by a cloud of nearby mutants, whose distribution changes in time (quasispecies), whereas in the second case the whole system is moving through the sequence space (random replication state). Recent works have proved that this behavior is the result of considering a simple fitness landscape. In more complicated landscape the system behavior can be considerably different[11,12]. In any case, the chain length is always limited by the mutation rate, which implies an information crisis.

Quasilinear models give rise to evolutive and selective processes, and the same to an optimitation process[13]. But, it's clear that emergence phenomena can't appear in this models where the only relationships among the species are mutation and those imposed by external constraints.

4 CATALYZED DYNAMICS: NON-LINEAR MODELS

In the original Eigen's paper[1], the catalytic activity was introduced in order to overcome the information crisis mentioned above, making species, otherwise competitive, cooperative. First analysis on the quasilinear model demonstrated that more sophisticated systems (systems composed of larger molecules, and therefore with the possibility of more information storage needed to carry out complex activities) only could be originated by means of a catayatic relationships among the network species[1]. Since then many studies have been carried out on this kind of systems[14,15].

In most non-linear network models the replication function expansion is cut after the second term. The dynamics that appears when these new terms are involved in the time evolution of the system is more complicated than in the quasilineal model. Different simplifications have been considered in order to get conclusions. Ones, at the level of catalytic realtionships; others at the level of the mutation matrix.

Considering catalytic relationships, several particular cases have been exhaustively analyzed. Many studies have been focussed on networks whose elements are cyclically coupled (hypercycle)[14]. From these studies, it was deduced that this kind of relationship allows the coexistence of all the species involved in the hypercyclic organization. Then, hypercycle was proposed as a general mechanism in order to assure the information content in each of its elements (therefore increasing the whole information content). In addition, this mechanism enables to take "once for all time" decisions. Other cases, as hypercycles with parasites, and connected or disconnected hypercycles, have been also analyzed as particular catalytic networks[8,14,15]. Another important case is the fully connected network. Similar conclusions can be obtained from the analysis of these models, although fundamental diferences exist[7,21]. Anyway, all these approaches can be derived from a general formulation of catalytic networks, setting up particular values of the catalytic constants R_{ij}[16].

Speaking of mutation matrix, two basic considerations may lead the simplifications to be carried out. If the catalytic network is formed from species involved in a stable quasispecies structure[11], it seems reasonable that a non null probability to reach every sequence from the erroneous self-replication of one network element should be considered [7]. In this approach all possible species of the sequence space may belong to the catalytic network. On the contrary, if the network is formed from species separated through large Hamming distance (species genetically disconnected), it seems that the probability of reaching a network species due to the erroneous self-replication of other network species must be negligible. This organization might be the result of catalytic stablishement after the evolution of a quasispecies system in a multipeak (but separated) landscape.

The model our group proposed some years ago takes into account these last assumptions[17], a general fully connected catalytic network formed by error-prone self-replicative molecules very separated genetically. In addittion, it is assumed that mutants have lost their catalytic activity.

In order to make this model mathematically tractable some additional assumptions are needed. The main one is that mutants are indistinguishable (error-tail) and that there is a null probability of getting a network species from the erroneous self-replication of error-tail species.

This model has been extensively analyzed for low dimension networks, both from a deterministic[18,20] and from a stochastic[17,19] point of view. The generalization to larger dimensions networks has been recently developed[21,22]. Obviously, after these simplifications the model fails to take into account important properties (e.g. evolutive properties, the distribution of species in the sequence space, the possibilty of changing the network dimension, etc...). Hovever, other relevant questions can be outlined from it: coexistence properties of units forming the network (e.g. symmetric networks have better stability properties than asymmetric networks); role of the error-tail in the behavior of catalytic networks (e.g. there is some parameter range where bistability between error-tail and network species exists); effect of superimposed competition on the network dynamics that appear when individual characteristics of each networks element are considered (e.g. in some circumstances low dimension network subclasses can be better that the whole network); decreasing of the error-threshold, etc,.... (For a complete review see the references above cited).

5 WHAT'S CATALYSIS GOOD FOR?

In this section we would like to pose some questions we think are basic within the framework of catalytic networks theory. Some of them are certainly able to be reasonably answered; others are still subject of scientific speculation.

- What's the contribution of catalytic activity to the evolution?

Evolution implies not only optimization (i.e. searching for the fittest species among the overall sequences space), but also emergence phenomena. Uncatalyzed replicator systems have been proved to be driven by a potencial function (cost function) which is optimized in the evolution process[13], and hardly yield more complicated behaviors. So, other kind of organization is needed in order to explain more sophisticated steps in biological evolution.

Catalysis originates different relationships among molecules from those derived from mutations and external constraints. This can be considered as a natural way of emerging complex structures. From a mathematical point of view this catalytic activity is displayed in intrinsic non-linear terms appearing in the time evolution equations. Then, both more complex behaviors and information increase emerge as a global characteristic.

- What kind of catalytic relationship fit better real biological systems?

Any kind of functional relation among the species forming a network must be supported by the physical properties of these molecules. So, catalysis of chemical reactions has been classically attributed to protein properties. The initial Eigen's model of catalytic network, implicitly or explicitly involved proteins[1]. But this implies the necessity of including a translation machinery following a genetic code well stablished. This model brought out this machinery as an emergent property of hypercycles. Since the catalytic activity of RNA molecules has been experimentally stablished[5,6], the formation of catalytic networks involving exclusively nucleic acids molecules appears more plausible in first steps of prebiotic evolution.

It seems that in these early stages an unspecific relation involving a large number of molecules might exist. However, hypercyclic relations must be developed under a very specific situation, which would be a very unlikely event to occur in these stages. In addition, in a recent study carried out on finite size systems[23], it has been proved that cyclic relation has advantages beside a general fully connected network for low dimension networks, but for high dimension networks, the hypercyclic organization becomes more unstable than fully connected networks. So, the first catalytic organization formed at the beginning of prebiotic evolution might be a fully connected network with non-uniform catalytic constants.

- Are simplifications so drastic that basic properties of catalytic networks are hidden?, and where is our contribution placed within the general framework about catalytic networks?

A model necessarily simplifies reality, trying to answer particular questions about its phenomenology. Very important problems are both to define the limits of the aproximation and the kind of phenomena that can be solved from the proposed model.

Our approach[17,18,19,20,21,22] introduces new aspects which are not so well analyzed in other approaches. These aspects can be summarized in the following two: firstly, the competitive effect of the error-tail, and secondly, the effect of a superimposed competition among the network elements. The latter is introduced at two levels: catalyzed ($R_{ii} \neq 0$) and uncatalyzed ($R_i \neq 0$). On the other hand, the possibility of reverse mutations from the error-tail species to the network is neglected. Even though information about evolutionary properties of the system are lost, a general picture of such events can be concluded from these studies as it has been previously remarked.

Other kind of simplifications are referred to the mechanism proposed for the catalytic activity. In our model, as well as in most of the catalytic network models, single molecular collisions between ribozyme and substrate are assumed. But the existence of intermediate ribozyme-substrate complexes is more realistic, likely giving rise to different behaviors. Stadler *et al.*[24] carried out a study on the dynamics of catalytic networks with more general relations among the elements.

Diffusion is another property that plays a very important role in real systems, although less treated in literature. For instance, it has been recently proved[25] that diffusion can stabilize hypercycles against the action of parasites. This can also be the origin of heterogeneity, even coexistence, in quasilinear models.

Finally, it's worth noticing that most of the models has been analyzed under the assumption of infinite population using a deterministic description. But, the role of fluctuations on the dynamics of catalytic networks only can be measured by means of others techniques[19,26,11]. Thus, a probabilistic description is needed in order to analyze deeply what the real behavior of catalytic nertwork is like.

After the considerations mentioned above, what is the role of catalytic activity in evolution of self-replicative molecules?

Taking into account the catalytic activity in ribozymes populations is suggested by experimental facts. It seems plausible that biological evolution pass through an stage in which some biological species acquired this activity. From the picture recently drawn in reference [11], a catalytic network could be originated after the formation of a quasistationary structure, allowing the system to survive for a long time. A recent work has proved that extinction time for particular quasilineal model is lower than for any kind of catalytic networks[23]. Therefore, in order to have a plausible biological system, lasting enough time to evolve to better forms, it seems a necessary condition the emergence of any kind of catalytic relationship (likely fully connected) among the system molecules.

6 REFERENCES

1. M. Eigen, *Naturwissenchaften* **58**, 465 (1971)

2. S. Spiegelman, *Quart.Rev.Biophys.* **4**, 213 (1971)

3. L.E. Orgel, *J.Mol.Biol.* **38**, 381, (1968)

4. F.H.C Crick, *J.Mol.Biol.* **38**, 367 (1968)

5. T.R. Cech, A.J. Zaug and P.J. Grobowski, *Cell* **27**, 487 (1981)

6. J.A. Doudna and J.W. Stostak, *Nature* **339**, 519 (1989)

7. P.F. Stadler and P. Schuster, *J. Math. Biol.* **30**, 597 (1992)

8. B.O. Küppers, 'Molecular Theory of Evolution'. Springer Verlag, Berlin (1983)

9. M. Eigen, J. McCaskill and P. Schuster, *J. Phys. Chem.* **92**, 6881 (1988)

10. M. Eigen, J. McCaskill and P. Schuster, *Adv. Chem. Phys.* **75**, 149 (1989)

11. P. Tarazona, *Phys.Rev A* **45**, 6038 (1992)

12. P. Tarazona, Contribution to this volume

13. P. Schuster, in 'Molecular Evolution on Rugged landscapes: Proteins, RNA and The Inmune System', Ed. A.S. Perelson and S.A. Kauffman, 47 (1991)

14. M. Eigen and P. Schuster, 'The hypercycle- A principle of natural self-organization', Springer Verlag, Berlin (1979)

15. J. Hofbauer and K. Sigmund, 'The Theory of Evolution and Dynamical Systems'. Cambridge University Press, Cambridge (1988)

16. P. Schuster, in 'Molecular Evolution on Rugged landscapes: Proteins, RNA and The Inmune System', Ed. A.S. Perelson and S.A. Kauffman, 281 (1991)

17. A. García-Tejedor, R. Castaño, F. Morán and F. Montero, *J.Mol.Evol.* **26**, 294 (1987)

18. A. García-Tejedor, F. Morán and F. Montero, *J.Theor.Biol.* **127**, 393 (1987)

19. A. García-Tejedor, J.C. Sanz-Nuño, J. Olarrea, F.J. de la Rubia and F. Montero, *J.Theor.Biol.* **134**, 431 (1988)

20. M. Andrade, A. García-Tejedor and F. Montero, *Biophys. Chem.* **40**, 43 (1991)

21. J.C. Nuño, M. Andrade, F. Morán and F. Montero, *Bull.Math.Biol.* **55**, 385 (1993)

22. J.C. Nuño, M. Andrade and F. Montero, *Bull.Math.Biol.* **55**, 417 (1993)

23. J.C. Nuño and P. Tarazona, submitted to *Theor.Pop.Biol.* (1992)

24. B.M.R. Stadler and P.F. Stadler, *Bull.Math.Biol.* **53**, 469 (1991)

25. M.C. Boerlijst and P. Hogeweg, *Physica D* **48**, 17 (1991)

26. M. Nowak and P. Schuster, *J.Theor.Biol.* **137**, 375 (1987)

MOLECULAR QUASI-SPECIES IN HOPFIELD
REPLICATION LANDSCAPES

P. Tarazona

Departamento de Fisica de la Materia Condensada
Universidad Autónoma de Madrid
E-28049 Madrid, Spain

Twenty years ago Eigen[1] proposed a model which, in the mathematical formulation given by Eigen and Schuster[2], links the dynamics of population genetic in biological systems with the dynamics of chemical reactions opening the field of molecular evolution. In its simplest realization the model describes a system of aperiodic polymers with a fixed number, N, of monomers which may be of ν different classes. For RNA we would have $\nu = 4$ to represent the four nucleotides or $\nu = 2$ if we discern only between purines and pyrimidines. In this case, which is assumed all over this article, each polymer is represented by a binary sequence $(s_1, s_2, ..., s_N)$, with s_i coded here as 1 or -1. The dynamics of the population follows from the (very far from equilibrium) chemical reactions describing the replication of these molecules in the presence of activated monomers and possibly of some catalyst, like in the flux reactor experiments for RNA replication in the presence of a replicase enzyme[3]. If we use I_k with $k = 1$ to 2^N) as a shorthand for each possible sequence, the replication reaction is:

$$I_k + [\, activated \;\; monomers \,] \longrightarrow I_k + I_j \tag{1}$$

with a polymer I_k producing a new polymer I_j out of activated monomers, which are kept in a concentration high enough to have the reaction flowing only in one direction, at a rate denoted by W_{jk}.

The two key elements to obtain a model with darwinian selection are accurate but not perfect replication of the polymers and the different rates of the replication reaction for different polymers in what is usually called the 'replication landscape'. Still in the simplest case the replication accuracy may be described by a single parameter q giving the probability of faithful replication of a single digit in the binary sequence. The replication of different digits if taken to be independent, representing point mutations without hot spots, insertions or deletions. Values of q near 1 correspond to direct replication with mutation rate $1 - q << 1$. The complementary replication of real RNA would correspond to $q << 1$ and the production of random polymers is described by $q = 1/2$. The rate for reaction (1) is then given by:

$$W_{jk} = A[I_k] \, q^N \left(\frac{1-q}{q} \right)^{d_H[I_j, I_k]} \tag{2}$$

where $d_H[I_j, I_k]$ is the Hamming distance between the two sequences, i.e. the number of digits which are different in the two binary sequences. $A[I_k]$ is the replication landscape: an arbitrary function defined over the set of all possible sequences (the N dimensions hypercube).

The model has been explored mainly in terms of the ordinary differential equation describing the chemical kinetics in the deterministic limit [4,5]. Very simple choices for the replication landscape which already show the some of the main features of the model and studies have been devoted to a model landscape with only two values: A_0 for a single 'master sequence', I_0, and $A_1 < A_0$ for any other possible sequence [4,5]. For high replication accuracy the steady state population in this model landscape is made of master sequences and a cloud of mutants at small Hamming distance, this population structure was called the 'quasi-species' by Eigen and Schuster [2] and since then it has proved to be a very useful concept to describe real populations of RNA molecules and viruses. If the mutation rate $1 - q$ increases (e.g. by adding a mutagen agent to the flux reactor) the quasi-species becomes broader, with more mutants further away from the master sequence, and there is a critical value,

$$(1 - q_c) \, N = log \left[\frac{A_0}{A_1} \right], \tag{3}$$

at which the population spreads over all the configuration space, loosing all the genetic information contained in the quasi- species. This information catastrophe is known as the 'error-threshold'.

An alternative formulation of the problem was recently proposed by Leuthäusser[6,7] following the earlier suggestion by Little[8] to interpret the reaction rate matrix W as a transfer matrix in statistical mechanics. The idea is to discretize the time in generations so that, representing the concentrations of each sequence in the population at time $t = i$ by a vector $X(i) = (x_1(i), x_2(i), ..., x_{2^N}(i))$, the population n generations after a initial set up $X(0)$ is given by

$$X(n) = W^n \, X(0).$$

The dynamics of the polymer population may be represented in a square lattice with Ising spin variables $s_j(i) = \pm 1$ at each node. One direction in the lattice represent the sequence of monomers along the chain, with the index j running from 1 to N. Along the other direction runs the index i to represent the generations, from a initial set up to the present. The spin lattice hamiltonian

$$-\beta H = \sum_{i=0}^{n-1} \left(\beta \sum_{j=1}^{N} s_j(i) \, s_j(i+1) + \ln A[I(i)] \right) + \frac{nN}{2} \ln[q(1-q)]. \tag{4}$$

and the inverse temperature is $\beta = \ln \sqrt{q/(1-q)}$ (for $q > 0$) produce a left row transfer matrix with the form given in eq.(2). [7,9].

This map of the population dynamics into a spin-lattice in thermodynamic equilibrium allows the characterization of the error threshold as a phase transition [9]. For the simple replication landscape described above, with a single sharp peak at the master sequence and flat everywhere else, the error threshold becomes a first order phase transition, rounded by the finite sequence size and with some characteristic surface effects reflecting the irreversibility of the replication reactions (1). Other simple landscape models have also been explored[9] giving again some kind of phase transition, of first or second order, between the quasi-species and the population of random polymers, with the error rate $1 - q$ playing the role of the temperature.

The main apparent limitation of Eigen's model, in the light of the analysis with simple replications landscapes, was the lack of diversity in the population. A model landscape with two different masters sequences having a replication rate higher than any other sequence, fails to stabilize a population with two quasi-species, each one around one of the master sequences. Instead the steady state population is fully attracted towards one of them in a relatively short transient time proportional to the population size[10]. However, the coexistence of sequences with genetic stability but also with a good amount of diversity seems to be the way towards the formation of cross catalytic networks, linking genetic information to some primitive kind of metabolism and opening the path to self-organised complexity[2]. A few years ago Anderson [11] proposed a spin-glass model to obtain both stability and diversity in pre-biotic evolution. The model used a complex replication mechanism based on the conjugation of sequences with variable sequence length and a Sherrinton-Kirpatrick spin-glass model for the degradation landscape. The complexity of the replication-mutation mechanism impose that the only way to analyze the model is through computer simulations [12].

Leuthäusser's map of the population dynamics as a problem in thermodynamic equilibrium provides a route to analyze spin-glass models for the replication landscape with the powerful theoretical tools developed in spin-glass theory and to check the effects of the simpler replication mechanism in Eigens's model with a 'complex' replication landscape. This possibility has been very recently explored [9] with landscape models based on the Hopfield hamiltonian for neural networks [7,13]. These replication landscapes depend on a set of p master sequences $S_0^\eta = (\xi_1^\eta, ..., \xi_N^\eta)$ with $\xi_i^\eta = \pm 1$ and $\eta = 1$ to p, which in the original context of neural networks represent the stored patterns. The replication landscape takes the form:

$$A[I] = \exp \left[\frac{K}{2N^2} \sum_{\eta=1}^{p} \sum_{j \neq j'}^{N} \xi_j^\eta \xi_{j'}^\eta s_j s_{j'} \right].$$ (5)

K being a constant to give the relative difference between the peaks and the valleys of $A[I_k]$. The equivalent hamiltonian (4) becomes now,

$$-\beta H = \sum_{i=0}^{n-1} \left(\beta \sum_{j=1}^{N} s_j(i) s_j(i+1) + \frac{K}{2N^2} \sum_{\eta=1}^{p} \sum_{j \neq j'}^{N} \xi_j^\eta \xi_{j'}^\eta s_j(i) s_{j'}(i) \right)$$ (6).

where the last term in (4) has been dropped as an irrelevant shift in the energy scale. This hamiltonian, although more cumbersome than the original Hopfield model, may still be analyzed with the standard techniques developed for fully connected spin-glass models[14], which become exact in the limit of infinite chain length.

The relevant parameter to measure the 'complexity' of the landscape is the ratio between the number of master sequences and the polymer length, $\alpha = p/N$. For $\alpha = 0$, which corresponds to any finite number of patterns in the limit of infinite N, the system is self-averaging, so the statistical properties do not depend on the particular set of patterns. The phase diagram, in terms of the effective temperature,

$$\theta = \frac{N(1-q)}{q\,K},$$

has only two regions and the error threshold is a second order phase transition at $\theta_c = 1$. For $\theta < 1$ the population forms a quasi-species around only one of the master sequences, while for $\theta > 1$ it spreads over the full hypercube. Thus, in this limit the model still fails to achieve the stability of the diversity in the genetic structure of the population. However, for $\alpha > 0$ the behaviour is richer[9]. The system is not any more self-averaging, so that the quenched averaged over the master sequences must be performed with the use of replicas. The solution within the replica symmetric approximation[15] gives most of the phase diagram and shows the existence of a spin-glass intermediate phase between the the single quasi-species and the fully random population. In this phase the population has small but non-zero overlaps with many master sequences because it is spread over the network formed by the secondary maxima of (5) which appear at the intersection of any odd number of patterns[15]. This network occupies still a very low fraction of the configuration space, but it contains very different sequences. Thus, the spin-glass phase provides the stability of the genetic diversity even within the simple replication mechanism of independent point mutations with fixed sequence length. The only essential requisite seems to be the 'complexity' of the replication landscape. The simple landscape models which have been thoroughly analyzed with the ordinary differential equations[4,5] seem to be very particular cases, rather than simple but generic models of the real replication landscapes. The actual replication rate of RNA chains depends on the secondary and tertiary structures which result from the folding of the chain over itself. The folding process depends in a very complex way on the sequence of nucleotides along the chain, so that the (still largely unknown) real replication landscapes are probably very complex indeed[16]. Work is in progress to study the effects of finite populations.

REFERENCES

1. M. Eigen, Naturwissenchaften **58**, 465 (1971).
2. M. Eigen and P. Schuster, Naturwissenchaften **64**, 541 (1977); ibid **65**, 7 and 341 (1978).
   These works are also published in : 'The hypercycle- A principle of natural self-organization', M. Eigen and P. Schuster, Springer Verlag, Berlin (1979).
3. C.K. Biebricher, M. Eigen and R. Luce, Nature **321**, 89 (1986).
4 M. Eigen, J. McCaskill and P. Schuster, J. Phys. Chem. **92**, 6881 (1988).
5 M. Eigen, J. McCaskill and P. Schuster, Adv. Chem. Phys. **75**, 149 (1989).
6. I. Leuthäusser, J. Chem. Phys. **84**, 1884 (1986).
7. I. Leuthäusser, J. Stat. Phys. **48**, 343 (1987).
8. W.A. Little, Math. Biosci. **19**, 101 (1974).
9. P. Tarazona, Phys. Rev. A (to be published, 1992).
10 J.C. Sanz Nuño and P. Tarazona, preliminary results.

11. P. W. Anderson, Proc. Nat. Acad. Sci. USA **80**, 3386 (1983).

12. D.S. Rokhsar, P. W. Anderson and D.L. Stein, J. Mol. Evol. **23**, 119 (1986).

13. J.J. Hopfield, Proc. Nat. Acad. Sci. USA **79**, 2554 (1982).

14. M.Mezard, G. Parisi and M.A. Virasolo, 'Spin Glass Theory and Beyond', World Scientific, Singapore (1987).

15. D.J. Amit, H. Gutfreund and H. Sompolinsky, Ann. Phys. **173**, 30 (1987).

16. W. Fontana, W. Schnabl and P. Schuster, Phys. Rev. A **40**, 3301 (1989).

EFFECTS OF NOISE ON SELF-ORGANIZED

CRITICAL PHENOMENA

Albert Díaz-Guilera

Departament de Física Fonamental
Universitat de Barcelona
Diagonal 647, 08028 Barcelona, Spain

INTRODUCTION

Recently much attention has been paid to the phenomenon known as self-organized criticality. Bak, Tang, and Wiesenfeld[1] studied a cellular automaton model in which the dynamic rules lead the system to a critical state without any characteristic time or length scales. This scale invariance suggests that the system is critical in analogy with classical equilibrium critical phenomena. The difference lies in the fact that in these models no fine tuning of external parameters is needed to reach the final critical state. Several aspects of self-organized criticality have been extensively treated in the literature.[2]

Here we focus our attention on numerical simulations and on an analytical description by means of a non-linear equation. Numerical simulations are performed on lattices with a continuously distributed variable; this model was introduced by Zhang[3] and has been used by other authors. This model is expected to be in the same universality class as the original sandpile model[1]. We want to check the universality classes of the Zhang model by changing the microscopic rules to see under which circumstances the macroscopic behavior (dynamic exponent) is modified.

From the dynamic rules one can obtain a coarse-grained version where the microscopic parameters enter into the transport coefficients. Different models then give rise to different macroscopic equations. From these nonlinear equations containing the threshold condition one usually builds up simple nonlinear equations that retain the underlying symmetries and conservation laws as well as the characteristics of the noise sources. Our aim here is to analyze the simplest nonlinear equations obtained from different microscopic rules and to compare these results with the numerical simulations. The models we have studied have different symmetries from the models treated in other similar approaches[4-8] and, hence, different macroscopic behavior is expected.

THE MODELS

The model originally proposed by Zhang[3] consists of a lattice in which any site

Growth Patterns in Physical Sciences and Biology, Edited
by J. M. Garcia-Ruiz *et al.*, Plenum Press, New York, 1993

can store an energy E continuously distributed between 0 and E_c. This variable which we call energy can have different physical interpretations[9]. The system is perturbed by choosing a site at random and adding an amount of energy δE which is also a random variable. Once a site reaches a value of energy greater than some critical value E_c, this site becomes active and isotropically transfers the full amount of energy to its nearest neighbors. At this point, the input of energy from the outside is turned off and the energy transferred to the neighboring sites can make them active, leading to new transfers of energy, giving rise to an activation cluster or avalanche that ends when all the sites have reached an energy less than E_c. It is only when the avalanche has stopped that energy is added again, otherwise the system remains quiescent. Once this procedure has been repeated a large enough number of times, there exists a well defined distribution of energies characterizing the dynamic equilibrium state, which is homogeneous and isotropic, on the average[9]. This distribution is equivalent to the original discrete sandpile model of Bak et al[1], in which the variable describing the system state is the slope. Actually both models share some of the critical exponents and seem to belong to the same universality class[3, 9].

The microscopic rules for this model can be written in the form of a set of algebraic equations, one for each site,

$$E(i, t+1) = [1 - \theta(E(i,t) - E_c)]E(i,t) + \sum_{nn} \theta(E(j,t) - E_c)E(j,t)/q + \eta_e(i,t) \quad (1)$$

where $\theta(x)$ is the Heaviside step function, q is the lattice coordination number, and $\eta_e(i,t)$ is the external noise which generates the dynamics of the system. The sum includes the nearest neighbors j to the lattice site i. Energy flows out freely through the boundaries to preserve overall energy conservation.

In the next section we will discuss in detail some of the variations one can introduce in this model in order to check the universality classes. However, one of them deserves some remarks. A simple change in the microscopic rules can be done in such a way that an amount of energy equal to E_c be transferred to the set of neighboring sites when a site becomes active instead of transferring its total energy $E > E_c$. This new model is closer in spirit to the original sandpile model of Bak et al.[1] where this is necessary due to the discrete nature of the critical variable, the slope. In this case the set of algebraic equations read

$$E(i, t+1) = E(i,t) - \theta(E(i,t) - E_c)E_c + \sum_{nn} \theta(E(j,t) - E_c)E_c/q + \eta_e(i,t). \quad (2)$$

The importance of this model lies in the fact that it introduces a new symmetry: the deterministic equations are invariant under the transformation $E - E_c \rightarrow -(E - E_c)$. There are other symmetries which are common to the original Zhang model discussed previously, both are invariant under translation, rotation and reflection.

COMPUTER SIMULATIONS

We have performed computer simulations of the model sketched in the previous section together with some variants in order to determine the universality classes they

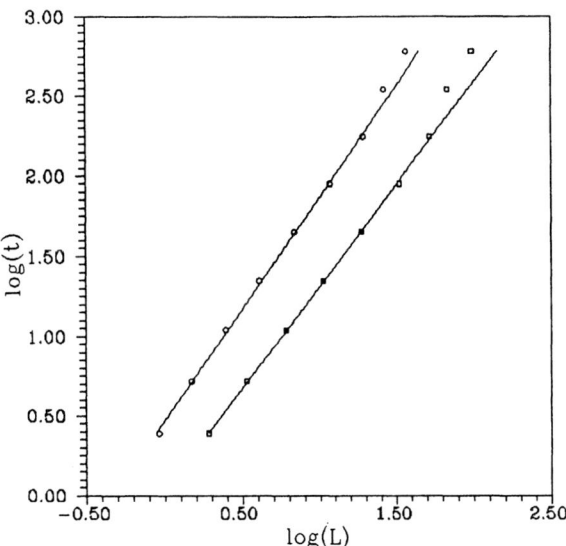

Figure 1. The duration of an avalanche versus its mean characteristic length in logarithmic scales; for two different definitions of the length: o) length is defined as the radius of gyration with respect to the center of mass, □) the length is the maximum distance to a perimeter site from the seed. Straight lines are best linear fits to the sets of data with slopes $z = 1.40$ and $z = 1.28$, respectively.

belong to. Our analysis does not pretend to be exhaustive since our main goal is to obtain one of the exponents, the dynamic exponent, which can be analytically calculated from a stochastic partial differential equation by means of the dynamic renormalization group (DRG) procedure[10-12], that will be the subject of the next section.

In our simulations we define the size of an avalanche as the number of sites that have become critical regardless of the number of times this has happened for a given site, while we define the time as the number of steps the avalanche takes and finally the characteristic length of the avalanche as the radius of gyration with respect to its center of mass. There are other choices in the literature, some of them are equivalent but others can lead to wrong conclusions, as we think it happens in Ref.[2], where the authors take the maximum distance to a perimeter site from the seed as a characteristic length of the avalanche. In our opinion this makes the dynamic exponent (the exponent relating the duration of the avalanche to its characteristic length) closer to unity since avalanches in a given direction make this choice of the length grow linearly in time. It is worth noting that this can modify the critical dimension of these systems.

Concerning the critical behavior, we have measured the dynamic exponent for a 128x128 square lattice with a starting configuration in which all sites are critical, taking $E_c = 1$ without loss of generality. We make 1000 runs for the system to get the dynamic equilibrium state and 10000 runs to get the dynamic statistical properties. In Fig. 1 we plot the duration of the avalanches as a function of their characteristic length for the different definitions mentioned above, each one extracted from the same set of simulations. The simulation data has been coarse-grained in order to get smoother curves. We consider time intervals of exponentially increasing amplitude 2,3-4,5-8,9-16,... and we associate the averaged lifetime and the averaged length of the avalanches

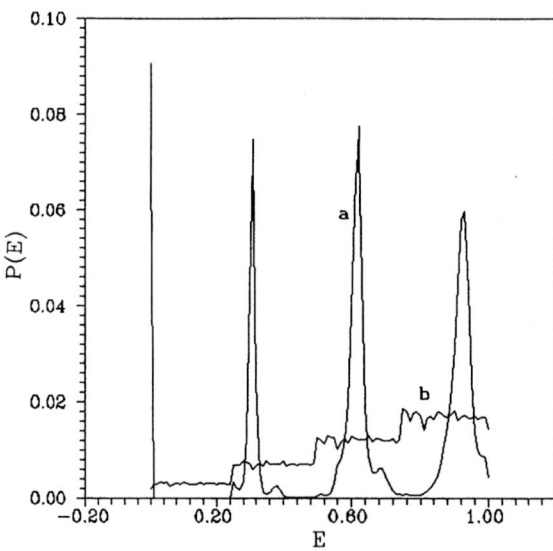

Figure 2. Distribution of energies for the microscopic dynamic rules described in the text: a) original Zhang model, b) introducing a new symmetry.

to this interval. For the definition of the characteristic length we have chosen, we get a dynamic exponent $z = 1.40 \pm 0.03$ whereas for the definition given in Ref.[2] one gets a smaller value $z = 1.28 \pm 0.03$, in agreement with our previous comments about this choice. In order to minimize the finite size effects, we have performed an extrapolation to infinite lattice size from which we get an estimate for the dynamic exponent $z = 1.36 \pm 0.03$.

We have checked universality by changing some of the parameters of the dynamic rule given by eq. (1). Instead of transferring the full amount of energy, we have considered that only a fixed fraction be isotropically transferred from a critical site to its four nearest neighbors. This pushes the system toward criticality, i.e. all sites are closer to the critical value of the energy E_c, and avalanches become smaller. We have computed the dynamic exponent for different values of the fraction of energy released. From there we get a set of dynamic exponents that do not change in comparison to the original model within numerical precision. Although some of the characteristic features of the model are changed (the distribution of energies still has four peaks but now they are closer to E_c and the average energy per site is different) the macroscopic behavior given by the dynamic exponent is unaltered. The same can be concluded when modifying the intensity of the noise, i.e. the amount of energy we add to perturb the system when it is at equilibrium.

In the previous section we presented an alternative to the original Zhang model that introduces a new symmetry. This new model seems to be closer to the discrete sandpile models that coined the term self-organized criticality. Although the energy is again a continuously distributed magnitude in each site, the distribution of energies is changed completely. In Fig. 2 we plot this distribution and the distribution obtained with Zhang model in order to compare them. When plotting the avalanche lifetime against the characteristic length for both models, it is difficult to distinguish between both sets of points and so, one can conclude that both models belong to the same universality class. This is confirmed by inspection of the distribution of lifetimes and sizes of the avalanches, and although the agreement is not as close, the curves have the same behavior within numerical accuracy.

DYNAMIC RENORMALIZATION GROUP APPROACH

From the microscopic rules (1) one can construct an effective medium equation in terms of a rescaled energy $E - E_c \rightarrow E$, in which the microscopic scales enter into the transport coefficients, as follows

$$\frac{\partial E(\vec{r}, t)}{\partial t} = \alpha \nabla^2 \left[\theta(E(\vec{r}, t)) (E(\vec{r}, t) + E_c) \right] + \eta_e(\vec{r}, t) + \eta_i(\vec{r}, t) \tag{3}$$

where α is related to the lattice spacing, to the unit time step, and to the fraction of energy released and plays the role of a diffusion constant. We have introduced $\eta_i(\vec{r}, t)$ as an internal noise which accounts for the removed microscopic degrees of freedom. This internal noise would obey a fluctuation-dissipation theorem linked to a conserved magnitude, in this case, the energy. Therefore, it has a zero mean and a correlation function given by

$$< \eta_i(\vec{r}, t) \eta_i(\vec{r'}, t') > \propto 2\alpha \nabla \nabla' \delta(\vec{r} - \vec{r'}) \delta(t - t'). \tag{4}$$

Since the external noise is fundamental for the dynamic properties of the model, this point deserves some discussion. The input of energy to the system is a random number between 0 and δE_{max} and the noise acts only between avalanches. This makes the external noise we write in (3) depend on the energy distribution in the lattice. However, in the previous section we reported numerical simulations in which the critical behavior (dynamic exponent) is not changed when δE_{max} is varied by orders of magnitude. For very small δE_{max}, the evolution is slowed down since many more inputs are needed to make a site critical and to start the avalanche. Moreover, for such a low intensity of noise, we have found that if noise is not turned off between avalanches, i.e. it is constant in time, the dynamic exponent is not modified, and interactions between avalanches remain negligible. Therefore we conclude that the external noise can be modelled by a time-independent stochastic process.

The noise character can also be analyzed from a different point of view. One can assume a gaussian process with a zero mean and a correlation function given by

$$< \eta(\vec{r}, t) \eta(\vec{r'}, t') > = \frac{2D}{\tau} e^{-|t-t'|/\tau} \delta^d(\vec{r} - \vec{r'}). \tag{5}$$

This is a Ornstein-Uhlenbeck process with τ being the correlation time. This general case can describe two very different limits. When dealing with internal noise which involves a microscopic time scale, τ should be very small and in this limit, we recover the usual white noise in space and time

$$< \eta(\vec{r}, t) \eta(\vec{r'}, t') > = 2D \delta(t - t') \delta^d(\vec{r} - \vec{r'}). \tag{6}$$

411

But in our problem, the external noise has no characteristic time scale. The only thing we know is that the noise acts once the avalanche is over, meaning that the only characteristic time would be a macroscopic time which scales with some power of the system size[13]. Then the limit $\tau \to \infty$ is appropriate and we can write

$$< \eta(\vec{r}, t)\eta(\vec{r'}, t') >= \frac{2D}{T}\delta^d(\vec{r} - \vec{r'}), \tag{7}$$

where T is a macroscopic time much greater than the unit time step. Therefore, both points of view are linked by the fact that there are only short-range spatial correlations and an intensity of the noise that scales with $1/T$ or $1/L^\mu (\mu > 0)$. Both types of noise (6) and (7) are nonconservative due to the way energy is added from the outside. This nonconservative nature breaks detailed balance. This is believed to be one of the main ingredients of self-organized criticality[6].

Clearly eq. (3) is a stochastic nonlinear differential equation from which one seeks to obtain the hydrodynamic (long-time and large-scale) behavior of the system. In order to make (3) manageable we choose one of the representations of the step function

$$\theta(x) = \lim_{\beta \to \infty} \frac{1}{2}[1 + \tanh \beta x] \tag{8}$$

and make a series expansion in powers of the argument. This can be performed by keeping β finite instead of $\beta \to \infty$. We can then write

$$\frac{\partial E(\vec{r}, t)}{\partial t} = \alpha \nabla^2 E(\vec{r}, t) + \sum_{n=2}^{\infty} \lambda_n \nabla^2 E^n(\vec{r}, t) + \eta_e(\vec{r}, t) + \eta_i(\vec{r}, t). \tag{9}$$

By simple dimensional analysis one can show that the internal noise with correlation function (4) makes the contribution of the nonlinearities irrelevant for any spatial dimensionality. This makes the internal noise itself irrelevant so from now on, we ignore this noise and discuss the effects of external noise as given by (7). Moreover, one realizes that all coupling constants λ_n are relevant when $d < 4$ and irrelevant when $d > 4$. Thus we can conclude that the upper critical dimension is four[14] and that below it, one needs the full set of nonlinearities to study the hydrodynamic behavior of this model, contrary to what happens, for instance, in surface growth, where the Kardar-Parisi-Zhang equation[15] has only one relevant nonlinearity, allowing the critical exponents to be computed from this equation. Nevertheless let us assume for the moment that some qualitative aspects of models exhibiting self-organized criticality can be obtained from the simplest nonlinear equation

$$\frac{\partial E(\vec{r}, t)}{\partial t} = \alpha \nabla^2 E(\vec{r}, t) + \lambda \nabla^2 E^2(\vec{r}, t) + \eta_e(\vec{r}, t), \tag{10}$$

which is consistent with conservation laws and with the symmetries of the problem: reflection, rotation, translation, and lack of any characteristic time or length scale.

Note that some of the symmetries obeyed by other continuum models such as Galilean invariance[5, 8] or discrete lattice structure[4] are lost.

We follow a DRG[11, 12] procedure to analyze the hydrodynamic behavior of the system given by eq. (10). The infrared divergencies of momentum integration are avoided by integrating out the fast modes with momenta in the range $\Lambda e^{-l} \leq k \leq \Lambda$, where l is the shell thickness and Λ is the short-distance cutoff. In order to recover the original Brillouin zone, one has to perform the following scaling transformation for the remaining short-wavelength modes

$$E(k,\omega) \longrightarrow e^{(\chi+d+z)l} E(ke^l, \omega e^{zl}), \tag{11}$$

where z and χ are the dynamic and the roughening exponent, respectively. These modes obey eqs. (7) and (10) with renormalized parameters which satisfy, under an infinitesimal RG transformation and in the hydrodynamic limit ($k \rightarrow 0, \omega \rightarrow 0$), the following recursion relations up to one-loop order

$$\frac{d\alpha(l)}{dl} = \alpha \left[(z-2) - \frac{D\lambda^2}{T\alpha^4} 16 A_d \Lambda^{d-4} \right] \tag{12}$$

$$\frac{d(D/T)(l)}{dl} = \frac{D}{T} [2z - d - 2\chi] \tag{13}$$

$$\frac{d\lambda(l)}{dl} = \lambda \left[(\chi + z - 2) + \frac{D\lambda^2}{T\alpha^4} 48 A_d \Lambda^{d-4} \right] \tag{14}$$

where $A_d = S_d/2(2\pi)^d$ with S_d being the surface area of a unit d-dimensional sphere. One can notice that $\bar{\lambda} = (16\Lambda^{d-4} A_d D\lambda^2/T\alpha^4)^{1/2}$ is the effective dimensionless coupling constant for which we can write the following RG recursion relation

$$\frac{d\bar{\lambda}(l)}{dl} = \bar{\lambda} \left[\frac{4-d}{2} + 5\bar{\lambda}^2 \right] \tag{15}$$

which enables one to evaluate fixed points and critical exponents up to the above mentioned one-loop order. One notices that the upper critical dimension is $d_{uc} = 4$ since for $d < 4$ there exists an unstable fixed point at $\bar{\lambda} = 0$ and for $d > 4$ there are three fixed points: $\bar{\lambda} = 0$ and $\bar{\lambda} = \pm\sqrt{(d-4)/10}$. The first one is stable whereas the other two are unstable. Under RG transformations, the flow in parameter space has the following behavior. For $d < 4$, a small nonlinearity flows away from the mean-field fixed point ($\bar{\lambda} = 0$) and a dynamic exponent z different from 2 is expected. On the other hand, for $d > 4$ there is a basin of attraction for the stable mean-field fixed point and purely diffusive behavior should be observed. However, for values of the effective coupling constant such as $|\bar{\lambda}| > \sqrt{(d-4)/10}$ the behavior is dominated by the strong-coupling unstable fixed point which gives rise to superdiffusive behavior ($z < 2$).

We conclude that it should be possible to observe a phase transition at $d \geq 4$ between a logarithmically rough phase with mean-field exponents and a smooth phase with a nondiffusive behavior[4, 5, 7, 8]. At this point, it is worth noting that with a correlation function for the noise as given by (6), the upper critical dimension is lowered to two. In the previous section, we reported extensive numerical simulations on Zhang model at $d = 2$ for different effective coupling constants (α is related to the fraction of energy released and D/T is linked to δE_{max} as the intensity of the noise) and no phase transition has been observed, thus providing some support to the assumptions about the external noise correlations we have made in the present analysis.

The same line of reasoning can be applied to the model described by microscopic rules (2). In this case, a continuum equation for a rescaled energy neglecting the effect of the internal noise is written

$$\frac{\partial E(\vec{r}, t)}{\partial t} = \alpha \nabla^2 [\theta(E(\vec{r}, t)) E_c] + \eta_e(\vec{r}, t),$$ (16)

which clearly shows the reflection invariance of the energy variable. Now the simplest nonlinear equation in agreement with symmetry rules is

$$\frac{\partial E(\vec{r}, t)}{\partial t} = \alpha \nabla^2 E(\vec{r}, t) + \lambda \nabla^2 E^3(\vec{r}, t) + \eta_e(\vec{r}, t)$$ (17)

giving rise to a different hydrodynamic behavior. This models has the same upper critical dimensionality $d_{uc} = 4$ but some qualitative differences appear. The RG recursion relation for the effective coupling constant $\bar{\lambda} = 12\Lambda^{d-4} A_d D\lambda / T\alpha^3$ is

$$\frac{d\bar{\lambda}(l)}{dl} = \bar{\lambda} \left[4 - d - 9\bar{\lambda} \right]$$ (18)

Now there are two fixed points above and below $d = 4$. For $d < 4$ the mean-field fixed point ($\bar{\lambda} = 0$) is unstable and the fixed point corresponding to $\bar{\lambda} = (4 - d)/9$ is stable whereas for $d > 4$, stability is exchanged. This enables us to obtain a dynamic exponent $z = (14 + d)/9$ below $d = 4$, which however, is far from the results obtained in the numerical simulations. Thus, this analytical approach would lead us to conclude that both models do not belong to the same universality class whereas from the simulations the conclusion is the opposite. All this leads us to believe that the approach based on simple nonlinear equations is incomplete and that one should study the full nonlinear equations (3) and (16), keeping the threshold condition.

CONCLUSIONS

In this paper we have performed numerical simulations in some models showing self-organized criticality in order to compute one of the exponents characterizing their critical behavior, the dynamic exponent. Starting with the dynamic microscopic rules proposed by Zhang[3] we have checked related models with different rules. We have emphasized that one must carefully define the characteristic length of the avalanche, since this can be relevant for the determination of the upper critical dimension.

When either the fraction of energy released at a critical site or the intensity of the noise are changed, the same dynamic exponent is obtained in a 2-d square lattice. The dynamic exponent is not modified when varying the effective coupling constant which is important for the appropriate choice of the noise correlations. We have also checked a different model which introduces a new symmetry but the dynamic exponent is not changed in this case, belonging then, in principle, to the same universality class.

We have analytically studied continuum models derived from the above microscopic dynamic rules. The stochastic differential equations satisfied by these models have two sources of noise: internal and external. Internal noise comes from the removed microscopic degrees of freedom and hence, is described by a fluctuation-dissipation theorem. On the other hand, external noise is specified by the model. We show that the internal noise turns out to be irrelevant and external noise with the appropriate correlation makes the critical dimension equal to four, in agreement with numerical simulations and other analytical approaches, and suggests that a phase transition, as a function of the bare coupling constant, above $d = 4$ should be observed. However, in the analytical approach, two models with different symmetries belong to different universality classes, in contrast with our numerical simulations. This suggests the need for a more complete analysis involving the full nonlinear equations describing these models.

ACKNOWLEDGMENTS

The author wishes to thank A.-M.S. Tremblay for suggesting this work and for many fruitful discussions and to the Spanish Ministry of Education for a Postdoctoral Fellowship at Université de Sherbrooke. This work has been partially supported by the CICyT of the Spanish Government (grant # PB89-0233).

REFERENCES

1. P. Bak, C. Tang and K. Wiesenfeld, Phys. Rev. A 38:364 (1988).

2. For a complete list of references, see K. Christensen, H.C. Fogedby and H.J. Jensen, J. Stat. Phys. 63:653 (1991).

3. Y.-C. Zhang, Phys. Rev. Lett. 63:470 (1989).

4. G. Grinstein and D.-H. Lee, Phys. Rev. Lett. 66:177 (1991).

5. T. Hwa and M. Kardar, Phys. Rev. Lett. 62:1813 (1989).

6. G. Grinstein, D.-H. Lee and S. Sachdev, Phys. Rev. Lett. 64:1927 (1990).

7. J. Toner, Phys. Rev. Lett. 66:679 (1991).

8. T. Hwa and M. Kardar, *Avalanches, Hydrodynamics, and Great Events in Models of Sand Piles*, Preprint.

9. L. Pietronero, P. Tartaglia and Y.-C. Zhang, Physica 173A:22 (1991).

10. S.-K. Ma, "Modern theory of critical phenomena", Benjamin, Reading (1976).

11. D. Forster, D.R. Nelson and M.J. Stephen, Phys. Rev. A 16:732 (1977).

12. E. Medina, T. Hwa, M. Kardar, Y.-C. Zhang, Phys. Rev. A 39:3053 (1990).

13. D. Dhar, Phys. Rev. Lett. 64:1613 (1990).

14. S.P. Obukhov, in: "Random Fluctuations and Pattern Growth", H.E. Stanley and N. Ostrowsky eds., Kluwer, Dordrecht (1988).

15. M. Kardar, G. Parisi and Y.-C. Zhang, Phys. Rev. Lett. 56:889 (1986).

THE PRACTICAL MEASUREMENT OF FRACTAL PARAMETERS

E.H. Dooijes, Z.R. Struzik

Department of Computer Science, University of Amsterdam
Kruislaan 403, 1098SJ Amsterdam, The Netherlands

1. INTRODUCTION

In our research group, we are investigating the applicability of fractal techniques for a variety of problems, ranging from the description and analysis of biological growth phenomena[1] to the design of systems for robot vision in natural terrain.

In this paper we address the following fundamental yet - to our knowledge - under-exposed questions: how can we determine, for a given object, the extent $\Delta\omega$ of the fractal scaling range, where a power law relation between scale parameter and measured size exists; how large should $\Delta\omega$ be at least that we may rightly qualify the object as a fractal, and that we can estimate its fractal dimension with the accuracy needed in the application?

The paper is organized as follows. In the next section we discuss the various factors which generally contribute to the inaccuracy in the determination of the fractal dimension and related parameters. A simple example, yet representative for many of the problems encountered is the determination of the divider (compass, coastline) dimension for self-similar objects with topological dimension 1. Details about the data sets used for experimentation are provided in section 3. The coastline measuring procedure is described in section 4. Section 5 addresses the generic problem of fitting an exponential function to a set of measurement data. Our main results are in section 6 where we propose a method for the determination of the location and extent of cross-over regions.

We will employ the following notation: $\Delta x = x_{max}-x_{min}$ denotes the 'yardstick' range corresponding to a given set of measurement data. $\Delta\omega = \omega_{max}-\omega_{min}$ is the range of scales where the object shows fractal behaviour; it is a subset of the scale-space Ω. We will assume that the object (usually refered to as 'coastline') is characterized by a single fractal dimension D, i.e. we are not dealing with conglomerates of fractals.

2. SOURCES OF UNCERTAINTY IN THE DETERMINATION OF FRACTAL PARAMETERS

In most applications of fractal geometry it is important to know the limits of confidence for an estimate D of the true value D^* of an object's fractal dimension. There are various reasons

Growth Patterns in Physical Sciences and Biology, Edited
by J. M. Garcia-Ruiz *et al.*, Plenum Press, New York, 1993

why the measured value of D should be suspected to differ from D*:

1. Any real-life object exhibits fractal behaviour, or self-similarity only over a finite range of scales $\Delta\omega$. It has been shown[2] that several definitions of fractal dimension and the related measuring methods are asymptotically equivalent; however the results from these methods may well diverge if we are dealing with fractal *segments*. Evidently, realizations of mathematical concepts like the Von Koch curve also belong to the class of real-life objects in this context.

2. Often, a fractal object should be regarded as a single realization of a stochastic process. Any dimension estimate computed from a finite segment of this realization will inherit the stochastic nature of the object.

3. The yardstick range Δx does not coincide, in most cases, with the self-similarity range $\Delta\omega$ (which is, in most practical cases, unknown *a priori*). Unless Δx is fully contained within $\Delta\omega$, values of the apparent coastline length incompatible with D* can be expected to arise, without being easily recognizable as such. In some cases assumptions about the behaviour of the data outside the self-similarity range can be put to use for detecting this situation, see section 6.

4. The range $\Delta\omega$ may be short, making it difficult to assign a D value with any reliability. This may signify that a description of the object as a fractal is inappropriate; but it can happen as well if the object really consists of a conglomerate of differently scaling fractal components.

5. Measurements on deterministic fractals (like the Von Koch fractal) may produce erratic results as explained in section 4.

6. For several measuring methods there exist both statistical (Monte-Carlo) and deterministic variants. Especially in the former case, the measurement procedure itself will introduce some (usually controllable) uncertainty in the measured value D.

7. Many measuring procedures (for instance, box counting based on a fixed grid) do not fully employ the information present in the data, and therefore introduce estimation errors. In this paper we deliberately avoid errors of this kind.

3. DESCRIPTION OF TEST DATA

We computed the divider dimension for a) the triadic Von Koch fractal curve, and b) trails of two-dimensional Brownian motion. In both cases the data are given as ordered sets of points. In case b the order can be interpreted as time order, whereas in case a the points are given in their natural order along the (polygonal) curve.

The triadic Von Koch fractal[2] can be constructed easily to any degree of precision, and has the dimension D* = log4/log3 = 1.2618.... Therefore it seemed useful to us as a test object for checking and calibrating our measuring routines. In reality, however, this turned out less straightforward than expected; see section 4. The fractal was constructed up to iteration level 6 (1024 polygon sides) by a simple recursion scheme.

Trails of two-dimensional classical and fractional Brownian motion (fBm) were constructed using the midpoint-displacement technique[3] up to the 10th generation (1024 data points). These curves act in our investigation as examples of fractals exhibiting self-similarity in the statistical sense.

The various test patterns have about the same linear size of 2 units of length. Divider dimensions are computed for a standard geometric sequence of divider lengths, starting from length 1, and with common ratio 2/3. The smallest of the 24 terms is 9.10^{-5} length units. The accuracy of our data is restricted only by their floating point representation. The random number generator needed for implementing the midpoint-displacement technique was taken from Wichmann and Hill[4].

4. MEASURING THE COASTLINE LENGTH

In the power law model $y = a. \, x^b$ the quantity y represents the effective coastline length and x the divider length. We want to infer the fractal dimension $D = 1 - b$ and the factor a from a collection of experimental (x,y) data.

Divider algorithm. Starting from the first data point, the polygonal curve is traversed in N full steps using a pair of dividers spanning length x. The last divider mark on the curve is connected then to the last data point (the endpoint of the curve), and a certain fraction μ of the length x' of this line segment is added to obtain the total length $y(x) = N.x + \mu.x'$. The correction factor μ is chosen as $\mu = (x / x')^{1-D}$ where D is the fractal dimension. It is introduced to cope with the fact that x' will overestimate the length of coastline it spans, as we are trying to determine the effective coastline length on the larger scale set by x. Of course, D is not known beforehand. As the subsequent computations are not very sensitive to the

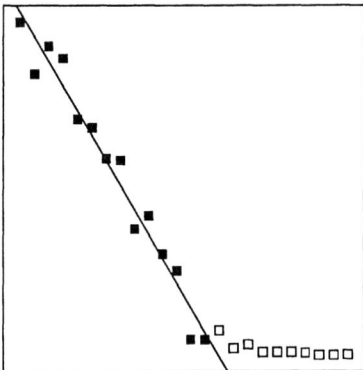

Figure 1.   log-log plot for Von Koch triadic fractal. Yardstick range (horizontal) is 4 decades. Deleted data points are shown as open squares.

particular choice of μ, we give it a value based on the *guessed* value of D. Obviously, $\mu.x' \leq x' < x$. This length measurement procedure is essentially free of errors (if we neglect the error remaining after the end effect correction). Notice that it is immaterial whether the divider circle is traversed in clockwise or counter-clockwise direction.

Modified divider algorithm. As shown in figure 1, the divider algorithm produces erratic results in the case of a deterministic fractal like the Von Koch curve. The fluctuations are caused by the jumps in apparent curve length each time the divider length approximates a submultiple of the generator's characteristic pattern length: a kind of 'beating' or 'moiré' phenomenon. A successful attempt to eliminate this problem was to introduce randomness in the measuring procedure, by replacing each equal-sided divider polygon by a 'random' polygon with the same number of vertices, thus making the measurement data amenable for statistical analysis. Details will be reported elsewhere[5].

5. FITTING A STRAIGHT LINE TO DATA POINTS IN A LOG-LOG GRAPH

In this section we address the problem of fitting a function of the form $y = a. x^b$ to a set of measurement data x_i, y_i, $i = 1......n$, ranging from x_{min} to x_{max}.
We start by assuming that the exponential relation is valid for the whole range of observed x-values:

$$y_i = a. x_i^b + d_i, \quad d_i = N(0, \sigma_i^2), \quad i = 1........n. \tag{1}$$

The x-data are assumed to have absolute accuracy whereas the y-data are expected to contain normally distributed 'errors' with zero mean. Generally with each measurement a different variance σ_i^2 is associated (which may not be known by the experimenter).
If we adopt the maximum-likelihood criterion for defining the error norm, (1) leads to a non-linear least-squares problem which can readily be solved using the Levenberg-Marquardt (LM) sum-of-squares minimizer[6]. However by taking logarithms the problem is reduced to a linear regression problem:

$$\log(y_i - d_i) = \log(a) + b.\log(x_i).$$

In the case of small errors the left hand logarithm can be approximated by $\log y_i - d_i/y_i$. After rearranging terms and substituting

$$\eta = \log y, \, \xi = \log x, \, \alpha = \log a,$$

we have the model

$$\eta_i = b\xi_i + \alpha + d_i/y_i \tag{2}$$

Minimizing the corresponding objective function

$$J(\alpha, b) = [w_i.(\eta_i - b\xi_i - \alpha)^2] \quad \text{with } w_i = (y_i/\sigma_i)^2$$

by setting the derivatives with respect to α and b to zero leads to a pair of equations linear in the parameters α and b. (Square brackets denote summing over the index i). Solving these equations yields the estimates $\underline{\alpha}$ and $\underline{b}$, the latter being

$$\underline{b} = ([w\xi\eta][w] - [w\xi][w\eta]) / \Delta \tag{3}$$

with variance $\quad \text{var}(\underline{b}) = [w] / \Delta \tag{4}$

where $\quad \Delta = [w][w\xi^2] - [w\xi]^2 .$

Notice that $\underline{b}$, in contrast with var($\underline{b}$), is invariant with respect to scaling of the weighting factors. In the absence of further information about the σ_i an intrinsic estimate for var($\underline{b}$) can be obtained by multiplying (4) with the correction factor $J(\underline{\alpha},\underline{b})/(N-2)$, after setting $w_i = 1$ for all i. See Numerical Recipes[6] for details.

An important peculiarity of this approach to fitting an exponential is that the weighting factors w are proportional to y^2. (The situation is not different if, instead of using linear regression, the parameter estimation is done by solving the original non-linear problem; in the latter case the data are weighted implicitly). In many physical situations it is quite plausible to assume that all y measurements are done with about the same *relative* precision (i.e. $\sigma(y)$ is proportional to y). In that case all data will have equal net weight for the linear regression. Thus, setting all weighting factors to 1 will produce correct values of the parameter estimates *provided that* $\sigma(y) \propto y$ *is true*; however our experiments indicate that this procedure is *not* justifiable in the present context (see section 7).

Moreover, the errors (i.e., the observed deviations from the model $y_i = a. x_i^b$) cannot always be adequately described by a zero-mean distribution. In other words, they contain a systematic component, which happens, for instance, in the cross-over regions (see section 6). This can alternatively be regarded as an inadequacy of the original model. Following this point of view, we will discuss in the following section how the model can be adapted to be able to deal with the cross-over behaviour of a fractal segment.

6. DETERMINATION OF CROSS-OVER POINTS

We have choosen our sequence of divider lengths in such a way, that the smallest one x_{min} is certainly below the lower boundary ω_{min} of the fractal scaling region of our samples. This has the important advantage that ω_{min} can be estimated with some accuracy, by putting to use our knowledge about the difference in nature of the data inside and outside the self-similarity range. Several authors have proposed so-called cross-over functions suitable for modelling this situation, sometimes making use of specific assumptions related to the physics of the system under study[7], and sometimes from a more general point of view. An interesting example of the latter class is the logistic-function approach[8], which accomodates the cross-over regions at ω_{min} and ω_{max} simultaneously, assuming that the system shows Euclidean behaviour beyond these points.

It can be made plausible with a geometric argument that in the neighbourhood of ω_{min} the coastline length tends to be underestimated. Also, this error tends to be larger (in absolute value) if the dimension D for the fractal regime is larger. Therefore, the extent of the cross-over region increases with D. To cover this phenomenon by extending the original power-law model, we introduce a *family* of cross-over functions parametrized by the quantity p:

$$y = k / (1 + (x/x_c)^{-bp})^{1/p} \tag{5}$$

The general idea is that, for x values significantly smaller than the cross-over point x_c, the function tends to the constant value k, whereas for large x the function is asymptotically reduced to $y = k.x_c^{-b}.x^b$; clearly b has the same interpretation as in (1), while $k.x_c^{-b}$ corresponds to the factor a.

The parameter p controls how closely the function follows its asymptotes in the vicinity of the cross-over point. This can be seen by considering the function value for $x = x_c$ which is

$y = k.2^{-1/p}$. For $p \to 0$ and $p \to \infty$ we have $y \to 0$ and $y \to k$, respectively. The parameter p may assume any positive real value; however in practice it is restricted to the interval (0.9, 5.5). We computed the parameters p, b, k and x_c using LM search for both Brown and Koch type data sets. As our error model predicts, objects with lower D need a higher value of p for a satisfactory fit to the cross-over function (3). Indeed, we found that corresponding values of b = 1 - D and p obeyed the simple relation b.p = -1 within about twenty percent. This is not good enough to justify reducing the parameter search space dimension to 3. However, it makes it possible to derive a simple expression for the width of the cross-over region. With p = -1/b, equation (3) reduces to

$$y(x) = k.x_c^{-b}.(x + x_c)^b.$$

The left- and right-hand branches of the asymptote are

$$y_-(x) = k, \quad y_+(x) = k.(x/x_c)^b. \tag{6}$$

We determine the factor λ for which $y(\lambda x_c) = q.y_+(\lambda x_c)$ where q is an arbitrarily chosen positive number < 1. A reasonable choice is q = 0.87; this is the value of $y(x_c)/k$ for D = 1.2. It is easy to see that q and λ are related by $\log q = -b \log(\lambda / (\lambda + 1))$. We find exactly the same relation if we look for λ satisfying $y(x_c/\lambda) = q.y_-(x_c/\lambda)$. This means that the cross-over region is symmetric with respect to $\log x_c$ and has the width (in decades)

$$2.^{10}\log \lambda = -2.^{10}\log (u - 1), \quad u = q^{1/b} = q^{1/(1-D)}. \tag{7}$$

For instance, for q = 0.87 and D = 1.26 (triadic Koch) we find a width of 0.3 decades, for D = 1.5, 1.3 decades; for D = 2 (Brownian motion), 1.6 decades. These numbers provide a realistic idea of the minimal extent of the fractal scaling region $\Delta\omega$. Pfeifer[9] has proposed a criterium stating that $\Delta\omega$ should be of order $2^{1/D}$, corresponding to 0.3/D decades on a logarithmic scale. However, this number decreases with increasing D, while our discussion above suggests that the reverse should be true.

We comment that the two-sided cross-over function[8] reduces for $\omega_{max} \to 0$ to a one-sided ('left-hand') cross-over function not identical to, but showing similar behaviour as our model (3) with p = 1; therefore it lacks the flexibility to explain the cross-over behaviour of fractals over the entire range $1 < D \leq 2$.

7. ANALYSIS OF THE TEST DATA

We present a few experimental results with the purpose of illustrating the points made in the previous sections. In all pictures (log-log plots, llp) the same range Δx is displayed, as explained in section 4. The vertical scale is chosen to fit the range of log(y) values for each particular example.

Von Koch fractal. The example (figure 1) shows the typical appearance of the llp for a deterministic fractal, in this case the Von Koch fractal segment. It was obtained using the basic divider algorithm discussed in section 3. Notice the fluctuations in the exponential scaling range. In figure 1, model (2) is fitted after deleting the data points identified by model (5) as 'Euclidean', resulting in D = 1.25(1). The error estimate, based on (3), is intrinsic in this case;

as explained in section 4 the observed fluctuations have no statistical background.

Trails of Brownian movement in the plane. Figure 2 shows the llp for a sample trail of classical Brownian motion. The σ_i corresponding to the length measurements y_i were estimated from an ensemble of 30 sample trails. We found that σ_i is roughly constant over the range of yardstick values (including the Euclidean region). On a logarithmic scale, this causes a tendency for the length fluctuations to increase with yardstick length; a phenomenon often found in published fractal measurement data. We found $p = 1.2$, $x_c = 0.05$, $D = 2.02(17)$. The availability of σ_i information makes it possible to derive explicit confidence limits from the covariance matrix computed by the LM search procedure. The vertical lines indicate the cross-over region (at the $q = 0.87$ level), symmetric around the cross-over point.

In figure 3 we show the result of averaging 30 sample trails of fBm with Hurst parameter[3] $H = 0.8$, illustrating the character of the sampling fluctuations. In this case $p = 5.0$, $x_c = 0.003$, $D = 1.260(34)$.

It should be remarked that some care is required in choosing the initial conditions for the Levenberg-Marquardt procedure, as the search space turns out to have many shallow local minima; only in the global minimum the objective function attains a sufficiently low value, however. We refer to Numerical Recipes[6] for details.

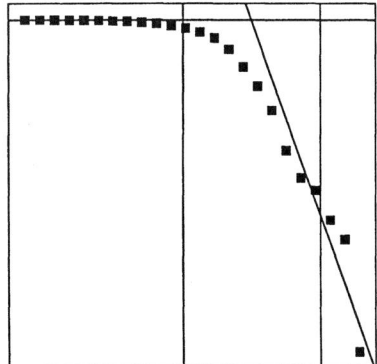 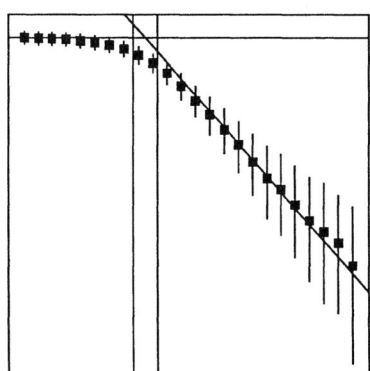

Figure 2 (left). log-log plot for one sample trail of Brownian motion.
Figure 3 (right). averages and standard deviations for 30 fBm trails, $H = 0.8$.
In both figures, yardstick range (horizontal) is 4 decades.

8. DISCUSSION AND FURTHER REMARKS

From our experiments we learn that the extent $\Delta\omega$ of the fractal scaling region, needed for the estimation of D, is determined to an appreciable amount by the system's behaviour in adjacent regions, as well as by the natural fluctuations of the length measurements around the value predicted by the true fractal dimension D^*. Fluctuations of the latter type can be reduced by combining the information of many fractal samples; however this can in practice only be realized in computer experiments. We speculate that for certain classes of fractals the statistics of the sampling fluctuations can be predicted theoretically, given D and $\Delta\omega$.

Sampling fluctuations are a natural consequence of the stochastic nature of real-world fractals, and should not be ascribed to local variations in D, as is proposed by Farin et al[10]. Moreover, it is difficult to find a useful interpretation for the concept of local fractal dimension, and its computation from measurement data is numerically unstable.

9. CONCLUSION

In an attempt to lay a basis for 'fractal metrology' we have identified in this paper some of the problems encountered in the determination of the fractal dimension of real-world objects. A parsimonious model has been proposed for the description of cross-over phenomena which seldomly can be neglected in realistic applications. This model has the interesting feature of working well with both deterministic and stochastic fractals. Probably, it can be generalized to deal with other situations than the transition from fractal to Euclidean behaviour.

REFERENCES

1. J. Kaandorp: Simulating radiate accretive growth using iterative geometric constructions. This volume.
2. K. Falconer, 1990, "Fractal Geometry - Mathematical Foundations and Applications", John Wiley & Sons.
3. H.O. Peitgen, D. Saupe (eds.), 1988, "The Science of Fractal Images", Springer Verlag.
4. B. Wichmann, D. Hill, An efficient and portable pseudo-random number generator, *Appl. Stat.* 31:188 (1982).
5. Z.R. Struzik, E.H. Dooijes: Experimental determination of the fractal dimension of a deterministic fractal. In preparation (1991).
6. W.H. Press, B.P. Flannery, S.A. Teukolsky, W.T. Vetterling, 1986, "Numerical Recipes - The Art of Scientific Computing", Cambridge University Press.
7. T. Freltoft, J.K. Kjems, S.K. Sinha, Power law correlations and finite-size effects in silica particle aggregates, *Phys. Rev.* B 33:269 (1986).
8. M. Sernetz, H.R. Bittner, H. Willems, C. Baumhoer, Chromatography. *In* "The Fractal Approach to Heterogeneous Chemistry", D. Avnir, ed., John Wiley & Sons (1989).
9. P. Pfeifer, Fractal dimension as working tool for surface-roughness problems, *Appl. of Surface Sci.* 18:146 (1984) .
10. D. Farin, S. Peleg, D. Yavin, D. Avnir, Applications and limitations of boundary-line fractal analysis of irregular surfaces: proteins, aggregates, and porous materials, *Langmuir* 1:399 (1985).

INDEX

Resistor networks, 258, 291
Reynolds number, 109-111
Richardson model, 148
RNA, 401-402, 404
Rough surfaces, 37-44
Roughness exponent, 32-35, 41-43, 45-55, 74, 77, 85-97, 107, 125, 149, 296

Saddle point, 227-228, 230
Saffman-Taylor instability, 186, 221, 225-231, 286-310
Samonella typhimurium, 8
Sand box analysis, 30-31
Sand pile models, 109, 112, 407-415
Scaling, 32, 37, 40, 45-55, 59, 70, 77-83, 85, 99, 103-107, 109-117, 130-133, 149, 166-168, 221-224, 239-241, 290-295, 308-311, 359
Scaling relation, 49-52, 70, 78, 81, 90, 107, 113-116
Scanning tunneling microscope, 37-44
Schlieren methods, 158
Screening, 3, 195-200, 203-213
Screening angle, 193, 195, 199-200
Screening length, 204
Self-affinity, 9, 29, 32-35, 37-44, 45, 65, 85, 110-115, 192, 246
Self-organized criticality, 110, 113, 213-219, 407-415
Self-similarity, 29, 37, 65, 175, 191, 214, 245, 255, 257, 273, 418
Self organization, 21, 341-372, 381
Self replication, 393-399
Semigroup, 145
Side branching, 16
Single step model, 100-106
Sivashinsky equation, 124-126
Slime mold, 21-27
Slit island analysis, 38-43
Solid-on-solid models, 59
Solidification, 163-169
Solitons, 390
Space charge, 156
Spin glass, 403
Spin models, 66
Spinodal decomposition, 354
Sponges, 332
Steady state, 238, 307
Stefan problem, 191
Strange attractor, 111
Stream function, 227-229
Streaming, 21-26
Stretched exponential, 251
Structure function, 359, 375
Subsets, 258
Sunflower, 341
Superconductivity, 323
Supersaturation, 163-169
Surface diffusion, 57-62

Surface growth, 45-55
Surface tension, 8, 11, 67, 109, 114, 116, 165-167, 169, 191, 221, 226
Surface width, 40-42
Surfactants, 320-323
Swift-Hohenberg equation, 373-377, 385-390
Symmetry, 195
Symmetry breaking, 223

Territories, 129
Thin films, 37, 61-62, 99
Tip splitting, 6, 11, 14, 17-18, 163, 178, 336
Time reversibility, 145
Topography, 37, 227
Topological entropy, 366
Transfer matrix, 402
Transients, 307-313, 344-346, 381
Trapping, 67
Trees, 29, 88, 191, 285
Topology, 37, 227
Turbulence, 48, 109
Turbulent viscosity, 115
Urn models, 143, 150
Universality, 50-54, 65, 70, 85, 263-264, 307-313

Vapor deposition, 39-43
Vascular networks, 267-276
Vascular tree, 269
Viscosity, 49, 109-112
Viscous fingering, 177, 183-189, 221, 225-231, 307-313
Von Neuman boundary conditions, 299
Vorticity, 119
Vortices, 159
Voter models, 146

Waiting time distribution, 82-83, 99-108
Waves, 21
Wave trains, 388
Wavelet transform, 191-200
Weierstrass equation, 327
Wetting, 67
Wetting fronts, 32-34
White noise, 47, 111
Williams-Bjerknes model, 148, 150
Williams domains, 353
Winding number, 227
Wings, 280, 283-284

Zeolites, 319